First International Conference on Electrical Energy and Power Engineering (ICEEPE 2020)

IOP Conference Series: Materials Science and Engineering
Volume 1045

Penang, Malaysia
25 - 26 August 2020

Editor:

Nor Azlan Othman

ISBN: 978-1-7138-3333-8
ISSN: 1757-8981

Printed from e-media with permission by:

Curran Associates, Inc.
57 Morehouse Lane
Red Hook, NY 12571

Some format issues inherent in the e-media version may also appear in this print version.

Institute of Physics
Dirac House, Temple Back
Bristol BS1 6BE UK

Phone: 44 1 17 929 7481
Fax: 44 1 17 920 0979

techtracking@iop.org

Additional copies of this publication are available from:

Curran Associates, Inc.
57 Morehouse Lane
Red Hook, NY 12571 USA
Phone: 845-758-0400
Fax: 845-758-2633
Email: curran@proceedings.com
Web: www.proceedings.com

TABLE OF CONTENTS

AUTOMATIC ROBOTIC SCANNING AND INSPECTION MECHANISM FOR MINES USING IOT. 1
Ibrahim Qadri, Amgad Muneer, Suliman Mohamed Fati

ELECTROCHEMICAL SELF-ASSEMBLY OF BUNCH-SHAPED ZNO NANOWIRE WITHOUT CATALYSTS 12
L S Chuah, M F Shafiee Haron, Y Yusof

NEURAL NETWORKS ALGORITHM FOR ARABIC LANGUAGE FEATURES-BASED TEXT MINING 21
Faiez Musa L Alrufaye, Hakeem Imad Mhaibes, Ali F Neamah

INFLUENCE OF NB AND ZR CO-DOPING ON THE STRUCTURAL, MORPHOLOGICAL AND DIELECTRIC PROPERTIES OF CACU3TI4O12 CERAMICS 28
M K Hasan, S H Shant, M Z Islam, M E Haque

DESCRIPTIVE ANALYSIS ON PRODUCTS/SERVICES DEMAND AND MARKET CONSPECTUS: A QUANTITATIVE APPROACH 37
Mohammed Afaounodden Ahmed, Adresya Suresh Azhakath, Raja Muthalagu

SYNTHESIS AND CHARACTERIZATION OF METAL-ORGANIC FRAMEWORK (MOF) COBTC AS A NON-ENZYMATIC ELECTROCHEMICAL BIOSENSOR FOR GLUCOSE 47
S A Abrori, M L A Trisno, R A Aritonang, I Anshori, Nugraha, Suyatman, B Yuliarto

RESIDUAL STRESS ANALYSIS OF FRICTION STIR WELDED AA6063 PIPE BUTT JOINT BY NEUTRON DIFFRACTION TECHNIQUE OF ENGIN-X 54
Azman Ismail, Tung Lik Lee, Mokhtar Awang, Fauziah Ab Rahman, Nabihah Salih, Bakhtiar Ariff Baharudin, Puteri Zarina Megat Khalid, Darulihsan Abdul Hamid

+2.5% IMPROVEMENT IN GENERATION UTILIZING ANTI SOILING MODULES BASED ON DSM TECHNOLOGY 59
Rajat Verma, Amresh Mahajan, Prem Pathak

SINGLE-MODE OPTICAL FIBERS COUPLING: STUDY OF THE FIELD OF VIEW 65
Lina M Shaker, Sara Osamah, Ahmed A Al-Amiery

A GRAPHENE-MODIFIED CO-BDC METAL-ORGANIC FRAMEWORKS (CO-MOF) FOR ELECTROCHEMICAL NON-ENZYMATIC GLUCOSE SENSING 71
R Yuniasari, F Amri, S A Abrori, N L W Septiani, M Rezki, Irzaman, M Z Fahmi, B Yuliarto

DESIGN, SIMULATION AND ANALYSIS OF MONOFACIAL SOLAR PV PANEL BASED ENERGY SYSTEM FOR UNIVERSITY RESIDENCE: A CASE STUDY 78
Abdullah Al Mehadi, Misbahul Alam Chowdhury, Mirza Muntasir Nishat, Fahim Faisal, Md Minhajul Islam

FABRICATION AND CHARACTERIZATION OF EPOXY RESIN–BARIUM TITANATE AT G-BAND USING WAVEGUIDE TECHNIQUE 89
Nurulfadzilah Hasan, Nur Shahira Mat Hussain, Nur Hazimah Syazana Abdul Razak, Ahmad Syahiman Mohd Shah, Mohamad Shaiful Abdul Karim

A STUDY ON SODIUM - THE FAST BREEDER REACTOR COOLANT 97
K K Rajan

PERFORMANCE EVALUATION OF A 200 KWP GRID TIED SOLAR POWER PLANT 109
K K Rajan, B Aruna

DESIGN OF PIEZO-RESISTIVE TYPE ACOUSTIC VECTOR SENSOR USING GRAPHENE FOR UNDERWATER APPLICATIONS 119
Smitha Prabhu, Amrita B Pai, Gurumukh Singh Arora, M R Kusshal, Veera Pandin, M A Goutham

IMPLEMENTATION OF SOBEL FILTER USING CUDA 129
Akshita Akasapu, V Sailaja, G Prasad

A BIBLIOMETRIC ANALYSIS ON PATH PLANNING METHODS FOR MOBILE ROBOTS PUBLISHED IN SCIENTIFIC CITATION INDEX-EXPANDED INDEXED JOURNALS BETWEEN 2000-2020 138
Aisha Muhammad, Mohammed AH Ali, Ibrahim Haruna Shanono

THERMAL ANALYSIS OF A THERMOELECTRIC GENERATOR (TEG) USING FEM TECHNIQUE 145
K T Prajwal, Pritam Bhat

TUNING AND EXPERIMENTAL ASSESSMENT OF SECOND-ORDER GENERALIZED INTEGRATOR –FREQUENCY LOCKED LOOP GRID SYNCHRONIZATION FOR SINGLE-PHASE GRID ASSISTED SYSTEM 152
Bhavik Brahmbhatt, Dr. Hina Chandwani

A REVIEW ON TRAINING AND LEADERSHIP DEVELOPMENT: ITS EFFECTIVENESS FOR ENHANCING EMPLOYEE PERFORMANCE IN INDIAN CONSTRUCTION INDUSTRY 162
Pramod Kumar Misra, Jitendra Mohanty

INVESTIGATING THE FAST COOLING PROCESS OF SILICENE BY MD SIMULATION 171
Vo Van On, Nguyen Thanh Hung, Dang Thi Khanh Huyen, Le Thi Phuong Trinh

IMPROVED PROGRESSIVE COLLAPSE RESISTANCE OF IRREGULAR REINFORCED CONCRETE FLAT SLAB BUILDINGS UNDER DIFFERENT CORNER COLUMN FAILURES 183
Suyash Garg, Vinay Agrawal, Ravindra Nagar

SMART AERIAL MONITORING SYSTEM TO PREVENT HUMAN WILDLIFE INTERVENTION 191
M Sri Harsha, P Sai Teja, Punyasloka Mahapatra, R Anand

PRODUCTION OF HIGH STRENGTH BIOLIQUID FROM MUNICIPAL SOLID WASTE (MSW) USING MIXED CULTURE 205
N Zainol, N S M Sidek, A Ya'acob

COSMIC ACCELERATION IN LRS BIANCHI TYPE I SPACE- TIME WITH BULK VISCOSITY IN GENERAL RELATIVITY 215
Sanjay Sharma, Laxmi Poonia

COLD FORGING OF SILICON CARBIDE PARTICULATE (SICP) REINFORCED ALUMINIUM METAL MATRIX COMPOSITE PREFORMS 222
M Agrawal, S Singh

COMPARISON OF REDUCING AGENT FOR REDUCED GRAPHENE OXIDE AS CATHODE FOR ZINC-ION HYBRID CAPACITORS 235
S Prasertkaew, T Dejthammathorn, P Pattananuwat

VIDEO QUALITY EVALUATION AND TESTING VERIFICATION OF H.264, HEVC, VVC AND EVC VIDEO COMPRESSION STANDARDS 240
Shreyanka Subbarayappa, K R Rao

DESIGN AND VLSI IMPLEMENTATION OF A DECIMATION FILTER FOR HEARING AID APPLICATIONS 251
V Purushothaman, R Rama devi, S Manjula

SIMULATION AND EXPERIMENTAL VALIDATION OF HYSTERESIS CURRENT CONTROL TECHNIQUE FOR SPEED CONTROL OF BRUSHLESS DC MOTOR 259
Hetal Patel, Hina Chandwani

SIMULATION OF EXHAUST GAS ENERGY RECOVERY SYSTEM USING TURBOCOMPOUNDING FOR A MOTORCYCLE ENGINE 271
Srivatsa S Joshi, Aditya Deshpande

THE INFLUENCE OF ELECTROMAGNETIC WAVES ON CONDUCTIVITY TENSOR IN THE PRESENCE OF LASER FIELD IN QUANTUM WELLS WITH PARABOLIC POTENTIAL FOR THE CASE OF ELECTRONS-ACOUSTIC PHONONSCATTERING 281
Hoang Van Ngoc

EFFECTIVE LOAD MANAGEMENT SYSTEM FOR SUDDEN SHUTDOWN AVOIDANCE OF STAND ALONE PHOTOVOLTAIC OPERATION 289
Nuratiqah Mohd Isa, Amirah Nurhafizah Abu Bakar, Nurul Husna Mohd Sarif, Mohamad Shaiful Abdul Karim, Ahmad Syahiman Mohd Shah

AE SIGNATURE ANALYSIS USING CONTINUOUS AND DISCRETE WAVELET TRANSFORMS TO PREDICT GRINDING WHEEL CONDITIONS 297
K Shivith, K Rameshkumar

LIFETIME STUDY OF WHITE ORGANIC LIGHT EMITTING DIODES FABRICATED BY BOTH SOLUTION AND VACUUM EVAPORATION PROCESSES 307
Fuh-Shyang Juang, Kuo Chun Huang, Jeng Yue Chen, Lin Kuo, Lin Ann Hong

SCPI BASED INTEGRATED TEST AND MEASUREMENT ENVIRONMENT USING LABVIEW 312
B Aravind Balaji, S Sasikumar, K Ramesh

EXTRACTION OF MICROBIAL CHITOSAN FOR PIEZOELECTRIC APPLICATION 319
M I Zamli, F B Ahmad, M H M Akmal

DESIGN AND IMPLEMENTATION OF FEC FOR CDMA SYSTEM 325
M N Smrithi, A R Abdul Rajak

CRYSTAL STRUCTURE AND OPTICAL PROPERTIES OF NON-VACUUM SOLUTION-BASED PROCESSED CU2ZNSNS4 (CZTS) THIN-FILM 337
H S Nugroho, G Refantero, E C Prima, C Panatarani, Suyatman, N Nugraha, B Yuliarto

INVESTIGATIVE ANALYSIS OF MATCH-LINE SENSING SCHEMES IN TCAM FOR HIGH-SPEED PERFORMANCE 343
Fahim Faisal, Mirza Muntasir Nishat, Sayka Afreen Mim, Hafsa Akter, Md. Rafid Kaysar Shagor

EVALUATING RMS BASED CONTINUOUS S-TRANSFORM WITH DEEP LEARNING FOR DETECTING AND CLASSIFYING VOLTAGE SAG AND SWELL 351
Kamarulazhar Daud, Syazreena Sarohe, Wan Salha Saidon, Saodah Omar, Nurlida Ismail, Nazirah Mohamat Kasim

A NON-UNIFORM EXCITATION ANTENNA ARRAY WITH DEFECTED GROUND STRUCTURES FOR DUAL-BAND 5G MM-WAVE APPLICATIONS 362
D Shashi Kumar, S Suganthi

HAND GESTURE RECOGNITION ON PYTHON AND OPENCV 369
Ahmad Puad Ismail, Farah Athirah Abd Aziz, Nazirah Mohamat Kasim, Kamarulazhar Daud

FAULT DIAGNOSIS AND TROUBLESHOOTING ON 2.1KW GRID-CONNECTED PHOTOVOLTAIC POWER SYSTEM IN UITM PULAU PINANG 379
Muhammad Faris Fahmi Mohd Fadzil, Ahmad Asri Abd Samat, A Idzwan Tajudin, S Omar, Kamarulazhar Daud

OPTIMUM LOCATION OF DG FOR LOSS REDUCTION WITH ANT COLONY ALGORITHM 389
Saodah Omar, Muhammad Nasrullah Abdul Manan, Muhammad Naqib Mat Siam, Ahmad Asri Abd Samat, Kamarul Azhar Daud

CONTROL STRUCTURES IMPLEMENTATION TO ALLOW HIGH PENETRATION OF A VSC WITHIN AN ISOLATED POWER SYSTEM 401
D Curto, S Favuzza, R Musca, M Navarro Navia, G Zizzo

SMART SOFT-SWITCHING SWITCHOVER CIRCUIT FOR DUAL-POWER PV-GRID SYSTEM ENERGY 414
I R Ibrahim, M A F M Nasir, M N M Hussain, A M Omar

Author Index

IOP Conf. Series: Materials Science and Engineering 1045 (2021) 011001 doi:10.1088/1757-899X/1045/1/011001

Preface

Being as one of the conference chairs for this 1st International Conference on Electrical Energy and Power Engineering (ICEEPE) 2020, it is my pleasure to welcome you all to this important and prestigious event. This conference goes to the heart of all matters relating to engineering as a whole, and it brings together the best research professionals and experience academicians from respective fields. There are delegates not only from Malaysia, but also delegates from around the globe.

This ICEEPE 2020 provides a platform to scientists, researchers, professors, scholars and academicians to share, discuss, debate, and dissect significant new developments and scientific advancements that will impact the future of electrical energy and power engineering or related fields. This ICEEPE 2020 is a unique and unmissable opportunity to meet face to face with colleagues from different parts of the world. It is a major contribution for the development and consilience of knowledge.

Although the world has been affected on the Covid -19 issue that has changed our daily life, ICEEPE 2020 has received tremendous supports and responses from researchers internationally and locally from various backgrounds in science and technology areas. The presentations of all research papers were successfully done via online plat form.

Due to the impact of Covid-19 pandemic, the conference has been converted into full virtual mode. The conference was held at Microsoft Teams platform where all the authors participated and presented their research works. Each presentation was scheduled for 15 minutes (12 minutes for presentation and 3 minutes for queries).

Once again, on behalf of the organizing committee, I am delighted to invite you all to take part in this conference and also to make the conference a grand success. I hope that the experiences in this event will be engraved in your memory.

Dr. Nor Azlan Othman
Chair & Editor
ICEEPE 2020
Contact: azlan253@uitm.edu.my

Committees

Committees

Conference Chair (s)
Dr. Nor Azlan Othman
Universiti Teknologi Mara, Malaysia

Dr. Senthil Kumar Sundaram
SVS College of Engineering, India

Organizing Chair
Dr. Ahmad Asri Bin Abd Samat
Universiti Teknologi Mara, Malaysia

Conference Secretary
Ir. Dr. Nor Salwa Damanhuri,
Universiti Teknologi Mara, Malaysia

Editors
Dr. Amjad Anvari, Aalborg University, Denmark
Dr. Nor Azlan Othman, Universiti Teknologi Mara, Malaysia

Scientific Committee
Dr. Osamah Ibrahim Khalaf, Al-Nahrain University – College of Information Engineering, Iraq
Dr. Faridun Naim Tajuddin, University Malaysia Perlis, Malaysia

Finance Chair
Ts Mohamad Adha Mohamad Idin
Universiti Teknologi Mara, Malaysia

Review Committee Chair(s)
Ts. Dr. Kamarulazhar Daud
Universiti Teknologi Mara, Malaysia

Publicity Chair(s)
Ts. Mohd Affandi Shafie,
Universiti Teknologi Mara, Malaysia
Technical / Local Arrangement Committee
Ir Aimi Idzwan Bin Tajudin, Universiti Teknologi Mara, Malaysia
Ts. Adi Izhar Che Ani, Universiti Teknologi Mara, Malaysia

Technical Program Committee
Dr. Mehmet DAL, Technical University of Munich, Munich, Germany
Dr. Lalit GOEL, Nanyang Technological University, Singapore
Dr. Sardar Ethem Hamamci, Inonu University, Turkey
Dr. Sanjay Misra, Covenant University, OTA, Nigeria
Dr. Omar Ellabban, Texas A&M University, Qatar
Dr. Antonella Ferrara, University of Pavia, Italy

Dr. Ali Emadi, McMaster University, Canada
Dr. Josep M. Guerrero, Aalborg University, Denmark
Dr. Riadh W. Y. Habash, University of Ottawa, Canada
Dr. Mani V N, Senior Scientist, DEIT, Hyderabad, Govt. of India
Dr. R. N. Patel, National Institute of Technology, Raipur
Dr. Mandal S K, NITTTR, Kolkatta
Dr. M.-R. Haghifam, Tarbiat Modares University, Iran
Dr. Krishnan A, KSR College of Engineering, India
Dr. Mohamed Ali Ismeil, Technical University of Munich, Munich, Germany.
Dr. Senthikumar S, SVS College of Engineering, India
Dr. Anand Nayyar, Duy Tan University, Vietnam
Dr. R. Murugappan, KIST, Kwait
Dr. M R Haghifam, Tarbiat Modares University, Iran
Dr. Chandima Gomes, University of the Witwatersrand, South Africa
Dr. Huai Wang, Aalborg University, Denmark
Dr. Albert Alexander S, Kongu Engineering College, Erode
Dr. Faridun Naim Tajuddin, University Malaysia Perlis, Malaysia
Dr. Umashankar S, USIT, Saudi Arabia
Dr. Mahmoud Fotuhi-Firuzabad, Sharif University of Technology, Dehran
Dr. Osamah Ibrahim Khalaf, Ai-Nahrain University, Iraq
Dr. Dr Fushuan Wen, Tallinn University of Technology (TalTech), Estonia
Dr. Ramani Kannan, University Technology Petronas, Malaysia
Dr. Abolfazl Vahedi, Iran University of Science & Technology, Iran
Dr. Gobbi Ramasamy, Multimedia University, Malaysia
Dr. Baharudhin Ismail, Universiti Malaysia Perlis, Malaysia
Dr. Hafizi, University Malaysia Perlis, Malaysia
Dr. Zhe CHEN, Aalborg University, Denmark
Dr. Hafizi, University Malaysia Perlis, Malaysia

IOP Conf. Series: Materials Science and Engineering 1045 (2021) 011002 doi:10.1088/1757-899X/1045/1/011002

Peer review declaration

All papers published in this volume of *IOP Conference Series: Materials Science and Engineering* have been peer reviewed through processes administered by the Editors. Reviews were conducted by expert referees to the professional and scientific standards expected of a proceedings journal published by IOP Publishing.

- **Type of peer review: Single-blind**
- **Conference submission management system: Email submission**
- **Number of submissions received: 128**
- **Number of submissions sent for review: 84**
- **Number of submissions accepted: 48**
- **Acceptance Rate (Number of Submissions Accepted / Number of Submissions Received X 100): 37.5**
- **Average number of reviews per paper: 2 (Internal screening + Technical review)**
- **Total number of reviewers involved: 17**
- **Contact person for queries (please include: name, affiliation, institutional email address)**

Dr. NOR AZLAN OTHMAN
Universiti Teknologi Mara
Penang, Malaysia
azlan253@uitm.edu.my

IOP Conf. Series: Materials Science and Engineering 1045 (2021) 012001 doi:10.1088/1757-899X/1045/1/012001

Automatic robotic scanning and inspection mechanism for mines using IoT.

Ibrahim Qadri[1], Amgad Muneer[2*] and Suliman Mohamed Fati[3]

[1,2]School of Engineering, Asia Pacific University of Technology & Innovation, Kuala Lumpur, Malaysia

[3]College of Computer and Information Sciences, Prince Sultan University, Saudi Arabia

E-mail: muneeramgad@gmail.com

Abstract. The mining industry has been recognized as one of the most hazardous industry. With the growth of the industry, the number of human casualties has increased. The mines can be very lethal for workers or in the event of an accident or with high humidity and temperature leading to workers fainting. Some of the problems that besets are of air blast, ground movement, dust explosion, inundation etc. This study reviews the common problems associated with Carbon Monoxide content with temperature and humidity. In this paper, we assemble an automatic robotic scanning and inspection mechanism for mines that is designed and assembled to recognize Carbon Monoxide (CO), humidity and temperature variance inside the mines. The proposed system employees a mobile robot that can be manually controlled by a self-developed mobile application and an Internet of Things (IoT) system. The sensors included take the input from the air and transmit them to the mobile application using Bluetooth module. The experimental results show that the IoT achieved an accuracy of 97.5 % for the mobile robot and the sensor system.

1. Introduction

With the latest advancement and the rapid progress of the technology, demand for automation of industries has spiked. These include the natural resource industries. Human workers have been replaced with big machines to increase the efficiency and the profits. The aftereffect of these machines is the unbalance of the natural environment at these extraction sites [1].

Mining industry is generally classified as the most hazardous industrial sector. Mining activities usually emit physical and chemical hazards in the workplaces and impose negative effects on the workers' health [2]. Many of incidents in the hazardous industry take place because of human error, the control of which would enhance safety levels in working sites to a considerable extent [3]. However, even with the introduction of various mobile robots by universities and companies, the industry still faces common issues like the design specification and functionality of the robot, in terms of size, weight [19] and the movement of the robot, including, the raw material of the robot and the sensors that were attached. Due to the air content inside the mines, the robots need to be dustproof and inflammable [18]. The robots did not do well with the still water in the mine and were stuck; and the functionality of the robot is compromised. The flammable gas content in the air needs flameproof protection for the components [4]. This paper is targeting some very important questions such as the safety of mine workers and the monitoring of mine air conditions.

The aim of this paper is to develop and use a robot to gather data inside a mine and transmit it wirelessly using IoT. The four sensors utilized in the proposed rescue robot are Carbon Monoxide (CO), humidity and temperature variance inside the mines. The core of the car is an Arduino MCU and the connection between the mobile app and the car is realized via Bluetooth.

The remainder of this paper is organized as follows: Section 2 discusses the related works. Section 3 introduces the proposed methodology of this work, while section 4 presents the experimental results. Finally, the paper has been concluded in section 5.

2. RELATED WORK

The fundamentals of this section are to identify the design specification and the working mechanism of a mobile robot for data acquisition inside a mine and the trends of robots used in the mining industry for the data acquisition. This allows the design specification of the robot to be scaled according to the requirements. It also investigates the different mobile robots developed by researchers for data gathering.

Rescue operation in the coalmine is extremely dangerous due to several factors. It is particularly very harmful to the rescuers to get into the coal mine tunnel in disaster without the prior knowledge of environment because the subsequent explosions may likely to occur at any time it is therefore essential to detect the explosive environment details such as toxic gases, high temperatures [5-17].

Added to the above reasons is the problem of spontaneous combustion, which is more common. Spontaneous combustion, if not detected in the early stages, could be fatal in terms of both material loss as well as the loss of life [6].

In one research [7] on the results of gases, emissions from the Lotts Creek coal mine fires. This study shows the analysis of the types of gases and their contents in the air after a coal fire. The environment effect of the coal fires is tremendous and thus does not lead to the halt of work. The results and analysis will show the need for an automated robot to analyze the environment for the health and safety of workers. Concentrations of CO_2, CO, CH_4, and H_2S in vent emissions were determined with an Industrial Scientific MX6 iBrid portable gas analyzer. The iBrid gas analyzer was calibrated with zero and 2000 ppmv standards prior to each field campaign.

Another research [8] on the Assessment of mine ventilation systems and air pollution impacts on artisanal tanzanite miners at Merelani, Tanzania, shows the study of air quality in five mines in Tanzania. The researcher set up the study for these five mines. A study done was by [9] on terahertz measurement of indicator gas emission from spontaneous coal combustion at low temperature. National Natural Science Foundation of China supported the study. The experiment was conducted on the temperature range between 0 to 200 degrees Celsius. The gas emissions in the mine were SO_2 and CO. In the experiment three samples of coal according to their ranking were studied and the. The researchers use gas coal, long flame coal and fat coal for the experiment to analyze and evaluate the emissions for coals of different quality ranks, with the increasing oxidizing temperature, the CO emission rise for any sample. The long flame coal releases a massive amount of CO in the air at higher temperature. The experiment is conducted to understand the effect of coal spontaneous combustion at low temperature, and the study was successful; however, the researchers could have employed more sensors to get the readings of other toxic gases.

In one research [10] on the study of unmanned vehicles (robot) for coal mines, talks about the deaths fatal accidents that happened in the mines in India. The Indian mining industry contributes to 8% of GDP and is in constant growth. The research suggests the use of the unmanned robot for the inspection of the geo-mining environments inside the mines through wireless communication. The aftermath of the attack on the world trade center lead to the introduction of robots for urban search and rescue. The study suggested different robots that have been developed and discusses about the problems that are faced by these robots. The problems addressed by the researchers are divided into three categories, addressed of searching, inspection and mapping and debris removal. In summary, their needs to be smarter software and hardware solutions for the robot to provide better inspection and analysis for the rescue teams, while maintaining better navigation through the complex mine structures.

IOP Conf. Series: Materials Science and Engineering 1045 (2021) 012001 doi:10.1088/1757-899X/1045/1/012001

According to [11] rescue robots can be classified into four major types as follows:

1. Unmanned Ground Vehicles (UGVs) - These mobile robots are best suited for the maneuver on ground.
2. Unmanned Surface Vehicles (USVs) - These robots are better suited for maneuver on water surface and are unsuitable for ground movement.
3. Unmanned Underwater Vehicles (UUVs) - These robots are designed and developed to be water proof and can move underwater.
4. Unmanned Aerial Vehicles (UAVs) - These robots can be called drones. They maneuver in the air and may meet ground for takeoff and landing.

An efficient robot can be either one of the above or a combination of any of the four. While they need to be developed according to the required factors

Table 1: List of Rescue Robot

Robot	Figure
Wolverine V2 Reddy, Kalyan & Murthy[12]	
Gemini scout Reddy, Kalyan & Murthy[12]	
Numbat Reddy, Kalyan & Murthy[12]	
Ground hog Reddy, Kalyan & Murthy [12]	
Sub-terranean Robot Ray et al., [13]	
GMRI Kasprzyczak, Trenczek and Cader,[14]	

In one study [11] compared the performance of multiple mine rescue robots. Some of these robots were designed for military purposes and later developed into rescue robots, while some were developed for research purpose only. The focus of this research was on the Unmanned Ground vehicles; the robots that took the focus of this research were highly functioning yet had many limitations.

Table 1 illustrates the comparison of the robots and their results. According to the findings, not many robots have been successful due to the complexity in the environment. While the only robot that had actual success and is still employed in the industry is the groundhog. However, GMRI and the Mine Bot are not industrial robot-like groundhog [14], wolverine V2, Gemini scout, they showed satisfactory result, and the worst performance was of Wolverine V2 as it did not serve the purpose it was designed for.

Another study conducted by [15] on the hazardous gas detecting rescue robot in coalmines. The instruments inside the robot helps to tolerate the rest of the subsystem and assists in control by the control system, the movement of the robot and a communication system that enables in the efficient transmission of data acquired by the camera and the sensors. In addition, it can carry some food and medicine to the miners trapped in the disaster. However, the robot was never deployed in a mine but tested inside a controlled environment. The limitations of this robot are that it is not flameproof, and that can be a big set point inside a mine. The robot also does not work efficiently in higher temperature; 40-degree Celsius is the temperature set point of the robot. Besides, the higher transmission is needed, as the robot cannot travel a greater distance. Changes to the camera also need to get better quality images inside the mine.

In one study by [3] presents a various robotic system for harsh environments and safety application. The robotic systems present in the study are developed to assist human to monitor and can be deployed in various types of places to measure different parameters. The applications are these robots at various harmful environments such as pipes, mines, and tube systems, for real-time monitoring the environment to reduce the economic and life risks. Each of these robots is equipped with different sensors/actuators, wireless communication networking and navigation system.

In the study [16] robot was deployed at the pyramid as while people survey the pyramid for evidence and research problems, damage can occur to the pyramid. The robot is deployed for video survey while protecting the pyramid from damage, and the robot was successful as no surface marks were left at the shaft walls. During the lab testing, the Minebot successfully performed its task; however, the drawback was its operational distance of 200m only. The other robot developed is LetterBot. The LetterBot was developed to access buildings with ease through the letterbox. The robot performance was successful; however, due to its complexity, it is not very reliable. The Minebot faced with the limitation due to its weight, so it cannot go more than 200m, and however, adding more weight will require the reconfiguration of the whole robot. The limitation for the Djedi robot is that the climbing speed is too long; in the experiment, it took the robot four hours for its mount.

According to the intensive research works done on mobile inspection robots, the robots are as good as their designer. They need to be designed and developed according to the factors that can affect its functionality and efficiency. The robots mentioned in Table 1 faced with several limitations. The wolverine v2 was too heavy and required too much power to operate. The robot transmits data through the fiber optic cable. It is due to the rather constant evolution of the mine and its unstructured nature that there are major limitations for the robot's application in the underground mines. The major robot, as addressed in Table 1, are heavy and bulky that limits their movement and are not suited for easy transportation. Their deployment in the mines may arise the risk of failure to maneuver and may lead to the risk of damage. The communication methods applied by many of the robots, as mentioned in this section, are not reliable and not efficient. The application of tethered communication method applied by these robots introduces different unnecessary complications for the maneuver. The cables prove data transmission, and the feed is efficient, but due to wear and tear, it compromises the robot's ability. The wolverine V2 was not a big success, and it faced many problems with the movement as the cables limited its range and were being overrun by the robot. The other problem was with the wireless transmission, as the mine structure is so complicated the major problem that arises for the developers is that of the communications. Hence, before the deployment of the robot, the structure of the specific mine should be studied and accordingly, the best solution needs to arise. This dilemma can be very challenging as the consistent wireless communication that was used by the robots were not reliable as it required the user to maintain the line of sight. The other factors concerning the limitations were of the presence of water, low roofs heights, and the wiring inside the mines.

3. RESEARCH METHODOLOGY

Figure 1 illustrates the block diagram of the proposed system. The system comprises of two sensors, MQ7 that gives the readings for the Carbon Monoxide content of the air and DHT11 gives the readings for the humidity and the temperature. The L293D Motor Controller IC is connected to the Arduino where it communicates with the microcontroller to output signals for the control of the DC motors. The Micro Controller controls the purposed system by acting as the channel for different components. The Mobile application is connected to the Arduino microcontroller using the HC06, and the power supply is connected to the DC motor and the Arduino.

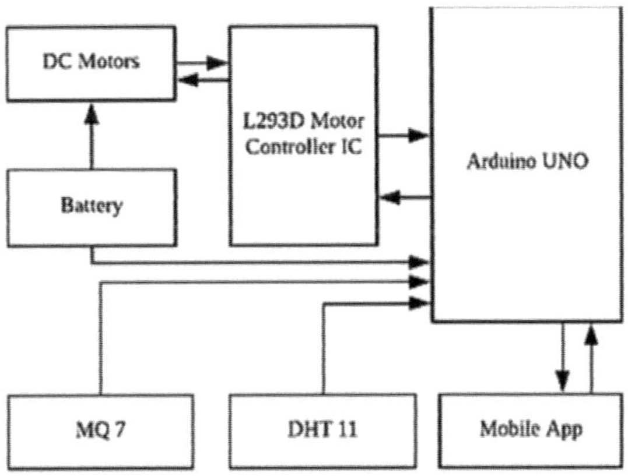

Figure 1. Methodology Schematic diagram

The mobile app's block diagram is illustrated in Figure 2; the mobile app works in coordination with the Arduino Uno using the Bluetooth to transfer data wirelessly. The application is developed using the MIT inventor application, an open-source web application and can be downloaded from MIT inventor or later from Google play store. The user must enable the Bluetooth of their cell phone and pair up with the HC-06 Bluetooth module, attached to the mobile robot. The mobile app allows the user to control the robot and get the data from the sensors. The objective of the mobile app is to allow the user to control the mobile app while simultaneously display the output data from the sensors. The data flows in two directions; one is the output for the movement of the prototype and inputs the data from the sensor system to display on the GUI.

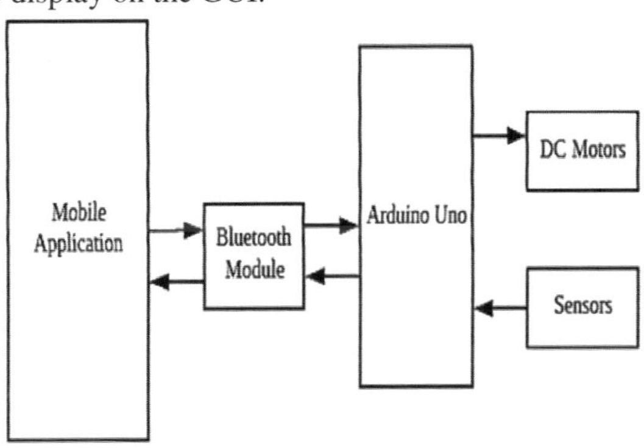

Figure 2. Signal flow for the Proposed System

Figure 3 illustrates the working principle of the proposed system and can be referred for visual understanding. According to Figure 3, the mobile robot and the mobile application is initialized. During the process, the built-in code is triggered; this is the internal self-check that illustrates the robot's movement capabilities. Following this process, open the mobile app on the smartphone and enable the Bluetooth on the device.

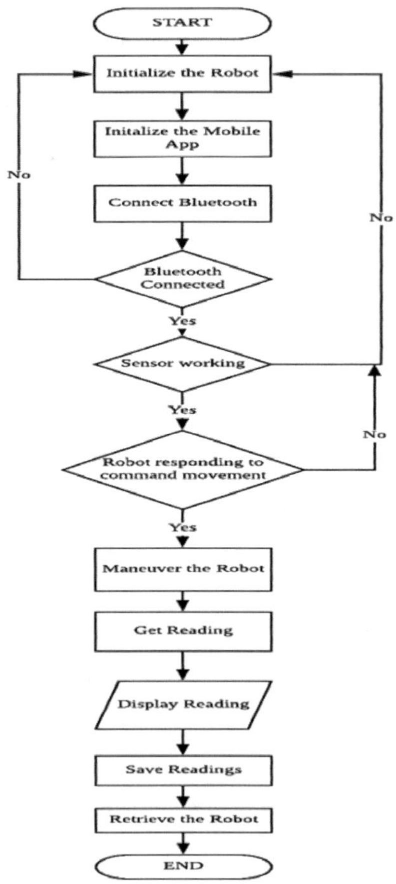

Figure 3. Working principle Flow Chart

The next stage is ensuring a wireless connection between the mobile robot and the mobile robot, which is done by connecting the Bluetooth on the prototype and the mobile app. The mobile application will illustrate four sensor readings; this is the self-test of the sensor system. If there are no readings illustrated in the GUI of the mobile app, it is recommended to initialize the mobile robot and the mobile application again. The user must send movement instruction to the mobile robot to test its response. If the prototype is not responding, once again a system initialization is needed due to connection issue.

Once the mobile robot and the sensor system are working fine, the proposed system can be deployed to carry out the task. The GUI's control panel can command the mobile robot for maneuver on the field and get the sensor readings on the mobile application. Once the task is finished, the readings can be saved and used for analysis. The robot should be retrieved from the site, and power should be offed.

4. EXPERIMENTAL RESULT

This section illustrates the hardware design and simulation results of the proposed system and the mobile app. Figure 4 shows the final design of the GUI developed for the mobile app. The GUI was designed to illustrate the data of the sensor and enable the user to control the robot wirelessly. The

GUI created consists of the Bluetooth connection button that enables the connection with the mobile robot. To control the mobile robot, a control panel was created in the GUI that consists of four buttons, illustrated with the arrow images. These buttons enable the movement control of the robot.
Moreover, the red octagon button in the center of the control panel acts as a safety button for the robot. This button stops any movement by the robot and halts the data output from the robot. The GUI also consists of the data display space labelled as the sensor system data. The sensor system in a mobile robot continuously transmits data to the mobile app, as illustrated in the bottom of the GUI. The data displayed is for the humidity and temperature levels. It also displays the temperature readings. All the readings are real-time. However, in the code, there is a time delay to ensure the GUI is not spammed with the readings, and the time delay is set to 2 seconds.

In addition, Figure 5 illustrates the real-life testing of the GUI. The proposed system connected to the mobile application and the sensor system is sending the data for the temperature and the humidity level. The results are shown at the bottom of the GUI, and the font is a yellow color. The readings are from an indoor location hence high humidity of 78% and 28 C temperature.

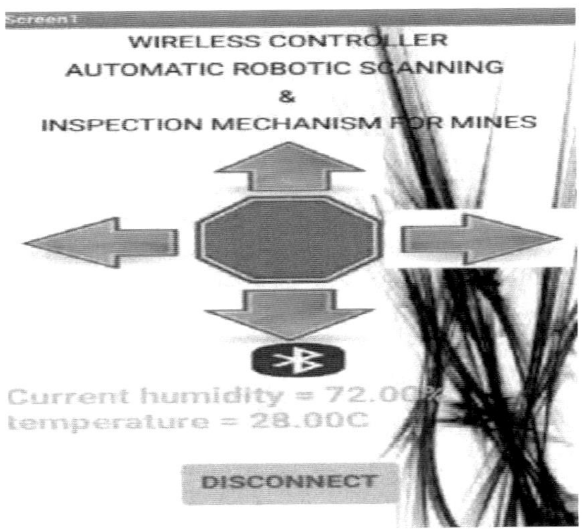

Figure 4. GUI: Humidity and Temperature results

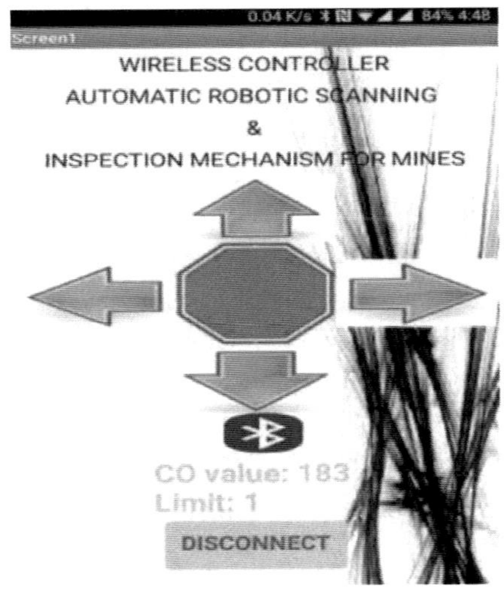

Figure 5. Carbon Monoxide results

Figure 4 illustrates the sensor data displayed on the GUI. It shows the result of the Co, temperature, and humidity together. If the level of Co is exceeding the safe amount, the font of the text with the change to red and the limit will show as one. Figure 6 illustrates the final design of the experimental mobile robot. The sensors and Bluetooth module are placed on the breadboard, while the Arduino Uno is wired with the breadboard and all the electrical components are attached on the top side of the acrylic body of the prototype. The Bluetooth module allows communication between the mobile application and the prototype. Moreover, the battery is connected towards the back of the main body. Lastly, the blinking LEDs for each component illustrates the working condition and that correct initiation has been achieved.

Figure 6. Experimental Mobile Robot

There are four types of testing conducted in order to analyses the accuracy of the proposed system. The testing as follow:

4.1. Carbon Monoxide

The sensor system is exposed to clean air in an indoor location and then placed in an open environment. The readings are conducted periodically. However, the last subtests are conducted with high exposure to CO gas using burning items such as near a cigarette, candle, and a burning paper. Moreover, for the exposure, the sensor system is placed inside a glass container with the burning item. To know the correct value of the air toxicity, a digital CO sensor was placed near the sensor system this ensures that the sensor is working according to the industrial standards and if there is a need for further calibration. There are 10 Sample tests conducted which consists of five indoor locations and five outdoor locations. The code is altered to get a single reading every 30 seconds for 50 samples for indoor locations and 30 seconds for 50 samples for outdoor. A carbon monoxide sensor was used to correlate with the readings from the sensor system, to understand the accuracy and the quality of results.

By measuring the carbon monoxide content in the indoor locations and the outdoor locations, the experiment illustrated that the results varied according to the locations. The indoor location had an accuracy of 94.4%, which is due to the limited air contents, and needed introduction of carbon monoxide contents in the air, Figure 7 illustrates the accuracy of the indoor results. The accuracy of outdoor locations was 94.8%, as illustrated in Figure 8. The number of gases present in the air is limited indoors. At the same time, outdoor there are different varying gases and includes higher carbon monoxide content due to the emissions from vehicles and other machines. However, the accuracy for the indoors is between 92% and 96%, while the accuracy for the outdoors is between 88% and 98%. Furthermore, according to the results, the accuracy of the indoors and the outdoor is nearly the same with indoor at 94.4 % and outdoor at 94.8%. Hence, it can be concluded that the accuracy of the system is useful in regarding the location.

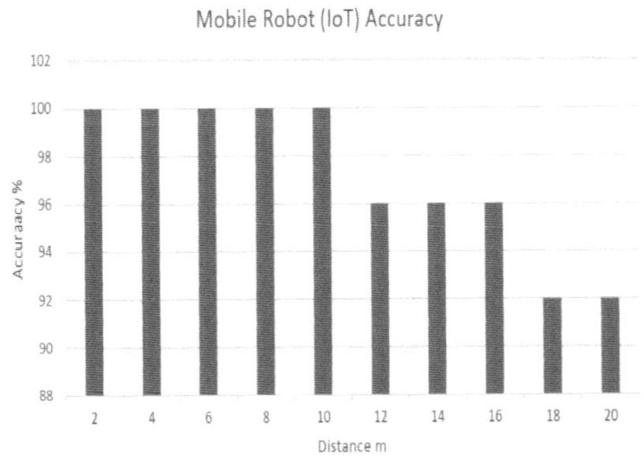

Figure 7. Co accuracy indoors

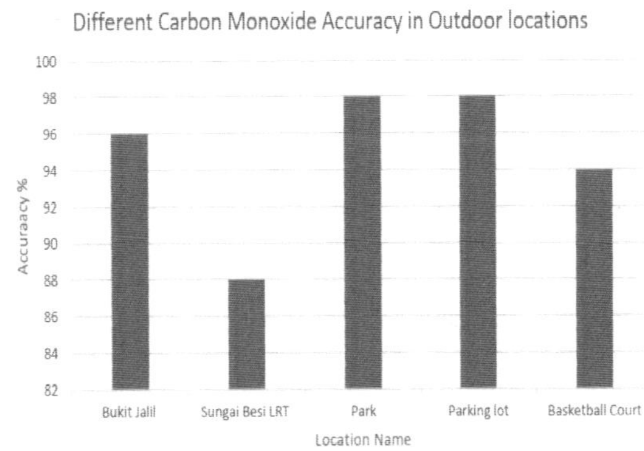

Figure 8. Co accuracy outdoors

4.2. Temperature Sensor

The sensor system was conveyed in a various environment with varying conditions. The readings were taken at an interim of 1 minute. A sum of 50 samples was collected at each location, and the outcomes acquired were noted down. Nevertheless, a digital thermometer was conveyed with the system with same exact conditions and time interims. The readings collected from the thermometer were noted down for correlation with the sensor system.

Figure 9 shows the results of testing the accuracy of the temperature sensor in the proposed system with a correlation with a digital temperature sensor. The results illustrate that the temperature sensing capability of the proposed system is excellent, with an accuracy of 97.80%. Figure 9 shows the highest accuracy of 100% at three locations, APU University, parking lot and south city, while the lowest accuracy of 92% at the Sungai Besi LRT. The highest being indoors while the lowest at the outdoor. However, with the lowest accuracy of 92% after taking 500 training samples at 10 different locations, it can be concluded that the temperature sensor is very reliable. While the reason for the lower accuracy of 92% at the LRT station can be assumed that due to the constant change in the environment that is due to the moving cars and trains, with many people moving around, these variables could have affected the accuracy of the sensor.

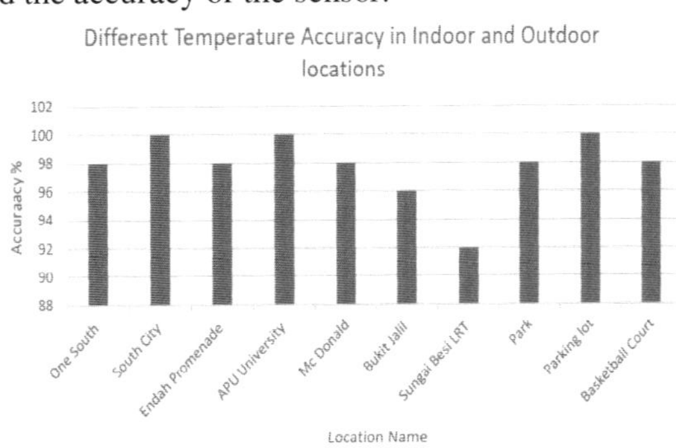

Figure 9. Temperature accuracy indoors and outdoors.

4.3. Humidity Sensor

The sensor system was placed in locations with high humidity and low humidity. Moreover, to measure the legitimacy and the accuracy of the humidity sensor in the proposed system, a digital humidity sensor was placed next to the system to get real-time values. Figure 10 illustrates the humidity accuracy of the proposed system. The results obtained show the minimum accuracy of 88% at South City with the highest accuracy of 100% at APU University, LRT, and Park. While, the accuracy results of the other locations are varying between 94% and 96%. The overall humidity accuracy is the lowest in the indoors as the humidity is constantly changing due to the human inputs as the area of testing is a small room a slight change can have a more significant impact on the results.

However, the accuracy of the humidity sensor of the proposed system is 96% with ±4 error; the results show that the humidity sensor is reliable and gives good results.

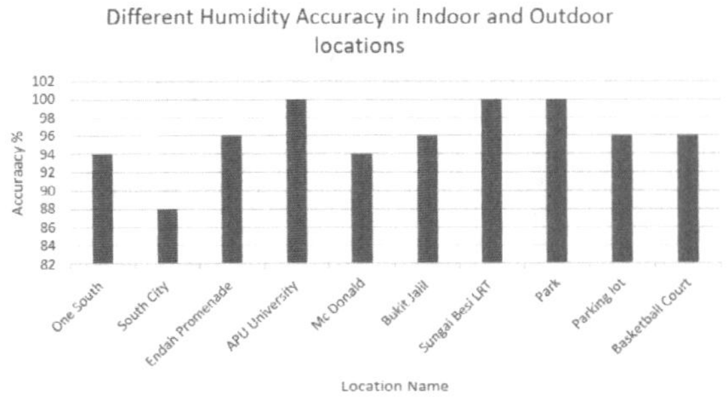

Figure 10. Humidity accuracy

4.4. IoT Experimental Results

The system was tested to know the accuracy of the response of the experimental robot over the distance of 2m to 20m; Figure 11 illustrates the results of 250 commands sent from the mobile app to the mobile robot and the correct response comprehended by the bot. Moreover, Figure 12 illustrates the time delay in response by the mobile robot over the varying distance from the smartphone.

The accuracy is varying from perfect 100% for 2m to 10m and then gradually decreasing to 92% at 20m. The decrease in the accuracy is due to the increasing distance and the loss of packets in the connection of Bluetooth. However, the accuracy is still decent, and the number of wrong output movement commands are two at 20m. Moreover, the maximum delay in response at 20m is of 1.00s, which is also acceptable; however, it can be improved. Hence, the initial delay is of 0.50s, which is due to the human error. This delay could be reduced using alternative automatic ways of calculating. However, the delay is also caused by the code in the mobile app and the communication between the two machines.

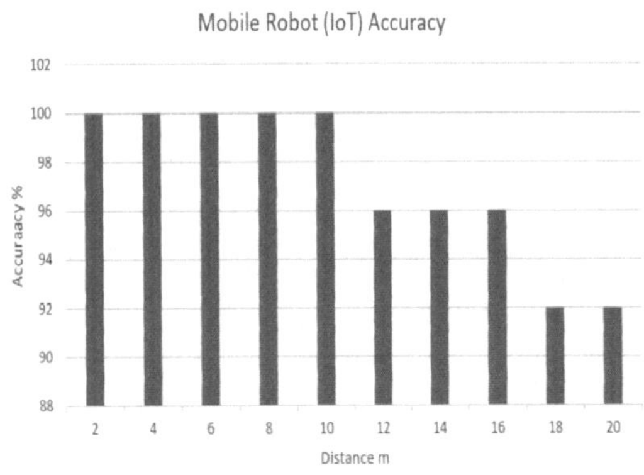

Figure 11. IoT experimental results

Figure 12. Response Time for Mobile Robot

5. Conclusion

To conclude, a mobile robot was designed using 3D printers and developed to go inside the mine. A sensor system was developed to detect the temperature, humidity, and carbon monoxide gas content in the air. For prototyping purpose, a mobile application was designed and developed to pilot the robot and to illustrate the data from the sensor system. The mobile application was developed to be deployed on the smartphone to be more sustainable.

For future work, Bluetooth 5.0 can be employed in the system that would give a greater range with improved speed. Increasing the price but simultaneously increasing the range to 60 meters and this would be a major upgrade to the system. Hence, the users can send and control the mobile robot from a higher distance. Moreover, the speed would be increased to two Mbps that would drastically reduce the system response delays, hence giving faster results and reducing the errors.

Furthermore, the proposed system could increase the number of gas sensors to detect more gases that are toxic. Hence, enabling more analysis and data content for the users. Moreover, use four dc motors instead of two as greater road grip and ease of maneuver. This would increase the movement speed of the robot also.

Lastly, developing a database for the sensor data and enabling sharing. Currently, the proposed system does not have the database for the results, and this limits the functionality, introducing it would be more user friendly and useful.

6. References

[1] Rabeiy, R.E., ElTahlawi, M.R. and Boghdady, G.Y., 2018. Occupational health hazards in the Sukari Gold Mine, Egypt. *Journal of African Earth Sciences*, 146, pp.209-216.

[2] Verma, S. and Chaudhari, S., 2017. Safety of workers in Indian mines: study, analysis, and prediction. Safety and health at work, 8(3), pp.267-275.

[3] Liu, J., Ma, B., Fry, N., Pickering, A., Whitehead, S., Somjit, N., Richardson, R. and Robertson, I. (2015). Exploration Robots for Harsh Environments and Safety. *IFAC-PapersOnLine*, 48(10), pp.41-45.

[4] Iwano, Y., Osuka, K. and Amano, H., 2004, December. Development of stretcher component robots for rescue activity. In *Robotics, Automation and Mechatronics, 2004 IEEE Conference* on (Vol. 2, pp. 915-920). IEEE.

[5] Hainsworth, D.W., 2001. Teleoperation user interfaces for mining robotics. Autonomous Robots, 11(1), pp.19-28.

[6] Green, J., 2013, October. Mine rescue robot's requirements outcomes from an industry workshop. In Robotics and Mechatronics Conference (RobMech), 2013 6th (pp. 111-116). IEEE.

[7] Dunnington, L. and Nakagawa, M. (2017). Fast and safe gas detection from underground coal fire by drone fly over.

[8] Mayala, L.P., Veiga, M.M. and Khorzoughi, M.B., 2016. Assessment of mine ventilation systems and air pollution impacts on artisanal tanzanite miners at Merelani, Tanzania. *Journal of Cleaner Production*, 116, pp.118-124.

[9] Wang, Y.M., Wang, W.Z., Shao, Z.L., Wang, D.M. and Shi, G.Q., 2013. Terahertz measurement of indicator gas emission from coal spontaneous combustion at low temperature. Ecological Chemistry and Engineering S, 20(4), pp.709-718.

[10] Subhan, M.A., Bhide, A.S. and SSGB COE, B., 2014. Study of unmanned vehicle (Robot) for coal Mines. International Journal of Innovative Research in Advanced Engineering (IJIRAE), 1(10), pp.116-120.

[11] Murphy, R. (2004). Activities of the rescue robots at the World Trade Center from 11-21 september 2001. IEEE Robotics & Automation Magazine, 11(3), pp.50-61.

[12] Reddy, A.H., Kalyan, B. and Murthy, C.S., 2015. Mine Rescue Robot System–A Review. Procedia Earth and Planetary Science, 11, pp.457-462.

[13] Ray, D.N., Dalui, R., Maity, A. and Majumder, S., 2009. Sub-terranean robot: A challenge for the Indian coal mines. The Online Journal on Electronics and Electrical Engineering, 2(2), pp.217-222.

[14] Kasprzyczak, L., Trenczek, S. and Cader, M., 2013. Pneumatic Robot for Monitoring Hazardous Environments of Coal Mines. In Solid State Phenomena (Vol. 198, pp. 120-125). Trans Tech Publications.

[15] Reddy, T.K. and Krishna, G.B.S., 2014, May. Hazardous gas detecting rescue robot in coal mines. In *Proceedings of IRF International Conference*, 13th April-2014, Chennai, India, ISBN (pp. 978-93).

[16] Narula, P., Gutheim, P., Rolnitzky, D., Kulkarni, A. and Hartmann, B., 2011. MobileWorks: A Mobile Crowdsourcing Platform for Workers at the Bottom of the Pyramid. *Human Computation*, 11(11), p.45.

[17] Kasprzyczak, L., Trenczek, S. and Cader, M. (2013). Pneumatic Robot for Monitoring Hazardous Environments of Coal Mines. *Solid State Phenomena*, 198, pp.120-125.

[18] Widiatmojo, A., Sasaki, K., Sugai, Y., Suzuki, Y., Tanaka, H., Uchida, K. and Matsumoto, H. (2015). Assessment of air dispersion characteristic in underground mine ventilation: Field measurement and numerical evaluation. *Process Safety and Environmental Protection*, 93, pp.173-181.

[19] Cheng, B., Cheng, X. and Chen, J. (2015). Lightweight monitoring and control system for coal mine safety using REST style. *ISA Transactions*, 54, pp.229-239.

Electrochemical self-assembly of bunch-shaped ZnO nanowire without catalysts

L S Chuah[1], M F Shafiee Haron [1] and Y Yusof[2]

[1]Physics Section, School of Distance Education, Universiti Sains Malaysia, 11800 Minden, Penang, Malaysia.
[2]School of Physics, Universiti Sains Malaysia, 11800 Minden, Penang, Malaysia.

Email: chuahleesiang@yahoo.com

Abstract:
There are many types of metal oxide nanostructure and one of them is zinc oxide (ZnO). Despite of vast field of application for this material, the fabrication process of nanostructure (ZnO) is rather costly or complicated due to exotic equipment or reagent required. To the leading of our understanding, there are no details about synthesizing the bunch-shaped ZnO nanowire patterns on silicon substrate. A series based on aqueous medium composition: (x) Zn $(NO_3)_2$ + 0.1 NaOH, where $0.005 \leq x \leq 0.2$ mol% is prepared using chemical bath method. The distribution and size of ZnO nanostructure was determined using FESEM spectroscopy. Meanwhile, EDX spectra show the presence of ZnO nanostructures in all samples. These researches generate new knowledge on embodiment of ZnO nanostructures on silicon thin film which may useful for finding material with high optical quality for diverse application in photonic devices.

Keywords: ZnO, nanostructured, thin film

1. Introduction

In the past few decades, there have been more studied in structural and optical behaviour of the ZnO nanostructures due to their technology and commercial applications. Thin film deposited with ZnO nanostructures presented unbelievable chances in photonic substances domain since of their vary of implementations by broaden from light energy conversion to detection, random laser, gas sensor, light emitting diode and solar cell technology [1-3]. Despite of that, the characterize of ZnO vigorously rely on its morphology and microstructure in those applications [4]. Hence, the investigation towards control the shape, size and microstructure of ZnO become crucial [5]. The deposited of ZnO material on substrates even though remain as a challenging issue due to the basic physics behind the properties of material deposition still are lacking.

Recently, there are many techniques available have been used to deposited the ZnO materials on the thin film substrates. In general, the depositions of material on the substrates are categorized in physical or chemical. However, the chemical bath deposition technique have attracts extensively attention of many researcher owing to its feature of the sort as low temperature, low cost, simple and easily coating of the substrate surface compare to the physical technique which using more sophisticated equipment [6-7]. Despite much research the mechanism of growth of ZnO nanostructure using chemical bath deposition is far away from being realize. Orderly sample emulsion with optimized the solution concentrations, controlled growth of nanostructures and particular characterization are required.

Synthesis and characterization of thin film deposited with nanostructure have been studied widely through optical, structural and morphology. ZnO materials become one of a good candidate due to their characteristic like outstanding chemical firmness, non-poisonous, good electrical, visual and electrifying property. However, the utilization of ZnO nanostructure materials applications which generally needs the crystalline morphology, direction and surface of nanostructures development need to be well controlled in order as optoelectronic material used in photonic and electronic application. To overcome this limitation, the use of optimization the solution concentration for aqueous medium in chemical bath deposition has been found a successful way to control the growing of ZnO nanoparticles stick on thin film substrate. Systematic experimental study will be made to analyze the structural and morphology of ZnO nanostructure. To recognize and ascertain the procedure of ZnO nanoparticles we will prepare a series of silicon substrate deposited using zinc nitrate and sodium hydroxide as aqueous medium in chemical bath deposition and characterize them.

In this study, the investigation will be carried by deposited ZnO nanostructures on silicon thin film. The morphology of ZnO nanostructures in substrates will be investigated to complete the samples with optimum utilization aqueous medium compositions applying chemical bath techniques. The interpretation on the process of ZnO nanostructures restoring the morphology and structural characteristics can be applied in the growth of photonic technology. This work will create the knowledge to enhance the morphology and structural properties of ZnO nanostructures deposited for wide application possibilities. Alert sample composition, controlled of aqueous medium composition, ZnO nanostructures growth of desired sizes, shape and pattern is the difficult piece of work that will be talented. Evaluation of morphology and structural aspect are the principal aim to be attained.

2. METHODOLOGY

The sequence of sample was conformation using chemical bath technique. The substance for the sample preparation of ZnO nanostructures are practically in using silicon as substrate meanwhile zinc nitrate as precursor and sodium hydroxide for the aquas medium. The samples were prepared with dissimilar concentration of zinc nitrate due to find the optimization of solution concentration for ZnO nanostructures fabrication.

First, the silicon substrates where the deposition of ZnO will take place were prepared respectively. Then, the aqueous medium which used in this chemical bath deposition was prepared using the reagent zinc nitrate and dissolve in common chemical reactive salts, sodium hydroxide. Following, 60 ml mixture of zinc nitrate and sodium hydroxide were put in a beaker. The silicon substrates which prepared under typical conditions were standing suspended in the beaker, meanwhile water bath was used at constant temperature of 85 - 90 ^{0}C. Generally, water bath is used to obtain the homogeneous temperature for this experiment and the deposition time was 90 minutes. The silicon substrates with deposited films were removed, rinsed with distilled water and pull out of dry after 90 minutes of deposition. The as-deposited ZnO thin films were also heated up at 200 ^{0}C and 300 ^{0}C in a furnace for 20 minutes.

All of the samples was fabricated by subsequent the same procedure and keep in vacuumed desiccators to delete any contamination and moisture. The characterization of prepared sample will be using various spectroscopy techniques. Figure 1 depicts the experimental of sample preparation meanwhile Figure 2 show the schematic diagram for preparation of sample.

IOP Conf. Series: Materials Science and Engineering 1045 (2021) 012002 doi:10.1088/1757-899X/1045/1/012002

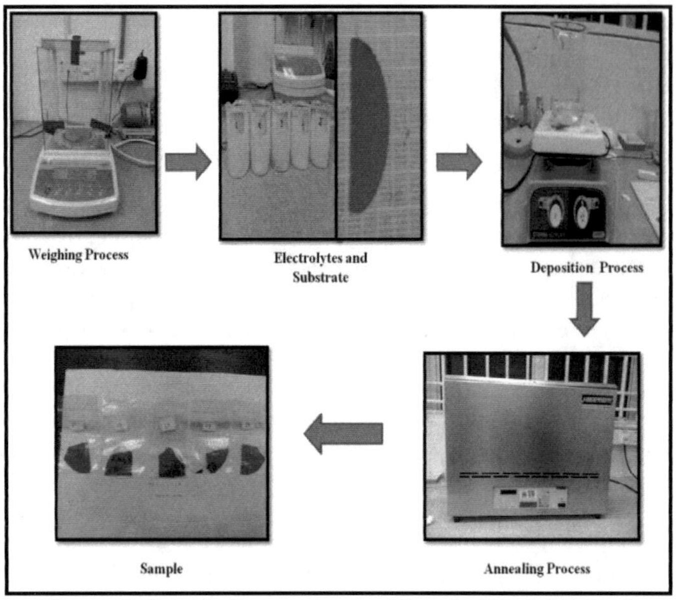

Figure 1. Sample Preparation by Chemical Bath Method.

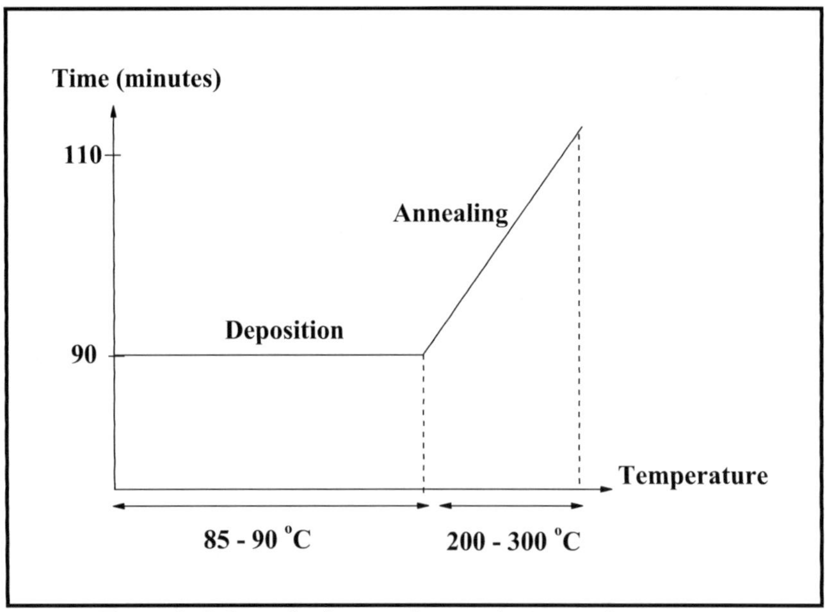

Figure 2. Schematic diagram for sample preparation.

This experiment was used analytical reagents zinc nitrate and sodium dioxide as aqueous medium in the chemical bath. The compositions for aquas medium in order to fabricate the ZnO nanostructures in the sample are:

$$(x)\ Zn(NO_3)_2 + 0.1\ NaOH$$

where $0.005 \leq x \leq 0.2$ mol%. The possible chemical reaction in the solution mixture in order to obtain the ZnO nanostructure is as follows:

$$NaOH \longrightarrow Na^{+} + OH^{-}$$

$$Zn(NO_3)_2 \longrightarrow Zn^{2+} + 2NO_3^{-}$$

$$Zn^{2+} + 2OH^{-} \longrightarrow Zn(OH)_2$$

$$Zn(OH)_2 \longrightarrow ZnO$$

During the growth process, first ZnO nucleus growth takes place which then aggregates and produces ZnO nanoparticles by Ostwald ripening. Nanoparticles crystallize and aggregate with each other through Van der Waals forces and hydrogen bonding and give ZnO nanostructures. Figure 3 depicts the schematic diagram for chemical bath deposition technique.

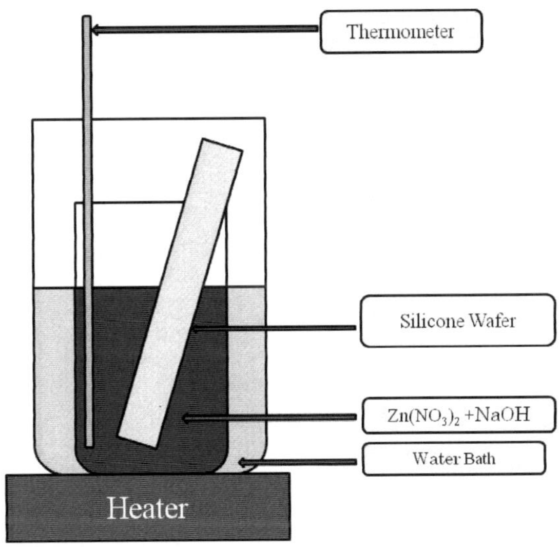

Figure 3. Schematic Diagram for Chemical Bath Deposition.

Generally, the study of this research is started by preparation of silicon (Si) wafer. Then, the deposition of ZnO are used an appropriate amount of zinc nitrate ($Zn(NO_3)_2$) (purity, 98.0%) and sodium hydroxide (NaOH) (99.0%). The series of sample was prepared and followed by the detail structural and morphology properties were investigated using several spectroscopic measurements.

The calculation to measure the analytical reagent for aqueous medium since the chemical bath deposition is a technique which control the homogenous precipitation of substance in the solution are used as equation 1:

IOP Conf. Series: Materials Science and Engineering 1045 (2021) 012002 doi:10.1088/1757-899X/1045/1/012002

$$\rho = mV \qquad (1)$$

Where, ρ is density of solution, m is mass of solution and V, is volume of solution. The nominal composition of aqueous medium for this study is shown Table 1 meanwhile Table 2 represent the experimental configuration.

Table 1. The composition of zinc nitrate concentration.

Sample	Concentration of $Zn(NO_3)_2$	Volume of Aquas Medium	Volume of $Zn(NO_3)_2$
Sample 1	0.005 M	60 ml	0.3 ml
Sample 2	0.01 M	60 ml	0.6 ml
Sample 3	0.05 M	60 ml	3.0 ml
Sample 4	0.10 M	60 ml	6.0 ml
Sample 5	0.20 M	60 ml	12.0 ml

Table 2. Experimental Configuration.

Sample	Temperature (°C)	Duration (minutes)	Volume (ml)	Concentration
Sample 1	80-90	90	60	0.005 M $Zn(No_3)_2$ + 0.1 M NaOH
Sample 2	80-90	90	60	0.01 M $Zn(No_3)_2$ + 0.1 M NaOH
Sample 3	80-90	90	60	0.05 M $Zn(No_3)_2$ + 0.1 M NaOH
Sample 4	80-90	90	60	0.10 M $Zn(No_3)_2$ + 0.1 M NaOH
Sample 5	80-90	90	60	0.20 M $Zn(No_3)_2$ + 0.1 M NaOH

3. RESULTS AND DISCUSSIONS

According to Urgessa, the ration of concentration in aqueous medium can affected the shape, size and distribution of ZnO deposited on substrates. From the Figure 4, the bunch-shaped ZnO nanowire arrays were observed. More nanowires intended to aggregate to form bunch shape with addition of zinc nitrate at the same temperature and deposition time. The presence of bunch-shaped nanowires can be well explained by a great amount of $Zn(OH)_2$ precipitations in the aqueous solution. The formation of $Zn(OH)_2$ precipitations in the solution prohibited the forced hydrolysis of anhydrous zinc acetate and further growth of ZnO.

The micrograph also shows that the ZnO cover the substrate more uniformly and are aligned predominantly perpendicular to the substrate. The increasing of zinc nitrate change the orientation, shape and size of ZnO nanostructure can be shown from the sample 1 until to sample 5. Figure 4 illustrate the ZnO nanostructures obtained from different composition of zinc nitrate.

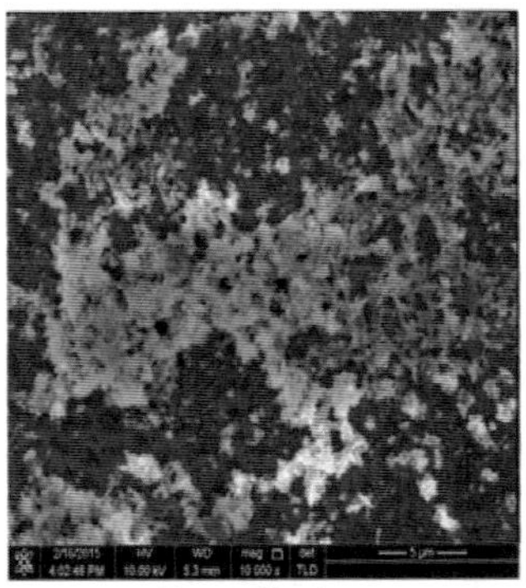
Sample 1

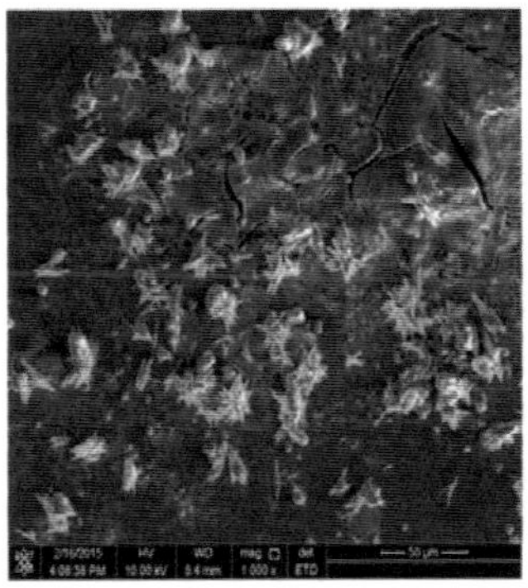
Sample 2

Sample 3

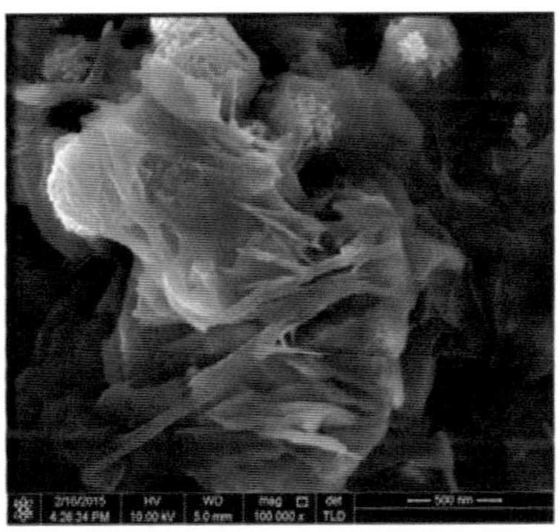
Sample 4

Sample 5

Figure 4. ZnO nanostructures obtained from different composition of zinc nitrate

Energy dispersive spectra confirm the presence of ZnO nanostructure in all samples since elements zinc and oxygen was detected. During deposition process, the oxygen molecular will be encounter chemical bonding with aqueous medium for shaping the ZnO nanostructures. The oxygen percentage increase proportional to the increment of zinc nitrate due to the interaction of hydroxyl group with crystallization of ZnO nanostructures [8-10]. Figure 5 displays the EDX spectra of the all sample where the elements (carbon, silicon, oxygen and zinc) present in the silicon thin film.

IOP Conf. Series: Materials Science and Engineering 1045 (2021) 012002 doi:10.1088/1757-899X/1045/1/012002

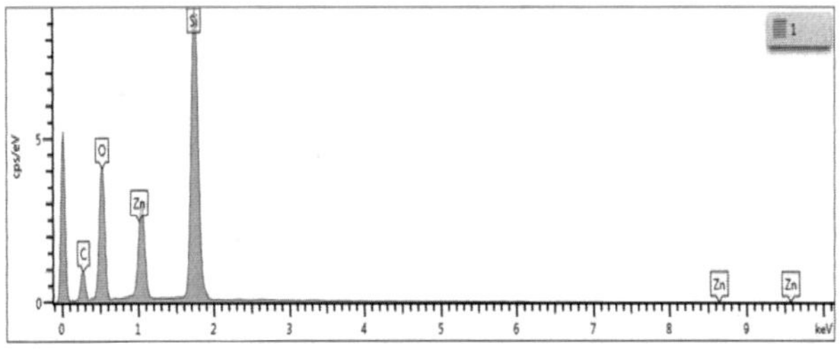

Sample 1

Element	Wt%	Wt% Sigma
C	17.28	0.64
O	25.24	0.41
Si	44.15	0.49
Zn	13.33	0.37
Total:	100	

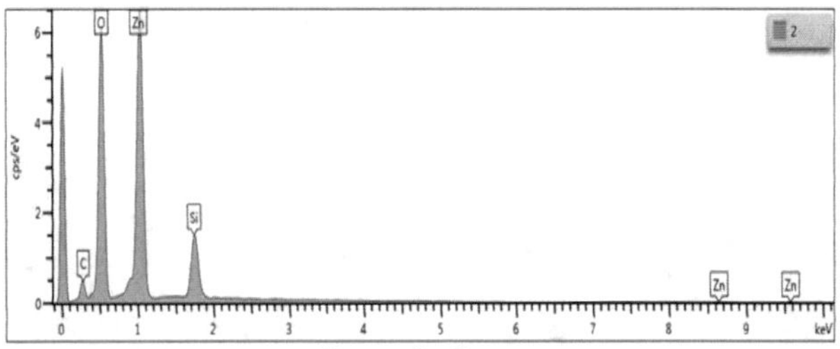

Sample 2

Element	Wt%	Wt% Sigma
C	7.46	0.49
O	34.31	0.46
Si	9.6	0.26
Zn	48.63	0.53
Total:	100	

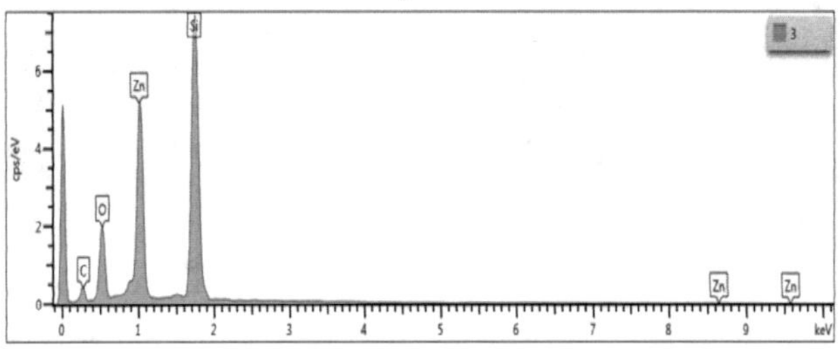

Sample 3

Element	Wt%	Wt% Sigma
C	8.93	0.65
O	12.53	0.33
Si	43.57	0.5
Zn	34.97	0.48
Total:	100	

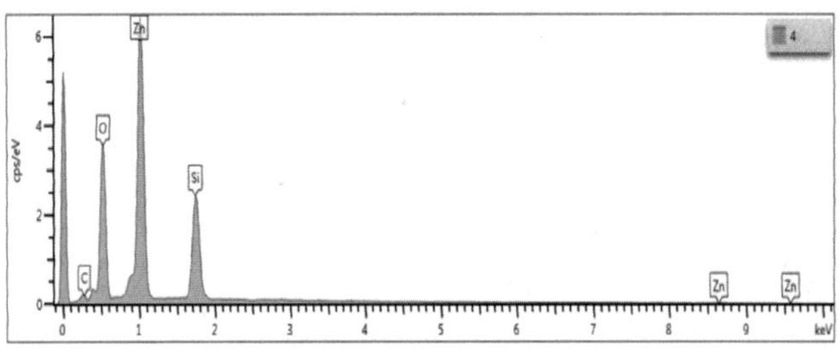

Sample 4

Element	Wt%	Wt% Sigma
C	3.6	0.53
O	23.85	0.43
Si	18.29	0.36
Zn	54.26	0.56
Total:	100	

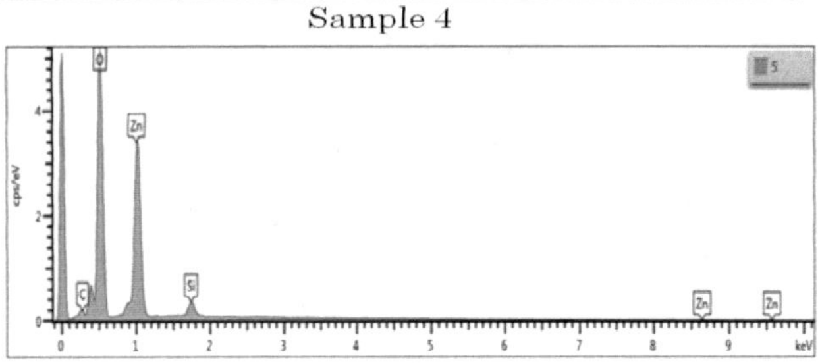

Sample 5

Element	Wt%	Wt% Sigma
C	3.78	0.58
O	44.91	0.64
Si	3.21	0.24
Zn	48.1	0.66
Total:	100	

Figure 5. EDX spectrums for all samples

4. CONCLUSION

Careful synthesis and thorough characterization of ZnO nanostructures was the focus task to achieve the objectives of this work. Firstly, the deposition was carefully prepared using chemical bath technique in order to control growth the crystallization ZnO. Incorporation of ZnO nanostructures was carried out to investigate the structural, morphology and optical properties. Consequently, the subsequent judgement is drawn establish on the analyzed data:

a. Series of silicon thin film deposited with ZnO based on aqueous medium composition: (x) $Zn(NO_3)_2$ + 0.1 NaOH, where $0.005 \leq x \leq 0.2$ mol% successfully prepared. All series of sample was prepared using different concentration of zinc nitrate in order to study the influence of composition on samples.

b. The micrograph obtained from FESEM spectroscopy shown the crystallization of ZnO nanostructure change by the increment of zinc nitrate. Sample 5 shown the ZnO nanorods array.

c. The EDX spectra showed the traces of all the elements (carbon, silicon, zinc and oxygen), present in the sample. Interaction of hydroxyl group with crystallization of ZnO nanostructures increases the oxygen percentage in samples.

Acknowledgments
This work was supported by the Bridging Incentive Grant (6316491) from Universiti Sains Malaysia.

References:

[1] K. Yu, Z. Jin, X. Liu, Z. Liu and Y. Fu, *Material Letters*, **61**, 2775, (2007).
[2] J. Deenathayalan, M.Saroja, M. Venkatachalam, P. Gowthaman, S. Shankar, "Natural dye sensitized solar cell using sol gel dip coated zinc oxide nanorods", *Elixir Thin Film Technology*, **46**, 8255-8258 (2012).
[3] J.Deenathayalan, M. Saroja, M. Venkatachalam, P. Gowthaman, T. S. Senthil, and S.Shankar "Fabrication of natural dye sensitized solar cell using hydrothermally grown zinc oxide nanorods", *International Journal of Nano science and Nanotechnology*, Vol **3** No:**2**, 81-89, (2012).
[4] L.S. Chuah, Z. Hassan, Nanoflakes of W/WO_3 Thin Films Grown by Electrochemical Anodization method, Nanoscience and Nanotechnology Letters, vol. **6**, num **6**, pp. 515-518, June (2014).
[5] L.S. Chuah, E. Kuang, S. S. Tneh, Z. Hassan, Synthesis, Characterization, and effect of concentration variation on metal oxide nanostructures, *Composite Interfaces*, vol. **21**, No. **3**, pp. 217-232 (2014).
[6] L.S. Chuah, Z. Mohamed, Z. Hassan, Microstructural and optical properties of SnO thin film by Thermal Evaporation, *Advanced Materials Research*, vol. 795, pp. 558-562, (2013).
[7] A. Mahmood, Z. Hassan, Y. Yusof, Y. F. Kwong, C. L. Siang, N. M. Ahmed, Structural and surface studies of undoped porous GaN grown on sapphire, *Advanced Materials Research*, vol. 620, pp.45-49 (2013).
[8] P. Hu, N. Han, D. Zhang, J.C. Ho, Y. Chen, Highly formaldehyde-sensitive, transition-metal doped ZnO nanorods prepared by plasma-enhanced chemical vapor deposition. *Sens Actuators B: Chem.*,169:74–80 (2012).
[9] M. Amirhoseiny, Z. Hassan, S.S. Ng, L.S. Chuah, M. A. Ahmad, Y. Yusof, InN Photoconductors on different orientations of Si substrates, *International Journal of Modern Physics* B, vol. **26,** no. **31**, 1250137-1250146 (2012).

[10] L.S. Chuah, M. S. Yaacob, Z. Hassan, Low temperature synthesis of Ni-doped SnO_2 thin films by spin coating route, *Optoelectronics and Advanced Materials Rapid Communications*, vol. 6, no. 1-2, 149-153 (2012).

IOP Conf. Series: Materials Science and Engineering 1045 (2021) 012003 doi:10.1088/1757-899X/1045/1/012003

Neural Networks Algorithm for Arabic Language Features-Based Text Mining

Faiez Musa L. Alrufaye[1,4], Hakeem Imad Mhaibes[2], Ali F. Neamah[3]

[1]Technical Institute/ Kut, Middle Technical University, Baghdad, Iraq
[2]Department of Accounting, Technical Institute/ Kut, Middle Technical University, Baghdad, Iraq
[3]Computer Science and IT Faculty, Wasit University, Wasit, Iraq
[4]Faiez.Alrufaye@mtu.edu.iq

Abstract. Text mining aims to understand texts correctly by utilising several phases to collect those features of Arabic words which are valuable and important to the applications mentioned above in making a correct decision. The technology then builds a strong system that relies on AI techniques, such as neural networks, to collect words in accordance with those features. An ANN is a collection of connected nodes called artificial neurons, which loosely model the neurons in a biological brain. Each connection, like the synapses in a biological brain, can transmit a signal to other neurons. An artificial neuron is one that receives a signal then processes it and can signal to neurons connected to it. The current study is concerned with building a system for analysing words in the Arabic language. This system can be included in any application to address the Arabic language, becoming part of it. The system generates strings for all names and pronouns appearing in the entered text and depends mainly on the automatic assembly of a set features by using neural networks. We implemented the system, with its two phases, on the documents in succession. The results were encouraging, ranging between 83% and 96%.

1. Introduction

Natural language processors have become one of the most important developments in computer science processing, as they provide a solution to people operating in many areas including Information Extraction, text summarization, machine answering, and other services[1].

Natural language processing includes two main phases; The first is the language understanding phase, which includes several sub-phases such as tokenize analysis, lexical analysis, morphological analysis, and syntactic analysis[2]. The second stage of natural language processing is the generation of language. In the first phase, researchers rely on several techniques to understand the language according to its required area[3].

The major challenge to the decision-making process in natural language processing is achieving accurate text understanding in the first stage (the stage of understanding natural language). If there is any incorrect understanding here, the decision will be wrong[4].

Text mining technology aims to understand texts correctly by utilising several phases to collect those features of Arabic words which are valuable and important to the applications mentioned above in making a correct decision. The technology then builds a strong system that relies on artificial intelligence techniques, such as neural networks, to collect words in accordance with those features[2]. The current study is concerned with building a system for analyzing words in the Arabic language. This system can then be included in any application to address the Arabic language, becoming part of it. The system generates strings for all names and pronouns appearing in the entered text and depends mainly on the automatic assembly of a set of morphological, grammatical, lexical, and semantic features by using neural networks.

2. Literature Survey

Many researchers have been interested in the field of natural language processing and especially in its integration with artificial intelligence technologies such as neural networks, fuzzy logic, decision trees, and other technologies to significantly improve performance and results.

In this part of the research, we will review previous work by researchers who focused their work on the technique of neural networks in text mining.

In contrast to conventional methods, S. Lai, L. Xu, K. Liu, and J. Zhao provided a frequent convolutional neural network to classify the text without human-designed features. The researchers applied a repetitive structure to capture contextual information as far as possible when learning word representations. To automatically control keywords in the text, researchers used the ultimate aggregation technique. They experimented with four commonly used data sets. Experimental results show that the proposed method outperforms modern methods in many data sets, especially in document-level data sets[5].

In 2017, M. Amajd, Zh. Kaimuldenov, and I. Voronkov used different neural networks to classify the text, where they used pre-filtered texts for the purpose of system conditioning, In light of that conditioning the system is able to classify vague texts later.

Researchers have proven that the bypass neuron network can work better at word level, and does not require knowledge of semantic or grammatical structure of language[6].

N. I. Widiastuti used a category of Neural Networks, called a Convolution Neural Network (CNN), that uses a multilayer perception variation that is designed for minimal preprocessing. We have surveyed more than 30 scientific articles obtained from scientific article portals where similarities and differences were detected depending on three parts: the data input preprocessing method, the problem solved, and the approach taken to achieve the goal[7].

Sh. Minaee, N. Kalchbrenner, E. Cambria, N. Nikzad, M. Chenaghlu, and J. Gao, also used a Convolution Neural Network with sets of deep learning technics. To strengthen their research and in documenting its results, they reviewed in detail more than 150 existing models of deep learning-based text classification developed in recent years. Similarities and strengths were identified, along with discussion of their technical contributions. They also provided models for more than 40 groups of widely used data for classifying text. Finally, to assess the performance of various deep learning models, they provided a quantitative analysis of the common criteria for text mining [8].

In our current work, we will rely on traditional neural networks for text mining, while controlling the layers of the neural network, especially the hidden layer, and increasing the number of nodes it contains, in order to obtain accurate and high results.

3. Artificial Neural Networks

The human mind was the basis for building societies. Intentionally, in order to benefit from its way of thinking, the human mind's way of working was simulated when building neural networks. The network neuron is similar to the human neuron in its work.

An ANN is a collection of connected nodes called artificial neurons, which loosely model the neurons in a biological brain. Each connection, like the synapses in a biological brain, can transmit a signal to other neurons. An artificial neuron is one that receives a signal then processes it and can signal to neurons connected to it[9].

Neural networks go through two stages: the first stage is the training stage, and the second stage is the testing stage. In the first stage the neurons are trained by giving them inputs and outputs as well as the way of working to reach those outputs by using a set of weights. It then matches between the outputs[10].

In the testing stage, we give only the inputs and weights. The system gives the outputs based on the knowledge base formed through the training stage. Figure 1 illustrates the training stage, and figure 2 illustrates the testing stage.

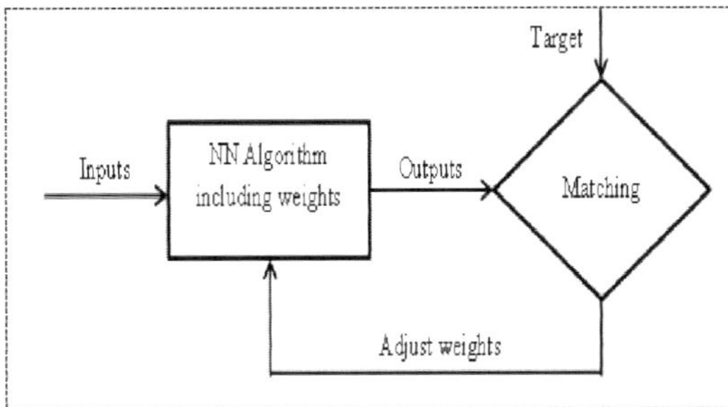

Figure 1. Training stage structure.

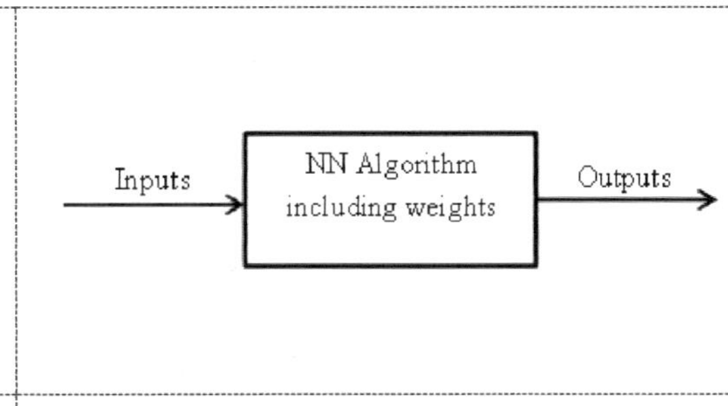

Figure 2. Testing stage structure

Neural networks are composed of three layers: The Input layer, which consists of a set of nodes, each node representing an input to the system. The second layer is the hidden layer, which includes the function of neural networks and the activation function,. The outputs of the data processing in the hidden layer, are outputs within the Output layer[11].

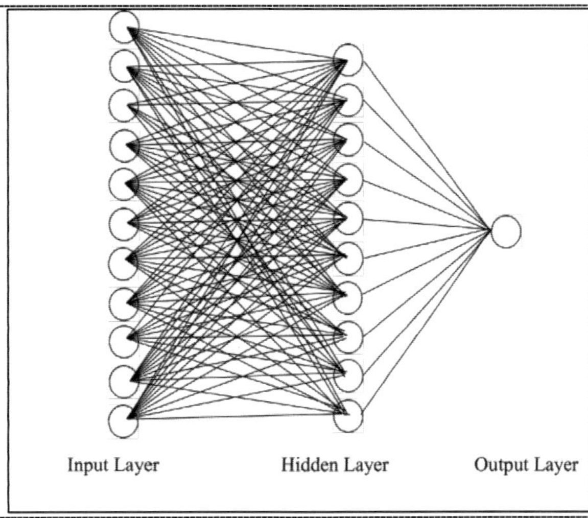

Figure 3. Neural networks layers

4. The Proposed System

This proposed system consists of two main phases. The first is the features collection, and the second is words collection by a neural networks algorithm. Each phase has an active role in the proposed system. In the first stage, the system seeks to divide the text in an entered document into its component words. It then collects the morphological, syntactic and semantic features of the text. These features are input to next phase as a number by which to compare it with others in same document, and to form a collection of words which have the same features. The performance and results of the second stage are entirely depended on the validity and accuracy of results from the first phase.

The first phase is the language understanding phase. It comprises several sub-phases, namely tokenize analysis to split the text in a document into a set of words; lexical analysis to classify the words as nouns or pronouns by looking at the lexicon; morphological analysis (which is an important sub-phase that ignores any prefixes and suffixes to the word when determining its class); and syntactic analysis to solve any ambiguity from the morphology analysis. By using ANN, the second phase aims to detect word-types in the text, that have the same features in the document. Figure 4 illustrates the proposed system phases.

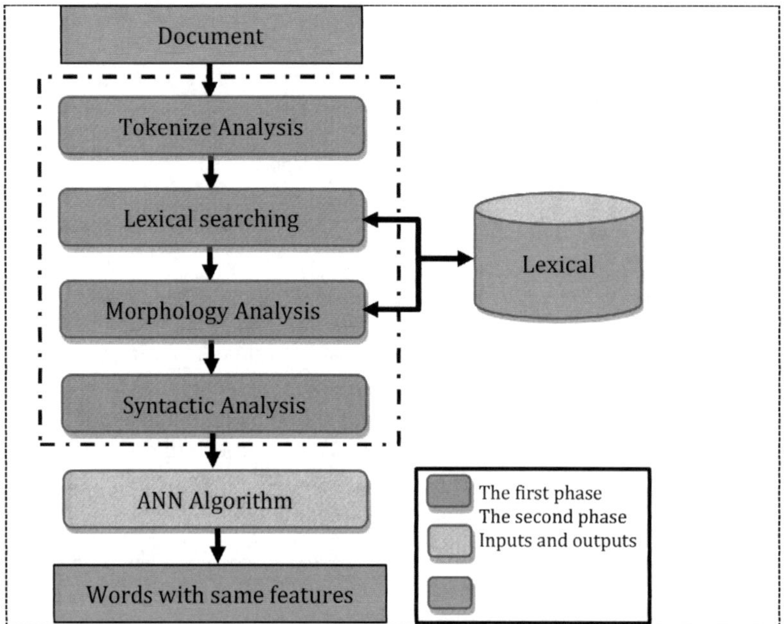

Figure 4. The proposed system phases.

4.1. The First Phase (Feature Collection)

This stage is vitally important, as the overall performance and results of the whole system depend mainly on the outputs from it. If the results of this phase are accurate, then the final results of the system will be accurate. There is a direct relationship.

This stage consists of a set of sub-phases, each of which has a job to perform, before handing over the work to subsequent sub-phases to produce the entire stage (the first phase) resulting in a set of properties that then are entered as inputs for the second phase (the word collection phase).

Given that most systems for natural language processing are similar in their early phases, it is felt unnecessary to further explain the sub-phases of the first stage.

4.2. The Second Phase (Words Collection by ANNs)

At this phase, the system will aggregate words that have the same features by applying the neural networks algorithm. In the previous stage, a set of properties was collected for each of the text words in the document, and those characteristics were coded into symbols for ease of handling by this stage. For example if the word was a single male human name symbolized by the symbol (1, 1, 2), where the symbol 2 symbols is for human being, symbol 1 for masculine, symbol 1 for singular, and so on. In the neural networks algorithm, we trained the system by entering a set of documents pre-classified by humans, so that the system would be able to group words in the light of that exercise.

The work of neural networks is formulated on three basic layers; Input Layer, Hidden Layer, and Output Layer. In the training stage we will feed the system with all the data for all of the layers, including the output layer. In the testing stage it will be sufficient to feed the system with the data of the input layer and the hidden layer equation only. Using the basis of the training we will then get the output.

In the neural network equation, inputs and their weights will be produced, and the results gathered, as per the following function:

$$y = \sum_{i=1}^{n}(x_i * w_i) + b_i \qquad (1)$$

When:

x_i is each word feature, w_i is a weight which computes as a random number smaller than 1, and bi is a threshold.

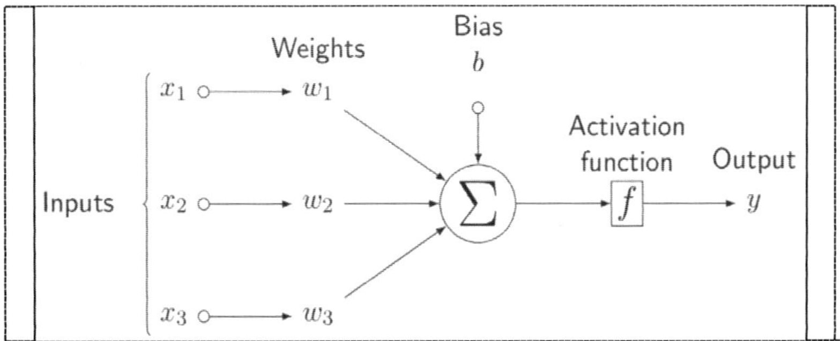

Figure 5. Arithmetic neural network algorithm.

Through the above equation, the system takes inputs to the input layer and generates a weight for each of them. The value of weights ranges between 0 and 1, where the value of those weights is multiplied by the value of the input and then cumulatively with the rest of the features to get a value that is close to the output value. Table1 shows an example of the features and their values in the input and output layers.

Table 1. Feature values in I/O layer.

Input1(Type)	Input1(Sex)	Input1(Number)
1	1	1
1	2	1
1	3	2
2	1	1

When calculating the outputs using neural networks, we may find a big gap between those results that we obtained in the testing stage and the original outputs. One of the most important units that can then be used with the neural network equation is the activation function, which reduces the gap between the outputs to reach the nearest value[12].

There are many activation functions. In our work, we used an activation function called Uni-Polar Sigmoid Function, for ease of handling and few transactions[13].

$$F(x) = \frac{1}{1+e^{-x}} \qquad (2)$$

The uni-polar sigmoid function looks like this, graphically:

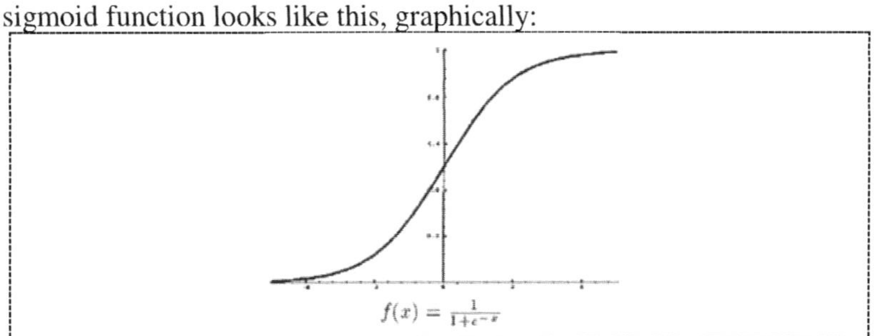

Figure 6. Uni- polar sigmoid function graph.

5. Results And Discussion

To evaluate the system used, we needed a corpus of 100 documents that contained various texts of news, stories, sports writings, and other texts. Almost all languages have standard corpus of texts. The Arabic language is exceptional in that one isn't included, so we built our own corpus. We implemented the system with its two phases on the documents in succession and the results were encouraging as the results achieved ranged between 77% and 86% without using the activation function. When using the activation function - a complementary function to the equation of the neural network algorithm - the results increased to between 83% and 96%.

The evaluation of any system or algorithm in natural language processing follows three methods to assess the results[14,15];

Precision method (P), which finds the ratio between the correct collection (CC) by the system to a sum of wrong collection and correct collection (WCC),

Recall Method (R), which finds the ratio between the correct collection (CC) to correct collection identified by the human (CH),

F-Measure Methods(F), which finds the ratio between multiplying twice the first method (P) and the second method (R), as (2P*R), and adding the two methods (P+R).

Table 2. ANNs system results.

Documents	Precision	Recall
Decument1	91%	92%
Document 2	92%	94%
Document 3	87%	79%
Document 4	95%	83%
Document 5	88%	86%
Document 6	89%	84%
Document 7	93%	93%
Document 8	96%	96%
Document 9	91%	95%
Document 10	87%	96%

To illustrate the above results, they can be represented as a graph as follows:

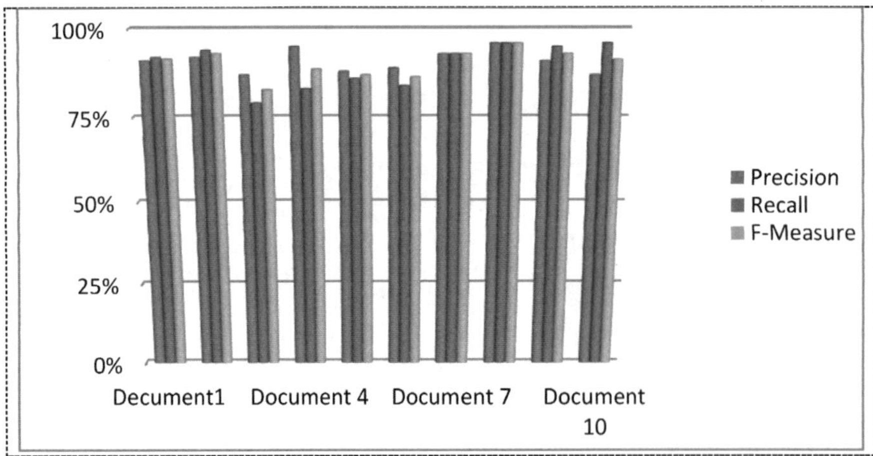

Figure 7. ANNs system results.

References

[1] V. Ng and C. Cardie, "Improving Machine Learning Approaches to Coreference Resolution", Proceedings of the 40th Annual Meeting of the Association for Computational Linguistics (ACL), Philadelphia, pp (104-111) , July 2002.
[2] S. Mahmood and F. M. L. Al-Rufaye, "Arabic text mining based on clustering and coreference resolution", 2017 International Conference on Current Research in Computer Science and Information Technology (ICCIT), Slemani, 2017, pp. (140-144), doi: 10.1109/CRCSIT.2017.7965549.
[3] Nitin Indurkhya and fred J. Damerau, "Handbook of Natural Language Processing" by Taylor and Francis Group, LLC, 2010.
[4] Heeyoung Lee and Sudarshan Rangarajan, "A Deterministic Coreference System with Rich Syntactic Features and Semantic Knowledge", Stanford University, 2009.
[5] Siwei Lai, Liheng Xu, Kang Liu, Jun Zhao, " Recurrent Convolutional Neural Networks for Text Classification ", Proceedings of the Twenty-Ninth AAAI Conference on Artificial Intelligence, PP. (2267,2273), 2015.
[6] Maaz Amajd, Zhanibek Kaimuldenov, Ilia Voronkov , " Text Classification with Deep Neural Networks ", Proceedings of the 5th International Conference on Actual Problems of System and Software Engineering, Supported by Russian Foundation for Basic Research, Moscow, Russia, PP. (362,370), November 14-16, 2017.
[7] N I Widiastuti, "Convolution Neural Network for Text Mining and Natural Language Processing", IOP Conf. Series: Materials Science and Engineering 662, 2019, doi:10.1088/1757-899X/662/5/052010.
[8] Shervin Minaee, Nal Kalchbrenner, Erik Cambria, Narjes Nikzad, Meysam Chenaghlu, Jianfeng Gao, "Deep Learning Based Text Classification: A Comprehensive Review" ,Cornell university, arXiv open-access archive, Vol. 1, No. 1, Article . Publication date: April 2020.
[9] V.N. Delgrange, N. Cabassud, M. Cabassud, L. Durand Bourlier, J.M. Laine, " Neural Networks for Prediction of Ultrafiltration Trans membrane Pressure: Application to Drinking Water Production ", Journal of Membrane Science 150, Pages 111–123, 1998.
[10] Harun Türkmenler, Murat Pala, " Performance Assessment of Advanced Biological Wastewater Treatment Plants Using Artificial Neural Networks ", INTERNATIONAL JOURNAL of ENGINEERING TECHNOLOGIES-IJET Harun Türkmenler et al., Vol.3, No.3, Pages 151-156, 2017.
[11] R. A. Naji, E. J. Khadum, " Efficiency of Using Artificial Neural Networks to Predict Child Mortality in Najaf Province and Neuroshell Predictor Program", AL GHAREE for Economics and Administration Sciences, Vol.14, No.38 Pages 311–320, 2016.
[12] G. Cybenko, "Approximation by Superposition of A Sigmoidal Function", Mathematics of Control, Signals, and Systems, Vol. 2, No. 4, PP. (303–314), 1987.
[13] R. P. Lippmann, "An Introduction to Computing with Neural Nets", IEEE Acoustics, Speech and Signal Processing, Vol. 2, No. 4, PP. (4–22), 1987.
[14] D. S. Al-Azzawy and F. M. L. Al-Rufaye, "Arabic words clustering by using K-means algorithm", 2017 Annual Conference on New Trends in Information & Communications Technology Applications (NTICT), Baghdad, 2017, pp. (263-267), doi: 10.1109/NTICT.2017.7976098.
[15] Catalina Barbu and RuslanMitkov, "Evaluation tool for rule-based anaphora resolution methods", Proceedings of the 39th Annual Meeting of the Association for Computational Linguistics, PP. (34-41), 2001.

Influence of Nb and Zr co-doping on the structural, morphological and dielectric properties of $CaCu_3Ti_4O_{12}$ ceramics

M K Hasan, S H Shant, M Z Islam and M E Haque

Department of Glass & Ceramic Engineering, Rajshahi University of Engineering & Technology, Rajshahi-6204, Bangladesh

*e-mail: khniloy786@gmail.com

Abstract. In this work, $CaCu_3Ti_{3.8-x}Nb_{0.2}Zr_xO_{12}$ (where x=0.0, 0.05, 0.1, 0.2) ceramics were made using the normal solid-state reaction method to assess the influence of Nb^{5+}& Zr^{4+} co-doping on the structure, morphology and electrical properties of $CaCu_3Ti_4O_{12}$ ceramics. Reitveld refinement by Fullprof package was done to achieve the structural information. No phase transformation was observed upon doping. The CCTNZO ceramics occupied a fine-grained microstructure with grain sizes ranges from 0.26-0.33 μm. Furthermore, grain size uniformity was enhanced with gradual doping of Zr^{4+} ion. within the meantime, the CCTNZO ceramics unfolds extensive durability of dielectric constant over a good range of frequency and also shows a lower loss factor. Doping resulted during a gradual decrease in grain sizes and a dielectric loss at low and high frequency. It had been found that 10% of Zr doped CCTNZO ceramics has the very best dielectric constant (3655) at 0.1 kHz. Meanwhile, rock bottom dielectric loss of 0.241 at 0.1 kHz was observed for 20% Zr doped CCTNZO ceramics.

***Keywords*:** Structural stability; Uniform grain size; Co-doping; Dielectric constant; Dielectric loss

1. Introduction

"Researchers have found that the CCTO ceramics showed tremendous lofty dielectric constant of 100,000 at low temperature and little temperature dependence from 100 to 400 K" [1]. "Moreover, CCTO ceramics has nearly frequency independent dielectric constant beneath 106 Hz, that is wished for several microelectronic appliances" [2]. Besides, materials having extensive dielectric constant over both wide temperature range and frequency range would extremely contribute to the miniaturization of memory storage devices and capacitors [3]. "The CCTO ceramics are much impressionable to several parameters, which can affect its dielectric properties. "Mass research has been equipped to define the premodial of the enormous dielectric property of CCTO ceramics" [4, 5]. CCTO belongs to the family of a double perovskite group with space group Im3 [6]. "The dielectric loss of CCTO ceramics is 0.5 at 1 kHz, which is extremely high" [7]. It's highly necessary to understand the explanation for high loss factor in CCTO ceramics and find an efficient way to reduce it. "Several well-known methods are established to mitigate the loss factor of Calcium copper Titanate ceramics by doping, substitution, two-phase composite processes" [7]. As of now, several researchist have effectively employed these methods like CCTO doped with Cr_2O_3 [8], CCTO doped with La [9], $CCTO/CaTiO_3$ composite [10], etc.

Elements with ionic charges +3 like Bi^{3+}, La^{3+}, Nd^{3+}, etc. are used as donors for A site, whereas elements having ionic charges +5 or +6 like Nb^{5+}, Ta^{5+}, Sb^{5+}, W^{6+}, etc. are often used as donor for Ti ions [7]. The surplus charge at A site creates (Ca^{2+}) vacancies and excess charges at B site (Ti^{4+}), generally creating vacancies. Sintering time increasing may create both more oxygen vacancies also as space charges [11]. Sulaiman, Muhammad Azwadi et al. reported that 1% of Nb-doped CCTO ceramics produced by a standard solid-state reaction method gives the very best dielectric constant with a high dielectric loss. The Dielectric loss was minimized when 3% of Nb was utilized in CCTO [7]. Hong et al. reported that 20% of Nb-doped CCTO ceramics has the very best dielectric constant and highest dielectric loss value [12]. W. L. Li et al. reported that doping with Zr^{4+} simultaneously decreases dielectric loss [13]. Single element doping may be a common approach, but the effect of doping is uncertain. A couple of works are found on co-doping in CCTO ceramics. H.A. Ardakani et al. investigated that Cr & La co-doped CCTO ceramics could reduce dielectric loss to an excellent extent [14], A.K Thomas et al. reported that dielectric constant increased with co-doping of La at A site and Nb at B site [6]. It is considered that doping simultaneously at A site and B site or only A site or only B site is worth it in improving the dielectric constant with a big reduction in dielectric loss.

In this present study, Nb & Zr co-doped CCTO ceramics were yielded using the traditional solid-state reaction method to enhance dielectric constant property. Several dopants percentages were used to minimize dielectric loss. The dielectric constant and dielectric loss were measured during a wide selection of frequencies. However, the effect of simultaneous partial substitution of Nb & Zr at B site has not been reported yet. This study will present the effect of partial substitution of Nb & Zr co-doped CCTO ceramics on the structural, morphological, and dielectric properties.

2. Materials and Method

Conventional solid-state reaction way was accustomed to produce monocrystalline samples of $CaCu_3Ti_{3.8-x}Nb_{0.2}Zr_xO_{12}$, where x = (0.00, 0.05, 0.10, 0.20) ceramics. The samples are named CCTNZO-0, CCTNZO-5, CCTNZO-10 & CCTNZO-20 because of the 0%, 5%, 10%, and 20% replacement of Ti-sites respectively. Pure raw materials of $CaCO_3$, CuO, TiO_2, Nb_2O_5, ZrO_2 were used with $\geq$ 99.9% purity and supplied by Merck Specialties, India. The desired powders were weighed with accurate ratio and circumstantially kneading by ball milling with high pure ethanol for 20 hrs. to keep up a consistent solution. After that, an oven was used to dry the liquid solution at 100° C for 24 hrs. to induce the powder, and then the powder was hand-milled thoroughly to avoid lump formation. Then calcination was administered at 860° C for two hrs. Then, the obtained powder was again hand-milled and pressed into pellet (13 mm diameter and 2 mm thickness) with an appropriate binder named PVA. Sintering was administered at 1020° C for 2 hrs. at ambient pressure. XRD analysis was performed using Bruker Advanced Diffractometer (Germany) to spot the possible synthesized phases. Reitveld refinement was done using Fullprof software to urge detailed the structural information. Scanning electron microscope was administered to watch surface morphology and particle sizes of the samples. The Linear intercept method was applied to calculate the grain size and Image J software was used to determine grain size distribution. Dielectric constant was measured with an Impedance analyzer.

3. Results and discussion

3.1 Structural analysis

X-ray diffraction analysis was performed to collect information about structure. Figure 1 exhibits the XRD pattern alongside the magnified pattern at 31-39 degrees. The XRD curves are well-matched with the peaks of $CaCu_3Ti_4O_{12}$ (CCTO). 0%, 5%, 10%, 20% Zr doped CCTNO (Nb-doped CCTO) are denoted by CCTNZO-0, CCTNZO-5, CCTNZO-10 & CCTNZO-20 respectively. The magnified portion in figure 1 starting from 31-39 degrees shows an entire merging of (202) peak from CCTNZO-0 to CCTNZO-20." All the peaks in figure 1 are indexed considering body-centered cubic with the space group Im3" [12].

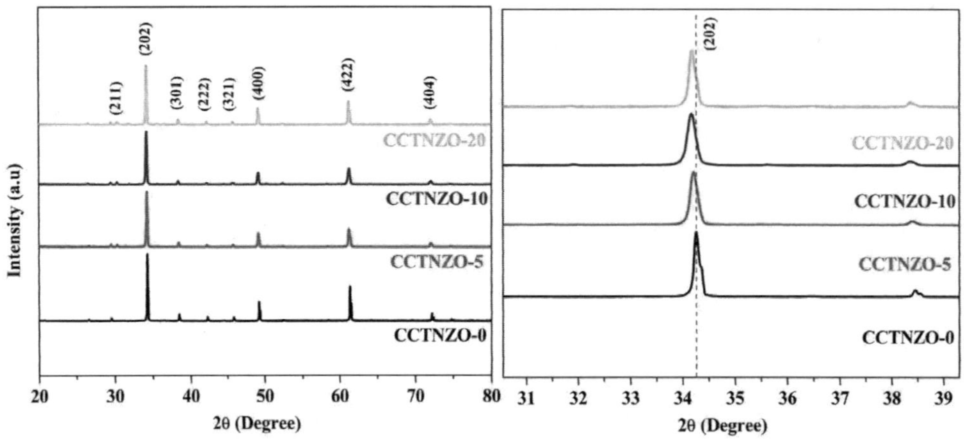

Figure 1: X-ray diffraction pattern of various percentage doped CCTNZO ceramics (left) at room temperature and magnified XRD pattern at 31-39° (right)

Fullprof software was used to further explore the information about structure. Figure 2 shows the possible structural refinement and curve fitting using the software. Hence, 20% substitution reveals no secondary phases. Refined structural information are enlisted in Table 1 which is in consistent with JCPDS #75-2188. Calculated lattice constant shows increment with increasing doping percentage.

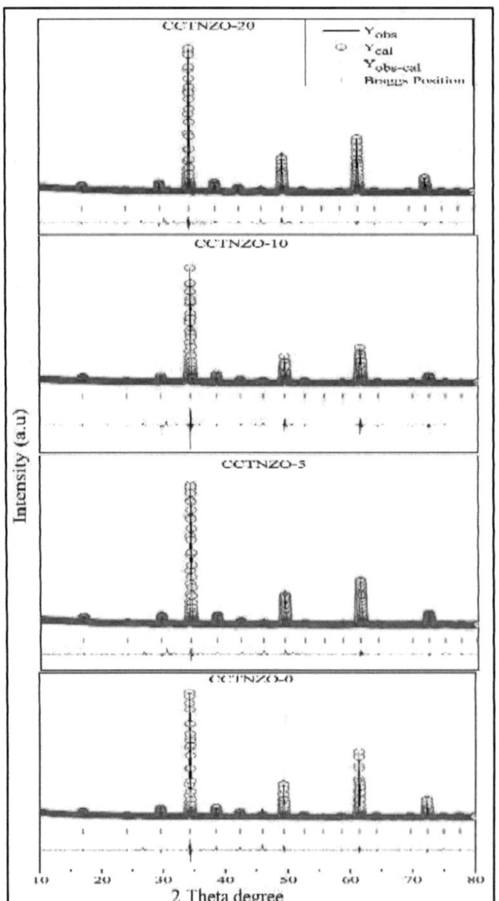

Figure 2: Stacking Reitveld refined graphs of CCTNZO ceramics

Table 1: Structural parameters of CCTNZO- (0, 5, 10, 20) ceramics obtained by Reitveld Refinement

Parameters	CCTNZO-0	CCTNZO-5	CCTNZO-10	CCTNZO-20
Lattice parameters	a = 7.3918Å V= 403.8709 $Å^3$	a = 7.4018Å V= 405.5152 $Å^3$	a = 7.4029Å V= 405.7047 $Å^3$	a = 7.4030Å V= 405.7239 $Å^3$
Density	6.225 g/cm^3	6.210 g/cm^3	5.5057 g/cm^3	5.866 g/cm^3
R_{exp}	13.7	13.6	7.34	9.86
GoF	1.3	1.6	1.6	1.4

From Figure 3 it is often seen that increasing dopant percentage causes a gradual increment in lattice parameters. "The explanation behind increasing lattice constant could also be ascribed to the difference in ionic radius between Ti and Zr ions ($r(Ti^{4+})$ =0.605 Å and ($r(Zr^{4+})$ =0.80 Å" [12]. Moreover, the Goodness of fit (GoF) indicating the reliability of our refinement.

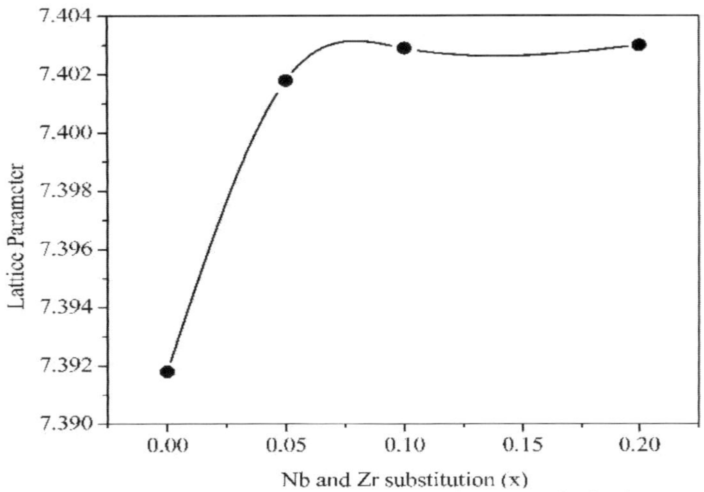

Figure 3: Lattice parameter Vs Co-dopants substitution

3.2 Morphological analysis

The Scanning electron microscope observed the Surface morphology of CCTNZO ceramics, and therefore the micro images are shown in Figure 4. Careful observation depicts that CCTNZO-0 has larger size grain.

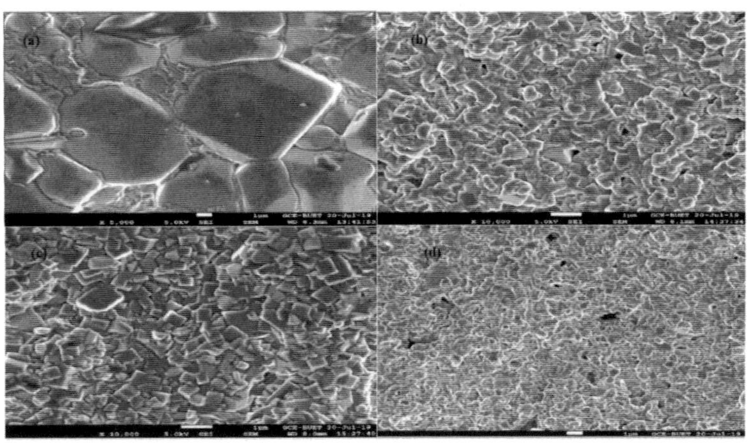

Figure 4: Scanning electron micrographs of CCTNZO (0, 5, 10, 20) ceramics respectively

Increasing the Zr percentage into CCTNZO ceramics causes grains to be core shell-like structure. Another clear visualization from the sem micrographs suggests that increasing dopant percentage causes a consecutive reduction in average grain size. Li, W. L et al. also reported that the Zr dopant reduces grain size [13]. Hence, the grain size of CCTNZO ceramics is minimized, and grain size similarity is attained by a gradual increase in Zr dopant percentage. Linear intercept method was used to calculate the average grain size [16]. The grain size for CCTNZO-0, CCTNZO-5, CCTNZO-10 and CCTNZO-20 ceramics are 4.82 μm, 0.33 μm, 0.29 μm and 0.26 μm respectively. Figure 5 depicts the grain size distribution of CCTNZO-0, CCTNZO-5, CCTNZO-10 and CCTNZO-20 ceramics. The grain size distribution becomes narrower with increasing the dopant percentage.

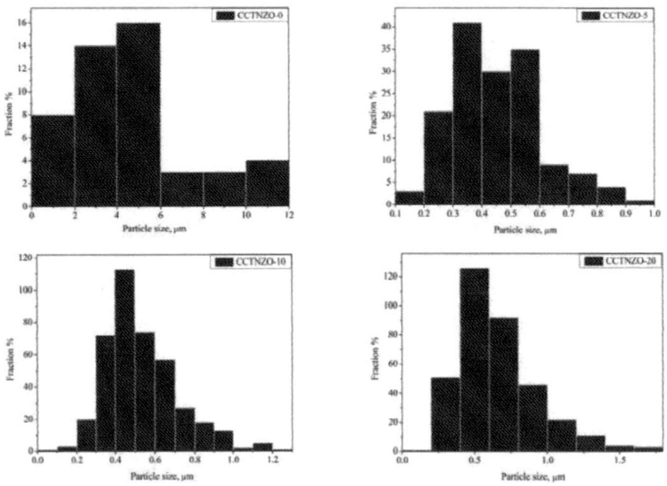

Figure 5: Grain size Distribution of CCTNZO ceramics respectively

The calculated average grain size for CCTNZO-0, CCTNZO-5, CCTNZO-10, and CCTNZO-20 ceramics are 4.82 μm, 0.33 μm, 0.29 μm, and 0.26 μm respectively. Figure 5 depicts the size distribution of CCTNZO-0, CCTNZO-5, CCTNZO-10, & CCTNZO-20 ceramics. The grain size distribution becomes narrower with increasing the dopant percentage. Crystallite sizes were enumerated using the Scherrer's formula:

$d={}^{k\gamma}/\beta cos\theta$ ………………………………………………………………………………… (1)

Where β denotes the FWHM of the main diffraction peak, k is that the dimensionless constant with a typical value ~0.9, λ is that the wavelength of Cu K_α radiation with the worth of 1.5406 Å, and θ stands for the Bragg angle of the main peak. Calculated parameters are enlisted in Table-2. As smaller crystallite size results in higher diffusion and agglomeration, a decrease in particle size with Zr doping is justified by the decrease in corresponding crystallite size [17, 18].

Table 2: Grain size and crystallite size of CCTNZO (0,5,10, 20) ceramics

Name of the element	Average grain size (μm)	Crystallite size (μm)
CCTNZO-0	4.82	0.058
CCTNZO-5	0.33	0.034
CCTNZO-10	0.29	0.031
CCTNZO-20	0.26	0.029

3.3 Dielectric property analysis

Figure 6 and Figure 7 exhibits the frequency depended dielectric constant (k) and dielectric loss (tanδ) respectively. The measured constant values of CCTNZO (0%, 5%, 10%, 20%) ceramics at 0.1 kHz are 7402, 3452, 3455, and 1597 respectively. Dielectric constant slowly reduces for all specimens with the increment of dopant percentage, and it becomes stable at a better frequency. This sort of phenomenon is often justified in 2 ways. "Firstly, among the four sorts of polarization, only interfacial and dipolar polarization contribute at lower frequencies. In this way, there are fewer polarization contributors at higher frequencies, leading to low dielectric constant" [19]. "Secondly, following Maxwell-Wagner theory, space charges move and coagulate at grain boundaries if sufficient resistance is present. Nevertheless, at a higher frequency, space charges don't receive adequate time to coagulate at boundaries of grain and so collapse the dielectric properties" [20, 22].The dielectric values are enlisted in Table 3. The grain size values can reveal the gradual decrement of dielectric constant values. Domain wall movement is simpler for larger grain, resulting in a rise in dielectric properties [21]. In other research, we've found this sort of property also [22]. Sem micrographs in figure 4 lucidly shows that partial substitution of Zr^{4+} into Ti^{4+} ion has an impression on decreasing grain size. Measured grain size is enlisted in Table 2 depicts, decreasing grain size with increasing Zr percentage causes a gradual decrease in dielectric constant. Table 2 depicts that CCTNZO-0 has the biggest grain size with the highest dielectric constant.

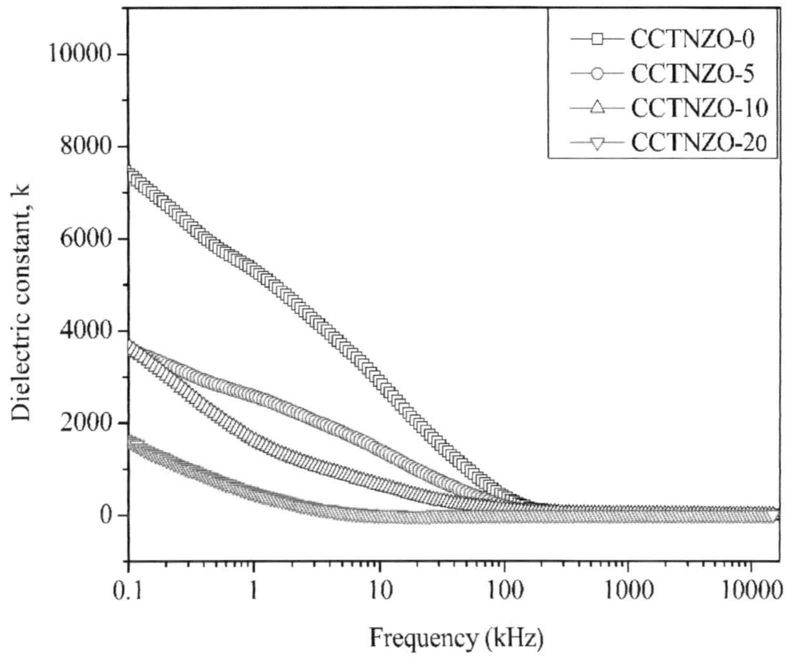

Figure 6: Frequency depended Dielectric constant values of CCTNZO ceramics at room temperature

Table 3: Dielectric constant and Dielectric loss of CCTNZO (0, 5, 10, 20) ceramics

Name of the element	Dielectric Constant (k)		Dielectric loss (Tanδ)	
	0.1 kHz	1 MHz	0.1 kHz	1 MHz
CCTNZO-0	7402	33	0.369	1.21
CCTNZO-5	3652	19	0.282	0.51
CCTNZO-10	3655	44	0.270	0.49
CCTNZO-20	1597	10	0.241	0.60

Whereas, the dielectric loss of CCTNZO (0%, 5%, 10%, 20%) ceramics are within the extent between 0.241-0.369 at 0.1kHz, which is within the tolerance limit. Figure 7 depicts that Nb-doped CCTO and

Nb & Zr doped CCTNZO ceramics show different defect chemistry. The sole exception is that CCTNZO-0 ceramics show an enormous dispersion at 10 kHz, as pointed that at this frequency extent sufficient volume of space charges did not ensue the channel of the electrical field [23]. Several mild arbors are seen within the range between 100-1000 kHz. Several other parameters such as internal stress, domain density, etc. needed to be taken into account for an extra detailed clarification of this dielectric loss tangent (Tanδ) behavior [21]. Above all, doping Zr ions into CCTNO ceramics can increase the frequency fixity also as decrease the loss factor, and this broaden fixity is conductive to be applied in microelectronics devices [24] ."Q.G Chi et al. reported that mixed valent structure and oxygen vacancies should be the elemental root of huge dielectric constant for CCTO ceramics. The dielectric relaxation mechanism of CCTO ceramics are often illustrated by losing of O_2 during its sintering at heat in accordance with reaction" [25],

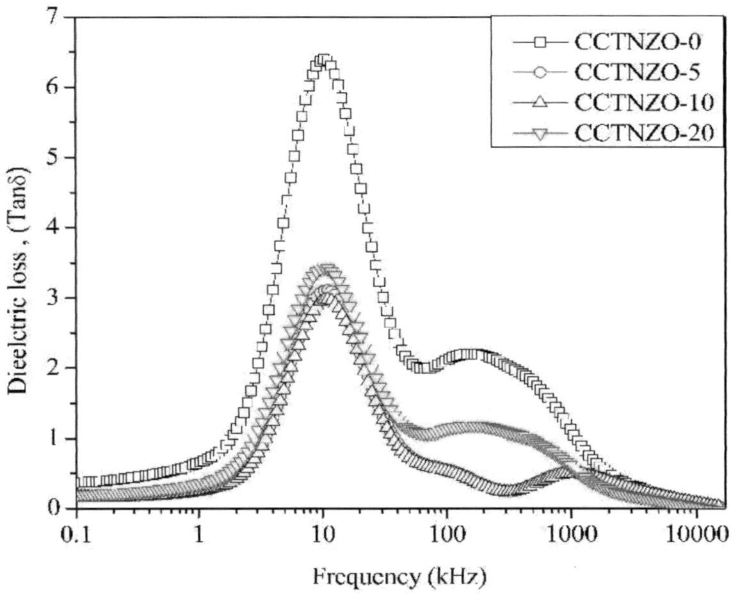

Figure 7: Frequency depended Dielectric loss of CCTNZO ceramics at room temperature

$O_0 = \frac{1}{2}O_2 + V_O^{\cdot\cdot} + 2e^-$ ……………………………………………………………………………… (2)

Cu^{2+} or Ti^{4+} capture those released electrons from equation 1 to generate Cu^+ or Ti^{3+} as follows,

$2Cu^{2+} + 2e^- = 2Cu^+$ ……………………………………………………………………………… (3)

$2Ti^{4+} + 2e^- = 2Ti^{3+}$ ……………………………………………………………………………… (4)

As Ti^{4+} with $3d^0$ configuration is in its most stable state, the probability of formation of Ti^{3+} is negligible. On the opposite hand, Cu^+ having $2d^{10}$ configuration is stable state of Cu due to noble gas configuration. Cu^+ ions on Cu^{2+} site will bear an efficient charge. These may produce dipoles with $V_O^{\cdot\cdot}$. "These dipoles can alteration their orientation because of either floating of electron among Cu^+ and Cu^{2+} or jumping of O^{2-} ions through vacant oxygen sites round the CuO_4 square or TiO_6 octahedral giving uprising to orientation polarization" [24]." "Q.G Chi et al. reported that prime frequency dielectric relaxation of CCTO ceramics is induced by mixed valent structure of Cu^+/ Cu^{2+} and Ti^{3+}/ Ti^{4+}" [24]. Furthermore, CCTNZO ceramics have an equivalent mechanism of dielectric relaxation. Doping Zr^{4+} on Ti^{4+} site, oxygen vacancy formation is suppressed as follows;

$ZrO_2 + 2V_{\ddot{o}TiO2\rightarrow ZrTi} + 2O_0$ ……………………………………………………………………………… (5)

This will alleviate the creation of Cu^+ and $V_O^{\cdot\cdot}$. The partial substitution of Zr^{4+} on Ti^{4+}curtails the coagulation of Cu^+ and $V_O^{\cdot\cdot}$ also as inhibits forming the mixed valent structure in CCTNZO ceramics and reduces the dipolar polarization leading to curtailing dielectric loss [24].

4. Conclusion

In precis,traditional solid-state reaction method was used to produce the Nb and Zr co-doped CCTO ceramics. Structural study reveals no accompanying phase has arrived with the rise of Zr^{4+} doping from 5% to 20%. Although Nb & Zr co-doping produced an equivalent cubic structure with space group Im3, Lattice parameters increased with co-doping. Reitveld refinement confirms the structural stability. Co- doping features a far-reaching effect on grain growth. The microstructural analysis predicts that Zr^{4+} might act as a grain growth inhibitor. Thus, diminished grain size and narrow size distribution are observed. Significant decrease in dielectric loss observed for CCTNZO ceramics with increasing doping percentage. 10% doping causes significant improvement both in dielectric constant & dielectric loss. Finally, though doping has made significant improvements within the dielectric constant and dielectric loss within the lower frequency, it's going to not be fit use in higher frequency devices.

Acknowledgement
The research project was fully supported by Rajshahi University of Engineering & Technology.

References

[1] Ramirez, A. P., Subramanian, M. A., Gardel, M., Blumberg, G., Li, D., Vogt, T., & Shapiro, S. M.2000 Giant dielectric constant response in a copper-titanate *Solid State Communications* **115**217-220.
[2] Homes, C. C., Vogt, T., Shapiro, S. M., Wakimoto, S., & Ramirez, A. P. 2001 Optical response of high-dielectric-constant perovskite-related oxide *science* **293**673-676.
[3] Renner, B., Lunkenheimer, P., Schetter, M., Loidl, A., Reller, A., & Ebbinghaus, S. G. 2004Dielectric behavior of copper tantalum oxide *Journal of applied physics* **96**4400-4404.
[4] Homes, C. C., Vogt, T., Shapiro, S. M., Wakimoto, S., Subramanian, M. A., & Ramirez, A. P. 2003 Charge transfer in the high dielectric constant materials CaCu 3 Ti 4 O 12 and CdCu 3 Ti 4 O 12 *Physical Review B* **67** 092106.
[5] Thongbai, P., Masingboon, C., Maensiri, S., Yamwong, T., Wongsaenmai, S., &Yimnirun, R. 2007 Giant dielectric behaviour of CaCu3Ti4O12 subjected to post-sintering annealing and uniaxial stress *Journal of Physics: Condensed Matter***19**236208.
[6] Thomas, A. K., Abraham, K., Thomas, J., &Saban, K. V. 2015 Structural and dielectric properties of A-and B-sites doped CaCu3Ti4O12 ceramics *Ceramics International* **41** 10250-10255.
[7] Sulaiman, M. A., Hutagalung, S. D., Ain, M. F., & Ahmad, Z. A. 2010 Dielectric properties of Nb-doped CaCu3Ti4O12 electroceramics measured at high frequencies *Journal of alloys and compounds* **493** 486-492.
[8] Kwon, S., Huang, C. C., Patterson, E. A., Cann, D. P., Alberta, E. F., Kwon, S., & Hackenberger, W. S. 2008 The effect of Cr2O3, Nb2O5 and ZrO2 doping on the dielectric properties of CaCu3Ti4O12*Materials letters* **62**633-636.
[9] Feng, L., Tang, X., Yan, Y., Chen, X., Jiao, Z., & Cao, G. 2006 Decrease of dielectric loss in CaCu3Ti4O12 ceramics by La doping. *physica status solidi (a)* **203** R22-R24.
[10] Yan, Y., Jin, L., Feng, L., & Cao, G. 2006 Decrease of dielectric loss in giant dielectric constant CaCu3 Ti4 O12 ceramics by adding CaTiO3 *Materials Science and Engineering: B* **130** 146-150.
[11] Patel, P. K., & Yadav, K. L. 2014 Extrinsic mechanism for colossal dielectric constant in CaCu3Ti4O12 ceramics evidenced by nanodomain *Materials Research Express* **1** 015037.
[12] Hong, S. H., Kim, D. Y., Park, H. M., & Kim, Y. M. 2007 Electric and dielectric properties of Nb-doped CaCu3Ti4O12 ceramics*Journal of the American Ceramic Society***90**2118-2121.
[13] Li, W. L., Zhao, Y., Chi, Q. G., Zhang, Z. G., & Fei, W. D. 2012 Enhanced performance of core-shell-like structure Zr-doped CaCu 3 Ti 4 O 12 ceramics prepared by a flame synthetic approach *RSC Advances* **2** 6073-6078.
[14] Ardakani, H. A., Alizadeh, M., Amini, R., &Ghazanfari, M. R. 2012 Dielectric properties of

CaCu3Ti4O12 improved by chromium/lanthanum co-doping *Ceramics International* **38**4217-4220.

[15] Denton, A. R., & Ashcroft, N. W. 1991 Vegard's law *Physical review A* **43** 3161.

[16] Abrams, H. 1971 Grain size measurement by the intercept method *Metallography***4** 59-78.

[17] Bellare, A., &Dadyburjor, D. B. 1987 Effect of the size dependence of particle diffusion and surface tension on the sintering of supported metal catalysts *AIChE journal* **33** 867-870.

[18] Billah, M., Ahamed, A. A., Sen, A., & Rahman, M. M. Influence of Lanthanum doping on the Crystallite Growth and Dielectric Properties of Barium Titanate (BaTiO3) Ceramics.

[19] Islam, M. R., Islam, M. S., Zubair, M. A., Usama, H. M., Azam, M. S., & Sharif, A. 2018 Evidence of superparamagnetism and improved electrical properties in Ba and Ta co-doped BiFeO3 ceramics *Journal of Alloys and Compounds* **735** 2584-2596.

[20] Wagner, KW 1913 On the theory of imperfect dielectrics *Annals of Physics* **345** 817-855.

[21] Zaman, T., Islam, M. K., Rahman, M. A., Hussain, A., Matin, M. A., & Rahman, M. S. 2019Mono and co-substitution of Sr2+ and Ca2+ on the structural, electrical and optical properties of barium titanate ceramics *Ceramics International* **45**10154-10162.

[22] Sen, A., Hasan, M. K., Islam, Z., Al Hassan, M. R., Zaman, T., Matin, M. A., & Gulshan, F. 2020 Influence of Ba and Mo co-doping on the structural, electrical, magnetic and opticalproperties of BiFeO3 ceramics *Materials Research Express* **7** 016312.

[23] Zheng, R. Y., Wang, J., & Ramakrishna, S. 2008 Electrical and magnetic properties of multiferroic BiFeO 3/CoFe 2 O 4 heterostructure *Journal of Applied Physics* **104** 034106.

[24] Chi, Q. G., Gao, L., Wang, X., Lin, J. Q., Sun, J., & Lei, Q. Q. 2013 Effects of Zr doping on the microstructures and dielectric properties of CaCu3Ti4O12 ceramics *Journal of alloys and compounds* **559** 45-48.

Descriptive Analysis on Products/Services demand and Market conspectus: A quantitative approach

Mohammed Afaounodden Ahmed, Adresya Suresh Azhakath and Raja Muthalagu

Department of Computer Science, Birla Institute of Technology and Science Pilani, Dubai Campus, Dubai, UAE

Email: f20170302@dubai.bits-pilani.ac.in, f20170131@dubai.bits-pilani.ac.in, raja.m@dubai.bits-pilani.ac.in

Abstract. The research conducted in this paper consists of finding out the shortcomings of Efficient Market Hypothesis as it is purely based on if information is immediately priced in the Trading markets. Efficient market hypothesis(EMH) also suggests that it is not possible for someone to give more returns than the market in long term , Which has been disproven in this research using multiple criteria decision making and taking the companies with Good Debt to Market cap ratio and check their aggregated outcome over a long period of time and the rate of return came out to be significantly higher than the growth in the market. One of the main assumptions of EMH also happens to be that fundamental analysis does not affect stock market or stock indexes but its proven to be false as in long term if indexes are selected based on good criterias. Efficient market hypothesis says that stock prices depend on news and the news of the next day is by default unpredictable, but it doesn't take into account the fact that not everyone interprets the news in the same way and even if they do they don't process the information on the same intensity.

Keywords: Prediction Analysis, Stock Market, Multiple-criteria decision analysis, Weights for parameters, Back Tracking, Correlation, PROMETHEE, AHP, technical analysis, stocks ranking.

1. Introduction

PROMETHEE is one of the latest MCDM strategies that was proposed by Brans et al. [7] and has progressively been applied in numerous fields [8], particularly in the speculation investigation and execution assessment. Mareschal and Brans, Vranegl et al., Babic and Plazibat[9], Bouri et al. what's more, Albadvi et al.[10] all applied PROMETHEE as a dynamic apparatus to take care of the various issues in the field of account. In this paper, we apply PROMETHEE II to choose firms with extraordinary execution in the choice cycle of significant worth contributing.
The proposed model is organized around two columns: Industry assessment and Company assessment. The inclination positioning association technique for enhancement assessment (PROMETHEE) has been utilized for tackling the issue [6]. As a genuine case and an overview from the specialists so as to decide the viable measures for industry assessment and friends assessment has been directed.[3] Nevertheless, it has long been criticized by academic researchers and users. [4] This criticism has been established based on two facts: the first fact which is the theory of labor markets argues that "prices always fully reflect available information . [5] The second fact is that technical analysis is based on weak fundamentals[6] . It appears that technical analysis can be a

compromising tool since it offers a relative composition of human, political and economic events. Like the companies before 9/11 and after 9/11 had the same fundamentals , but the prices dropped 50% .

2. The Efficient Market Hypothesis

2.1 Weak Form of Efficient market Hypothesis
On the off chance that a market is regarded to be 'frail structure proficient', it would imply that no relationship exists between chronicled costs and progressive costs.

2.2 Semi Strong Form of Efficient market Hypothesis
This would imply that principal examination may assist dealers with gathering data and produce better than expected returns. Chance that a market is regarded to be 'frail structure proficient', it would imply that no relationship exists between chronicled costs and progressive costs [11].

2.3 Strong form of Efficient market Hypothesis
Solid structure EMH states that all accessible data, both open and private, is valued into the cost of a security.[12] Strong Market Hypothesis expects that the market is great, thus the main way an individual could make an extreme return is by utilizing insider data [13]. Both specialized and central examinations would be rendered disputable, as neither could give profitable Information. Table 1 shows the six type of generalized criteria.

Table 1. Mathematical Model of PROMETHEE

Table 1
The six types of generalized criteria

Types of generalized criteria		Param -eters
I. Usual criterion	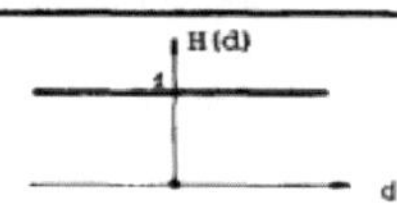	-
II. Quasi-criterion	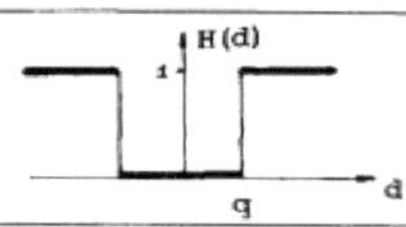	q
III. Criterion with linear preference	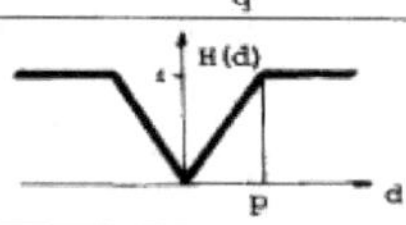	p
IV. Level criterion	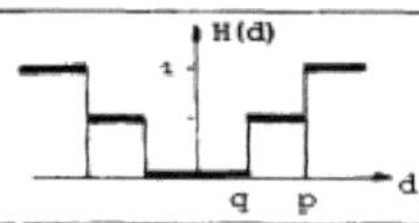	q,p
V. Criterion with linear preference and indifference area	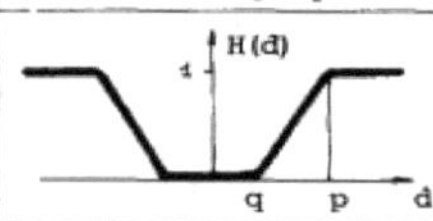	q,p
VI. Gaussian criterion	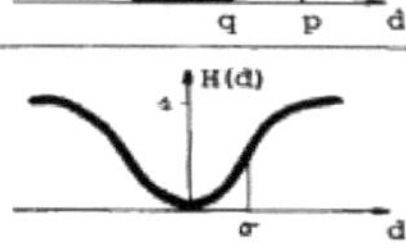	σ

3. Technical Analysis
Stock price data for all the stocks is taken and totals about to be more than a Million rows, and many many stock symbols [7]. Then another dataset is taken with all the fundamental values of the companies, then newer columns are created requiring the values like P/E ratio, normal ratio from Debt to Equity of the given company and taking all the companies of good Market cap, Ratio, Profit %, dividend. Through filtering using pandas Dataframe and after filtering the final names of the companies who fulfill all the criterias are kept in a separate Data Frame and all the names of the company values are extracted from the dataframe by using Unique Ticker , store them in a dataframe the 10 year stock prices for all the stocks [11]. After storing the stock prices of good stocks in the

variable then group all the values by Date column then sum it up , then plot it In line plot as an index ,so this will not be weight adjusted Weight or prices adjusted [6].

This is the Aggregate of all the stocks with the Assets to Liability Ratio above 10 .

Aggregate of all the stocks with the Assets to Liability Ratio above 10

As seen in the figure above the rate of return for Aggregated Index is around 1000% for a period of approx. 10 years.

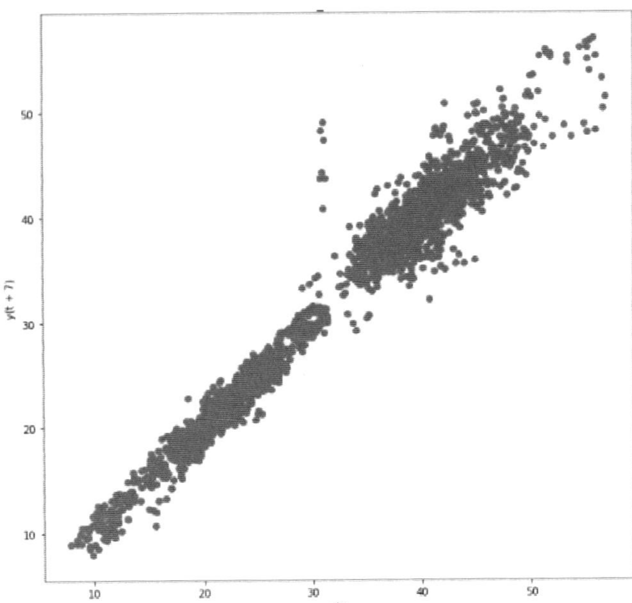

Figure 1. Correlation for the Created Index

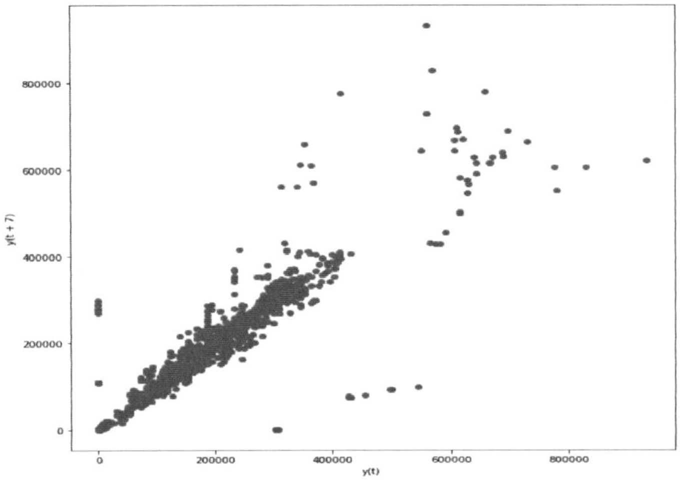

Figure 2. Correlation for the Aggregated Stock Market

Figure 1 Is a Scatter Plot showing Positive correlation for the selected Stock Index using PROMETHEE and Figure 2 is the Scatter plot which gives the correlation which is not ideal i.e. it shows less Correlation. Hence the Scatter Plot shows more Positive Correlation for Aggregated market's Index. As observed in Figures the Scattered matrix for Figure 2 is much more scattered and for the Selected index S&P 500.

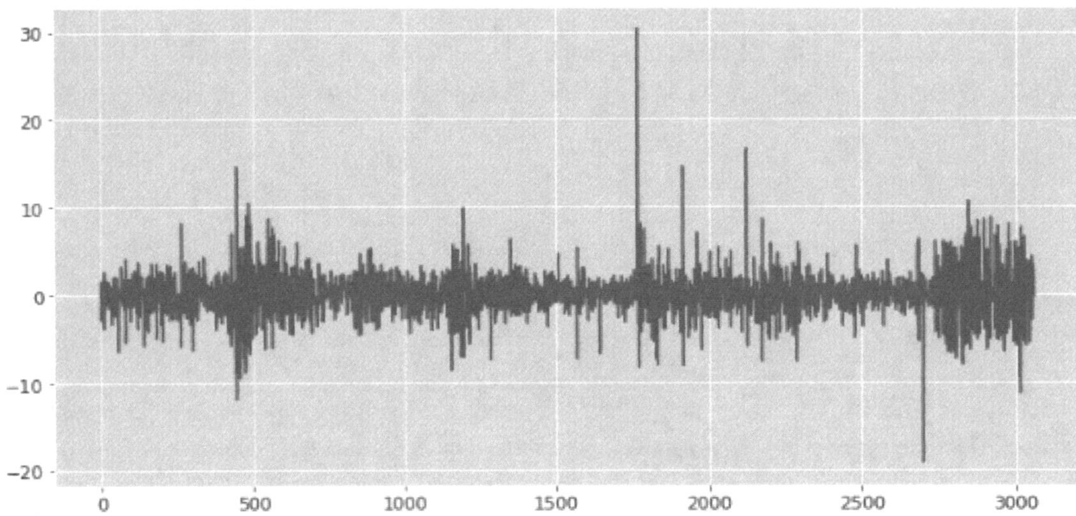

Figure 3. Mean Deviation for the Selected Index

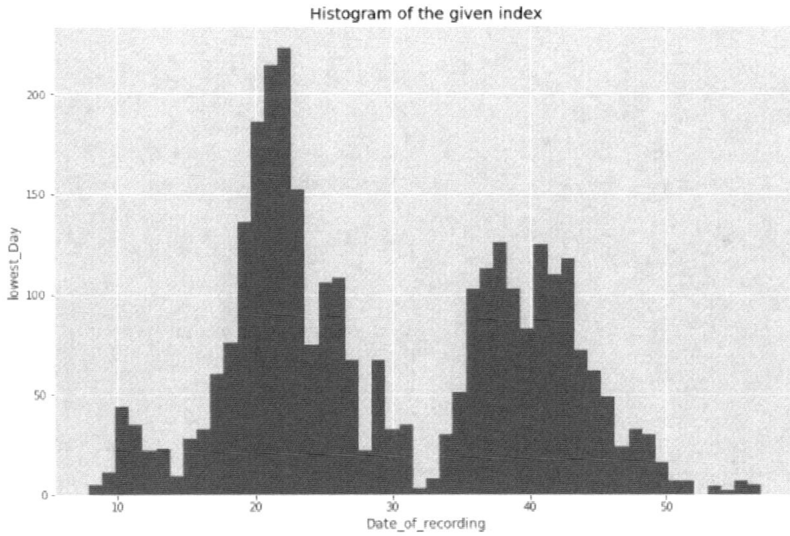

Figreu 4. Standard Deviation for the Selected Index

Fig 3 gives the Percentage change in daily values which gives a wide variation in percentage changes for the index which indicated high volatility on Daily Bases which proves the hypothesis that the market is Volatile and can change Positive 10 to Negative 10 percentage on daily basis but as we see over the long term it follows a uptrend and the Figure 4 gives the a double bell curve for the our selected Index of stocks using PROMETHEE indicates Bimodal Distribution also indicating percentage change is spread out into Two different group.

3.1 *Comparing the Method by Traditional methods*

<matplotlib.axes._subplots.AxesSubplot at 0x7f2b339e0cf8>

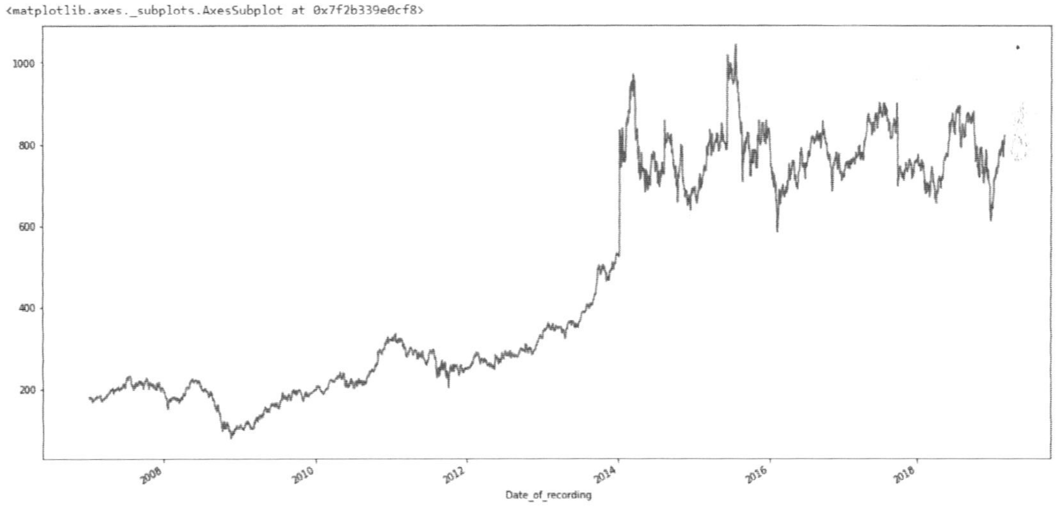

Figure 5. Time Series Aggregate for 23 Stocks

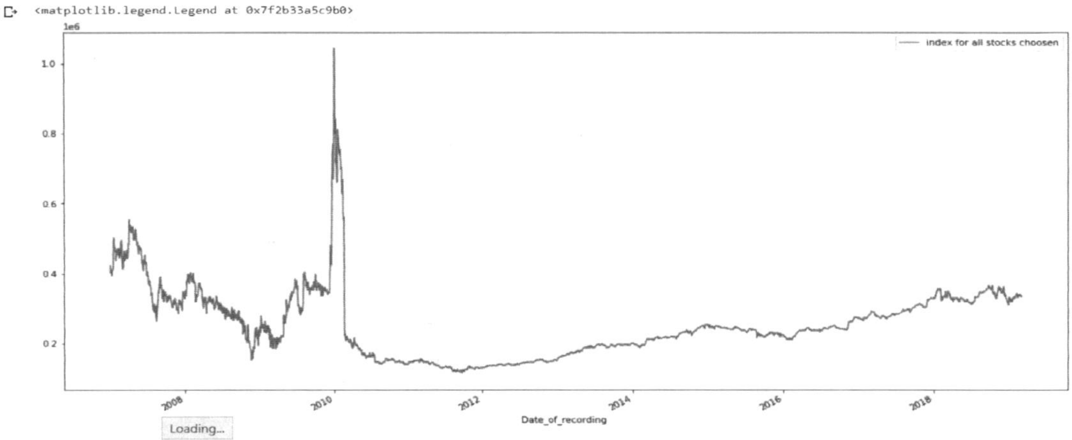

Figure 6. Time series aggregate of the Entire Stock List

Figure 5 contains Cumulative Index of stocks above Asset to Debt Ratio above 10 and Figure 6 Contains Cumulative return of the stocks Randomised stock collective rates of return for the aggregate for 23 stocks randomly picked in the market from 2009 to 2019 the return for the 10 year period is merely 200% , Fig.5 Looking at the Graph and taking the returns from 2008 - 2019 the average return is around 1000% but considering the return for S&p 500 or the market it comes out to be around 350% . Therefore using actual good parameters and ranking criterias it is possible to have a high rate of return compared to the market in the long run , Stated otherwise by the Efficient market Hypothesis[8] .

3.2 Increasing Gains Using Leverage

At that point file finances dependent on the specific boundaries are entered and dependent on those boundaries (like greatest market top, enchantment equation, all stocks with the most reduced P/E) TOP 100/500 STOCKS are chosen and dependent on those boundaries they are totaled and afterward contributed similarly [4]. The outcomes fluctuate enormously dependent on the measure of time you took for backtesting like the blend of those 3-4 models boundaries may really be acceptable however they probably won't function admirably for 2-3 years. for the others expectation may be high for a very long time. in any case, in some other mix of boundaries it can give a totally unique outcome dependent on the time span we pick. At that point the following is absolutely specialized dependent on that utilizing the past model we can plot the future stock cost in and afterward contribute it daily or two, yet the minutes after the specific episodes are not so much so we use influence to expand the benefit [9]. So assume we get a model right 80% of h time and we influence 1:1000 on a 1% sure increment what can happen is the multiple times you cash will go to zero however the rest times it will build multiple times. So even 1 time win can recoup your whole portfolio.

4. Fundamentals

1. In the securities exchange the measure of rate increment diminishes, not an issue the extent that it is certain we can use and get the necessary outcomes.

2. Discover the connection between's two distinct boundaries and discover what befalls this when this is taken [7]. As per Joel Greenblatt there is no relationship between's the past presentation of a chief and his exhibition in the following 3-4 years.

3. In the momentary the market runs on news and feelings in the drawn out it runs on basics.

Pick the stocks positioning savvy or make the total of the chose boundaries like the main 500 organizations or distinction between similarly weighted files. lists which gauge the boundaries considered similarly [8]. Like to pick the stocks positioning savvy (in view of 2-3 unique boundaries).

Let gj(a) be the estimation of a measure j for a choice a. We note dj (a, b), the distinction of estimation of a rule j for two choices an and b.

$$d_j(a,b) = g_j(a) - g_j(b) \qquad (1)$$

$P_j(a,b)$ a,b is the estimation of the inclination level of a standard j for two choices an and b. The inclination capacities used to figure these inclination degrees are characterized, for example

$$P_j(a,b) = F(d_j(a,b))\ with\ \forall x\ \in [-\infty,\infty], 0 \leq F(x) \leq 1 \qquad (2)$$

Leave C alone the arrangement of thought about standards and wj the weight related to the basis j. The worldwide inclination record for a couple of conceivable choice an and b is registered as follows:

$$\pi(a,b) = \sum_{j=C} \quad w_j \ \times \ P_j(a,b) \qquad (3)$$

For each possible decision a, we compute the positive outranking flow ɸ +(a) and the negative outranking flow ɸ −(a) [7]. Leave An alone the arrangement of potential choices and n the quantity of potential choices. The positive outclassing stream of a potential choice an is processed by the accompanying formulae:

$$\emptyset^+(a) = \frac{1}{n-1} = \sum_{x\in A} \quad \pi(a,x) \qquad (4)$$

The negative outranking flow of a possible decision computed by the following formulae:

$$\emptyset^-(a) = \frac{1}{n-1} = \sum_{x\in A} \quad \pi(x,a) \qquad (5)$$

The PROMETHEE I partial preorder (P~, It, R) is then obtained by considering the intersection of these two preorders:

$$\begin{cases} aP_1b & (a \text{ outranks } b) & \text{if } aP^+b \text{ and } aP^-b, \\ & & \text{or } aP^+b \text{ and } aI^-b, \\ & & \text{or } aI^+b \text{ and } aP^-b; \\ aI_1b & (a \text{ is indifferent to } b) & \text{iff } aI^+b \text{ and } aI^-b; \\ aRb & (a \text{ and } b \text{ are incomparable}) & \text{otherwise.} \end{cases}$$

5. Industries and Companies Ranking (Fundamental Analysis)

The required data in the PROMETHEE method are: Effective criteria in evaluation: these criteria are the foundation of ranking and have been extracted through the study of literature and surveying decision-making process in one of the Tehran investment and stock exchange companies [5]. Then, using a questionnaire, the necessity of using expert views is asked. Weight or relative importance of criteria: this weight will be provided with the help of the information obtained from the questionnaire where weights of criteria are calculated using the effects of their interaction on each other.
Type of criterion (max / min): other required information about the criteria, is its type which was considered using questionnaire and integrating experts views for each criterion. Preference function for each criterion: preference function in The PROMETHEE method depends much on the nature of the criterion and on the view of the decision maker [8]. For this purpose there are six standard preference functions that take into consideration most applications.

6. Methodology

- A dataset containing all the fundamental analysis like the dept of the company and the market capitalization , the long term assets , cash and cash equivalents are taken out from that Dataset all the company names which fit the criterias are filtered using the if else statement or the criterias set in place in multiple steps[15] . And each step filtering for a new criteria and storing into the same table.
- We want to extract the dataset out of all this for all the criterias one by one and the final dataset to meet all the criterias above like for example all the stocks where the market capitalization is more than 10 Billion , the P/E ratio is less than 5 , is only in industry with a lot of stability,
- It is only a famous name and last one if it has to be included (This wasn't included for the thought), it has huge protection in its industry (referred to as moat) and a good balance sheet .

- But what happens sometimes is a stock when its price is 1000 and another stock with equal market cap but with a price of 10 and kept in an index a 10% rise in stock A will give a total of 1110 or almost a 10% return but when Stock B rises in the index for 10 percent it becomes 1011 or a total rise of almost 0.001% .

7. Proposed system and proposed algorithm

In the event that leaders were dubious of standards loads and/or foresee a danger factor of high seriousness that could influence rules loads. On the off chance that chiefs were questionable of execution measures and/or foresee a danger factor of high seriousness that could influence execution measures. In the event that chiefs were unsure and/or foresee a danger factor of high seriousness that could influence the two standards loads and execution measures [9]. At whatever point two choices were impassive from one another in the ultimate result, they had distinctive execution measures as for standards. In the event that an elective scores most noteworthy on the most significant measure or rules and didn't have horrible showings on others. For a choice to be positioned first in an equivalent rules weight decisional issue [12].
The proposed model is organized around two columns: Industry assessment and Company assessment. The inclination positioning association technique for enhancement assessment (PROMETHEE) has been utilized for tackling the issue [6]. As a genuine case and an overview from the specialists so as to decide the viable measures for industry assessment and friends assessment has been directed. Theoretically, technical analysis tries to predict the trend of stock price using data on prices and trading volume over the last. The main problem with this approach is that it highly relies on empirical regularities based on price and volume movements [3]. The weight of the remaining indicators of distribution and finally the weights were calculated using paired comparisons method (improvement in AHP).

There are many other ways in which higher returns can be achieved rather than just risking everything. Stock market also doesn't need Any news Good or bad to affect its price as suggested because there has been countless cases where company prices have crashed despite showing high rates of return , because there was negative sentiment in market , and pump and dump schemes do exist where good news has nothing to do with good news but with the volume . As observed in the recent Covid-19 Pandemic the Stock market rose 44 % from the period of march to june 2020 when all businesses were going bankrupt, Travel was banned, consumer spending was the least and the unemployment rate was the highest it has ever been. [5].The proposed system suggests that investing in a stock market can be done on fundamental analysis and altho the market is Volatile in the short term, in the long term it returns to the fundamentals as the Stocks with the best fundamentals seem to return higher percentage gain overtime than the Standard benchmark S&P 500 . Hence, **Disproving the Efficient Market Hypothesis** according to which stocks can return a higher percentage over the market in the long term and are dependent on pure chance of luck as the market depends on News and News by definition is Unpredictable. Filtering out stocks by selecting the good fundamentals and ranking them using Multiple Criteria Decision making methods like PROMETHEE returns a higher percentage return over the Market in long term. Various types for Multiple-criteria decision analysis are shown in 'Fig 7'.

Types of MCDM

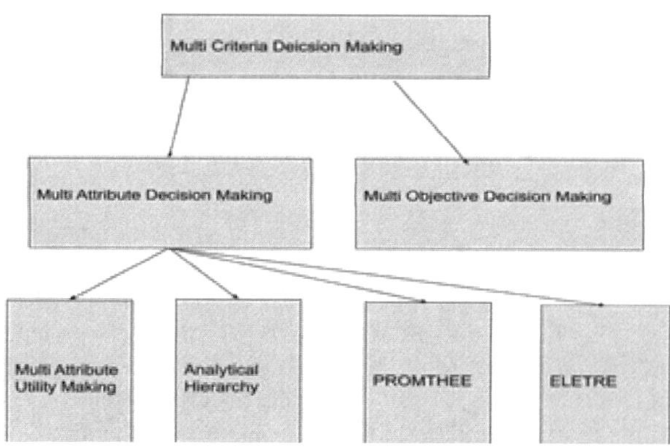

Figure 7. Proposed Working of the Model

8. Result and analysis

The Model gave following result after Testing:

Testing Mean Squared Error: 1.278
Symmetric mean absolute percentage error: 10.110

The Higher Percentage return over 10 years for the Stock Index chosen by observing the Fundamentals of the stocks and aggregating them to find index (i.e around 1000% rise over 10 Years) being 2-3 times higher than the Market proves our Hypothesis that EMH is faulty by negating its clause that One can't outperform the market over a long period of time unless provided with some sort of Insider information and also Fundamentals of Stocks are not of much use as the price of stocks depend on the new news with by definition can't be predicted .

9. Conclusion

Decision-making process involves setting goals, gathering relevant information and reviewing and evaluating information related to alternatives. After performing factor analysis on the first questionnaire (evaluating industries criteria) weights of the remaining criteria should be calculated based on their mutual influence on each other [14].

There are many things which are assumed as EMH says there is no way to outperform the market but as we know for a fact using the given index that the indexes with good filtering criteria can outperform the market in long term. According to the theory everything is efficient then bubbles shouldn't exist, but as we know there have been many many cases when the stock market crashed and recessions have hit, and it also has been proven that undervalued stocks can be picked.

10. References

[1] *Using the PROMETHEE Multi-Criteria Decision-Making Method to Define New Exploration Strategies for Rescue Robots.*
[2] *Decision making in stock trading: an application of PROMETHEE*
[3] *Application of Interval-PROMETHEE Method for Decision Making in Investing*
[4] Ranking and Managing Stocks in the Stock Market Using Fundamental and Technical Analyses.
[5] N. Basilico and F. Amigoni, "*Exploration strategies based on multicriteria decision making for search and rescue autonomous robots*," in Proc. of 10th Int. Conf. on Autonomous Agents and Multiagent Systems (AAMAS 2011), pp. 99–106, 2011.
[6] C. Stachniss and W. Burgard, "*Exploring unknown environments with mobile robots using coverage maps*," in IJCAI, 2003.
[7] F. Amigoni, V. Caglioti, and U. Galtarossa, "*A mobile robot mapping system with an information-based exploration strategy*," in Proceedings of the International Conference on Informatics in Control, Automation and Robotics (ICINCO 2004), pp. 71–78, 2004.
[8] W. Burgard, M. Moors, C. Stachniss, and F. Schneider, "Coordinated multi-robot exploration," *IEEE Transactions on Robotics*, vol. 21, no. 3, pp. 376–386, 2005.
[9] B. Roy, Méthodologie multicritère d'aide ` a la d ` ecision. Paris : Economica, 1985.
[10] B. Roy, "*The outranking approach and the foundations of electre methods*," Theory and Decision, vol. 31, pp. 141–160, 1991.
[11] G. Siskos, J. Wascher and H. Winkels, "*A bibliography of outranking ¨ approaches*," tech. rep., Cahiers du Lamsade, Paris Dauphine, 1983.
[12] Brans, J.P., Vincke Ph., Mareschal B. How to select and how to rank projects: the promethee method. *European Journal of Operational Research*, 1986, **24(2)**: 228-238.
[13] Babic ,Z., Plazibat,N.. Ranking of enterprises based on multi criteria analysis. *International Journal of Production Economics*, 1998: 29-35,56-57.
[14] Bouri,A., Martel,J.M., Chabchoub. A multi-criterion approach for selecting attractive portfolio. *Journal of Multi-criteria Decision Analysis*, 2002,11: 269-277.
[15] Albadvi,A., Chaharsooghi,S.K., Esfahanipour, A.. Decision making in stock trading: An application of PROMETHEE. *European Journal of Operational Research*, 2007,177:673-683.

IOP Conf. Series: Materials Science and Engineering 1045 (2021) 012005 doi:10.1088/1757-899X/1045/1/012005

[16] Li, Z-L. *A new improved ranking approach for interval numbers.* Mathematics in practice and theory (In Chinese) 2004, 34 (6): 124-127.
[17] *A multi-criteria decision model for selecting project portfolio with consideration being given to a new concept for synergies*

IOP Conf. Series: Materials Science and Engineering 1045 (2021) 012006 doi:10.1088/1757-899X/1045/1/012006

Synthesis and Characterization of Metal-Organic Framework (MOF) CoBTC as Non-Enzymatic Electrochemical Biosensor for Glucose

S A Abrori[1], M L A Trisno[1], R A Aritonang[1], I Anshori[2,3], Nugraha[1,2], Suyatman[1]*, B Yuliarto[1,2]*

[1] Advanced Functional Materials Laboratory, Engineering Physics Department, Faculty of Industrial Technology, Bandung Institute of Technology, Bandung 40132, Indonesia
[2] Research Center for Nanosciences and Nanotechnology (RCNN), Bandung Institute of Technology, Bandung 40132, Indonesia
[3] Lab-on-Chip Group, Biomedical Engineering Department, Bandung Institute of Technology, Bandung 40132, Indonesia
* Correspondence: yatman@tf.itb.ac.id Tel: +62 812-2101-205, brian@tf.itb.ac.id; Tel.: +6281312090020

Abstract. Diabetes mellitus is one of the highest causes of death in the world caused by high levels of glucose in the blood. Thus, monitoring glucose levels in the blood becomes very important. Non-enzymatic glucose detection is currently an emerging research area, one of which is to use Metal-Organic Material (MOF), which has advantages such as having a large surface area and having a large amount of micropore. The purpose of this study was to synthesize MOF CoBTC and see its performance as a glucose sensor. The synthesis method used is the solvothermal method, while the sensor performance testing method uses an electrochemical method. The solvothermal method of the CoBTC was carried out at a temperature of 100°C and duration of synthesis of 10 hours CoBTC materials. The structural characterization of the synthesized samples by X-Ray Diffraction (XRD), Scanning Elektron Microscopy (SEM), Fourier Transform Infrared (FTIR) Spectroscopy, Brunauer Emmett Teller (BET) Surface Area Analyzer showed that we successfully synthesized CoBTC. Furthermore, the electrochemical characterization via cyclic voltammetry (CV) and amperometry results showed the optimal working potential of the sensor at 0.4 V, the sensitivity of 53.455 μA mM-1 cm-2 and the limit of detection (LOD) at 5.49 μM.

1. Introduction

One of the diseases suffered by humans today is diabetes mellitus. Diabetes mellitus is a condition in which the glucose content in the blood exceeds the average threshold [1]. Diabetes mellitus is a metabolic disease caused by the inability of the pancreas to produce enough insulin or the inability of the human body to use the insulin effectively, resulting in an uncontrollable increase in glucose levels in the blood [2]. The International Diabetes Federation (IDF) estimates that in 2013 there were over 380 million people with diabetes, and it is estimated that this number will increase to 592 million people in 2035, wherein 2013, 175 million of them have not been diagnosed, so they are threatened with progressive development into complications without realizing it [3]. Furthermore, without prevention, Diabetes mortality alone reaches 80% in countries with lower-middle-income levels. Thus, the detection of diabetes mellitus is very much needed to be able to prevent and avoid developing diabetes. Until recently, diabetes detection was carried out using enzymatic electrochemical glucose biosensors, where

IOP Conf. Series: Materials Science and Engineering 1045 (2021) 012006 doi:10.1088/1757-899X/1045/1/012006

85% of the glucose sensors currently on the market are used for diabetes detection, and the remainder is used for the detection of glucose content in medicines, foods, and others[4].

Since fast and reliable glucose detection has become essential in many fields such as biotechnology, clinical diagnosis, and the food industry, the development of electrochemical based glucose sensors has attracted much attention. In the past, research on enzymatic glucose biosensors based on the oxidation reaction of glucose compounds was prevalent because of its sensitivity [5]. However, testing using modified enzyme electrodes has disadvantages such as relatively low stability due to environmental conditions, high cost of the enzymes, critical operational conditions, and complicated immobilization procedures. Therefore, glucose detection without the use of enzymes or non-enzymatic detection is highly desirable with the advantages of stability, simple, sustainable production, and low cost [6]. Currently, the development of non-enzymatic electrodes has achieved the modification of precious metal compounds, metal alloys, and metal nanoparticles. However, suitable detection materials are required for glucose detection.

One of the materials that can be used as active material in biosensors is material based on the metal-organic framework. Metal-organic frameworks (MOF) or coordination polymers are metal-ligand compounds whose structures will expand or extend in one, two, or three dimensions and where the ligands will become connecting organic groups [7]. MOF compounds that have microporous properties are interesting material to develop. The space between metals connected to organic ligands makes MOF material have a large number of micropores and a large surface area. This large surface area causes MOF to have the potential to be developed into a biosensor material. Therefore, in this study, the development of a non-enzyme sensor based on Metal-Organic Framework (MOF) was chosen, namely the Co material with BTC. Co is used because it has strong magnetic properties, does not readily react with oxygen in the air and does not dissolve easily in water, and is very popular as a glucose biosensor in its oxide formation. MOF with a Co metal base will be of particular interest due to the Co metal and the MOF material itself so that the material is considered to have excellent performance as an active material for glucose detection biosensors. Therefore, further study is needed on the performance of CoBTC and how the synthesis formulation of this material is to produce active glucose detection biosensor material with optimum performance so that it can be used as a glucose detection biosensor.

2. Experimental

2.1. Materials

Cobalt(II) nitrate hexahydrate ($Co(NO_3)_2.6H_2O$), trimesic Acid (H_3BTC), dimethylformamide (DMF), glucose, Nafion, and other common reagents were acquired from Sigma-Aldrich, Singapore. The organic solutions such as ethanol were purchased from Merck, Singapore. All chemical materials were of analytical grade and used without any further purification.

2.2. Synthesis of CoBTC

The synthesis of CoBTC was carried out by mixing metal precursors and linkers at heating temperature or what can be called the solvothermal method. First, we dissolved the cobalt dinitrate hexahydrate (1.454 grams) in 10 mL distilled water and trimesic acid precursor (0.525 grams) in 10 mL DMF and 10 mL ethanol, then stirred separately for ± 20 minutes. Next, the two solutions were mixed by dropping the trimesic acid solution drop by drop into the cobalt metal solution and stirred for ± 45 minutes. After the mixture is homogeneous, the mixture was heated thermally in the autoclave at 100 °for 10 hours. The mixture will produce a product in the form of powder. The as-collected powder then washed several times with ethanol while sonicated. Then the product was dried at a temperature of 100 ° C for 10 hours. Finally, the sample is ready to be used and characterized. The synthesis of CoBTC will follow the following formula:

$$3Co(NO_3)_2.6H_2O + 2H_3BTC \rightarrow Co_3(BTC)_2.12H_2O + 6NO_3^- + 6H_2O + H^+ \qquad (1)$$

2.3. Material Characterization

Bruker D8 Advanced collected the X-Ray Diffraction (XRD) patterns with CuKα x-ray generator, $\lambda = 1.5418$ Å. The morphologies property of the CoBTC were analyzed using Hitachi Scanning Electron Microscopy (SEM) SU3500 series. The surface area of the CoBTC was analyzed using a Surface Area Analyzer with the Brunauer, Emmet, and Teller (BET) analysis method and Quantachrome Instrument Nova 1000, while Fourier Transform Infrared (FTIR) spectrum was collected by Prestige 21 Shimadzu.

We employed CorrTest CS Series Electrochemical Station to carry out electrochemical measurements. All electrochemical measurements were carried out in a standard three-electrode system in an aerated condition at room temperature. In this work, we employed a glassy carbon electrode (GCE, 3 mm in diameter) modified with the CoBTC samples as the working electrode. Meanwhile, we used an Ag/AgCl standard electrode (3 M KCl) as the reference electrode and a Pt wire as the counter electrode. NaOH solution 0.1M was employed as the supporting electrolyte for the electrochemical characterization. The Cyclic Voltammograms (CVs) test of the samples was carried out in the potential range of -0.2-0.8 V with a scan rate of 50 mV/s. Meanwhile, the amperometry test was carried out at 0.4 V. All potential are versus the Ag/AgCl standard reference electrode.

3. Results and Discussion

3.1. Structural Characterization

The graph of the synthesized CoBTC diffraction peaks compared to the reference can be seen in Figure 1 (a). Graph of FTIR spectroscopy results are shown in Figure 1 (b) shows the classification of the analysis from the top of the graph to the range of the infrared spectrum absorbed from the covalent bond. Based on Figure 1(a), the synthesized CoBTC has peaks that are the same as the CoBTC synthesized by Crane et al., 2014 [8], and Aguiar et al. 2017 [9], namely at 17.3, 18.4, and 26.7 2θ (deg), which indicates the material is $Co_3(BTC)_2.12H_2O$ material. In Figure 1 (b), it can be seen that all forms of chemical bonds that make up CoBTC compounds based on its chemical structure can be detected, which also confirms that the CoBTC material is perfectly formed.

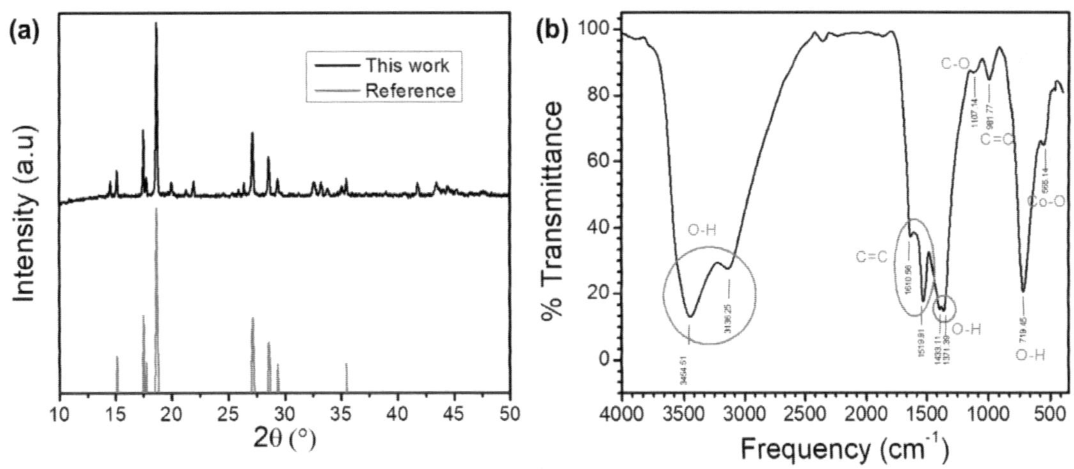

Figure 1 (a) XRD patterns of the CoBTC in this work and its reference (b) FTIR spectrum of the CoBTC

SEM on the synthesized CoBTC material was carried out at two different magnifications, namely 2,000 times and 15,000 times. The result of magnification of 2,000 times can be seen in Figure 2 (a), while for magnification of 15,000 times can be seen in Figure 2 (b). Based on SEM results at a magnification of 2,000 times, it is known that the morphological formation of the CoBTC material synthesized tends to clump. At a magnification of 15,000 times, it appears that CoBTC has a cubical shape, and it appears that CoBTC has a length between 25-30 μm and a width of 21-24 μm. Based on the references found, CoBTC has various forms of morphology, depending on the synthesis method used and the variations in the method concerned. The morphological forms of CoBTC include a distorted octahedral form [10] and rod [9]. The synthesized CoBTC has a cubical shape that has six surfaces with uneven conditions, and the resulting CoBTC tends to be agglomerated, and the size is not homogeneous.

The large surface area is one of the advantages of MOF materials. MOF is a porous material that has so far been used and holds great promise for storing hydrogen and methane, which has already been

Figure 2. SEM images of the CoBTC (a) at 2k magnification (b) at 15k magnification

applied to fill car tanks. As a sensor, a large surface area is an advantage of MOF material, because the more significant the surface area, the higher the active area of interaction between MOF as a sensor with glucose molecules. CoBTC itself, which has been characterized by a BET Surface Area Analyzer, has a surface area of 557.862 m^2/g. However, these results are much better than the synthesis results from Aguiar et al. 2017 [9], which is only 57 m^2/g with the same XRD characterization results and higher than Co in its oxidation form, Co_3O_4 which is only 260 m^2/g [11].

3.2. *Electrochemical Characterization*

The process of cyclic voltammetry testing is carried out in alkaline conditions where a 0.1 M NaOH solution is used as the analysis solvent. The use of an alkaline environment is because glucose is not reactive in an acidic environment and the instability of the transition metal/metal oxide-based electrode materials in acidic conditions. This instability can cause the oxidation process not to run optimally, and the resulting voltammetry peak will be small. Glucose in a neutral environment tends to form a double layer region where the oxidation peak will be reformed so that the peak from the maximum oxidation period of glucose is challenging to know. The ideal alkaline environment becomes a detection environment because glucose will be easily oxidized and become gluconic acid, which is indicated by the peak value of oxidation from the cyclic voltammetry test.

This voltammetry test is carried out in a voltage range from -0.2 - 0.8V with a positive test direction and a scan rate of 50 mV / s. The oxidation equation that occurs is as follows:

$$CoBTC + 2OH^- \rightarrow [CoBTC\,(OH)_2] \quad (2)$$

$$[CoBTC(OH)_2]^{-2} + e^- \rightarrow Co(OH)_2 + BTC \quad (3)$$

$$Co(OH)_2 + OH^- \rightarrow CoOOH + e^- + H_20 \quad (4)$$

$$CoOOH + e^- + \textit{glucose} \rightarrow CoO + OH^- + \textit{gluconic acid} \quad (5)$$

$$CoO + OH^- \rightarrow CoOOH + e^- \quad (6)$$

Equations (2), (3), and (4) are the material activation process. In this process, the material is tested in 0.1 M NaOH solution without glucose so that the material is active. Equations (3) and (4) also prove that the sensor material will produce current even in the absence of glucose. After the material is activated, it is tested with glucose. During the testing process with glucose, reactions such as equations

(2), (3), and (4) are added with the reactions according to equations (5) and (6), which will cause an increase in the peak flow when without glucose and when tested at the glucose concentration higher. This process occurs when the oxidation of Co becomes a catalyst for the oxidation of glucose to gluconic acid. The cyclic voltammetry testing process can be seen in Figure 3. It can be seen that CoMOF can increase glucose oxidation current compared to bare GCE. The oxidation of glucose by CoMOF occurs

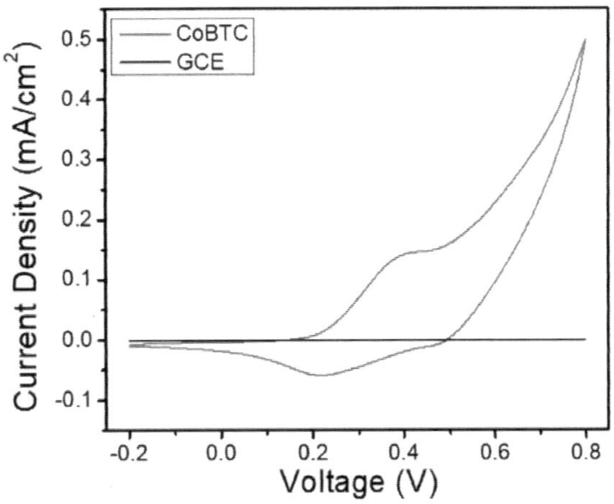

Figure 3. Cyclic Voltammetry curves of CoBTC and bare GCE on glucose 5mM and NaOH 0.1M in potential rage -0.2-0.8V and scan rate 50mV.s^{-1}

at a potential of 0.4 V.

Furthermore, amperometric testing was carried out to determine the sensitivity and limit of detection (LOD) of the selected CoBTC material to glucose. In this study, a concentration variation from 0 to 5 mM of glucose was used, and potential value of 0.4V was used where this potential value was obtained from cyclic voltammetry experiments because, at a potential of 0.4V, the material was able to oxidize glucose completely. The graph of the results of this amperometric test can be seen in Figure 4.

From Figure 4, it can be seen that the increase in the value of glucose concentration causes an increase in the resulting current value, where the difference in the increase in flow is directly proportional to the difference in the increase in glucose. A graph of the relationship between concentration and the resulting current is drawn, as shown in figure 4(b), to find out quantitatively about this comparison. From the graph, it can be seen the linearity of the material in detecting glucose from a range of 0 to 5 mM with a

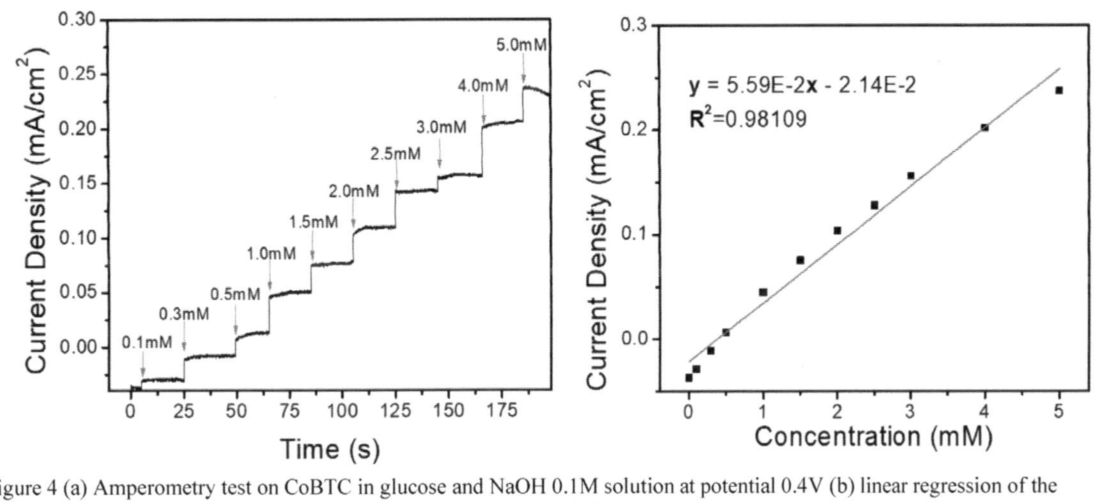

Figure 4 (a) Amperometry test on CoBTC in glucose and NaOH 0.1M solution at potential 0.4V (b) linear regression of the amperometry test

value of $R^2 = 0.9859$. The sensitivity value can also be seen from the slope value of the line on the graph of the ratio of concentration to current. The sensitivity value obtained was 53.455 μA $mM^{-1}cm^{-2}$, and the LOD was 5.49 μM. The sensitivity of the CoBTC biosensor of 53.455 μA $mM^{-1}.cm^{-2}$ was still lower than the non-enzymatic biosensor based on Co_3O_4, which was 1377 μA $mM^{-1}cm^{-2}$ by Wang et al., 2017 [12] so that the modification of this material was to exceed the sensitivity of Co_3O_4 will need to be done in the future.

4. Conclusions

In this work, we successfully synthesized CoBTC material using the solvothermal method. The CoBTC was characterized using XRD, SEM, FTIR, and BET and showed similar features to those previously synthesized. The CoBTC was then used as a sensing material for non-enzymatic electrochemical detection of glucose. The results showed that the optimal working potential of the sensor at 0.4 V, the sensitivity of 53.455 μA mM-1 cm-2 and the limit of detection (LOD) at 5.49 μM. This result indicates that CoBTC is promising to be further studied as a non-enzymatic electrochemical sensor material.

Acknowledgments

This work is also partially supported by the Ministry of Education and Culture and the Ministry of Research and Technology under the grant scheme of the World Class University (WCU) Program managed by Institut Teknologi Bandung. The authors also acknowledge financial grants provided by the Institute of Technology Bandung (ITB), the Ministry of Education and Culture and the Ministry of Research and Technology of Indonesia

References

[1] Habibi, S.; Ahmadi, M.; Alizadeh, S. Type 2 Diabetes Mellitus Screening and Risk Factors Using Decision Tree: Results of Data Mining. *Glob. J. Health Sci.* **2015**, *7*, 304–310, DOI:10.5539/gjhs.v7n5p304.

[2] Li, R.J.W.; Zhang, S.-Y.; Lam, T.K.T. Interaction of glucose sensing and leptin action in the brain. *Mol. Metab.* **2020**, *39*, 101011, doi:https://doi.org/10.1016/j.molmet.2020.101011.

[3] Glasson, N.M.; Crossland, L.J.; Larkins, S.L. An Innovative Australian Outreach Model of Diabetic Retinopathy Screening in Remote Communities. *J. Diabetes Res.* **2016**, *2016*, 1267215, DOI:10.1155/2016/1267215.

[4] Bruen, D.; Delaney, C.; Florea, L.; Diamond, D. Glucose Sensing for Diabetes Monitoring: Recent Developments. *Sensors (Basel).* **2017**, *17*, 1866, DOI:10.3390/s17081866.

[5] Hsu, Y.-W.; Hsu, T.-K.; Sun, C.-L.; Nien, Y.-T.; Pu, N.-W.; Ger, M.-D. Synthesis of CuO/graphene nanocomposites for non-enzymatic electrochemical glucose biosensor applications. *Electrochim. Acta* **2012**, *82*, 152–157, doi:https://doi.org/10.1016/j.electacta.2012.03.094.

[6] Luo, J.; Jiang, S.; Zhang, H.; Jiang, J.; Liu, X. A novel non-enzymatic glucose sensor based on Cu nanoparticle modified graphene sheets electrode. *Anal. Chim. Acta* **2012**, *709*, 47–53, doi:https://doi.org/10.1016/j.aca.2011.10.025.

[7] Liu, S.; Lai, C.; Liu, X.; Li, B.; Zhang, C.; Qin, L.; Huang, D.; Yi, H.; Zhang, M.; Li, L.; et al. Metal-organic frameworks and their derivatives as signal amplification elements for electrochemical sensing. *Coord. Chem. Rev.* **2020**, *424*, 213520, doi:https://doi.org/10.1016/j.ccr.2020.213520.

[8] Crane, J.L.; Anderson, K.E.; Conway, S.G. Hydrothermal Synthesis and Characterization of a Metal-Organic Framework by Thermogravimetric Analysis, Powder X-ray Diffraction, and Infrared Spectroscopy: An Integrative Inorganic Chemistry Experiment. *J. Chem. Educ.* **2015**, *92*, 373–377, DOI:10.1021/ed5000839.

[9] Aguiar, L.W.; da Silva, C.T.P.; de Lima, H.H.C.; Moises, M.P.; Rinaldi, A.W. Evaluation of the synthetic methods for preparing metal-organic frameworks with transition metals. *AIMS Mater. Sci.* **2018**, *5*, 467–478, DOI:10.3934/matersci.2018.3.467.

[10] Du, Y.; Thompson, A.L.; O'Hare, D. Resin-assisted solvothermal synthesis of metal-organic frameworks. *Chem. Commun.* **2008**, 5987–5989, DOI:10.1039/B814180K.

[11] Hoa, L.T.; Chung, J.S.; Hur, S.H. A highly sensitive enzyme-free glucose sensor based on Co3O4 nanoflowers and 3D graphene oxide hydrogel fabricated via hydrothermal synthesis. *Sensors Actuators B Chem.* **2016**, *223*, 76–82, doi:https://doi.org/10.1016/j.snb.2015.09.009.

[12] Wang, L.; Zhang, Y.; Xie, Y.; Yu, J.; Yang, H.; Miao, L.; Song, Y. Three-dimensional macroporous carbon/hierarchical Co3O4 nanoclusters for non-enzymatic electrochemical glucose sensors. *Appl. Surf. Sci.* **2017**, *402*, 47–52, doi:https://doi.org/10.1016/j.apsusc.2017.01.062.

Residual stress analysis of friction stir welded AA6063 pipe butt joint by neutron diffraction technique of ENGIN-X

Azman Ismail[1,a], Tung Lik Lee[2], Mokhtar Awang[3], Fauziah Ab Rahman[1], Nabihah Salih[3], Bakhtiar Ariff Baharudin[1], Puteri Zarina Megat Khalid[1] and Darulihsan Abdul Hamid[4]

[1]Universiti Kuala Lumpur Malaysian Institute of Marine Engineering Technology, Perak, Malaysia.
[2]Science and Technology Facilities Council, Oxford, United Kingdom.
[3]Department of Mechanical Engineering, Universiti Teknologi PETRONAS, Perak, Malaysia.
[4]Kolej Universiti Poly-Tech MARA, Kuala Lumpur, Malaysia.

[a]Corresponding author: azman@unikl.edu.my

Abstract. Residual stress refers to stress that remains in a structure when all applied stresses have been removed. It is important because the combination with applied stress can cause failure at the level below at which the failure could occur. A study has been carried out using the non-destructive neutron diffraction (ND) technique of ENGIN-X to quantify the redistribution of residual stresses in a pipe. A three-dimensional residual stress-measurements test is fully performed on a 89 mm (3.5 in) diameter pipe in AA6063 grade aluminium (5 mm wall thickness) containing a girth friction stir weld. The results have shown an opposite residual stress effect due to tool rotational speed increment especially on the retreating side of the weld. This weld residual stress analysis is an important factor for Fitness-For-Purpose Assessments. This residual stress measurement data provides a basis for structural integrity assessment and inspection planning.

1. Introduction

Aluminum alloys are commonly used in railway transport, automobile industries, aerospace, bridges and high speed vessels. It is due to lightweight, with a high strength which is relative to weight ratio, corrosion resistance and ductility. The joining of aluminum is commonly performed in the industry using the arc welding technique but the technique would easily induce severe weld defects and distortion. The friction stir welding (FSW) technique offers excellent mechanical properties and safe to the environment as it produces no fumes and arc, requires no shielding gas and filler metal. In 1991, FSW was invented and introduced by The Welding Institute and was mainly used in welding flat and tubular structures.

Residual stress is the stress that still persists in the structure after all the added load are removed. Stresses due to residues are easily generated in welded structures. The material's yield robustness and level of residual stresses may vary. This condition may cause an adverse effect on structural integrity and can cause premature failure [1]. This premature failure may result from the residual stresses when combined with the applied stresses which occur due to the applied stresses. This weld residual stress analysis is an important factor in Fitness-For-Purpose assessments. Similar to the applied stresses, this residual stress can also be determined by the use of proper techniques i.e ENGIN-X [2,3].

A deterioration in the integrity of the structure and performance of the components may persist due to residual stresses in welded structures. Neutron diffraction (ND) is an authenticated approach for the measurement of residual stresses occurring in welds due to its peculiar ability to profoundly penetrate, and map three-dimensional elements. In addition, the scattering neutron beam with bulk dimensions are also averaged on volume. The variations of the residual stress within the nugget zone (NZ), thermo-mechanically affected zone (TMAZ), and heat-affected zone (HAZ) can be measured using the ND [4-6]. The present study focuses on how the friction stir welding parameters affecting the residual stress of tubular component was measured and analyzed. Figure 1 shows the dedicated ENGIN-X facility at ISIS, UK.

Figure 1. ENGIN-X Facility

2. Methodology

FSW of relatively small diameter of 89mm aluminum pipes is a relatively new application thus presents a unique challenge which requires the use of specialized techniques and equipment [2,3].The lab test was conducted at the ISIS STFC neutron diffraction facility (ENGIN-X) in UK. The welding residual stress measurement was characterized in the as-welded pipe specimen (in the hoop, longitudinal and radial directions relative to the pipe), measuring residual stresses at the 45° angle position. The test was performed on full three-dimensional residual stress measurements on 89mm diameter pipe in AA6063 grade aluminium with 5-mm wall thickness containing a girth friction stir weld and weld width of 20mm. The single line scans were performed at this location at 2 mm below the cap level. There were 12 measurement positions along each measurement line extending -20 mm and +20 mm from the weld center as shown in Figure 2.

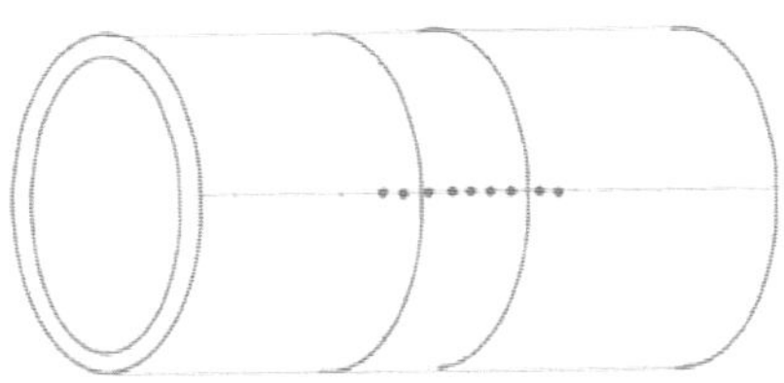

Figure 2. ND measurement points on pipe specimen

A set of stress-free (d_0) reference combs from similar locations on a pipe containing a girth weld manufactured using the same welding parameters were prepared as shown in Figure 3 . Cross-sectional samples of 4 mm x 5 mm x 100 mm were wire cut along the length of the pipe specimen to serve as d_0 samples [2,3]. The positions where the d_0 samples were corresponded to the ND measurement position made on the pipe specimen in order to maintain comparable microstructure distribution. These matchsticks were focused on weld and heat-affected zone regions as well as the base material. The same number of positions were measured on these stress-free combs specimen. The two sets of welding parameters used for these specimens were shown in Table 1 below. The setup of the specimen on the ENGIN-X is shown in Figure 4.

Figure 3. A stress-free comb specimen (d_0) of pipe

Table 1. Friction stir welding parameters

Specimen No. (SN)	Rotational Speed (rpm)	Travel Speed (mm/s)
SN.1	1000	5
SN.2	1300	5

Figure 4. As-welded pipe specimen and comb specimen setup at the ENGIN-X facility

The experiment was carried out using the time-of-flight neutron diffractometer, ENGIN-X at RAL. These 3 strain components (hoop, longitudinal and radial) were measured via the neutron diffraction technique across the thickness of the two (2) pipe specimens using 2×2×2 mm^3 gauge volume at ambient conditions. The aforementioned procedures were repeated for the d_0 samples cut to obtain the stress-free lattice spacing at the corresponding scanned positions. The total estimated scan time required for the study was 3 days. The data analysis was conducted using ENGIN-X OpenGenie software with GSAS (peak positions analyzed with Pawley refinement).

3. Results and Analysis

All three residual stress components were measured in this study. Based on Figures 5 and 6, the residual stresses in the hoop direction were maintained low after the increment of rotational speed. In the vicinity of the weld (±4mm from the weld centreline), the residual stresses dropped to a minimum value from yield magnitude. However, the residual stress was quite high on the retreating side, approximately 15-20mm away from weld centreline. The residual stresses in the longitudinal direction decreased in general with the increment of rotational speed except on the retreating side which gave the opposite effect. However, the residual stress on the retreating side was lesser than the hoop direction.

The residual stresses in the radial direction were also low but limited to the vicinity of the weld zone. The distribution of residual stresses before and after straining follows the same trend in the far field especially on the retreating side, approximately 15-20 mm away from the weld centreline. The retreating side gave higher but opposite readings with increment of tool rotational speed that was due to cold stirring process as compared to the advancing side which gives more heat [7,8].

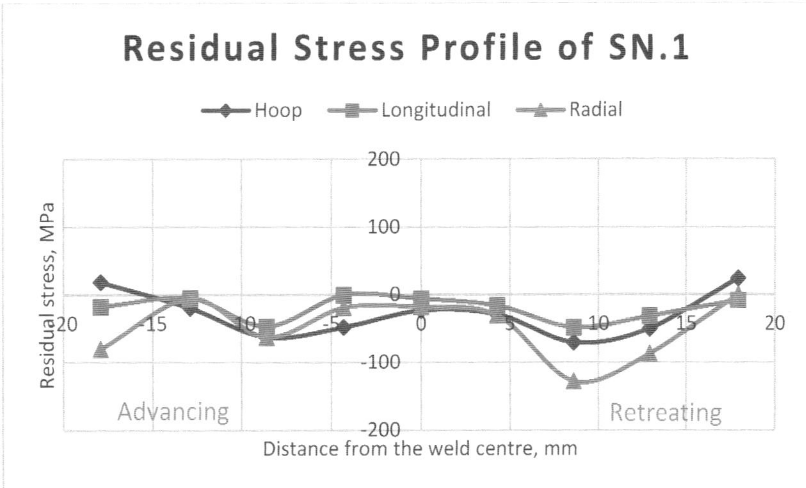

Figure 5. Residual stress for tool rotational speed at 1000 rpm.

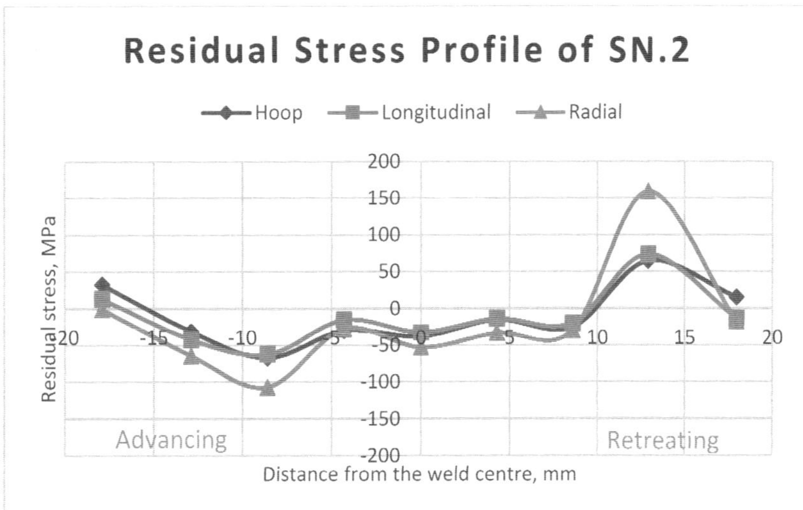

Figure 6. Residual stress for tool rotational speed at 1300 rpm

4. Conclusion

This present study reports on the comprehensive 3D measurements of residual stresses in full scale girth welds of pipes. The following specific conclusions are made. With the increment of tool rotational speeds:

a. Residual stresses in the hoop, longitudinal and radial direction were measured at low value in friction stir welded zone.
b. These residual stresses gave a higher magnitude on the retreating side were due to cold stirring effect and uneven heat distribution between these two zones, as the advancing side gave more heat as compared to the retreating side.

Acknowledgement

The authors acknowledge the support of the Science and Technology Facilities Council (STFC) in United Kingdom for their neutron diffraction facilities of ENGINE-X provided and sponsored for this lab testing, RB 1810192 and the travel grant provided by the Universiti Kuala Lumpur (UniKL), PEM No.80 (3/2018).

References

[1]. Yan-Hui Zhang, Simon Smith, Liwu Wei and Carol Johnston, "*Residual stress measurements and modelling*", TWI HSE Books, pp.1-80, 2012.

[2]. P.J Withers, M.Preuss, A.Steuwer and J.W.L Pang, "Methods for obtaining the strain-free lattice parameter when using diffraction to determine residual stress", *Journal of Applied Crystallography*, 40, pp 891-904, 2007.

[3]. J.R. Santisteban, M.R Daymond, J.A James and L. Edwards, "ENGIN-X: a third generation neutron strain scanner", *Journal of Applied Crystallography*, 39, 2006, pp.812-825.

[4]. W. Woo, Z. Feng, X.L. Wang and S. A. David, *Neutron diffraction measurements of residual stresses in friction stir welding: A review,* Science and Technology of Welding and Joining, Vol 16, No 1, pp 23-32, 2011.

[5]. Meysam Haghshenas, M. A. Gharghouri, V. Bhakhri, R. J. Klassen, and A. P. Gerlich, Assessing residual stresses in friction stir welding: neutron diffraction and nanoindentation methods, *Int J Adv Manuf Technol*, pp1-15, 2017.

[6]. Luciano Buglioni, Leonardo N. Tufaro, and Hernán G. Svoboda, "Thermal cycles and residual stresses in FSW of aluminum alloys: experimental measurements and numerical models", *Procedia Materials Science*, 9, pp. 87-96, 2015.

[7]. A Ismail, M Awang, M. A. Rojan and S. H. Samsudin, "The Characteristic of Temperature Curves for Friction Stir Welding of Aluminium Alloy 6063-T6 Pipe During Tool Plunging Stage", *ARPN Journal of Engineering and Applied Sciences*, Vol.11, No.1, pp.277-280, 2016.

[8]. M. P. Iqbal, R K Vishwakarma et al. Influence of plunge depth during friction stir welding of aluminum pipes, Proc IMechE Part B: *J Engineering Manufacture*, pp.1–12, 2020.

+2.5% Improvement in generation utilizing Anti soiling modules based on DSM technology

Rajat Verma[1], **Amresh Mahajan**[2] **and Prem Pathak**[3]
[1, 2] Quality Department, ACME Cleantech Solutions Private Limited, Gurugram, India
[3] DSM Engineering Plastics, Pune, India

Corresponding author: Email: rajat.verma@acme.in

Abstract. Accumulation of dust on solar modules decreases the transmittance of photovoltaic glazing, resulting in both deteriorated efficiency and daily energy losses. Actually, the soiling phenomenon in PV systems has been shown to be responsible for a medium- and long-term yield reduction of up to 16% over the first year of outdoor exposure [1].The most prominent origin of dust is driven by wind and soil particles from the ground surface. All over the planet, the properties of dust due to geological, topographical, and environmental ambiences such as relative density, particle structure, surface effects, and vapour content differ from one location to another. In the present study, an exploration of the energy gain observed by Photovoltaic modules coated with DSM Anti-Soiling coating (ASC) installed at 150 MW solar power plant in Andhra Pradesh is presented. The experiment for the analysis of the generation data started on 6 September 2018. Four different slots, one with DSM ASC and the rest with an anti-reflective coating (ARC), with equivalent DC connected load are chosen for analysis. The washing cycle was maintained at an interval of 10 days. The study shows that the modules with ASC generate +2.5% more power than the modules with ARC.

1. Introduction

The mass and structure of dust granule have a considerable effect on their sedimentation manner and photovoltaic (PV) efficiency. These days, the decreased generation output of PV modules due to the impact of soiling on the glass surface is receiving a lot of attention. The agglomeration of the dust on the plane of the PV modules decreases the transmittance and results power losses. Previous studies show that these losses could reach 15% in dry areas [1], while another study in a desert location reports losses greater than 20% [2]. In such events, the remedy is to wash the modules with water and dry cleaning with robots [3].

The accretion of dirt on the frontal glass of PV systems continuously interferes with power generation phases. In fact, PV module energy generation is proportional to the accumulation of incident photos reaching the PV cell surface of the modules. In practical terms, the amalgamation of self-cleaning coatings enables a remarkable decrease in maintenance expenses associated to wiping activities and in the quantity of water required for solar module washing. The application of radiation by any PV cell or solar thermal appliance is limited by arrival losses due to backscattering, reflection, or exterior absorption. Thus, anti-soiling surfaces and anti-reflecting coatings can make a remarkable contribution to the respective overall process production.

The industrial science of renewable power, in specific photovoltaic's (PVs), is troubled by diverse specifications (examples, technology, environment, region, and assimilation). The environmental criterion has a broad influence on the power generation and effectiveness of PVs. Some environmental criterion includes direction, moisture, solar radiation and temperature, wind speed, and the presence of dirt.

The soiling on the sun-facing area of installed PV modules is a well-known obstacle, generally resulting in reductions of about 5% to 25% in power output, depending on environmental conditions and maintenance and cleaning proceedings [4], [5] & [6]. Soiling happens very quickly, so losses due to soiling can be severe even with continued pattern of cleaning. In different circumstances, soiling fluctuates tremendously, even in a specific region, as the effects of climate, microclimate, atmosphere, the configuration of wind-borne dust, local dust and other contaminants, and many other characteristics determine the soiling that happens in any time span.

2. DSM Technology

DSM Anti-Soiling coating (ASC) consists of anti-reflective and anti-soiling properties designed specifically for dry environments. It improves the internal rate of return by requiring fewer cleaning cycles and reducing performance loss. It comprises of 100% inorganic material, which provides an excellent UV resistance under high irradiance and its anti-reflective properties are comparable to regular anti-reflective coatings (ARCs). It reduces the soiling rate by preventing dust and sand from sticking to the module, while using wind and gravity as cleaning processes. This also reduces cleaning costs by increasing the intervals between cleaning cycles. The Taber Oscillation test was designed to test coatings in lenses and windows (according to ASTM F735), and DSM adapts such a setup for soiling tests by using standardized 'Arizona Sand'. The ASC shows virtually no loss in optical performance in such tests, while ARC and uncoated glass exhibit a loss of roughly 2% to 4%.

Traditional anti-soiling coatings are based on two physical principles. These effects are based on hydrophobic or hydrophilic properties, which are, respectively, "water-repellent" or "water-attracting". "Water-repellent" hydrophobic coatings have a low surface energy and a contact angle between the surfaces and water above 120 degrees (super-hydrophobic), which is relevant for enhanced cleaning. Materials used for hydrophobic coatings are mainly "fluoropolymers".

"Water-attracting" hydrophilic coatings have high surface energies, leading to low contact angles and, assuming the presence of abundance of water, to a closed water film on the surface. With contact angles of less than 10 degrees (super-hydrophilic), moving of dirt particle of the surface is assured under terms of surface inclination. Titanium oxides (TiOx) and silicon oxides (SiOx) materials are used for such coatings. Also, some hydrophilic coatings have the photocatalytic functionality, being able to precisely decompose organic element after solar activation [7].

However, there is hardly any rain to remove the dust in dry areas. Sometimes dew can enhance cleaning, but it can also have an adverse effect (e.g., stripes, and concentration of dirt). Another problem is that water doesn't run off easily at low installation angles. Furthermore, coatings containing fluorine or organic components are sensitive to UV degradation, leading to a limited lifetime.

DSM's ASC is based on our proprietary core-shell particle technology and designed to reduce soil dust bond and maximize elimination of dust. It reduces the soiling rate by preventing dust and sand from sticking to the module while using wind and gravity as the "cleaning process", thus reducing the output loss in between cleaning cycles and making it easier to clean the modules.

3. Methodology

All the assessments defined in this paper are designed at demonstrating the expected impact of ASC application at solar power plants. Figure 1 shows the actual three different plots, each with 50 MW capacities, were installed, and each plot was distributed in six blocks. Four blocks were installed with 10 MW capacity and the other two were installed with 5 MW capacity. Actual site location is mentioned in the Figure 2. Three inverters were selected for analysing the generation. For precise results, the comparison was done in terms of slots.

Figure 1. Aerial drone view of M/s ACME 150 MW SPV Plant, A.P.

Figure 2. The site is located at Galiveedu, Anantapur, Andhra Pradesh State, India (14° 01' N, 78° 30' E) about 200km north-east of Bangalore.

Figure 3. Anti-Soling coated (ASC) modules installed at the Galiveedu site.

Table 1. Details of connected DC load for ARC & ASC slots

Inverter Details	Capacity(KW)
P6 B6 I2-Slot 2 (ASC)	793.8
P6 B1 I1-Slot 2 (ARC)	819.2
P6 B2 I1-Slot 1 (ARC)	819.2
P6 B2 I1-Slot 2 (ARC)	857.6

The measured generation variation of different slots mentioned in Table 1 was collected and analyzed on a daily basis between 6 September 2018 and 31 March 2020. Of the 19 months observed, the data of 4 months was disrupted due to communication problems and therefore omitted from the analysis. These modules were installed during July and August 2018 at the site as shown in Figure 3. The washing cycle was maintained at an interval of 10 days throughout the study cycles. The comparison was done between the ASC and ARC modules.

4. Analysis and Discussion

As per the standard test conditions, the output of maximum power generated from installed modules helps in monitoring the rate of soiling. The difference in between the generating units of the ARC and ASC modules authenticates that the implementation of such anti soiling coatings is able to reduce soiling deposition and accumulation on the front glass of the installed modules.

The reference ARC modules demonstrate noteworthy loss directly linked to soiling phenomenon, more than +2.5% of the nominal generating units. Investigational activities have proven that a self-cleaning coating exercise can eradicate soiling-related losses during the year of installation, allowing for consistent savings in terms of the operation and maintenance of solar power plants. The monthly gain and accumulated specific energy difference between AS and AR coated modules are shown in Figure 4.

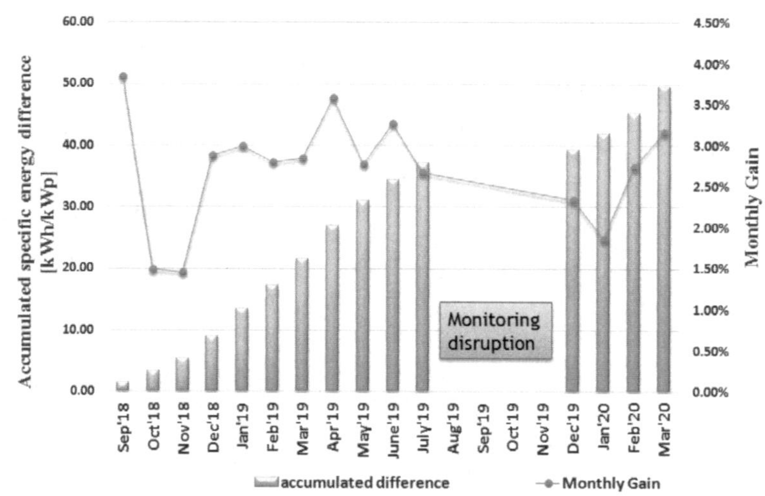

Figure 4. This bar graph shows the monthly gain (green line) from September 2018 to March 2020. The accumulated specific energy difference (kWh/kWp) between AS and AR coated modules is represented with yellow bars on a monthly basis.

The generation measured on a daily basis was captured and its equivalent performance ratio (PR) was calculated with the specific insolation of that particular day. The respective PR% results are mentioned in the Table 2 and shown in Figure 5.

Table 2. Details of calculated PR% for ARC & ASC slots

Inverter Detail	Performance Ratio (%)
P6 B6 I2-Slot 2 (AS)	82.86
P6 B1 I1-Slot 2 (ARC)	80.74
P6 B2 I1-Slot 1 (ARC)	80.01
P6 B2 I1-Slot 2 (ARC)	80.71

The PR is calculated based on the following formula:

$$\textit{Performance Ratio \%} = \frac{\textit{Slot-wise Generation (kWh)}}{\textit{DC capacity (kW) * Plant Availability * Grid Availability * Insolation (kWh/m}^2\textit{/day)}}$$

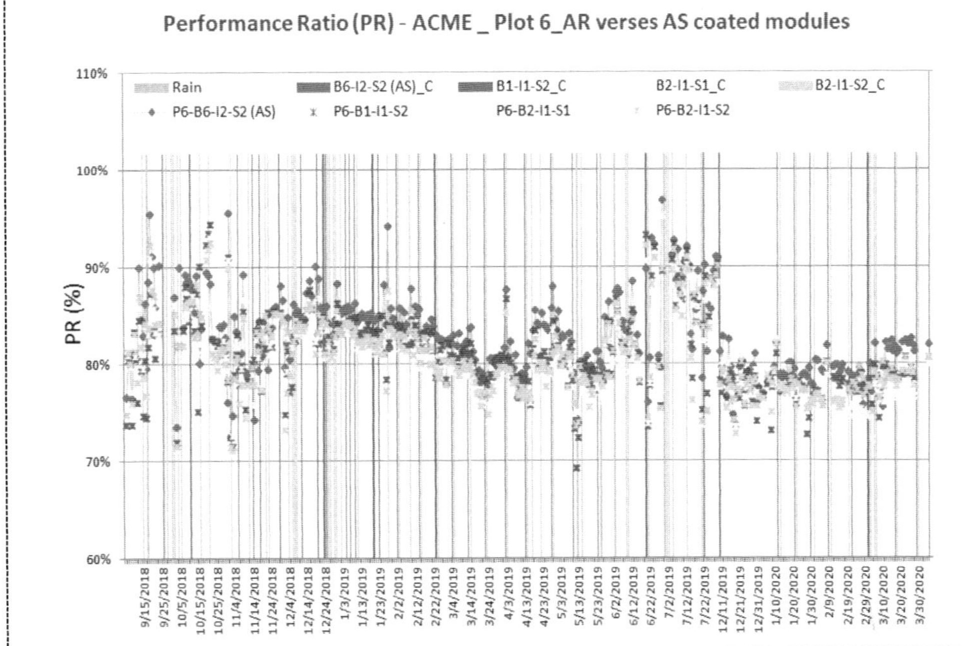

Figure 5. The analytics are done on the calculated PR% of each slot. The PR% (ASC) is higher than the other PR% (ARC).

5. Conclusion

A habitual occurrence is observed on the installed modules due to aggregation of surrounding dust. Collected soil curtails performance of installed modules. The results show in Figure 4, depicts a +2.5% average increase in Photovoltaic module power generation for the anti-soiling modules, recommending that the preferable washing cycle should be 10th day after every washing cycle.

The activity furnished in said research was focused on validating the utilization of ASC being a feasible option for Photovoltaic module preservation. As a matter of fact, the main concerns of PV plant managers and owners are soiling-related losses, the losses are accountable for increasing the return time of investment in case of noteworthy power generation losses throughout the viability of the plant. Furthermore, module soiling can develop confined hindrance on the front surface of glass, which can expedite to electrical delay.

The effect of ASC would be even durable for PV techniques installed at dusty sites, and the earnings in PV yield would be enormous.

In addition, during the soiling evaluation, the transmittance values were remarkably decreased. Finally, the AS coated glass shows a higher gain in transmittance after soiling than the ARC glass substrates.

6. References

[1] Piliougine M, Carretero J, Sidrach-de-Cardona M, Montiel D, Sanchez-Friera P., Comparative analysis of the dust losses in photovoltaic modules with different cover glasses, 2008, *23rd European photovoltaic solar energy conference and exhibition*, Valencia, Spain **p 2698-700**

[2] Alajlan S., Photovoltaic grid-connection system as load-shaving tool in Riyadh, Saudi Arabia, 1999 *Appl Energy 63(2):***91-9**

[3] M. Piliougine, C. Canete, R. Moreno, J. Carretero, J. Hirose, S. Ogawa, M. Sidrach-de-Cardona., Comparative analysis of energy produced by photovoltaic modules with anti-soiling coated surface in arid climates.

IOP Conf. Series: Materials Science and Engineering 1045 (2021) 012008 doi:10.1088/1757-899X/1045/1/012008

[4] A. Kimber, The Effect of Soiling on Photovoltaic Systems Located in Arid Climates, 2007 *Proceedings of the 22nd European Photovoltaic Solar Energy Conference*, Milan, Italy.

[5] http://rredc.nrel.gov/solar/calculators/PVWATTS/system.html, 2012.

[6] M. Garcia, L. Marroyo, D. Lorenzo, M. Perez, Soiling and other optical losses in solar-tracking PV plants in navarra, 2011 *Progress in Photovoltaics: Research and Applications 19*, **p 211-217**.

[7] Can anti-soiling coating on solar glass influence the degree of performance loss over time of PV modules drastically, 2013, *28th European PV solar energy conference and exhibition.*

Acknowledgement

This work was supported throughout by Mr Bikas Kumar for resolving the complications of the calculated trends due to non-washing cycles and rainfall days. He also helped in mapping the different slots at respective site for accurate conclusions.

IOP Conf. Series: Materials Science and Engineering 1045 (2021) 012009 doi:10.1088/1757-899X/1045/1/012009

Single-mode optical fibers coupling: Study of the field of view

Lina M Shaker[1*], Sara Osamah[1] and Ahmed A Al-Amiery[2]

[1]Laser and Optoelectronics Engineering Department, University of Technology, Baghdad, IRAQ.
[2]Energy& Renewable Energy Technology Center, University of Technology, Baghdad, IRAQ.

Correspond: Email: linamohmmed91@gmail.com

Abstract. Single mode Fiber (SMF) is commonly used in long distance applications as in optical communication. Coupling of the laser beam into SMF core to get high coupling efficiency is a challenging task in practice. The aim of our work is to study the impact of the field of view on the coupled light beam, taking into account well designed coupling lens to be sure that the focal spot has matched to the mode of propagation.Here in this article, a conical lens of 1.5 refractive index was used for light coupling process. The used criteria at HeNe laser of 0.5438 nm wavelength showed a degradation in the output signal due to generated distortion as the field of view increased from 0 to 6 degrees.

1. Introduction

1.1. Fiber Coupling

Optical fiber misalignment most often occurs when optical fibers are attached, when coupling the laser with the fiber, or when using conductors. The transmission loss between the connected fibers can be due to intrinsic losses that result from parameter mismatch between the fibers, for example, core diameter, digital aperture (NA) and so on. External losses occur due to the optical and physical misalignment of the ends of the fibers such as the presence of air gaps between the ends of the fibers [1,2], the tilt of the fibers [3] or the radial displacement [4]. One of the conditions for achieving power and information transfer is that the attenuation be limited when connecting plastic optical fibers.The received laser is coupled to a single-mode fiber (SMF) in optical fiber device applications such as interferometers, Sensors, fiber amplifiers and others. Hence it is essential to maximize the coupling efficiency in optical communication systems [5]. The coupling efficiency is affected by the same misalignment conditions mentioned above. In addition, the aberrations caused by the optical systems are important factors affecting the coupling efficiency and should be carefully eliminated [6-8].Beyond telecommunications, optical fibers can also transport optical energy to powering electric or electronic devices remotely.

1.2. Optical Aberrations

Imperfections in any optical component may cause deviation of light from its path. Such deviations are called optical aberrations, causes blurred image and reduces the optical system performance. Optical aberrations comes into two kinds; monochromatic aberrations (Spherical, coma, astigmatism, image curvature and distortion) and chromatic aberrations. Off-axis aberrations can be eliminated by concentering the optical and detecter surfaces with a common curvature center at the stop aperture.The Seidel polynomial uses the polar coordinate system and described mathematically as follows [9]:

$$W = \sum_{i,j,k} W_{ijk} \bar{H}^i \rho^j \cos^k \phi............(1)$$

W_{ijk} is the wavefront aberration coefficient whose value can be positive or negative (in units of wavelength). The subscripts i, j and k refer to the powers on the other factors that indicate the aberration kind. The factor H is the fractional image height. Its value ranges between 0 and 1. The fractional pupil radius ρ also ranges between 0 and 1. Another pupil coordinate is denoted by ϕ in the cosine term (has values between -1 and 1).

2. Method and Analytical Work

For optical modelling, ZEMAX 13 premium-34900software is used. The designed model works at far distances as illustrated in Figure (1). This model is composed mainly of a laser source (HeNe laser of 0.5438 nm), entrance pupil diameter (stop aperture) of 4 mm, the exit pupil position of -65.217 mm included the glass lens (conic of -2.250 and a refractive index of 1.50) and the last part is focusing at the fiber.

At a fixed aperture diameter, field of view (FOV) was changed from 0-10 increased by step 2.Table (1) summarizesthe input parameters of each component.An efficient plano-convex lens is used in this study to get higher light intensity deliver and higher coupling efficiency as the field of view is varied. Because such lens eliminates the spherical aberration that may be caused due to the applied monochromatic laser beam because it collects and collimates the light rays into a small spot.

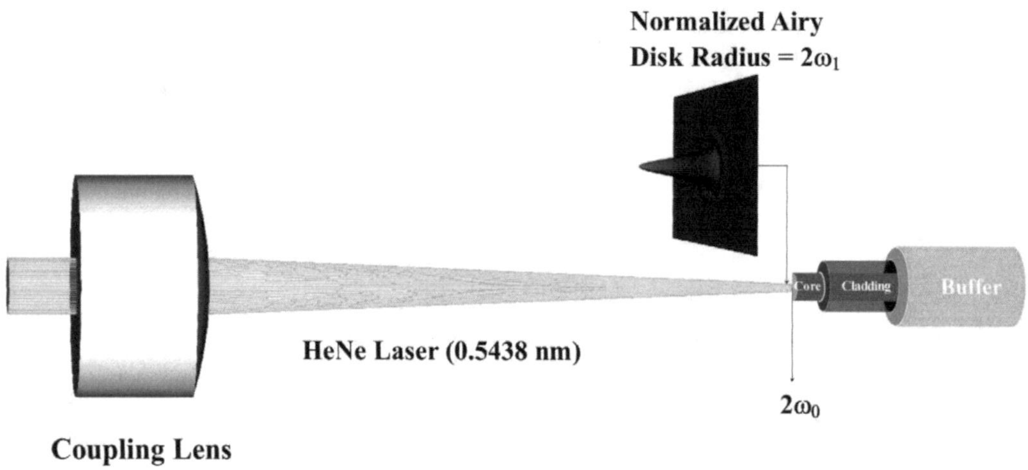

Figure 1. The designed single mode fiber coupling

Table 1. Input parameters of the designed single mode fiber coupling model. All measured in (mm).

Surface	Surface Type	Radius	Thickness	E.P.D.	Conic
Light Beam	Standard Surface	Infinity	Infinity	Infinity	-
Aperture Diameter	Standard Surface	Infinity	5.00	4	-
Lens-Front	Standard Surface	Infinity	10.00	8	-
Lens-Back	Standard Surface	-25.00	50.00	8	-2.25
Focus at Fiber	Standard Surface	Infinity	-	2	-

IOP Conf. Series: Materials Science and Engineering 1045 (2021) 012009 doi:10.1088/1757-899X/1045/1/012009

3. Results and Discussion

3.1. PSF

The Point Spread Function (PSF) of a converging lens is known as the Airy disk. The Airy disk is then defined as $J_1(v)$ for the order (1) of the Bessel functionwhere v is the optical distance.In turn, v is defined as a function of r (the radial distance). PSF is also defined as the square of the modulus of the point spread function amplitude, h(v), as indicated in the following equations [10]:

$$h(v) = \left[\frac{2J_1 v}{v} \right] \text{..............................} (2)$$

$$v = \frac{2\pi}{\lambda} \left(\frac{a}{f} \right) r \text{....................................} (3)$$

Where a is the lens radius, λ is the laser wavelength and f is the lens focal length.

Briefly, when the laser beam is well focused by the optical system, an Airy pattern of distributed light obtained at the fiber end. This can be found in on-axis (FOV = 0) optical systems as shown in Figure (2.a). From this figureand the relative values of intensity, the system is perfectly alignedwith a Strehl ratio of 1. Significantly, the peak intensity diminished when the FOV has changed by 6 degrees as shown in Figure (2.b). As a result, the Strehl ratio decreased to 0.568. In addition, small protrusionshave appeared and the peak intensity has shifted slightly. This is due to the generated aberrations through the coupling process. The Total irradiance surface graphs facilitate the way to know the generated aberration type to overcome such issues

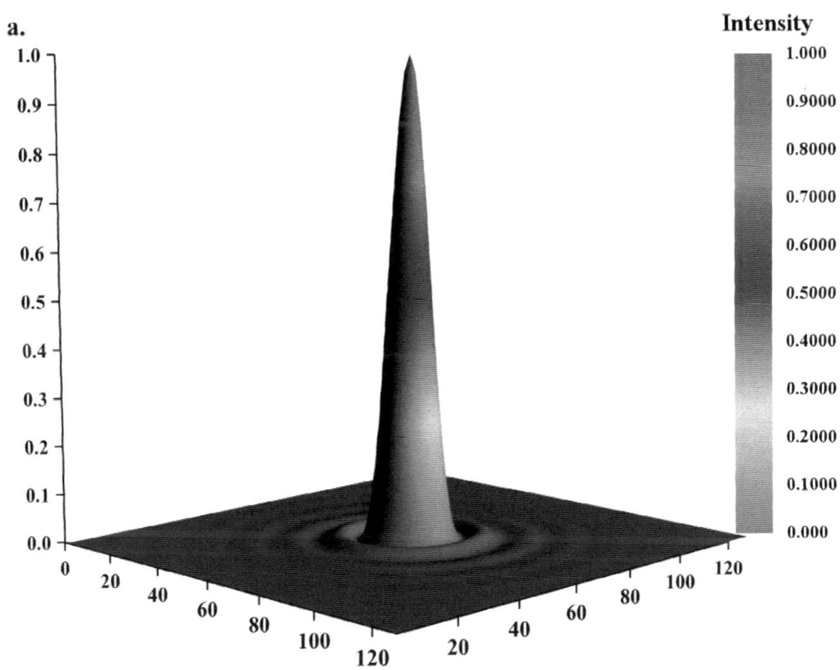

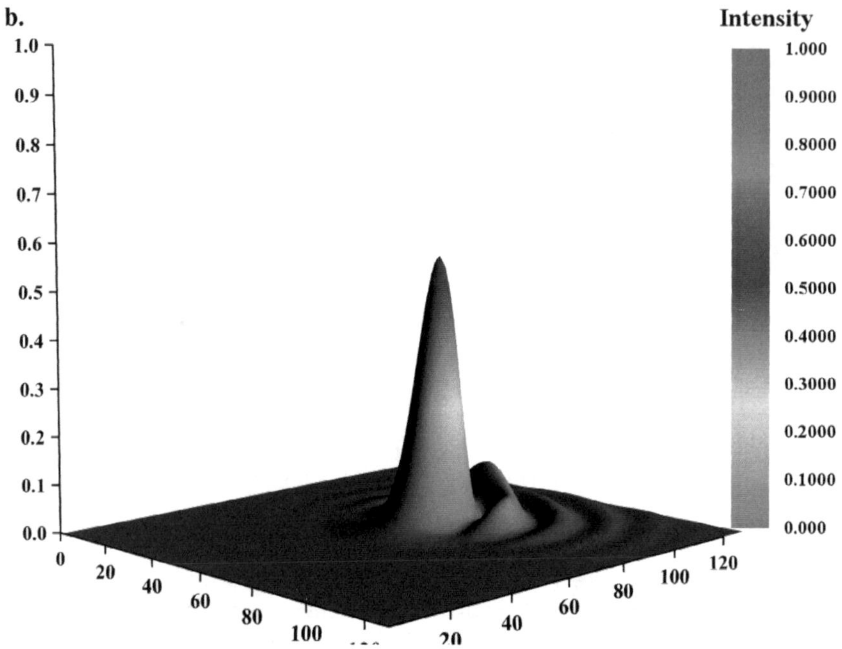

Figure 2. PSF at FOV of**: a.** 0 degrees and **b.** 6 degrees

3.2. Total Irradiance Surface

The system efficiency is the fraction of the energy in the source beam that exits the optical system. This value is determined by the input numerical aperture (NA), entrance pupil size and position, apodization, transmission of the optics, and vignetting. While The receiver efficiency is the fraction of the transmitted energy that couples from the exit pupil to the receiving fiber. This value is determined by aberrations and the NA of the receiving fiber. The resultant coupling efficiency is the fraction of energy radiated by the source that couples into the receiver. This is the product of the system and receiver efficiencies.The spatial HeNe laser beam to single mode fiber coupling efficiency denoted by (η) is a function of phase(A(r)) and Gaussian distribution amplitude (M(r)). The mathematical expression of each term is indicated below [11,12]:

$$\eta = \frac{\left|\int A(r) M(r) dr\right|}{\int \left|A(r)\right|^2 dr} \quad (4)$$

$$A(r) = \frac{\pi D^2}{4\lambda f}\left|2\frac{J_1(3.83/\omega_1)}{3.83r/\omega_1}\right| \quad (5)$$

$$\mathrm{M(r)} = \sqrt{\frac{2}{\pi\omega_0^2}}\exp(-\frac{r^2}{\omega_0^2}) \quad (6)$$

$$\omega_a = \frac{\lambda f}{\pi\omega_0} \quad (7)$$

Where D is the lens diameter, ω_1 and ω_0 are the Ariy disk radius and the mode-field radius of the single mode fiber.The Irradiance graph represented by Figure 3.a. is the ideal system as seen in the PSF Figures. While figure 3.b.introduces the reason behind the appearance of the protuberances and the decrease in intensity, which is due to distortion and coma aberration. As a result of the widen in FOV, which in turn led to the system misalignment.

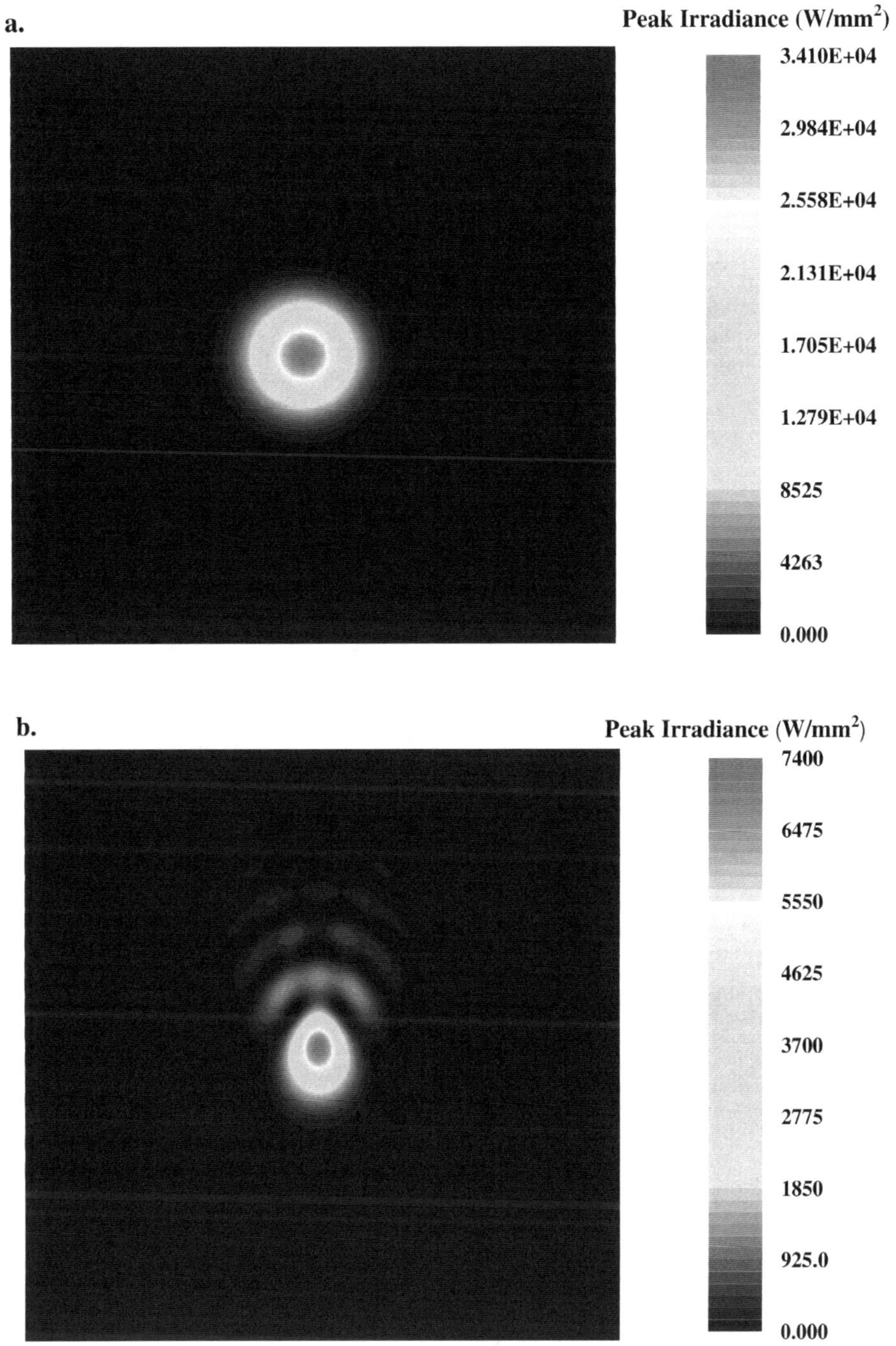

Figure 3. Irradiance measured in watts/millimetre^2 focused at the fiber at: **a.** 0 degrees and **b.** 6 degrees.

From the Irradiance graphs (Figures 3.a and 3.b) at 0 and 6 degrees FOV, respectively, the fiber efficiency components' values are listed in Table 2 below:

Table 2. Fiber efficiency of both models 0 degrees and 6 degrees FOV.

Fiber Efficiency	FOV= 0 degrees	FOV = 6 degrees
System Efficiency	1	1
Receiver Efficiency	0.699	0.425
Coupling Efficiency	0.699	0.425

4. Conclusion

In the event that the laser beam is not completely centered, there are two solutions. First, if the offset is less than 1 mm, compensation can be made easily by adjusting the coupling optics with respect to the laser head transformer. This capability is built into the source coupler design. For the second solution, if the displacement is more than 1 mm, then an adapter with inserted holes must be constructed to compensate for the displacement. This work demonstrated that the coupling efficiency response to the established pairing models changed with the change of FOV (misalignment) values. This is attributed to the amount of coma and distortion aberrations generated within the system which affect the total transmitted power.

References

[1]. Wagner R E and Sandahl C 1982 *Appl. Opt.***21** 1381–1385.
[2]. Van Etten W, Lambo W and Simons P 1985*Appl. Opt.***24** 970–976.
[3]. Gloge, D. 1976 *Tech. J.***55** 905–916.
[4]. Young, M. 1973*Opt. Commun.***7** 253–255.
[5]. Takenaka H, Toyoshima M and Takayama Y 2012*Opt. Express***20** 15301–15308.
[6]. Ma J, Zhao F, Tan L Y, Yu SY and Yang YQ 2010 *Opt. Eng.***49** 6.
[7]. Yin X H, Wang R, Wang S X, Wang Y K, Jin C B, Cao Z L and Xuan L 2018*Appl. Opt.***57** 573–581.
[8]. Zhu XL, Wang KL, Wang F, Zhao CL and Cai YJ 2018*Appl. Sci.***8** 1313.
[9]. Joseph M. Geary 2002*Introduction To Lens Design With Zemax.*
[10]. Wagner R. E. and Tomlinson W. J. 1982*Appl. Opt.***21** 2671-2688 .
[11]. Toyoshima M J 2006 *Opt Soc Am A***23**(9) 2246-2250.
[12]. Wallner O, Winzer P J and Leeb W R 2002*Appl. Opt.***41** 637–643.

A graphene-modified Co-BDC metal-organic frameworks (Co-MOF) for electrochemical non-enzymatic glucose sensing

R Yuniasari[1], F Amri[1], S A Abrori[1], N L W Septiani[1], M Rezki[1], Irzaman[2], M Z Fahmi[3] and B Yuliarto*[,1,4]

[1] Department of Engineering Physics, Faculty of Industrial Technology, Institute of Technology Bandung, Bandung 40132, Indonesia.
[2] Department of Physics, Faculty of Mathematics and Natural Sciences, Bogor Agricultural University, Bogor, Indonesia.
[3] Department of Chemistry, Faculty of Science and Technology, Universitas Airlangga, Surabaya 60115, Indonesia.
[4] Research Center for Nanoscience and Nanotechnology (RCNN), Institute of Technology Bandung, Bandung 40132, Indonesia.

* *Corresponding Author: brian@tf.itb.ac.id*

Abstract. In this work, a metal-organic framework (MOF) based on cobalt was decorated with graphene and used as a sensing material for glucose determination with electrochemical principles. The selection of Co-MOF material is based on its porous nature, large surface area, and excellent electrochemical properties. The combination of Co-MOF with graphene (high conductivity) effectively increased the electrochemical sensor current. The fabricated composite owned the good crystallinity with graphene particles attached to the Co-MOF surface. The biosensing performance was evaluated by cyclic voltammetry (CV) with 0.1 M NaOH solution as the bolstering electrolyte. The electrochemical measurement indicated that the prepared materials possessed a well-moved transfer electron between the electrode surface and electrolyte solution. The Co-BDC-3Gr sample obtained the best electrochemical performance with the lowest limit of detection (LOD) of 5.39 μM and the highest sensitivity of 100.49 $\mu A\ mM^{-1}\ cm^{-2}$. The selectivity test of the modified Co-MOF was done by comparing the response with other compounds such as dopamine, uric acid, and NaCl. The acquired biosensor had excellent stability, with 93% of the initial response after 30 days of storage.

Keywords. Cobalt-metal organic framework, graphene, electrochemical, glucose sensing.

1. Introduction

Non-enzymatic glucose sensors have been recently developed to replace an enzymatic glucose sensor. Enzymatic-based sensors are considered less effective and efficient due to many factors that affect the quality of the enzyme and their sensing performance such as temperature, humidity, and pH conditions. Those factors cause several problems including short lifetime, instability, complicated immobilization procedure, high cost, and low reproducibility [1-4]. That technique generally requires glucose oxidase, acting as oxidation of glucose to gluconolactone. In contrast, non-enzymatic glucose sensor becomes popular due to its ability to convert glucose to gluconolactone directly [3-5]. Moreover, this technique also offers high stability, excellent repeatability, and relatively low cost [6].

According to several studies that have been reported, materials containing transition metal catalysts such as cobalt, nickel, iron, noble metals such as palladium, platinum, gold, and carbon nanomaterial show relatively good performance as a non-enzymatic glucose biosensor [7,8]. Among the types of material, three-dimensional MOF containing metal catalyst provides the large surface area, tunable pore, and plentiful active sites for catalytic activity [9-11]. Cobalt-based MOF (Co-MOF) possesses high catalytic activity in the identification of the chemical and biological compounds. For instance, Li *et al.* developed Co-MOF nanosheet array to identify glucose with great sensitivity of 10,886 μA mM^{-1} cm^{-2} and low LOD of 1.3 nM [12]. Co-MOF developed by Yang *et al.* was reported to show high performance as an H_2O_2 sensor with high sensitivity and low LOD of 83.10 μA mM^{-1} cm^{-2} and 3.76 μM, respectively [9]. Moreover, L-cysteine can also be appropriately detected by Co-MOF with a sensitivity and LOD of 0.0032 μA μM^{-1} and 1.85 μM, respectively [13].

In this study, Co-MOF was synthesized using a solvothermal technique in the presence of graphene as a modifier to be applied as a glucose sensor. Graphene is one of the carbon materials in the form of sheets and has a honeycomb structure of graphite [14]. Graphene owns the high potential to be applied as a sensor material because it has a large surface area, good electrical conductivity, high thermal, and chemical stability [14-17]. Moreover, the high conductivity of graphene supplies more electron channels promoting high electron transfer [18,19]. Alizadeh and co-worker synthesized metal oxide combined with graphene as a non-enzymatic glucose biosensor [20]. CuO/graphene showed a high sensitivity and low LOD of 2939.24 μA mM^{-1} cm^{-2} and 0.09 μM, respectively. Improvement of charge transfer capability provided by a combination of CuO and graphene sheet was believed as a reason for its high performance of glucose sensor. Another study was carried out by Mehek *et al.,* who combined Co-MOF with graphene to be applied as a methanol sensor [21]. The work reported that graphene could increase the surface area (contact area) and electrochemical response. In this study, graphene-modified Co-MOF was applied as electrochemically non-enzymatic glucose biosensor with excellent sensitivity, selectivity, and stability.

2. Experimental Section

2.1 Materials

Cobalt (II) nitrate hexahydrate ($Co(NO_3)_2.6H_2O$, 99.5%), D(+)-glucose, *N, N*-dimethylformamide (DMF) were purchased from Fujifilm Wako. Benzene-1,4-dicarboxylic acid or terephthalic acid (H_2BDC, 98%), uric acid ($C_5H_4N_4O_3$, ≥99%), and dopamine were purchased from Sigma-Aldrich. All chemicals were employed as gained without any purification.

2.2 Synthesis of Co-BDC MOF

Co-MOF was synthesized by employing a solvothermal method at 120°C for 24 hours. Typically, 0.28 grams of $Co(NO_3)_2.6H_2O$ was added in a solvent which contains 20 ml of DMF and 20 ml of ethanol. Subsequently, 0.5 gram of 1,4-benzene dicarboxylic acid was mixed to the solution and then stirred for 1 h. After dissolving, the mixture solution was poured into 100 mL Teflon autoclave and heated at the destined temperature. The precipitated product obtained after cooling down was laundered employing DMF and ethanol three times, respectively, then was dried overnight.

2.3 Synthesis of graphene-modified Co-BDC

The graphene-modified Co-BDC was prepared with a similar procedure as Co-BDC preparation. The dissolving process of $Co(NO_3)_2.6H_2O$ and 1,4-benzene dicarboxylic acid was followed by the addition of graphene to the mixture with various concentrations of 1, 3, and 5%wt, which then in this report are noted as Co-BDC-1Gr, Co-BDC-3Gr, and Co-BDC-5Gr, respectively. After stirring for an hour, the mixture solution was transferred into 100 mL Teflon autoclave and heated at 120°C for 24 h. The precipitated product obtained after cooling down was washed using DMF and ethanol three times, respectively, subsequently was dried overnight.

2.4 Characterization

The phase and crystal structure of all products were characterized using X-ray diffraction Bruker D8 Advance. The morphological of the products was observed using a scanning electron microscope (SEM)

Hitachi SU3500. The infrared spectra were collected by Fourier transform infrared (FTIR) Shimadzu - Prestige 21. Nitrogen (N_2) adsorption-desorption characteristic was performed using Quantachrome Instruments (USA) via the Brunauer-Emmett-Teller (BET) approach.

2.5 Electrochemical measurements

There are three electrodes employed to measure the electrochemical properties of all samples. The measurements, including CV and amperometry, were performed using Corrtest 350 with glassy carbon electrode (GCE), platinum wire, and Ag/AgCl acted as the working, counter, and reference electrodes, respectively. Before the measurement, 5 mg of each sample was added in 0.5 mL of distilled water and sonicated to reach a homogenous state. 5 μL of the solution was then dropped onto GCE and dried naturally. After completely dried, 2 μL of Nafion was added on the surface of the layer and was dried naturally. In this research, 0.1 M of NaOH was emploted as an electrolyte, and the glucose solutions with varying concentrations (1-10 mM) were prepared in 0.1 M NaOH.

3. Results and Discussion

3.1 Synthesis and characterization of hierarchical Co-BDC MOF

The diffraction pattern of graphene-assisted Co-BDC can be observed in figure 1. The MOF crystal formed from the bonding between cobalt metals and BDC ligands are marked by fingerprints peak at 2θ of 8.8°, 11.5°, 12.0°, and 15.9° originating from the diffraction planes of (100), (110), (101), and (200), respectively. The diffraction pattern shown by the Co-BDC-graphene composite is very similar to the pattern of pure Co-BDC. The presence of graphene does not interfere with the crystallinity of Co-BDC. Also, graphene owing a diffraction peak at 2θ of 26.3°does not indicate its presence in the prepared composite system. It also can be reported that 1 wt%, 3 wt%, and 5 wt% graphene added to the Co-BDC system are small enough to be detected by XRD.

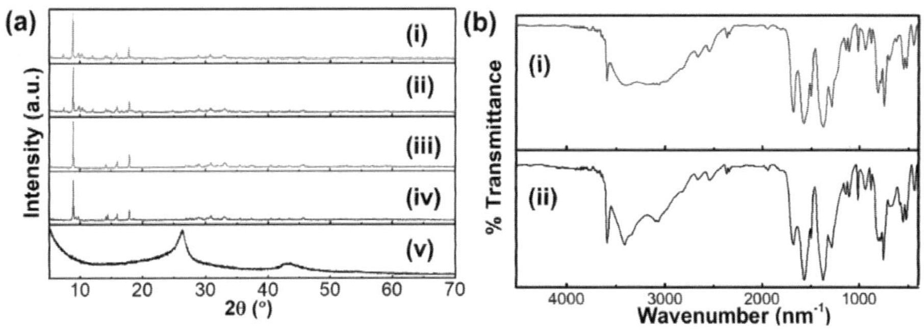

Figure 1. (a) Diffraction pattern of Co-BDC MOF (i) and its modification with 1 wt% (ii), 3 wt% (iii), 5 wt% (iv) of graphene in comparison with pure graphene (v). (b) infrared spectrum of Co-BDC (i) in comparison with Co-BDC-3Gr (ii).

Co-BDC or also known as MOF-71, is formed from CoO_6 octahedra chains, and each chain is connected to four other parallel chains by a linker BDC, while the two other sites connect chains in series through DMF [11,22]. Coordination between cobalt and BDC is confirmed through the FTIR spectrum displayed in figure 1b. The infrared spectrum produced by Co-BDC and Co-BDC-graphene composites is relatively similar and showing that the presence of graphene does not interfere with the Co-BDC framework. The sharp peak at 3700 cm^{-1} originates from OH bound to BDC ligands [23,24]. The broad peak centered at the wavenumber 3300 nm^{-1} originates from the stretching vibration OH group of water vapor adsorbed on the surface. In contrast, the sharp peak at 1690 cm^{-1} is derived from stretching vibration group of C = O. The bond between Co metal and BDC ligands is characterized by the presence of asymmetric and symmetric vibrations -COO-, respectively in the wavenumber range of 1575-1501 cm^{-1} and 1380 cm^{-1} [25,26]. Hu *et al.* found a decrease in the intensity of the C = O peak when compared with the same peak in the IR H_2BDC spectrum showing the interaction between Co^{2+} and deprotonated H_2BDC [11]. Co-bonding and BDC ligand are also confirmed by the appearance of peak at 437 cm^{-1}

originating from stretching vibration Co-O. The 3D framework of Co-BDC observed via SEM can be seen in figure 2. Pure Co-BDC has a spindle-like shape with length and diameter of 5 μm and 2.5 μm, respectively.

Figure 2. SEM images and XRD patterns of obtained Co-MOF and Co-MOF/Gr with the optimized ratios of graphene to Co-BDC: (a,b) Co-BDC, (c,d) Co-BDC-1Gr, (e,f) Co-BDC-3Gr, and (g,h) Co-BDC-5Gr.

Based on the analysis results, the surface of pure Co-BDC looks smooth. This behavior is different from the Co-BDC sample, which has been added by graphene. The surface of the modified Co-BDC looks rougher because there are several particles attached to the surface of the Co-BDC spindle, which is believed to be crumpled of graphene. This structure is reinforced by the increasing number of particles on the Co-BDC surface, along with the increasing amount of graphene given.

N_2 adsorption-desorption tecnique were employed to calculate the surface area of the as-prepared Co-BDC/Gr samples. From figure 3, the obtained surface area is 36.2 m^2 g^{-1} and 36.7 m^2 g^{-1} for Co-BDC and Co-BDC-3Gr, respectively. These results indicate that the addition of graphene does not give a significant effect on the surface area of Co-BDC MOF, even though the graphene has high surface area. This phenomenon strengthens the results of the FTIR analysis earlier.

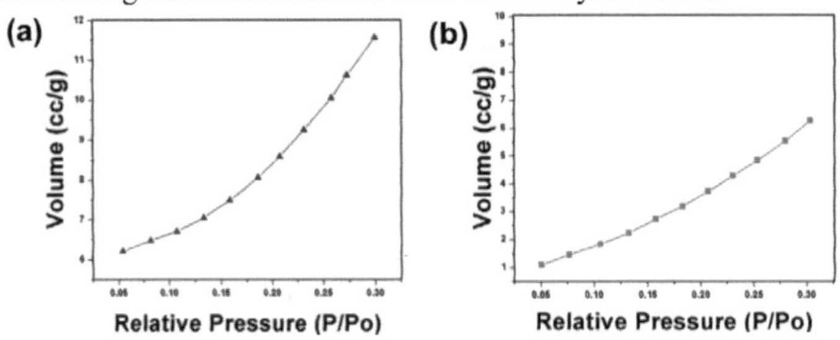

Figure 3. Nitrogen adsorption measurement of (a) Co-MOF and (b) Co-MOF-3Gr

3.2 Non-enzymatic glucose biosensor performance of hierarchical Co-BDC MOF

The electrochemical property of all samples was evaluated using CV, as shown in figure 4. Oxidation and reduction process of the electrode surface was observed in the potential range of 0-0.8 V in 5 mM glucose solution. There are no oxidation and reduction peaks taking place in the surface of bare GCE and graphene, while oxidation and reduction peaks around 0.4 V and 0.3 V, respectively were observed during the CV process of Co-BDC and Co-BDC-Gr. These peaks also show the increasing of electron transfer, so it can be concluded that the modified electrodes have better electrochemical catalytic properties. The increase of currents in the case of Co-BDC and Co-BDC-Gr are attributed to the change of Co oxidation state from Co^{2+} to Co^{3+} or vice versa in alkaline media [27]. Addition of graphene to Co-MOF shows an increase in electrochemical sensor activity. Co-MOF can effectively be the active site of the electrode, while graphene increases the electrical conductivity of the sensor. Graphene acts as a supporting material that increases the electronic conductivity in Co-MOF. Combining the roles and benefits of the two materials causes the Co-MOF/Gr material to have better electrocatalytic properties compared to each material. Furthermore, it is found that the current response peak produced by redox reaction increases along with the increase in the amount of graphene. This situation continues until it reaches the maximum current response in the addition of 3 wt% graphene. At the addition of 5wt% graphene, the resulting current response decreased (figure 4a). Therefore, the addition of 3wt% graphene is the best amount in the electrode system to generate the best performance of the glucose sensor. The effect of scan rates on the CV curve behavior was analyzed mainly for the Co-BDC-3Gr sample (figure 4b). The anodic and cathodic current peaks shift to the right and left side, respectively with increasing of scan rate indicating diffusion-controlled process involved in this glucose detection [28,29].

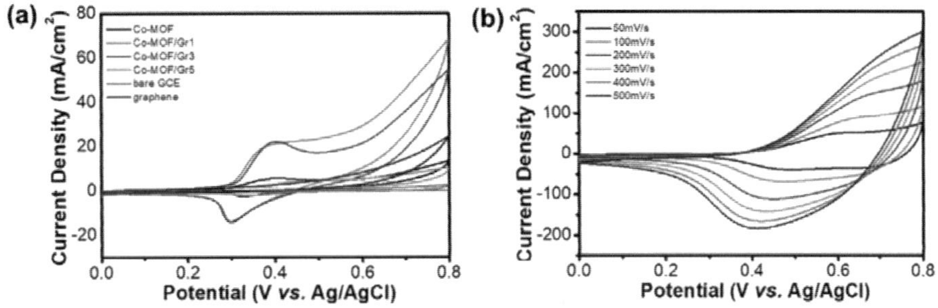

Figure 4. Cyclic voltammogram of (a) Co-MOF and Co-MOF/Gr in 5 mM glucose solution with a scan rate of 50 mV.s^{-1}. (b) Co-BDC-3Gr with various scan rates.

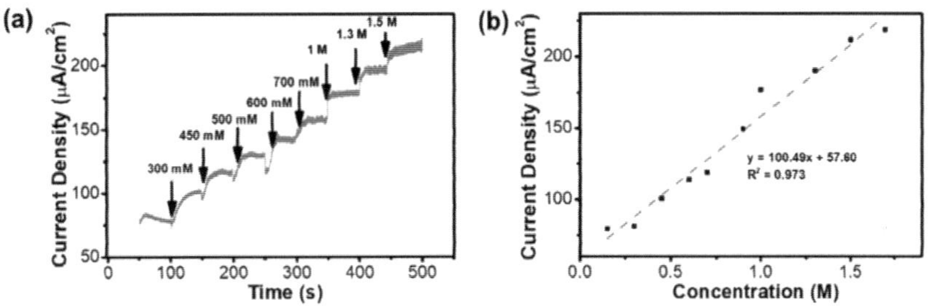

Figure 5. (a) Amperometric response and (b) the change in current density of Co-MOF-3Gr.

The amperometric technique can be used to observe the ability of a sensor to produce a direct response to the concentration of analytes. In this case, linearity in sensitivity was determined by making a series of analyte solutions in different concentrations. Figures 5a and 5b show that the current response enhances with increasing concentration of the given glucose solution in the electrolyte solution. This phenomenon causes more ion diffusion to the electrode so that the detected current also increases. The hierarchical Co-BDC MOF sensor shows a sensitivity of 100.49 μA mM^{-1} cm^{-2} with a correlation coefficient of 0.973. LOD obtained for biosensor is 5.39 μM (S/N = 3). The results of biosensor selectivity exhibit excellent performance as indicated by a significant change in current response when injected with glucose compared to other solutions such as dopamine, uric acid, and NaCl. The stability of the Co-BDC-3Gr was evaluated by observing its current response toward 5 mM glucose for several cycles after long-time storage (30 days). The obtained stability of biosensor shows excellent stability, with 93% of the initial response.

4. Conclusion

The Co-MOF-based biosensor platform, combined with graphene for glucose detection, was successfully prepared. The fabricated biosensor has an excellent performance in terms of sensitivity, LOD, selectivity, and stability. Based on the analysis of electrochemical results, the Co-BDC-3Gr sample has the lowest LOD of 5.39 μM with a sensitivity of 100.49 μA mM^{-1} cm^{-2}. The biosensor selectivity was tested by comparing the current response between glucose with uric acid, dopamine, and NaCl. The analysis result shows that the biosensor possesses a good selectivity, which is indicated by the most significant current response when injected with a glucose solution. Also, the acquired stability of biosensor exhibits excellent stability, with 93% of the initial response.

Acknowledgements

This work is partially bolstered by Ministry of Education and Culture, and Ministry of Research and Technology under grant scheme of World Class University (WCU) Program managed by Institute of Technology Bandung (ITB). The authors also acknowledge financial grants provided by ITB, Ministry of Education and Culture, and Ministry of Research and Technology of Indonesia.

References

[1] Khosroshahi, Z.; Karimzadeh, F.; Kharaziha, M.; Allafchian, A., A non-enzymatic sensor based on three-dimensional graphene foam decorated with Cu-xCu2O nanoparticles for electrochemical detection of glucose and its application in human serum. *Materials Science and Engineering: C* **2020,** *108*, 110216.

[2] Lin, X.; Wang, Y.; Zou, M.; Lan, T.; Ni, Y., Electrochemical non-enzymatic glucose sensors based on nanocomposite of Co3O4 and multiwalled carbon nanotube. *Chinese Chemical Letters* **2019,** *30* (6), 1157-1160.

[3] Hwang, D.-W.; Lee, S.; Seo, M.; Chung, T. D., Recent advances in electrochemical non-enzymatic glucose sensors – A review. *Analytica Chimica Acta* **2018,** *1033*, 1-34.

[4] Zhang, L.; Liang, H.; Ma, X.; Ye, C.; Zhao, G., A vertically aligned CuO nanosheet film prepared by electrochemical conversion on Cu-based metal-organic framework for non-enzymatic glucose sensors. *Microchemical Journal* **2019,** *146*, 479-485.

[5] Guo, S.; Zhang, C.; Yang, M.; Zhou, Y.; Bi, C.; Lv, Q.; Ma, N., A facile and sensitive electrochemical sensor for non-enzymatic glucose detection based on three-dimensional flexible polyurethane sponge decorated with nickel hydroxide. *Analytica Chimica Acta* **2020,** *1109*, 130-139.

[6] Ahmad, R.; Tripathy, N.; Ahn, M.-S.; Bhat, K. S.; Mahmoudi, T.; Wang, Y.; Yoo, J.-Y.; Kwon, D.-W.; Yang, H.-Y.; Hahn, Y.-B., Highly Efficient Non-Enzymatic Glucose Sensor Based on CuO Modified Vertically-Grown ZnO Nanorods on Electrode. *Scientific Reports* **2017,** *7* (1), 5715.

[7] Zhuang, Z.; Su, X.; Yuan, H.; Sun, Q.; Xiao, D.; Choi, M. M. F., An improved sensitivity non-enzymatic glucose sensor based on a CuO nanowire modified Cu electrode. *Analyst* **2008,** *133* (1), 126-132.

[8] Baghayeri, M.; Amiri, A.; Farhadi, S., Development of non-enzymatic glucose sensor based on efficient loading Ag nanoparticles on functionalized carbon nanotubes. *Sensors and Actuators B: Chemical* **2016,** *225*, 354-362.

[9] Yang, L.; Xu, C.; Ye, W.; Liu, W., An electrochemical sensor for H2O2 based on a new Co-metal-organic framework modified electrode. *Sensors and Actuators B: Chemical* **2015,** *215*, 489-496.

[10] Furukawa, H.; Cordova, K. E.; O'Keeffe, M.; Yaghi, O. M., The Chemistry and Applications of Metal-Organic Frameworks. *Science* **2013,** *341* (6149), 1230444.

[11] Hu, X.; Hu, H.; Li, C.; Li, T.; Lou, X.; Chen, Q.; Hu, B., Cobalt-based metal organic framework with superior lithium anodic performance. *Journal of Solid State Chemistry* **2016,** *242*, 71-76.

[12] Li, Y.; Xie, M.; Zhang, X.; Liu, Q.; Lin, D.; Xu, C.; Xie, F.; Sun, X., Co-MOF nanosheet array: A high-performance electrochemical sensor for non-enzymatic glucose detection. *Sensors and Actuators B: Chemical* **2019,** *278*, 126-132.

[13] Murinzi, T. W.; Hosten, E.; Watkins, G. M., Synthesis and characterization of a cobalt-2,6-pyridinedicarboxylate MOF with potential application in electrochemical sensing. *Polyhedron* **2017,** *137*, 188-196.

[14] Kang, X.; Wang, J.; Wu, H.; Aksay, I. A.; Liu, J.; Lin, Y., Glucose Oxidase–graphene–chitosan modified electrode for direct electrochemistry and glucose sensing. *Biosensors and Bioelectronics* **2009,** *25* (4), 901-905.

[15] Misra, A., Carbon nanotubes and graphene-based chemical sensors. *Current Science* **2014,** *107* (3), 419-429.

[16] Debataraja, A.; Muchtar, A. R.; Septiani, N. L. W.; Yuliarto, B.; Nugraha; Sunendar, B., High Performance Carbon Monoxide Sensor Based on Nano Composite of SnO2-Graphene. *IEEE Sensors Journal* **2017,** *17* (24), 8297-8305.

[17] Debataraja, A.; Septiani, N. L. W.; Yuliarto, B.; Nugraha; Sunendar, B.; Abdullah, H., High performance of a carbon monoxide sensor based on a Pd-doped graphene-tin oxide nanostructure composite. *Ionics* **2019,** *25* (9), 4459-4468.

[18] Luo, J.; Jiang, S.; Zhang, H.; Jiang, J.; Liu, X., A novel non-enzymatic glucose sensor based on Cu nanoparticle modified graphene sheets electrode. *Analytica Chimica Acta* **2012,** *709*, 47-53.

[19] Lv, W.; Jin, F.-M.; Guo, Q.; Yang, Q.-H.; Kang, F., DNA-dispersed graphene/NiO hybrid materials for highly sensitive non-enzymatic glucose sensor. *Electrochimica Acta* **2012,** *73*, 129-135.

[20] Alizadeh, T.; Mirzagholipur, S., A Nafion-free non-enzymatic amperometric glucose sensor based on copper oxide nanoparticles–graphene nanocomposite. *Sensors and Actuators B: Chemical* **2014,** *198*, 438-447.

[21] Mehek, R.; Iqbal, N.; Noor, T.; Nasir, H.; Mehmood, Y.; Ahmed, S., Novel Co-MOF/Graphene Oxide Electrocatalyst for Methanol Oxidation. *Electrochimica Acta* **2017,** *255*, 195-204.

[22] Miles, D. O.; Jiang, D.; Burrows, A. D.; Halls, J. E.; Marken, F., Conformal transformation of [Co(bdc)(DMF)] (Co-MOF-71, bdc = 1,4-benzenedicarboxylate, DMF = N,N-dimethylformamide) into porous electrochemically active cobalt hydroxide. *Electrochemistry Communications* **2013,** *27*, 9-13.

[23] Li, L.-X.; Xu, D.; Li, X.-Q.; Liu, W.-C.; Jia, Y., Excellent fluoride removal properties of porous hollow MgO microspheres. *New Journal of Chemistry* **2014,** *38* (11), 5445-5452.

[24] Ansari, A.; Ali, A.; Asif, M.; Shamsuzzaman, Microwave-assisted MgO NP catalyzed one-pot multicomponent synthesis of polysubstituted steroidal pyridines. *New Journal of Chemistry* **2018,** *42* (1), 184-197.

[25] Lou, X.; Chen, M.; Hu, B., The effect of nitrogen and oxygen coordination: toward a stable anode for reversible lithium storage. *New Journal of Chemistry* **2018,** *42* (19), 15698-15704.

[26] Kurisingal, J. F.; Babu, R.; Kim, S.-H.; Li, Y. X.; Chang, J.-S.; Cho, S. J.; Park, D.-W., Microwave-induced synthesis of a bimetallic charge-transfer metal organic framework: a promising host for the chemical fixation of CO2. *Catalysis Science & Technology* **2018,** *8* (2), 591-600.

[27] Shi, L.; Li, Y.; Cai, X.; Zhao, H.; Lan, M., ZIF-67 derived cobalt-based nanomaterials for electrocatalysis and non-enzymatic detection of glucose: Difference between the calcination atmosphere of nitrogen and air. *Journal of Electroanalytical Chemistry* **2017,** *799*, 512-518.

[28] Yin, H.; Zhu, J.; Chen, J.; Gong, J.; Nie, Q., MOF-derived in situ growth of carbon nanotubes entangled Ni/NiO porous polyhedrons for high performance glucose sensor. *Materials Letters* **2018,** *221*, 267-270.

[29] Zhang, X.; Xu, Y.; Ye, B., An efficient electrochemical glucose sensor based on porous nickel-based metal organic framework/carbon nanotubes composite (Ni-MOF/CNTs). *Journal of Alloys and Compounds* **2018,** *767*, 651-656.

Design, simulation and analysis of monofacial solar pv panel based energy system for university residence: a case study

Abdullah Al Mehadi, Misbahul Alam Chowdhury, Mirza Muntasir Nishat, Fahim Faisal and Md Minhajul Islam

Department of Electrical and Electronic Engineering (EEE), Islamic University of Technology, Dhaka, Bangladesh

{abdullahmehadi, misbahulalam, mirzamuntasir, faisaleee, minhajulislam40} @iut-dhaka.edu

Abstract. This paper proposes a design and software simulation of monofacial solar photovoltaic panel based energy harvesting system for university residence. As the site of the proposed system, the rooftop of North Hall of Residence of the Islamic University of Technology (IUT) is considered. The 3D model of the system is demonstrated by utilizing PVSOL software. The monthly energy production is obtained by performing the simulation in three software named PVSOL, PVsyst, and System Advisor Model (SAM). However, the monthly consumption of the residence is tabulated and graphical presentation is depicted. Hence, the energy flow diagram is illustrated showing that the proposed PV system can contribute 18.4% in the annual demand of the site. The performance of each of the software is evaluated and deviation analysis is performed so that more insight can be achieved in terms of harvesting solar energy. Hence, a cost-effective and self-reliant solar energy model for the rooftop of a university residence is proposed by utilizing monofacial solar PV panels.

Keywords. Monofacial PV panel, PVSOL, PVsyst, SAM, Solar Energy Model

1. Introduction

For the past few decades, most of the global energy consumption is heavily dependent on fossil fuels which causes severe carbon emission and leads to environmental pollution and ultimately global warming [1-5]. Considering all the negative aspects of fossil fuels and environmental concerns, renewable energy is a promising alternative, and it has a massive prospect in addressing real-world issues. The mainly used renewables are solar and wind energy, possessing the immense potential to be more reliable and efficient than coal-fired power plants in the context of a developing country like Bangladesh [6-8].

One of the most reliable forms of renewable energy around the world is solar energy which can be harnessed using Photovoltaic (PV) technology. Generally, two types of PV configurations are in use which are on-grid and off-grid systems [9]. On-grid PV system, being a grid-connected array is capable of converting DC energy to AC energy with the help of suitable power conditioning technology and synchronizing the AC energy to the grid [10]. By generating power during daytime when sunlight is abundant, this system also ensures that the average daily PV output matches the utility's peak demand period [11]. Combining on-grid and off-grid capacity, the world saw a total production of 627 GW, which is substantially high in comparison to the production of the last ten years [12].

Hence, PV demand is increasing significantly in residential, commercial, and utility-scale applications. Rooftop based off-grid PV system is an effective method of utilizing solar energy. The installation of the solar panels is quite a challenge, especially in the crowded cities of Bangladesh. But open space, especially rooftops and highways, can be utilized for the purpose. *Jamal et al.* reported the potential rooftop distribution for solar installation using GIS mapping in the Dhanmondi area [13]. *Mahmud et al.* worked on a bifacial module based solar installation in some of the main highways of Bangladesh [14]. *Shuvho et al.* predicted the solar irradiation and evaluated the performance of an 80 kWp on-grid solar plant in Dhaka using fuzzy logic [15]. *Kazem et al.* conducted a case study on the techno-economic feasibility of a 1 MW on-grid connected photovoltaic system in Adam, Oman [16]. *Dondariya et al.* has reported a performance simulation for an on-grid solar panel system for small households in Ujjain, India, using several PV simulation-tools [17]. Therefore, it is observed that the obstacles of available land area for PV installation can be mitigated by utilizing the open highways and the available rooftop areas in the urban areas.

In this paper, we present the Islamic University of Technology (IUT) as the site of our research, which is situated in the Board-Bazar of Gazipur district of Dhaka division. The area is still quite underdeveloped, and load shedding is a common problem. The University itself has a huge electricity demand. There are two residential halls, two academic buildings, and many laboratories and machine workshops, which require a continuous power supply for most of the time of the day. Our main objective is to design and develop an off-grid monofacial photovoltaic panel on the rooftop of one of the residential halls (North Hall) using three prominent PV simulation tools namely PVSOL, PVsyst, and SAM (System Advisor Model).

In Section 2, system design is illustrated where all the site details, meteorological data, load profile, solar components as well as system orientation are depicted. Afterward, in Section 3, the results of the software simulation of the system are shown and discussed. Lastly, the overall comparison and deviation analysis is presented. A rigorous software-based investigation and feasibility analysis are carried out.

2. Design of the System

2.1. Site Details

Islamic University of Technology (IUT), located in Board-Bazar, Gazipur, on the outskirts of Dhaka city, Bangladesh, is chosen as the site of our project. It is situated at a latitude of 23.9481° N and a longitude of 90.3793° E.

Figure 1. Imported 3-D model on PVSOL

The area selected for the installation of PV panels is the North Hall of Residence with a large space available on the rooftop, which can be effectively utilized for PV installation. The hall has approximately 503 square meters available mounting surface on it. The building is south facing and there is no external shading on the rooftop by trees or any other obstacle [18]. The 3-D model of the entire building is shown in Figure 1.

2.2. Meteorological Data

After site selection, necessary meteorological data is collected from various online databases like Meteonorm, NASA, PVGIS, SOLARGIS, SOLCAST, NREL which will act as key factors in designing the system [19]. The required data is reported in Table 1, utilized for further design of the system in PVSOL, PVsyst, and SAM.

Table 1. Required Data for the Design.

Site Name	Islamic University of Technology Board-Bazar, Dhaka
Annual global irradiance	1793 kWh/m^2
Horizontal Diffuse Irradiance	864 kWh/m^2
Average temperature	24.9 °C
Wind Velocity	0.8 m/s
Humidity	81%

2.3. Load Profile Analysis

An approximate monthly consumption for one hall (North Hall of residence) was collected from the engineering section of IUT, from which the estimated annual load was found to be 444733.5 kWh. The monthly load profile is tabulated in Table 2.

Table 2. Tabulated Monthly Load Profile.

Month	Consumption (kWh)
January	24797.52
February	30173.52
March	47799.1
April	47799.1
May	23983.9
June	47799.1
July	47799.1
August	23881.9
September	47799.1
October	48100.2
November	30003.5
December	24797.5
Total	444733.5

As we can see from the load profile, the electricity consumption is naturally high in March, April, June, July, September, and October due to the summer and autumn seasons. The electrical demand is the lowest during the winter season in December and January. The electrical consumption is in the medium range in the early part of winter in November and towards the end of winter in January.

2.4. Solar Components

For any photovoltaic system, PV array is considered the heart of the system, which determines how much solar energy can be received from the sun [19]. LG modules have been chosen as they have relatively low degradation and longer life. For optimum operation, three inverter models from Huawei technologies have been selected. They are SUN2000L-3KTL, SUN2000L-4KTL, and SUN2000L-5KTL. The parameters of the PV system and their corresponding values are presented in Table 3.

Table 3. Parameters of the proposed PV system

Parameters	Values
PV Model	LG340N1K-v5
Power	340W
Module Type	Monofacial
Nominal Efficiency	20.6
Maximum Power	340.6 W
Maximum Power Voltage	34.9 V
Maximum Power Current	9.8 A
Open Circuit Voltage	41.2 V
Short Circuit Current	10.3 V
Nominal Module Operating Temperature	42 ± 3 °C

2.5. System Operation

Monofacial modules were placed at an optimum tilt angle of 24° which is depicted in Figure 2 [20]. A total of 174 modules were utilized where all modules were set at 2 m height above the mounting [21]. A summary of the parameters necessary for our simulations and orientation is given in Table 4.

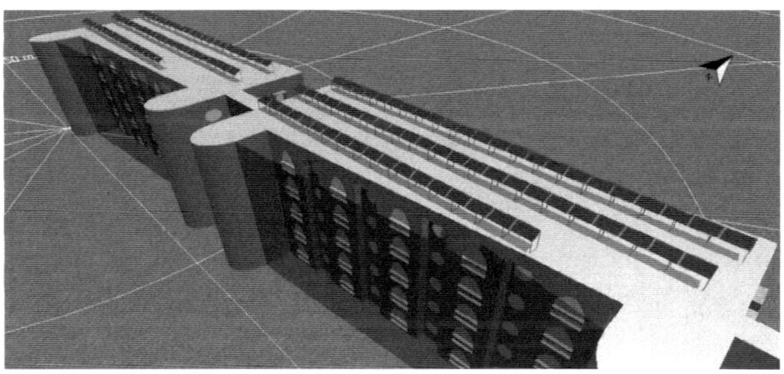

Figure 2. Monofacial modules tilted at 24°

3. Software Simulation and Result Analysis

Three popular software namely PVSOL, PVsyst, and SAM are utilized for the simulation purpose. The parameters taken for simulation are tabulated in Table 4. By utilizing these parameters, rigorous simulation is carried out and a monthly profile is obtained from each of the software. However, each of the software showed quite a promising result in terms of energy production which is tabulated in Table 5 and the graphical representation is depicted in Figure 3. The annual yield was 82,076 kWh, 83,038 kWh, and 80,401 kWh in PVSOL, PVsyst, and SAM respectively. It is observed that maximum yield was obtained during March due to high global horizontal irradiation. However, the lowest yield was during June when the irradiation was low due to cloudy sky and rain.

Table 4. Simulation parameters.

Parameters	**Values**
Panel rating	340 W
No. of Panels	174
Total PV Generator Output	59.16KW
Tilt	24°
Azimuth	180°
Albedo	0.65
Soiling loss	15%
Mismatch Loss	2%
Diode Loss	0.5%
Module Gap	0.1 m
Row Gap	3.5 m
No. of Inverters	11
Mounting Surface	2 m

Table 5. Monthly energy production in kWh.

Month	**PVSOL**	**PVsyst**	**SAM**
January	7369.4	7481	6168.08
February	6989.7	7089	6690.24
March	7868.6	7969	8139.5
April	7404.6	7458	7527.78
May	7036.5	6978	7334.4
June	5642.9	5664	5728.9
July	5923.9	5867	6144.68
August	5884.5	5969	6253.18
September	6156.5	6344	6172.1
October	6751.6	6702	7007.99
November	7641.4	7700	6822.54
December	7406.6	7817	6411.84
Total	**82076.2**	**83038**	**80401.23**

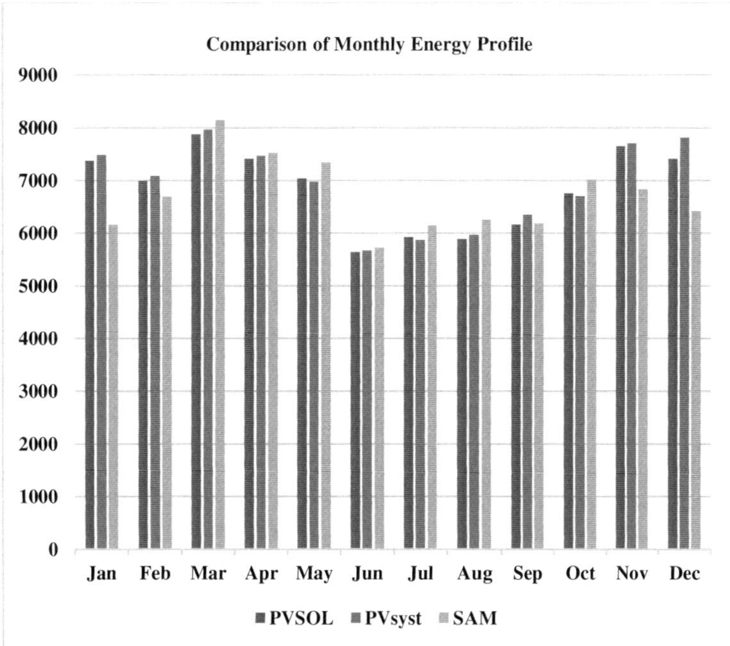

Figure 3. Comparison of monthly energy profile among three software (PVSOL, PVsyst, and SAM)

For North Hall of residence of IUT, the annual energy demand was 444,733.5 kWh approximately. Considering the mean of three software, it can be presumed that the proposed PV system would be supplying 81,838.5 kWh (18.4 %), and the rest of the 362,895 kWh would be taken from the grid. The monthly energy consumption from the PV system and the grid are illustrated graphically in Figure 4 and tabulated in Table 6. Hence, the energy flow diagram is displayed in Figure 5.

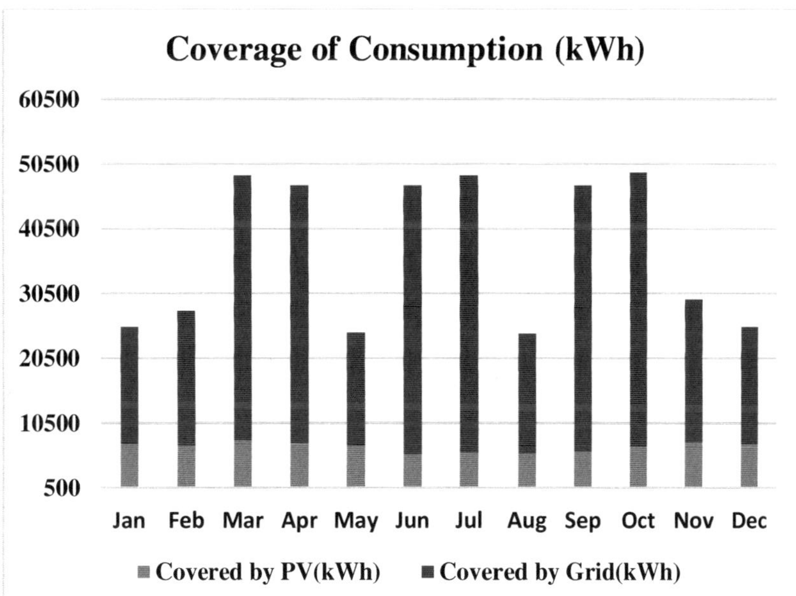

Figure 4. Coverage of Consumption (kWh)

Table 6. Coverage of Consumption (kWh).

Month	Covered by PV	Covered by grid
January	7218.7	18101.5
February	6989.7	20793.2
March	7868.6	40887.7
April	7404.6	39776.1
May	6969.1	17485.3
June	5642.9	41536.9
July	5923.9	42828.5
August	5774.7	18589.1
September	6156.5	41025.7
October	6751.6	42322.2
November	7641.4	21996.1
December	7296.6	18024.8

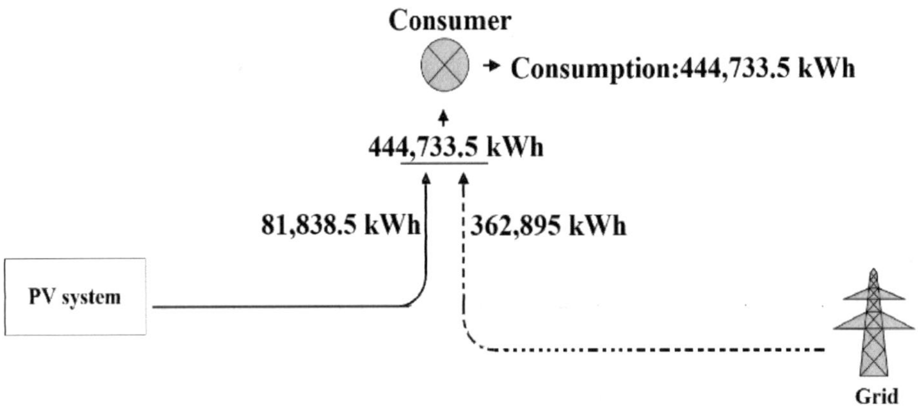

Figure 5. Energy Flow Diagram

Deviation analysis is performed among the obtained results of the software which will help to evaluate each software performance and aid in further estimation and planning for rooftop monofacial setup [22]. There is a difference in the system production between the three software. In the case of PVsyst vs. SAM and PVSOL vs. SAM, the monthly production difference is quite high, because SAM takes a few more loss parameters into account during the simulation. On the other hand, for the instance of PVSOL vs. PVsyst, there is a lesser deviation since they use pretty much the same parameters during the simulation process. Besides, there is a slight difference in the mathematical model used in each of the software and how irradiation value and yield are being calculated. Moreover, there is a dissimilarity in the weather profile used in each of the software which also affects annual yield or system production. The monthly deviation analysis among the three PV simulation software is tabulated in Table 7, Table 8, and Table 9. In Figure 6, the graphical representation is depicted.

Table 7. Deviation Analysis between PVSOL and PVsyst.

Month	Deviation
January	-111.6
February	-99.3
March	-100.4
April	-53.4
May	58.5
June	-21.1
July	56.9
August	-84.5
September	-187.5
October	49.6
November	-58.6
December	-410.4

Table 8. Deviation Analysis between PVSOL and SAM.

Month	Deviation
January	1201.32
February	299.46
March	-270.9
April	-123.18
May	-297.9
June	-86
July	-220.78
August	-368.68
September	-15.6
October	-256.39
November	818.86
December	994.76

Table 9. Deviation Analysis between PVsyst and SAM.

Month	Deviation
January	1312.92
February	398.76
March	-170.5
April	-69.78
May	-356.4
June	-64.9
July	-277.68
August	-284.18
September	171.9
October	-305.99
November	877.46
December	1405.16

IOP Conf. Series: Materials Science and Engineering 1045 (2021) 012011 doi:10.1088/1757-899X/1045/1/012011

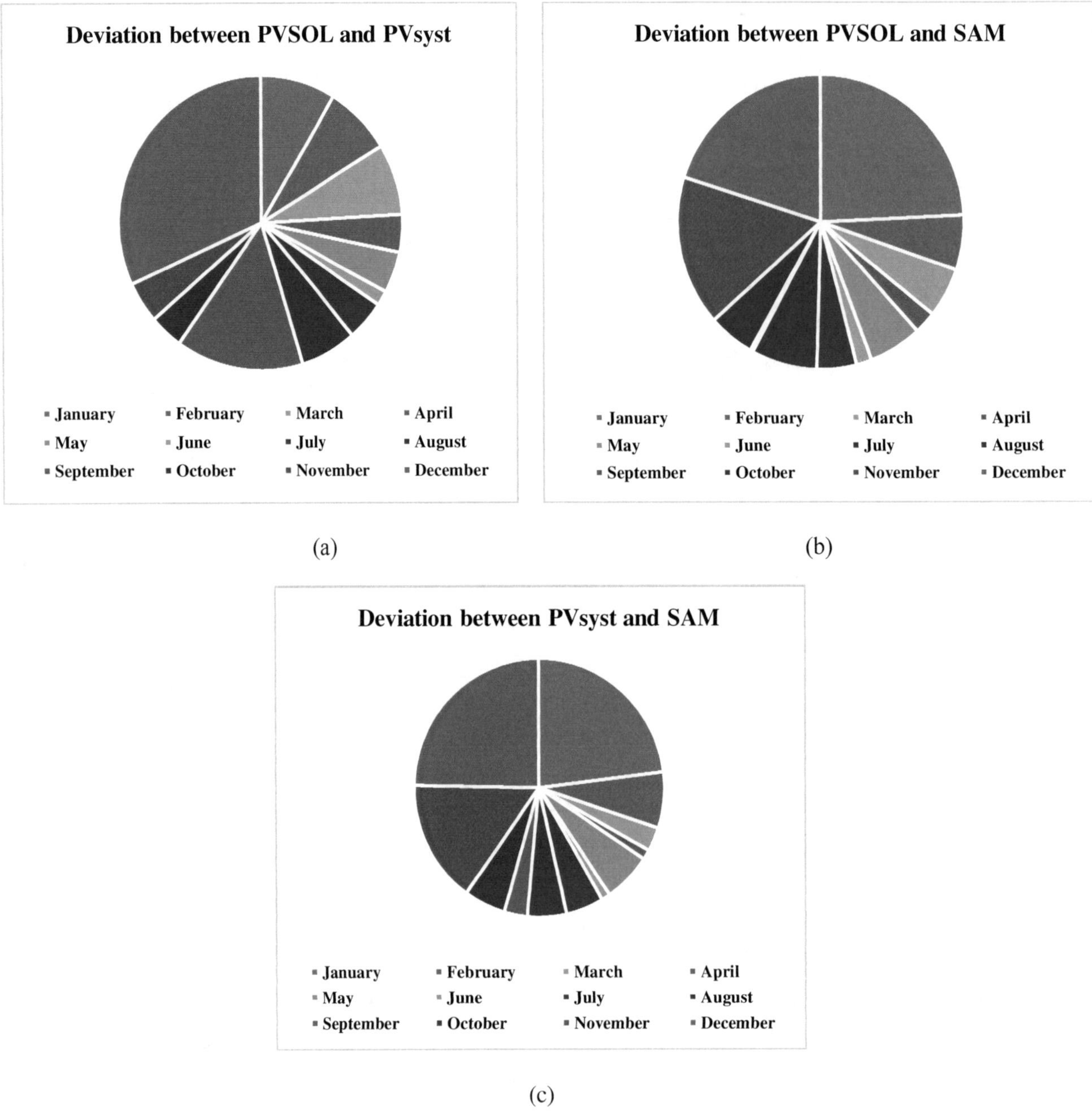

Figure 6. Deviation Analysis between (a) PVSOL and PVsyst (b) PVSOL and SAM and (c) PVsyst and SAM

4. Conclusion

The design, simulation, and analysis of monofacial PV module on the roof of the North-hall of residence in IUT have been presented in this paper. A detailed 3D design was created and analysed using three different software. The 3D design of the rooftop simulation provided a promising yield and shading profile. To meet the annual energy demand of 444,733.5 kWh, a total of 174 modules rated 340W have been used to generate 81,838 kWh approximately in a year. Hence, 18.4% of energy demand is met using this PV setup which implicates that university can be self-reliant in energy consumption by adopting solar energy and installing it in different open areas and rooftops. As a result, huge load shedding during peak hours can be undertaken and the diverse international community and stakeholders of the university will be benefitted. On the other hand, this attempt of clean energy would be able to save up-to USD 5810.5$ annually considering the grid price to be 0.071 $/kWh. Therefore, a cost-effective and self-reliant energy system can be designed for such universities situated in load shedding prone industrial areas so that the academic and research activities can be carried out smoothly, without any interruption of power. In the future, there is scope to analyse the performance of the PV panels by utilizing various tilt angles and changing other parameters so that a more optimum energy production system can be designed and implemented in educational and healthcare institutions.

5. References

[1] Dey, Diptiman, and Bidyadhar Subudhi "Design, simulation and economic evaluation of 90 kW grid connected Photovoltaic system." *Energy Reports* 6 (2020): 1778-1787

[2] Arent, Douglas J., Alison Wise, and Rachel Gelman. "The status and prospects of renewable energy for combating global warming." *Energy Economics* 33.4 (2011): 584-593.

[3] Bilgen, Selçuk, et al. "Global warming and renewable energy sources for sustainable development: a case study in Turkey." *Renewable and sustainable energy reviews* 12.2 (2008): 372-396.

[4] Lau, Lee Chung, Keat Teong Lee, and Abdul Rahman Mohamed. "Global warming mitigation and renewable energy policy development from the Kyoto Protocol to the Copenhagen Accord—A comment." *Renewable and Sustainable Energy Reviews* 16.7 (2012): 5280-5284.

[5] Yüksel, Ibrahim. "Global warming and renewable energy sources for sustainable development in Turkey." *Renewable energy* 33.4 (2008): 802-812.

[6] Khan, M. J., M. T. Iqbal, and S. Mahboob. "A wind map of Bangladesh." *Renewable energy* 29.5 (2004): 643-660.

[7] Nandi, Sanjoy Kumar, and Himangshu Ranjan Ghosh. "A wind–PV-battery hybrid power system at Sitakunda in Bangladesh." *Energy Policy* 37.9 (2009): 3659-3664.

[8] Islam, AKM Sadrul, Mazharul Islam, and Tazmilur Rahman. "Effective renewable energy activities in Bangladesh." *Renewable energy* 31.5 (2006): 677-688.

[9] Menconi, Maria Elena, et al. "Energy sovereignty in Italian inner areas: Off-grid renewable solutions for isolated systems and rural buildings." *Renewable energy* 93 (2016): 14-26.

[10] Mukherjee, D., and S. Chakrabarti. *Fundamentals of renewable energy systems*. New Age International, 2004

[11] Natural Resources Canada. "Clean energy project analysis: RET Screen Engineering & Cases textbook." (2005).

[12] *"Renewables 2020 Global Status Report"* https://www.ren21.net/wp-content/uploads/2019/05/gsr_2020_full_report_en.pdf

[13] Jamal, Taskin, et al. "Potential rooftop distribution mapping using Geographic Information Systems (GIS) for Solar PV Installation: A case study for Dhaka, Bangladesh." *2014 3rd international conference on the developments in renewable energy technology (ICDRET)*. IEEE, 2014.

[14] Mahmud, Md Sultan, et al. "Solar Highway in Bangladesh Using Bifacial PV." *2018 IEEE International Conference on System, Computation, Automation and Networking (ICSCA)*. IEEE, 2018.
[15] Shuvho, Md Bengir Ahmed, et al. "Prediction of solar irradiation and performance evaluation of grid connected solar 80KWp PV plant in Bangladesh." *Energy Reports* 5 (2019): 714-722.
[16] Kazem, Hussein A., et al. "Techno-economic feasibility analysis of 1 MW photovoltaic grid connected system in Oman." *Case Studies in Thermal Engineering* 10 (2017): 131-141
[17] Dondariya, Chandrakant, et al. "Performance simulation of grid-connected rooftop solar PV system for small households: A case study of Ujjain, India." *Energy Reports* 4 (2018): 546-553.
[18] Lodhi, Ehtisham, et al. "Application of particle swarm optimization for extracting global maximum power point in PV system under partial shadow conditions." *Int J Electron Electr Eng* 5.3 (2017): 223-229.
[19] Şenol, M., et al. "A guide in installing large-scale PV power plant for self consumption mechanism." *Solar Energy* 132 (2016): 518-537.
[20] Kalogirou, S. A. "Solar Energy Engineering: Processes and Systems, Else-vier." (2009).
[21] Yusufoglu, Ufuk Alper, et al. "Simulation of energy production by bifacial modules with revision of ground reflection." *Energy Procedia* 55 (2014): 389-395.
[22] Axaopoulos, Petros J., Emmanouil D. Fylladitakis, and Konstantinos Gkarakis. "Accuracy analysis of software for the estimation and planning of photovoltaic installations." *International Journal of Energy and Environmental Engineering* 5.1 (2014): 1.

Fabrication and characterization of epoxy resin–barium titanate at G-band using waveguide technique

Nurulfadzilah Hasan, Nur Shahira Mat Hussain, Nur Hazimah Syazana Abdul Razak, Ahmad Syahiman Mohd Shah, and Mohamad Shaiful Abdul Karim

College of Engineering, Universiti Malaysia Pahang, 26300 Gambang Pahang, Malaysia
mshaiful@ump.edu.my

Abstract. In this paper, fabrication process of epoxy resin-barium titanate nanocomposite and measurement of its complex permittivity are presented. The material is prepared by mechanical mixing of epoxy resin and barium titanate nanopowder. The nanocomposite is intended to be used as high permittivity microstrip antenna substrate, which requires accurate measurement of its electrical characteristics. Thus, characterization of materials is done using waveguide technique, which does not require a precise machining of sample's width and thickness, and does not utilize small reflection coefficient, which can cause error in measurement. The complex permittivity of the nanocomposite is measured in G-band (4 to 6 GHz). Then, the measured values are compared with prediction method, Lichtnecker and Maxwell-Garnet method. The results show that the measured permittivity of composite materials are in good range with prediction method, while the measurements of loss tangent show that the developed materials are low-loss and suitable to be used as substrate of antenna.

Keywords: antenna substrate, barium titanate, epoxy resin, material characterization, material fabrication

1. Introduction

In our modern society, microwave materials have been used widely by researcher in a lot of advanced communication devices. For this reason, a wide variety of new materials have been developed especially for communication system [1], radar system [2], energy storage device [3], and industrial waste [4]. These devices require high permittivity of microwave material to develop a small, lightweight and robust design [5]. The dielectric permittivity of the composite can be enhanced by adding ferroelectric ceramics, such as barium titanate ($BaTiO_3$) and lead magnesium niobate ($MgNb_2O_9Pb_3$)[6]. They are used as fillers in dielectric composites.

In this study, a technique to prepare and characterize composite substrate of antenna, which is made of epoxy resin and barium titanate is presented. Epoxy resin has been widely used due to their low cost, high moisture, and high chemical resistance [7], while barium titanate is well known as high permittivity material [8]. Thus, the mix of these two materials can afford high value of permittivity in order to be used in compact and small antenna design.

Prediction methods such as Lichtnecker [9]-[11], Maxwell-Garnett [9]–[12], Jayasundere-Smith [11], effective medium theory [11], and Yamada model [10] have been used to guess the value of permittivity for composite material involving volume fraction of base matrix and ceramic material. In microwave region, the permittivity is decreasing as the frequency increases. Hence, these methods are not applicable to estimate permittivity for antenna applications, as the methods not consider frequency changes but yet important to get initial idea of permittivity during fabrication of material. In the past, many techniques have been carried out to determine complex permittivity (permittivity and loss tangent) of material in microwave measurement such as free-space method [13], resonant method [14]-[15], and transmission-line method [16]–[18]. Measurement for free-space method is among the easiest setup, which is typically, consists of two horn antennas as transmitter and receiver while the sample is placed between the antennas. However, the measurement using this method is easily affected by surrounding and cause inaccuracy determination of complex permittivity. Alternatively, resonant method able to provide high accuracy of measurement, but the measurement can be done only at one point of frequency. In this paper, the fabricated substrate is measured using transmission-line method in order to provide wide-frequency measurement while the accuracy is comparable with resonant method. G-band waveguide, frequency range from 4 to 6 GHz, is used in this method to measure scattering parameters. Then the complex permittivity will be determined from scattering parameters using an accurate and effective hybrid electromagnetic method as proposed by [19].

The rest of this paper is organized as follows. Section 2 explains the prediction of permittivity and sample fabrication of nanocomposite materials. Then, the measurement and complex permittivity estimation of fabricated materials is described in Section 3. Then, the cumulative results are presented in Section 4. Finally, Section 5 concludes our works.

2. Nanocomposite Sample Preparation

2.1. Permittivity Prediction Method

In general, adding high permittivity materials such as barium titanate to a low permittivity material, such as epoxy resin will increase the permittivity of the composite. Furthermore, the permittivity increases as the concentration of filler is increase. Thus, the permittivity of a composite material can be predicted by using theoretical models such as Lichtnecker, Maxwell-Garnett, Jayasundere-Smith, effective medium theory, and Yamada model. The theoretical models relate the permittivity of the new composite material, with the permittivity and the volume fractions of base matrix and filler. In this work, commonly used Lichtnecker and Maxwell-Garnett model are adopted to verify our permittivity's measurement. The relationship between volume fraction of base matrix and ceramic material, and permittivity in Lichtnecker model [9] is given as following;

$$\ln \varepsilon_{reff} = v_1 \ln \varepsilon_{r1} + v_2 \ln \varepsilon_{r2} \quad (1)$$

where, ε_{reff}, ε_{r1} and ε_{r2} are the permittivity of the composite, base, and ceramic material, respectively. v_1 and v_2 are the volume fraction of the base and ceramic material, respectively.

Meanwhile, according to Maxwell-Garnett model [9], the relationship can be represented as;

$$\frac{\varepsilon_{reff} - \varepsilon_{r2}}{\varepsilon_{reff} + 2\varepsilon_{r2}} = v_1 \frac{\varepsilon_{r1} - \varepsilon_{r2}}{\varepsilon_{r1} + 2\varepsilon_{r2}} \quad (2)$$

2.2. Sample Fabrication Technique

Base matrix of the composite is EpoxAmite™, epoxy resin system with medium hardener. The epoxy resin system has density of 1.25 g/cm^3 and permittivity between 2.7 and 3 [16]. Filler of the composite is barium titanate nanopowder, with particle size of less than 3 μm and 98.0% purity. The density is 6 g/cm^3, while the permittivity is 3279 [20].

To prepare composite material with 5 vol.% filler, first the weight ratio (wt.%) for base matrix and filler is calculated based on the filler volume ratio and density of the materials. Fabrication process starts with adding 12.63 g barium titanate nanopowder to 50 g epoxy resin system. The mixture is then stirred thoroughly using an overhead stirrer (WiseStir HS-30D) at 500 rpm for two minutes. Stirring at high speed is necessary to ensure the filler particles are well dispersed within the base matrix. However, it

IOP Conf. Series: Materials Science and Engineering 1045 (2021) 012012 doi:10.1088/1757-899X/1045/1/012012

must be done within a short period to minimize trapped air bubbles in the mixture. Next, the mixture is placed inside a vacuum chamber (30 in.-Hg) for 2 min to remove trapped air bubbles. Then mixture is carefully poured into mold with dimension of 47.55 mm × 22.15 mm and thickness of 22.10 mm. Curing of the composite is done in two stages. First, the mixture is let to cure inside vacuum oven at 60°C for

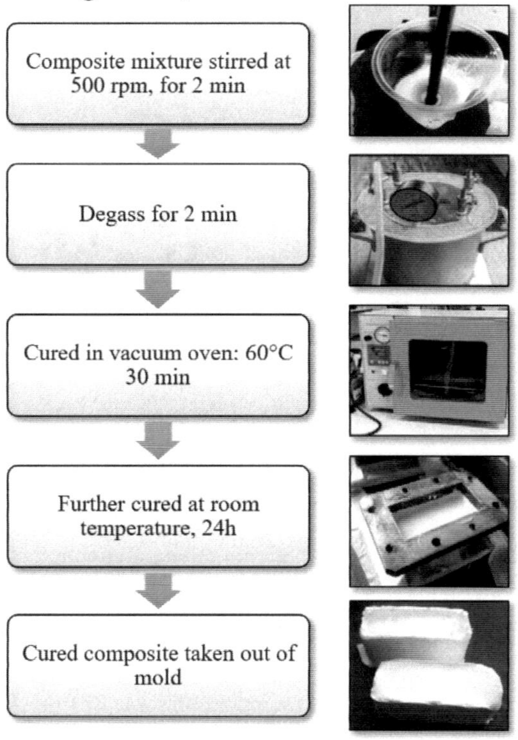

Figure 1 Material preparation process.

30 minutes. This is done to remove trapped air in the mixture and improves the mechanical properties of composite [21]. The composite mixture is then further cured at room temperature for 24 hours. Finally, the cured composite is taken out of the mold and prepared for permittivity measurement. Composite preparation process is depicted in Figure 1.

3. Material Characterization Based on Waveguide Technique

3.1. Sample Preparation and Measurement

Three fabricated nanocomposite samples are machined to fit the height of G-band waveguide, i.e., 22.15 mm, while the width and length can be varied in order to ease the processes of sample preparation and measurement. Figure 2 shows the sample material under test (MUT). The dimensions for MUT 1, MUT 2 and MUT 3 are 20.94 mm × 21.96 mm, 21.12 mm × 21.77 mm, and 21.15 mm × 16.57 mm, respectively.

Figure 3 shows the experimental setup to measure the scattering parameters. First, MUT is partially placed inside the waveguide. Then, the measurement is done by connecting the waveguide to the network analyzer using low-noise RF cable and waveguide-to-coax adapters. Partially loading MUT into the waveguide is beneficial, since TE_{10} mode characteristic; where the electric field is stronger at the center of the waveguide can be fully utilized. For instance, if the sample absorb high level of energy, the transmission parameter will be very low and lead to inaccurate measurement. Thus, to overcome this problem, the sample might be placed at the sidewall of the waveguide where the electric field is weak so that less energy will be absorbed by the sample.

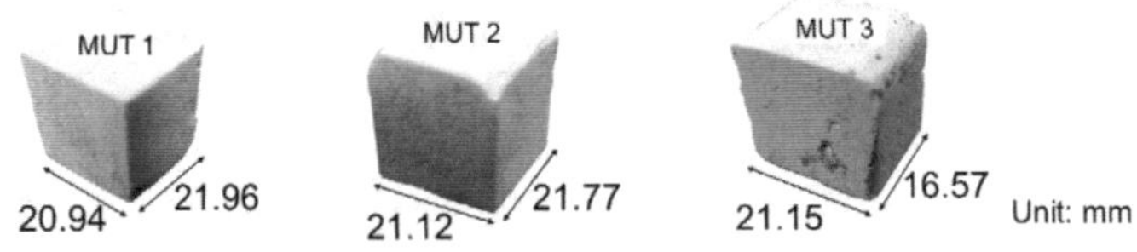

Figure 2 Material under test

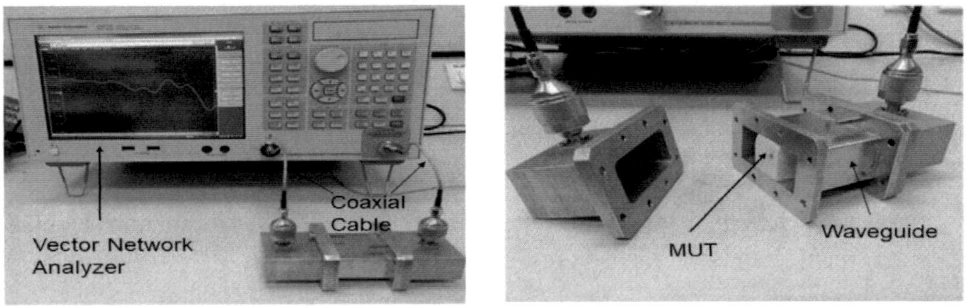

Figure 3 Measurement setup

3.2. Material Characterization Technique

To enhance the accuracy of material characterization, systematic errors is suppressed by utilizing calibration-free technique of waveguide measurement [22]. It is done by taking difference of complex transmission parameters (magnitude and phase) of waveguide measurement with and without MUT. Hence, this technique is able to eliminate electrical disturbances from measurement setup such as noise of cable and vector network analyzer (VNA).

Electrical properties in terms of complex permittivity are derived by resorting inverse technique [14] where the difference of measured and calculated complex transmission parameters are minimized by changing the complex permittivity. Once the difference is below than tolerance value, the last changed value of complex permittivity is adopted as the electrical properties of MUT. The full illustration for the material characterization technique is shown in Figure 4.

4. Results and Discussions

Permittivity of composite materials are measured using waveguide technique and the compared with prediction methods as mentioned above. Meanwhile, the measurement of loss (tangent delta) will be presented even the developed composite material is low loss, yet important for an accurate antenna design. However, there is no comparison between measurement and prediction data for tangent delta since the prediction value will lead to almost zero value due to low loss properties.

Figure 5 shows the measured complex transmission parameters (magnitude and phase), S_{21}, for all three MUTs within frequency range of 4 to 6 GHz. There is no significant resonance of magnitude of S_{21}, so it is very promising to obtain an accurate measurement of complex permittivity. Complex permittivity is determined from complex transmission parameter by resorting the inverse technique explained above (section 3).

4.1. Permittivity/Dielectric Constant

In this section, the prediction value of permittivity of epoxy resin with 5% filler of $BaTiO_3$ is compared with the measurement value done using waveguide technique. Figure 6 depicts the measured permittivity of all nanocomposite materials. The prediction values using Lichtnecker and Maxwell-Garnett model are included for comparison. All measured values are in good range with the predicted value. However, the measured value for each MUT is slightly different due to the difficulty to keep the

IOP Conf. Series: Materials Science and Engineering 1045 (2021) 012012 doi:10.1088/1757-899X/1045/1/012012

same conditions during fabricating and percentage of trapped air bubbles might different. Hence, each sample has its own characteristics and measurement for each sample is necessary before using it for design purposes. It should be noted that the predicted values are not frequency-dependent. Therefore, it can be a reference for developing material but cannot be used for design purposes.

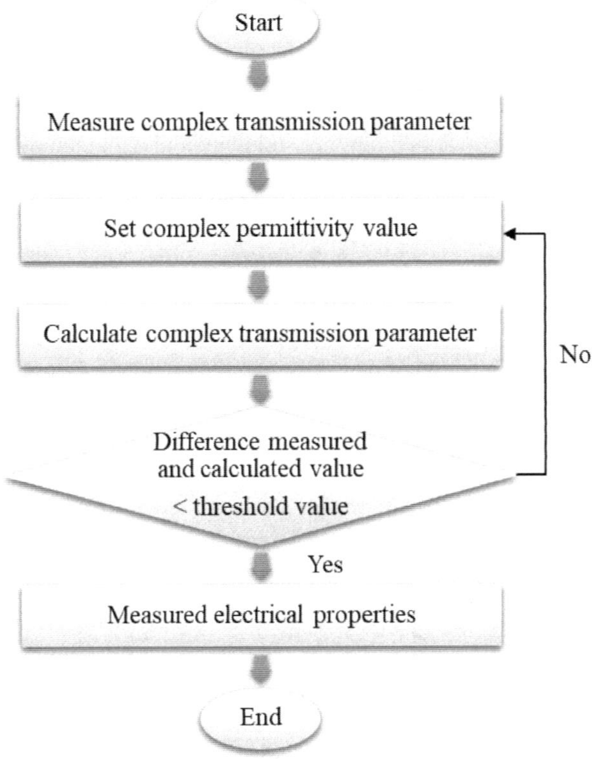

Figure 4 Material characterization flow.

4.2. Tangent Delta

Value of tangent delta for these nanocomposites are small, yet important for design purposes. Figure 7 shows the measured value of tangent delta. All three samples have tangent delta values that are consistently close to zero. Small value of tangent delta is desirable for an antenna substrate, because the loss is small.

5. Conclusions

Nanocomposite materials made of epoxy resin system as base matrix and barium titanate nanopowder as filler were fabricated for antenna applications. The necessary electrical properties for antenna designing were measured using waveguide technique at 4 to 6 GHz. This measurement method was promising since there is no resonant for measurement of scattering parameters. The measured values of permittivity were compared with two prediction methods, Lichtnecker, and Maxwell-Garnett model. They are in good range with the predicted value. Meanwhile, the measured values of loss tangent were almost zero, which is indicated that the fabricated nanocomposite material is low loss and suitable to be used as substrate of antenna.

IOP Conf. Series: Materials Science and Engineering 1045 (2021) 012012 doi:10.1088/1757-899X/1045/1/012012

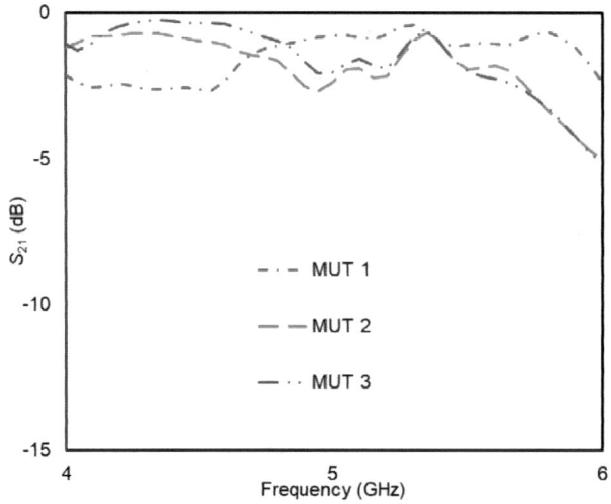

(a) Magnitude of transmission parameter, S_{21}

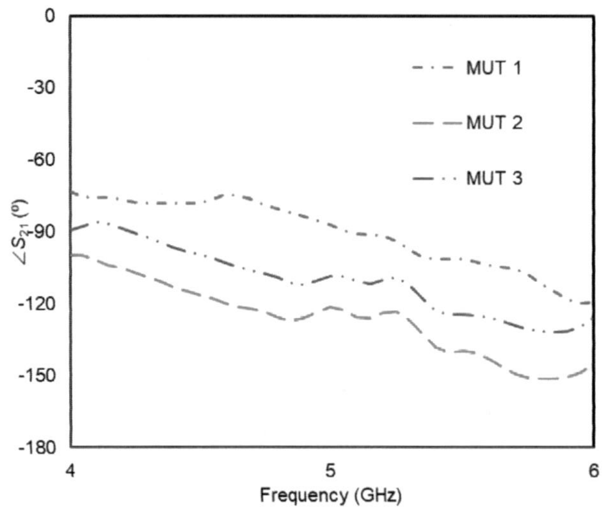

(b) Phase of transmission parameter, S_{21}

Figure 5 Measured complex transmission parameter.

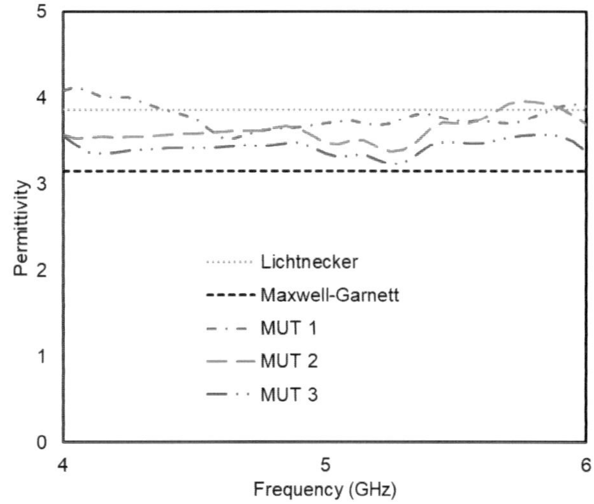

Figure 6 Comparison of measured and predicted permittivity.

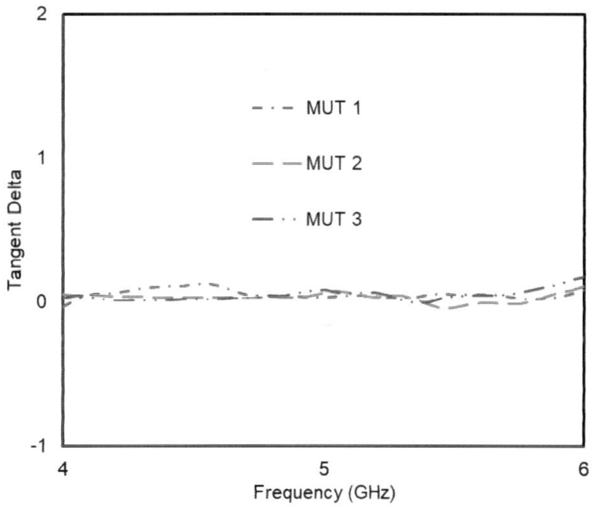

Figure 7 Measured values of tangent delta (loss tangent).

6. Reference

Guociang P, et. al, 2018 "Investigation of the composites of epoxy and micro - scale $BaTi_4O_9$ ceramic powder as the substrate of microwave communication circuit," Microsyst. Technol., vol. 24, no. 1, pp. 343–349

Yuri A L, et. al, 2018 "A Synthetic Aperture Radar (SAR) -based technique for microwave imaging and material characterization," *Electronics*, vol. 7, no. 373, pp. 1–15

Georgia C M and Georgios C P 2016 "Energy storage and harvesting in $BaTiO_3$ / epoxy nanodielectrics," *Inst. Eng. Technol.*, vol. 1, no. 4, pp. 151–157

Azmi E, et. al, 2019 "Friction and wear behavior of epoxy composite filled with industrial wastes," J. Clean. Prod., vol. 237

Nurulfadzilah H, et. al, 2020 "Dielectric properties of epoxy – barium titanate composite for 5 GHz microstrip antenna design," *SN Appl. Sci.*, vol. 2, no. 62

Kuo C C, et. al, 2007 "Dielectric properties of epoxy resin – barium titanate composites at high frequency," Mater. Lett., vol. 61, pp. 757–760
Pei L T, et. al, 2007 "The properties of epoxy resin coated silica fillers composites," Mater. Lett., vol. 61, pp. 2156–2158
Edward K N, et. al, 2008 "Effect of microwave frequency on hydrothermal synthesis of nanocrystalline tetragonal barium titanate," J. Phys. Chem. C, vol. 112, no. 26, pp. 9659–9667
Dong S K, et. al, 2008 "Effect of microwave frequency on hydrothermal synthesis of nanocrystalline tetragonal barium titanate," *J. Phys. Chem. C*, vol. 112, no. 26, pp. 9659–9667
Shaohui L, et. al, 2016 "Dielectric properties and energy storage densities of poly(vinylidenefluoride) nanosomposite with surface hydroxylated cube shaped Ba0.6Sr0.4TiO3 nanoparticles," *Polymers (Basel)*, vol. 45, no. 8, pp. 10–14
Bingcheng L, et. al, 2015 "Synthesis, characterization and dielectric properties of surface functionalized ferroelectric ceramic / epoxy resin composites with high dielectric permittivity," *Compos. Sci. Technol*, vol. 112, pp. 1–7
Francesco D P, et. al, 2011 "From maxwell garnett to debye model for electromagnetic simulation of composite dielectrics part I : random spherical inclusions," *IEEE Trans. Electromagn. Compat*, vol. 53, no. 4, pp. 933–942
Deepak K G, Vasundara V V and Vijay K V 1989 "A free-space method for measurement of dielectric constants and loss tangents at microwave frequencies," *IEEE Trans. Instrum. Meas.*, vol. 37, no. 8926909, pp. 789–793
Salwa A A, et. al, 2019 "An accurate characterization of different water properties using resonant method for underwater communication activity," *Springer Nat. Singapore Pte Ltd.*, pp. 113–120
Mohamad S A K, Nurhafizah A T @ Y and Toshihide K 2017 "Scattering analysis of rectangular cavity with input and output waveguides and its application to material characterization," *Proc. 2017 Asia Pacific Microw. Conf.*, pp. 5–8
Nurulfadzilah H, et. al, 2019 "Cured epoxy resin dielectric characterization based on accurate waveguide technique," *AIP Conf. Proc. 2129*, pp. 1-5
Teppei K, et. al, 2013 "Determination of complex permittivity of materials in rectangular waveguides using a hybrid electromagnetic method," *CEM'13 Comput. Electromagn. Int. Work.*, pp. 56–57
Mohamad S A K, Yuki K and Toshihide K 2014 "Robustness analysis of simultaneous determination method of complex permittivity and permeability," *2014 Int. Conf. Numer. Electromagn. Model. Optim. RF, Microwave, Terahertz App.*, pp. 1–4
Mohamad S A K and Toshihide K 2015 "Determination of material parameters based on hybrid numerical methods using complementary source quantities," *IEICE Trans. Electron.* vol. J98-C, no. 12, pp. 356–365
Burcu E 2013 "The overview of the electrical properties of barium titanate," *Am. J. Eng. Res.*, no. 08, pp. 1–7
Devalingam S K, et. al, 2015 "Effect of post-curing on thermal and mechanical behavior of GFRP composites," *IOP Conf. Ser. Mater. Sci. Eng.*, vol. 75, no. 1
Mohamad S A K, et. al, 2104"Determination of complex permittivities of layered materials using waveguide measurements," *IEEE Trans. Microw. Theory Tech.*, vol. 62, no. 9, pp. 2140–2148

Acknowledgments

This work was supported in part by the Universiti Malaysia Pahang and Ministry of Education (MOE) Malaysia under Fundamental Research Grant Scheme (FRGS) with reference no. FRGS/1/2018/STG02/UMP/02/1 or RDU190140. This work also was partly supported by the Research and Innovation Department, Universiti Malaysia Pahang under Postgraduate Research Grants Scheme (PGRS) with reference no. PGRS200315.

IOP Conf. Series: Materials Science and Engineering 1045 (2021) 012013 doi:10.1088/1757-899X/1045/1/012013

A study on sodium - the fast breeder reactor coolant

K K Rajan

Viswajyothi College of Engineering and Technology,Vazhakulam, Muvattupuzha, Kerala, India,PIN-686670

kkrajan56@gmail.com

Abstract. Sodium, the second alkali metal in the first group of periodic table is highly chemically reactive. Sodium is widely used in pharmaceutical industries to manufacture lifesaving medicines. Liquid sodium is an excellent reducing agent and heat transport medium, which makes it an important industrial material. The application of this element in liquid form at high temperature as a coolant in fast breeder nuclear reactors necessitated development of sodium technology. Physical, chemical and nuclear properties of sodium led to the choice of sodium as the universally accepted fast breeder reactor (FBR) coolant. The manufacturing of sodium is by electrolysis process. The design and manufacturing requirements of high temperature liquid sodium system components are successfully addressed. Special type of sensors needed for high temperature liquid sodium system are specifically designed and developed. Indira Gandhi Centre for Atomic Research (IGCAR), Kalpakkam is involved in the development of sodium technology in India. Sodium technology is matured to an extent and around twenty FBRs with sodium as the coolant were constructed, commissioned and successfully operated all over the world. Five sodium cooled fast reactors are currently in operation including fast breeder test reactor (FBTR) in India. Prototype fast breeder reactor (PFBR) is in the advanced stage of commissioning at Kalpakkam. This review paper explains the choice of sodium as FBR coolant and gives highlights on the research and development took place in sodium technology all over the world especially in India with an aim to give an exposure to this technology to the academic community.

1. Introduction

All must have heard of sodium and studied about its peculiar properties. Every one also might have seen solid sodium pieces stored in bottles under kerosene, in high school laboratories. A student would be considering it as a dangerous element, which will catch fire when exposed to atmosphere. If it is exposed to water severe reaction and explosion would also take place. It is sixth element in the order of abundance and forms 2.8% of earth's crust. However this element is finding application in nuclear engineering and is used as the coolant in fast breeder type of nuclear reactors. Because of this, a branch of science and technology by name sodium technology has emerged in India and abroad. This paper describes basic properties of sodium metal, why sodium is selected as the fast breeder reactor coolant, the challenges involved while using it as the coolant, highlights of sodium technology developed and the opportunities for scientific community in research and development. This paper intends to give an exposure to academic community about sodium technology.

2. Physical and chemical properties

Sodium is a group 1A element in the periodic table and is an alkali metal. In the periodic table other alkali metals are Lithium, Potassium, Rubidium, Cesium and Francium. These elements have only one electron in the outermost shell. If we remove this electron, the electronic structure will become that of aninert gas. So the natural tendency of alkali metals is to release electron in the outer most shell and attain a state of inert gas electronic structure. Because of this inherent tendency to release electron, alkali metals are chemically very reactiveand attack most of the elements and forms compounds. The atomic number of sodium is11 and atomic weight is 21. Figure 1 shows the atomic structure of sodium atom. Table 1 gives the physical properties of sodium.

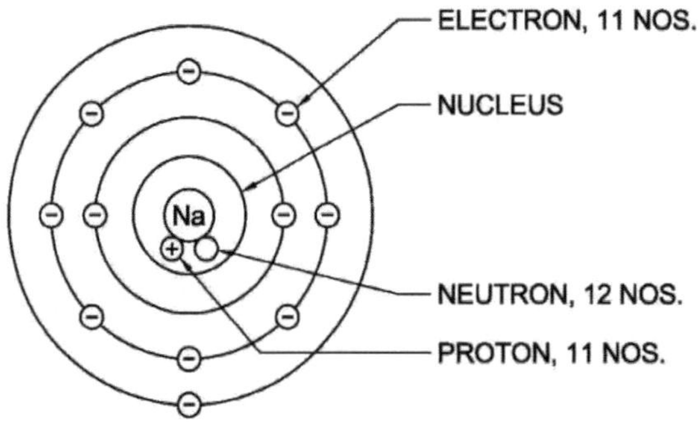

Figure 1. Atomic structure of sodium atom

Table 1. Physical Properties of Sodium

Density of liquid sodium	0.927 g cc^{-1}
Melting point	97.8°C
Boiling point	883 °C
Thermal conductivity	76 W m^{-1} °C^{-1}
Specific heat capacity	1.38 kJ kg^{-1} °C^{-1}
Viscosity	0.670 cps
Electrical conductivity	9.67 x 10^{-8} Ω m

3. Why sodium as a coolant

This highly reactive metal, which is solid at room temperature is widely used in pharmaceutical applications for the manufacture of life saving medicines. The element sodium is a universally accepted fast breeder reactor (FBR) coolant. Fast breeder reactors [1,2] will produce power and it also converts fertile nuclear materials such as U238 and Th232 to fissile nuclear materials or nuclear fuels such as Pu 239 and U233 respectively. By proper configuration of core, with fuel and surrounding blanket with fertile elements, and by maintaining the neutron energy spectrum high (neutrons at high speed) the conversion ratio can be made more than one. When the ratio is more than one the reactor produces more fuel than it consumes and it will breed nuclear fuel. In fast breeder reactors, the abundant amountof fertile elements available in nature can be converted into fissile elements and the shortage of nuclear fuel can be effectively solved. Because of this special advantage,along with other developing countries, India is also developing fast breeder reactortechnology in the second stage of our nuclear power programme.

In fast breeder type of nuclear reactors, heat is generated in the core of the reactor where fission chain reaction takesplace. Core is made of fuel pins of around 6 mm diameter. This fuel pin contains pellets, made from a mixture Uranium oxides/carbides and Plutonium oxides/carbides. They are concealed in stainless steel tubes with ends closed. The heat generated in the fuel pin due to fission chain reaction will be enormous and the power density will be very high. Thepower density is maintained high to achieve desired economic benefit. The temperatureof the fuel pin should be limited by transferring and transporting the heat. For transferring and transporting the heat, the coolant is circulated through the fuel pin bundle which is known as fuel sub assembly. Figure 2 shows schematic of fuel subassembly with sodium flow direction. The economic and safe operation of the fast breeder reactor depends on effective transfer of heat and limiting the hot spot temperature of fuel pins within the design limit. The physical and heat transfer property of the coolant plays a major role here. Any circulating fluid will transfer heat from the fuel pin by convection mode.

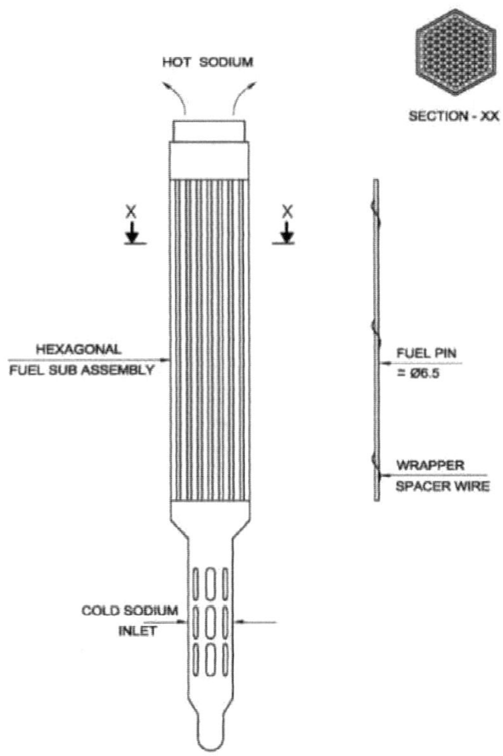

Figure 2. Schematicof fuel subassembly with sodium flow direction

A metal in contact with fuel pin will transfer the heat by conduction mode. If a liquid metal is used as the coolant and circulated through the fuel subassembly heat transfer will be in convection and conduction mode, which will elevate the overall heat transfer coefficient to a very high value compared to water, steam or any gaseous coolant. As the boiling point of sodium is high, 882.8° C, the liquid need not be pressurized like water to avoid boiling at the operating temperature. So the sodium which is circulated through the fuel subassembly of fast breeder reactor can remain at low pressure and at high temperature. The high temperature is a challenge but low pressure is an added advantage for the mechanical design of the system. Low pressure leads to low thickness of structural material used in the sodium system components. Since we are able to operate the sodium system at high temperature the sodium heated steam generators can deliver super-heated steam at high temperature and high pressure. High temperature and high pressuresteam will increase the thermodynamic efficiency of the steam cycle and hence sodium cooled fast breeder reactors have high thermal efficiency. Figure 3 shows the heat transport system of a FBR [1]. In addition the sodium atoms will absorb less number of neutrons in the fast breeder reactor neutron energy spectrum. If at all it absorbs neutronsand formsradioactive isotopes of sodium, they will have lower halflife which enhances radioactive safety of the fast breeder reactor systems.

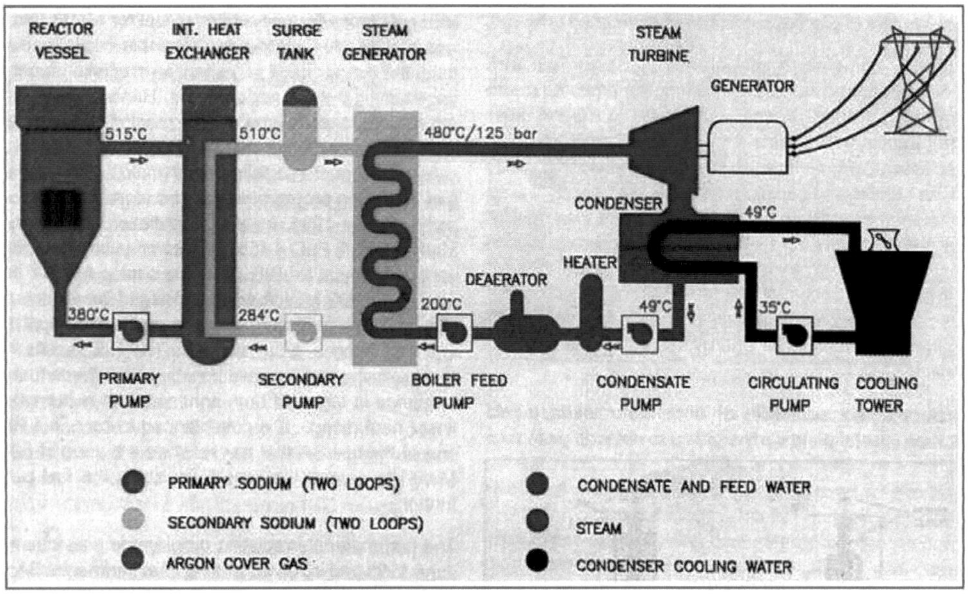

Figure 3. Heat transport system of a FBR.

4. Manufacturing, transportation and storage

Sodium is manufactured using Downs process by electrolysis of sodium chloride [3]. The mixture of sodium chloride and calcium chloride is used as the electrolyte. Graphite is the anode and the Cast iron vessel act as the cathode. Chlorine gas formed at the anode is collected, treated and used for production of bleaching powder. Sodium element deposited at the cast iron vessel float on the surface of electrolyte and collect in a container located at one side of the cell. Cell operates at around 8 V DC and the current will be of the order of thousands of amperes and it depends on cell capacity. Operating temperature of cell is around 600°C. Addition of Calcium chloride to sodium chloride will reduce the melting point below 600°C as it become a eutectic alloy. A metallic wire mesh diaphragm is inserted in between the anode and cathode which is normally separated by around 25 mm gap prevents recombination reaction which will lead to better cell efficiency. Pure sodium chloride is added periodically to the cell as production of sodium progresses. The sodium production is energy intensive and the minimum specific energy consumption reported all over the world is around 15 kWh/kg. Figure 4 shows the schematic of a Downs cell.

For pharmaceutical applications purity of sodium is not a limitation. For this purpose oil coated sodium bricks are stored in steel drums and are closed with rubber gaskets. These sodium drums are normally transported in trucks, through road under water tight double tarpaulin cover. For nuclear application filtered sodium is filled in sealed steel tankers, frozen and an inert gas atmosphere is maintained in the gas space[4]. These tankers with frozen sodium are transported by road, rail or ship to the destination. At the reactor site, sodium is melted and transferred to conditioned storage tanks through fine filters by differential pressure [5]. Storage tanks with sodium will be under inert atmosphere and under positive pressure to avoid any ingress of air. Sodium is frozen and stored if filling of reactors system is delayed. Cover gas pressure is monitored during storage to ensure system integrity. The sodium inventory in the fast breeder test reactor (FBTR) under operation at IGCAR Kalpakkam is around 150 tones. For the 500 MWe capacity Prototype Fast Breeder Reactor under construction at Kalpakkam the sodium inventory is around 1700 tones.

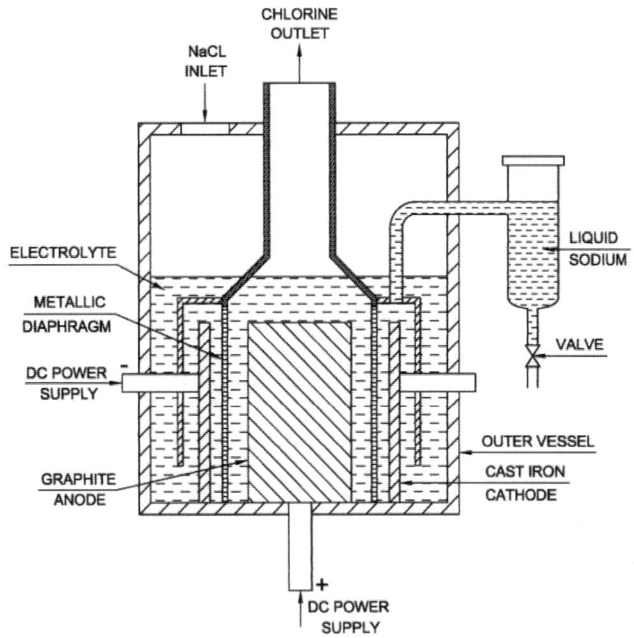

Figure 4. Schematic of a downs cell

5. Challenges in sodium technology

Since sodium is a highly reactive metal and is solid at room temperature most of the conventional systems used in high temperature process industry to handle fluids are not adoptable for sodium systems [6]. Selected materials compatible with high temperature liquid sodium only can be used for sodium system construction. Special instrumentation system is required as most of the conventional sensors are not readily usable. Preheating system is needed as sodium is solid at room temperature. Leak detection and fire fighting is an added requirement. Sodium purity has to be maintained at around 99.95 % to limit blockage of flow passages and structural material corrosion. Special purification methods and purity monitoring techniques have to be adopted to maintain sodium purity. Inert cover gas system is required to provide inert atmosphere over free space of sodium tanks, vessels and components. Operation and maintenance techniques are unique for liquid sodium systems. Economic and safe realization of all these special requirements are the major challenges in sodium technology.

6. Sodium system design features

Sodium is solid at room temperature but under operating conditions sodium is brought to liquid state. The material used for construction of sodium system is stainless steel. Because of the favorable mechanical properties at high temperature SS 316 LN grade stainless steel is recommended for use. The specifications are based on ASTM standards. The mechanical design, quality assurance and quality control of sodium systems are carried out as per ASME section VIII division I and associated international codes.

6.1 Special features

Sodium system should not have any flanged joints to avoid leakage. Hence all the joints in the piping and joints in the piping to vessels of sodium systems are of welded construction. Nitrogen or argon is used as the cover gas for sodium system and all free space of vessel and tanks is filled with cover gas. Separate cover gas system is a part of the auxiliary systems for any sodium system. Purified cover gas at desired pressure will be admitted to the free space of vessels and tanks. Provision for venting the gas and evacuating the vessels and tanks are also provided as part of the cover gas system. On any emergency condition the sodium in the system has to be drained or dumped into a storage tank. To ensure draining of sodium in a passive way, gravity dumping is resorted in sodium systems. Hence the storage tank or the dump tank will be positioned at lower most elevation. In order to avoid trapping of sodium in the pipe lines and to accelerate the flow of sodium to the storage tank a slope of around 3 % is provided for pipelines. Conventional valves cannot be used in sodium systems as the gland packing will be a source of leak. Bellows sealed valves made up of SS316 LN material is used in sodium systems. Sodium systems are allowed to operate upto a temperature of 600°C. The piping system should have adequate number of bends to have structural flexibility to accommodate thermal expansion during operation. Vapor traps are required in the cover gas line close to the interface with sodium system to avoid transport of sodium vapor to gas line and solidification. Solidification of gas line will lead to lack of gas communication which will loose the dumpability of the system.

6.2 Sodium instrumentation

Sodium is a very good conductor of electricity and this property is successfully used for developing sensors for sodium systems. Special type of sensors are required for measuring sodium level in tanks and vessels. Resistance type and mutual inductance type level sensors are used or continuous and discontinuous level measurements [7]. Alnico based permanent magnet type electromagnetic flow meters are used for sodium flow measurement [8]. Figure 5 shows the schematic.

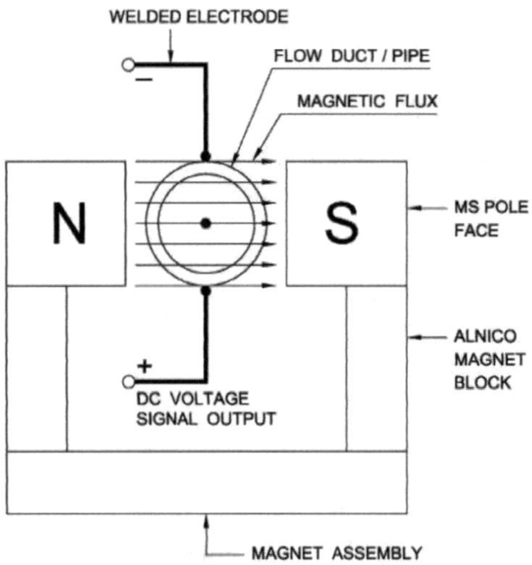

Figure 5. Alnico based permanent magnet flow meters

6.3 Sodium pumps

As liquid sodium is a good electrical conductor, for pumping sodium in the system electromagnetic pumps can be used [9]. Because of its low efficiency and associated design difficulties for large sodium systems centrifugal mechanical sodium pumps are used. Figure 6 shows schematic of DC Electromagnetic pump.

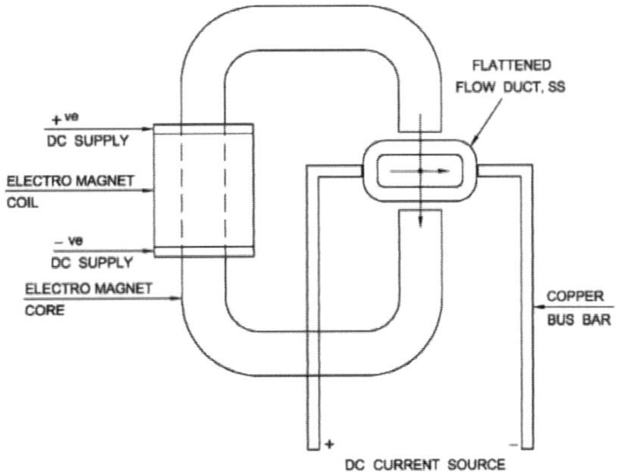

Figure 6. Schematic of DC electromagnetic pump

Centrifugal sodium pumps are vertical type with shaft sealing only at one end that too in the cover gas. These pumps have hydrostatic bearing at bottom which is submerged in sodium and tapper roller bearings at the top with oil cooled mechanical seals. Figure 7 shows schematic of a centrifugal sodium pump.

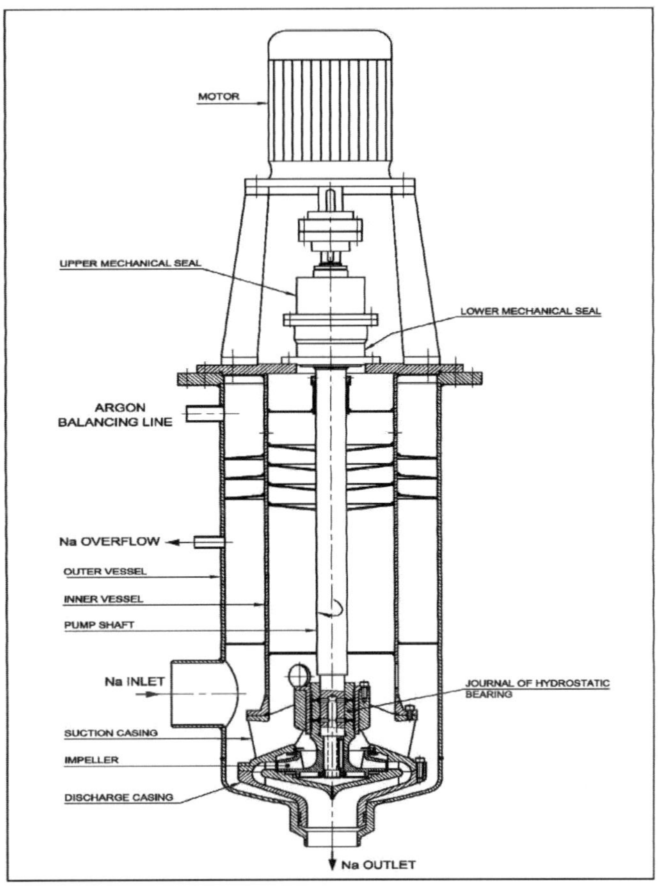

Figure 7. Schematic of a centrifugal sodium pump.

6.4 Heating System
Since sodium is solid at room temperature solidification of sodium in a system has to be avoided. The system has to be maintained well above the melting point of sodium which is 98.2 °C. This necessitated a preheating system for the sodium system. All pipe lines, tanks and vessels are provided with electrical surface heaters and thermal insulation. The preheating system allows the sodium system to be heated up around 200°C before filling sodium and also helps to maintain at this temperature after filling with sodium. The temperature of the piping and components are measured by means of Chromel-Alumel thermocouples and the heaters are controlled based on temperature readings. Mineral wool thermal insulation of sufficient density and thickness would be provided over the surface to limit heat loss also and maintain the surface skin temperature below 70°C. In order to raise the temperature of sodium in experimental facilities to desired value, immersion type heaters in heater vessels are provided in sodium systems. These heaters can be used for raising sodium temperature in the system at upto 600 °C. In nuclear reactor heat generated in the reactor core will be used to raise the temperature to the rated conditions.

7. Sodium system operation techniques
Safe operation of sodium system is also a challenge. Minor errors in design, construction, quality control, commissioning, operation and an maintenance can lead to major incidents of sodium leak, fire and damage to material, equipment, and injury to human beings.

7.1 Safe operation

Major measures to be adopted for safe operation of sodium system are

- Integrity of sodium system shall be maintained throughout the life of the system and periodically tested at desired and specified conditions.
- Sodium in the system shall be purified and purity of sodium in the system should be monitored periodically.
- Temperature of all parts of the system shall be maintained above 200°C during operation. This will avoid freezing of sodium at any part of the system. Freezing of sodium in dump lines should be avoided at any cost as it will lead to loss of system dumpablity.
- Availability of normal power supply and standby power supply shall be ensured
- Remote dumping facility which should be operable even after complete power failure, is needed. Dump valve actuation shall be failsafe.
- Purified Argon cover gas shall be used to avoid system contamination
- Automatic temperature controller and stand by controllers are provided to limit system temperature below the design limits
- Sensitive local and global sodium leak detection system is required. Healthiness of leak detection system should be checked periodically
- Dumping time should as per design intent and shall be checked periodically by actual dumping operation.
- Availability of adequate number of quality fire extinguishers and its operability should be checked
- In-service inspection on the structural material of components and piping at high stressed location shall be carried out at specific intervals.

7.2 Sodium purification
Reactor grade sodium is 99.95 % pure[10]. Impurities in sodium are oxygen, Hydrogen, Carbon, Calcium and Magnesium. Oxygen level should be maintained below 10 ppm. Sodium is purified by two stages of filtration and then by cold trapping. Molten sodium is coarse filtered using wire mesh filter at 125 °C. Bulk sodium oxide and hydroxide suspended over liquid sodium is removed by this process. Coarse filtered sodium is aged at 300°C to convert metallic calcium present in sodium to calcium oxide. Fine filtering of sodium using sintered stainless steel filter elements of 14 microns

pores size, will remove calcium oxide and precipitated sodium oxide. Further purification of sodium is carried out by a process called cold trapping. In the cold trap liquid sodium is cooled to a temperature below the saturation value of impurities and the dissolved impurities allowed to precipitate and deposit on the surface of wire mesh present in the cold trap. Figure 8 shows schematic of a cold trap.

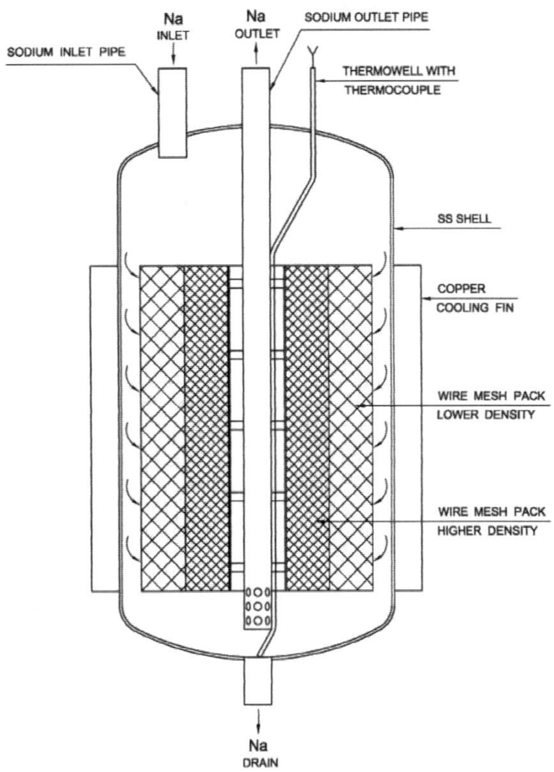

Figure 8. Schematic of cold trap

Impurities having temperature dependent solubility such as sodium oxide and sodium hydride only are removed by this method. Purity level of sodium in systems are monitored by a device by name plugging indicator which measures the saturation temperature of dissolved impurities indirectly.

7.3 Sodium leak detection

Leakage of sodium from the system is avoided by selecting suitable construction materials, adopting all welded construction and stringent quality control measures. The sodium systems are operated in round the clock shift under enforced technical specifications by deploying trained and qualified manpower. Limiting conditions of operating parameters are defined and are not allowed to violate. However minute probability of barrier failure exists and if it fails, liquid sodium may leak to atmosphere. Detection of sodium leak at the incipient stage by local and global leak detection is the first step of defense followed. Sodium leaks from vessels, tanks and pipes are detected at the incipient stage by wire type leak detectors. These detectors have capability detect leakage of 200 grams of sodium leak within 2 hours. The detection capability of sodium leak detecting system were experimentally demonstrated. Nickel or stainless steel wires are positioned over the vessel tanks and pipes at regular intervals and they are electrically insulated from the system components by means of insulating ceramic beads. In case of a sodium leak, leaked sodium being a good electrical conductor, bridge the gap between the nickel/SS wire and electrical circuit is made to close. This signal is sensed

IOP Conf. Series: Materials Science and Engineering 1045 (2021) 012013 doi:10.1088/1757-899X/1045/1/012013

by proper electronic circuits and alarm indication appears in control room and alert the operator[11,12]. Figure 9 shows sodium leak detectors fixed on pipe line. Sodium ionization detector based system is used for global leak detection.

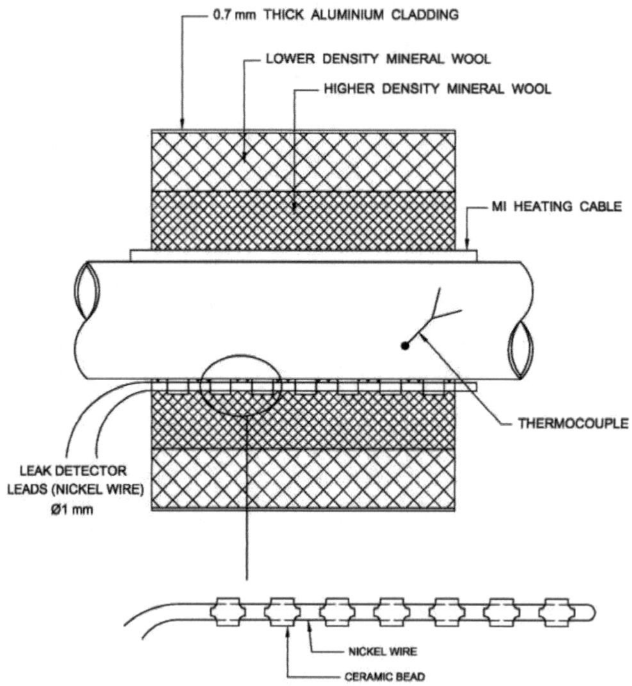

Figure 9.Sodium leak detector on pipe line

7.4 Sodium fire fighting

Mitigation measures envisaged are collection of leaked sodium in specially designed leak collection trays, which has provision to cut off air supply to the collected sodium and provision for sodium fire fighting. Figure 10 shows schematic sodium of leak collection tray.

Sodium can leak and form a spray fire or pool fire depending on the process condition of the leaking system and ambient conditions. Sodium fire fighting is complicated and risky because of the emission of large quantity of smoke containing of sodium oxide aerosol, which will completely block the vision. On inhalation of sodium oxide aerosols, respiratory system of human beings will be badly affected. Blanketing is the principle adopted for sodium fire fighting. Dry chemical powder (DCP)made of sodium bicarbonate is sprayed over the sodium pool under fire and oxygen supply is cut off. The low density moisture free sodium bicarbonate powder will float on the burning sodium surface and prevent the fire. Fire fighting systems with automatic dispersal of DCP were used in large sodium facilities which enables remote operation.

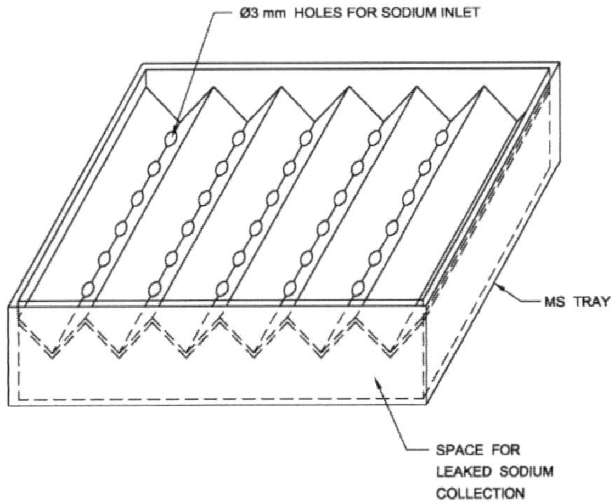

Figure 10. Schematic sodium of leak collection tray

7.5 Sodium disposal

Sodium is not consumed in a nuclear reactor or in an experimental facility. It is only circulated in the primary and secondary circuits of the heat transport system of a reactor. However relatively small quantity waste solid sodium is formed during modifications and maintenance of sodium system. These waste sodium may have lot of impurities such as oxides and hydroxides and is not worth to purify and recharge into the system. This waste sodium has to be converted into sodium carbonate or chloride. Disposal of solid sodium waste is by allowing it to burn in air by applying finewater spray. The sodium hydroxide formed is collected in tanks and is neutralized by acidic solutions. The pH of discharge to the environment is kept within the allowable limits. In order to avoid release of sodium oxide toambient, wastesodium is burned in chemical reactors at controlled air atmosphere and the smoke emitted is admitted to water bubblers for conversion to sodium hydroxide solution.

8. Technology spinoff

The possibility of application of liquid sodium as a heat transfer medium and as a reducing agent in other process industries may be of worth consideration[13]. The equipments, components and systems developed for liquid sodium may find application in other process industries with suitable modifications. Similarly many modern smart sensors, devices and technologies developed elsewhere for high temperature process systems may find application in sodium systems with appropriate modifications. IGCAR has not given priority in this area as the centre was focusing to meet the technology requirements of PFBR and future FBRs. Removal of Sulphur content from petroleum isessential for limiting SO_2 emission from automobiles. Sodium may be an effective reducing agent for removal of residual sulphurfrom petroleum products. The technology is yet to be developed. There are many more areas where technology development took place in IGCAR and results are published and are available in open domain.

9. Conclusion

Sodium which is stored in small bottles in laboratories have wide application in fast breeder type of nuclear reactor as coolant. Thousands of tons of liquid sodium is circulated through primary and secondary heat transport system up to an operating temperature of 550 °C. Sodium technology is branch of fast breeder reactor technology and is around 60 years old. Sodium technology is matured enough to meet the requirement ofa fast breeder reactor andaround 20 reactors are commissioned and operated with sodium as coolant all over the world. However thereare gap areas in sodium technology which has to be addressed. Performance enhancement, increased safety and economic measures are to be taken, to make energy from fast breeder reactor economically competitive and safe. In addition the application of this excellent heat transfer medium and reducing agent in other process industry also has to be considered. Extension of technologies developed for sodium to other electrically conductive high temperature fluids in industrial and process application is also possible. Hence, though this hazardous element poses many challenges in handling, it gives immense opportunity for scientific and academic community for application oriented research.

10. References

[1] Srinivasan, G., Suresh Kumar, K.V., Rajendran, B., and Ramalingam, P. V., 2006, *Nuclear Engineering and Design*, Volume 236.

[2] Chetal,S.C.,Balasubramaniyan,V.,Chellapandi,P.,Mohanakrishnan,P.,Puthiyavinayagam,P., Pillai, C.P. and Pillai, C.S, 2006, *Nuclear Engineering and Design, Volume* 236.

[3] Down's Cell, Available : https://prezi.com/ 0cz8uzrqe5jd/downs-cell/.(accessed on 10 August 2020)

[4] MetauxSpeciuax product catalog, Available: https://www. metauxspeciaux.fr/sellers-services/, (accessed on 10 August 2020)

[5] Rajan, K. K., Rajan, M. and Kale, R. D., 1995, *Indian Chemical Engineering Congress*, Kalpakkam.

[6] Kale, R.D., and Rajan. M., 2004, *Current Science*, Vol. 86**,** No. 5, p 668.

[7] Sylvia, J.I., Vijayamohana Rao, P., Babu, B., and Madhusoodanan, K., 2013, *Nuclear Engineering and Design*, Volume 262, p 219.

[8] Rajan, K. K., 2015 Ph D Thesis,Homi Baba, National Institute, Mumbai.

[9] Prashant Sharma, Sivakumar, L.S., Rajendra Prasad, R.,Saxena, D.K. Suresh Kumar, V.A., Nashine, B.K and Kalyanasundaram P, *2010, Energy Procedia Volume* ***7,****Asian Nuclear Prospects,* p 622.

[10] Rajan, M., Rajan K. K., Gurumoorthy, K., Prabhakar, R., Kale, R. D., Rajamani, P and Subramanian, R., 1985, *Indian Chemical Engineering Congress*, Kolkata.

[11] Sylvia, J I, Vijayamohana Rao P., Babu, B., Madhusoodanan, K., and Rajan, K. K, 2012, *Nuclear Engineering and Design*, Volume 249, p 419.

[12] Vijayakumar,G., Chandramouli, S., Nashine,B.K., Selvaraj, P. and Rajan, K. K., 2017, *Annals of Nuclear Energy*, Volume 103, p 326.

[13] Rajan, M. and Kale, R. D., 2000, *IndianChemical*Rajan, M. and Kale, R. D., 2000, *IndianChemical Engineering Congress*, Kolkata, p 18.

Acknowledgement

Entire sodium technology development work in India was carried out in different groups of IGCAR Kalpakkam especially in Fast Reactor Technology Group. Author express his sincere gratitude and acknowledges the help rendered by IGCAR officials in sharing the resources for carrying out this review.

Performance evaluation of a 200 kWp grid tied solar power plant

K K Rajan and B Aruna

Viswajyothi College of Engineering and Technology, Vazhakulam, Muvattupuzha, Kerala, India, PIN-686670

kkrajan56@gmail.com

Abstract. Grid tied solar power plants are found to be very adaptable for academic institutions as the load profile almost matches with the generation profile. For the academic institution under consideration, optimum capacity of the solar power plant is arrived at as 250 kWp, based on studies carried out. Since the institution was already having a plant of 50 kWp, it is decided to augment it with another solar plant of 200 kW capacity. The normal power supply available from the utility was not very reliable, which leads to frequent power failures. Hence a 200 kVA capacity Diesel Generator (DG) was provided for standby power requirement. An automatic power factor control (APFC) unit is also provided to improve the power factor of the system. During commissioning, the performance of the 200 kWp solar power plant was assessed based on the measurement of various parameters at site and were compared with the standard specifications. Integrated operation of the system was carried out and few operating difficulties were faced. In this paper, performance evaluation test results of 200 kWp solar plant, experience in integrated operation, the difficulties faced and the measures adopted to rectify them are discussed.

1. Introduction

In India, solar power generation is growing at an accelerated pace. Solar power is green and clean power. Use of grid tied solar power plants are gaining acceptance and are widely used in domestic and industrial applications[1]. Grid tied solar power plants are found to be very suitable for academic institutions as the load profile is generally in coherence with the generation profile. Studies have been carried out on the basis of the load profile of the academic institution under consideration and an optimum capacity of the solar power plant for the institution is arrived at as 250 kWp. Since the institution was already having a plant of 50 kW it was decided to augment it with another plant of 200 kW capacity. As per the rules and regulation of the utility providing power supply to institution, the maximum capacity of solar plant permitted is based on the feeder capacity and the contract demand. The contract demand of the Institution was 280 kW and feeder capacity was much higher. The Institution is fed with single circuit, three phase feeder by the utility without any redundant power supply provision. Hence, the power supply was not very reliable, leading to frequent power failure. Therefore, a 200 kVA capacity Diesel Generator (DG) was installed for standby power. The operation of the solar plant along with Diesel Generator set was considered in the design. Automatic power factor control (APFC) unit, using capacitor banks have been connected in order to improve the power factor of the system and are automatically switched on/off based on variation in power factor. During commissioning, the performance of the 200 kWp solar power plant was assessed based on measurements of various parameters at site and were compared with specifications. Integrated operation of the system was carried out and few operating difficulties were faced. The difficulties were overcome by adopting appropriate measures. This paper describes the methodology adopted for plant capacity selection, configuration details of the plant, specifications of the plant components, commissioning experience, performance evaluation, experience in integrated operation, the difficulties faced and the measures adopted to tackle them.

2. Plant capacity selection

The power demand of the institution varies with respect to time on each day and also with respect to season in each year. The power demand on a working day and a holiday will be entirely different. To determine the optimum capacity of the solar plant, the power demand on a working day in the peak summer season is taken as the reference. Figure 1 shows power demand curve of the institution. Area under the power demand curve is 3705 kWh, which gives the energy demand of the day. The power generation curve of a PV solar power plant for a day will be parabolic in nature and area under the curve would be the total energy generated. To make the Institution self-sufficient in electrical energy, the area under power demand curve should be made equal to the power generation curve assumed. To achieve that, plant capacity should be very much higher than the peak demand of 310 kW, which is not economical. A plant capacity equal to peak demand is also found to be uneconomical. Average demand of the entire day is only 154 kW. The average demand during day time (6 AM to 6 PM), when solar plant generation is taking place is 213 kW. The optimum capacity needed for solar power plant is taken as the mean of day time demand and peak demand which is 261.5 kW. A 50 kW plant was already installed and functioning in the institution. Hence it was decided to go for a 200 kWp capacity solar plant (211.5 kWp approximated to 200 kWp). Considering economic factors and as the load profile is generally matches with the generation profile, grid tied system is selected.

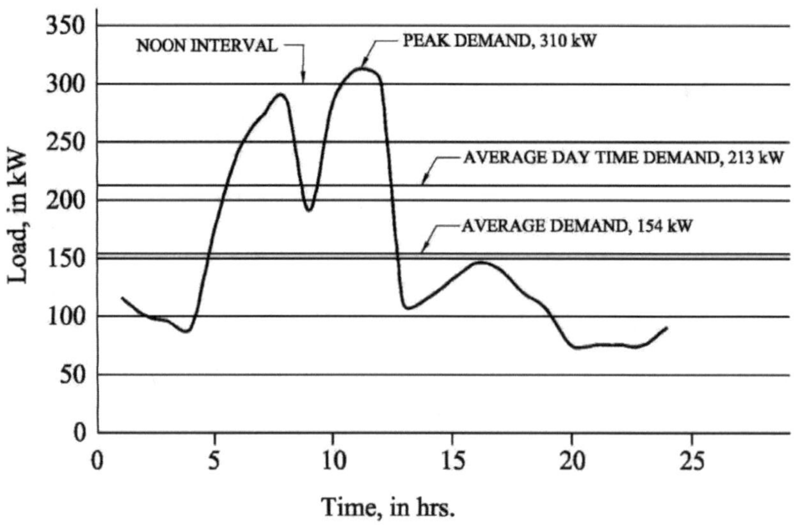

Figure 1. Power demand curve of the institution

3. Solar plant configuration

The location for the 200 kWp solar plant was selected at the roof top of the D block of the institution which was shadow free and the highest elevation in the campus The distance from the plant to the substation is around 250 m. In order to ensure availability of the plant and reduce the DC cabling it was decided to go for two independent 100 kWp units. It was also decided to go for inverters of 100 kW capacity and solar panels of the same rating. Based on market survey, the state-of-art poly crystalline solar modules of 350Wp capacity was selected. Sixteen modules are connected in series to form a string. Three strings are connected in parallel to a Maximum Power Point Tracker (MPPT) of the inverter [2]. So there are 48 modules connected to an MPPT. An inverter has 6 MPPTs and total number of modules connected to the inverter is 288. There are two inverters and the total number of modules are 576. Modules are wired in series using 4 sq.mm copper cable and it forms strings. Inverters have DC and AC wiring box. Two numbers 250 A DC isolator and 48 numbers of 15 A DC fuse is provided in the DC input side of inverter. The inverters are connected to the LT main switch board (MSB) in the substation using two UG cables in parallel. Output from both the plant is joined at the substation and connected to MSB bus through switch gears and energy meter. The capacitor banks for power factor improvement are connected to the MSB through SFU and switching contactors. The panel receives AC power from the utility at 415 V, 50 Hz. In addition, a 200 kVA capacity DG is linked to MSB through auto main failure panel. Figure 2 shows the schematic layout of the integrated power distribution system of the institution.

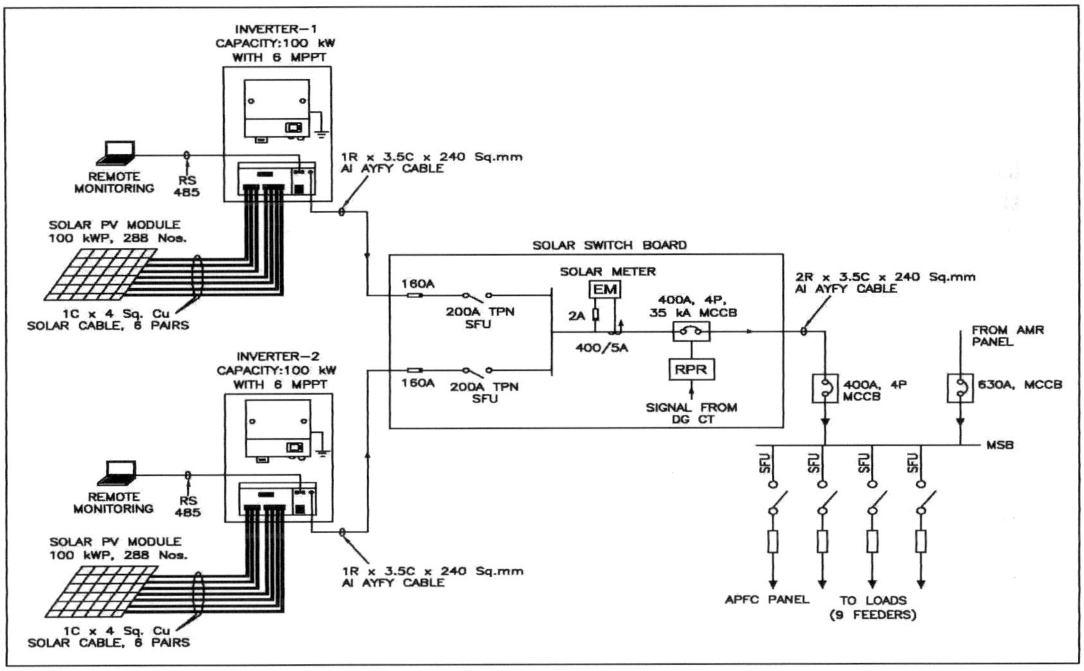

Figure 2. Layout of the integrated power distribution system of the institution

4. Specifications of plant components

The PV modules, Inverters and the interconnecting cables are the main components in the power generation and transmission chain.

4.1 PV modules

Polycrystalline cell is used to form the modules. 72 cells are used in a module. PV cells have generic characteristics of diminishing efficiency with increasing temperature. Specifications of the module are given below. The nominal cell temperature and the temperature coefficient at rated power are as specified below.

Maximum power (P_{max})	:	350 W
Maximum power Voltage (V_{mp})	:	38.1 V
Open circuit voltage V_{oc}	:	46.39 V
Short circuit current I_{sc}	:	9.79 A
Module efficiency	:	18.00

Nominal operating temperature	:	47 ± 2°C
Temperature coefficient of V_{oc}	:	-0.35 %/°C
Temperature coefficient of I_s	:	+0.05 %/°C
Temperature coefficient of Power	:	- 0.45 %/°C
Module surface area	:	19404 cm^2
Est. shunt resistance	:	69.2447 ohm
Est. series resistance	:	0.1971 ohm

4.2 Inverters

Two inverters of 100 kW capacity delivering power at unity power factor was used for converting DC to 3 phase AC at 415 V and 50 Hz [3,4]. The specifications of inverter is given below.

4.2.1 Input side.

Absolute maximum DC input voltage ($V_{max,abs}$)	:	1000 V
Start-up DC input voltage (V_{start})	:	420 V (400…500 V)
Operating DC input voltage range ($V_{dcmin}...V_{dcmax}$)	:	360…1000 V
Rated DC input voltage (V_{dcr})	:	620 V
Rated DC input power (P_{dcr})	:	102 000 W
Number of independent MPPT	:	6
MPPT input DC voltage range at ($V_{MPPTmin}...V_{MPPTma}$x) Pacr	:	480…850V
Maximum DC input power for each MPPT ($P_{MPPT,max}$)	:	17500 W
Maximum DC input current for each MPPT (I_{dcmax})	:	36 A
Maximum input short circuit current (I_{scmax}) for each MPPT	:	50 A
Maximum Number of DC input pairs for each MPPT	:	4

4.2.2 Output side.

AC Grid connection type three phase	:	4W+PE
Rated AC power (P_{acr} @cosφ=1)	:	100 000 W
Maximum AC output power (P_{acmax} @cosφ=1)	:	100 000 W
Maximum apparent power (S_{max})	:	100 000 VA
Rated AC grid voltage ($V_{ac,r}$)	:	400 V
AC voltage range	:	320...480 V
Maximum AC output current ($I_{ac,max}$)	:	145 A
Rated output frequency (f_r)	:	50 Hz
Output frequency range ($f_{min}...f_{max}$)	:	45...55 Hz
Total current harmonic distortion	:	< 3%
Nominal power factor and adjustable range	:	> 0.995

Inverter has RS 485 communication port and the plant data from the inverter is linked to PC and the campus network for remote monitoring.

4.3 Under Ground Cable

Two runs of 240 sq.mm, three and a half core, Aluminium conductor, PVC insulated, steel strip armoured and PVC sheathed cable (AYFY) is used for power transmission from plant to the substation. The length of the cable is around 250 m and one end is connected to a SFU installed in the switch board near inverter. From inverter copper cable is laid and connected to the switch board. Other end of the cable is connected to another SFU at the substation.

5. Installation and commissioning

Solar Modules are installed at an angle of 11 degree in the east-west direction. Installation of module structure had been done based on analysis carried out with geographical coordinates as input and ensured that there will not be any shade on any of the panels between 6 AM to 6 PM on any day in the year. Galvanized structural steel is used for support structural works. Columns of the support structure is made of 120 x 50 x 2.5 mm size, lipped C channels and are welded to the base plate of 120 x 120 x 6 mm size. These base plates are anchored to the roof slab. Concrete blocks are provided surrounding the column over the base plate. Rafters used for support structure is also lipped C channels of size 100 x 40 x 2 mm size. Purlin used for fixing the module is 80 x 40 x 2 mm size lipped C channels. Entire support structure with module is analyzed for a wind speed of 150 km/h and found fit. Pathway is provided in between arrays of modules to ensure approach for water spray and cleaning of the modules. Entire Structure and solar modules are electrically earthed at different locations to two separate earth pits using GI flats of 3 mm thickness and 25 mm width. Switch gears and surge protection devices (SPD) are provided wherever needed as per standards. Lightning arrestors provided for the building were adequate for the solar panel too.

Cables from the plant are laid underground to the substation. The cables are terminated to SFU at substation using suitable legs and glands. Energy meter is provided to read the amount of energy generated. In the common output line, a moulded case circuit breaker (MCCB) is provided. Revere Power Relay (RPR) installed in DG power line circuit is linked to the MCCB. On reverse power to DG, the MCCB will trip and isolate solar plant. MCCB output is connected to MSB where loads are connected. Parallel to MSB six capacitor banks of total capacity 125 kVAR is connected in parallel. Capacity of banks ranges from 25 kVAR to 15 kVAR. These capacitor banks are automatically switched on/off based on the value of power factor which is monitored real time. After completing the installation, inspection and after receiving clearance from the state electrical inspectorate, the plant was commissioned by switching on the inverters. The voltage and current at different points of the plant were monitored to ensure that the functioning is as per design. Energy meter reading was observed and confirmed the feeding of power to the MSB panel. The remote monitoring system was activated and plant parameters could be obtained through mobile phones.

6. Plant performance evaluation

In order to evaluate the performance of the plant various plant parameters were measured. Measurements were carried out at two conditions of irradiance of 578 W/m^2 and 1088 W/m^2. The first one corresponding to evening and mornings and the second one corresponds to peak power density during noon period. Solar 100 Irradiance Meter, Fluke 438-II Power Quality Analyzer and Fluke 62 MAX IR thermometer are used for the measurements. The accuracy of the instruments used were better than 1 %.

6.1 PV module efficiency evaluation

Measurements are carried out for an array of 48 modules which are connected to one MPPT of the inverter. Details of measurement carried out are given in Table 1.

Table 1. PV module efficiency evaluation

Parameter	Measured Value	Measured Value
Solar Irradiance(W/m^2)	578	1088
PV Module Area (m^2)	1.9404	1.9404
Total No of PV Modules	48	48
Total PV Array Area (m^2)	93.1392	93.1392
I_{DC} (A)	17.5	24.33
V_{DC} (V	520	482
PV Input Power(W)	53834.46	101335.45
PV Output Power(W)	9100	11727.06
Efficiency (%)	16.90	11.57
Module temperature	62.2	71.2
Extrapolated power of the plant kW(576 modules)	109.2	140.7

Expected power of plant as specs	-	159.52
Percentage reduction	-	13.4

From the observations, the actual value of power per module at 1088 W/m^2 irradiance is estimated as 244 W. The expected value of power per module as per specification is 277 W. Expected efficiency is 13.12 % against an actual efficiency of 11.57%. This is due to high value of temperature coefficient. The actual temperature coefficient is estimated as - 0.66%/°C against the specification of - 0.45%/°C.

6.2 Inverter Efficiency

Efficiency of the inverter 1 is computed here for two different loading levels as given in Table 2. Inverter 2 is identical to inverter 1.

Table 2. Inverter efficiency evaluation

Parameter	Measured Value	Measured Value
Solar Irradiance(W/m^2)	578	1088
I_{DC} (A)	17.5	24.33
V_{DC}(V)	520	482
Input Power To PCU (W)	54506.25	69629.41
V_{RN} (V)	241.02	245.92
V_{YN} (V)	239.02	243.68
V_{BN} (V)	240.26	243.26
I_R (A)	66.5	93.8
I_Y (A)	66.8	93.9
I_B (A)	67	94.3
P_{Total} (W)	47940	69629.41
Power Factor	0.99	1
Efficiency (%)	87.95	99

6.3 Overall efficiency

Overall efficiency of one string of 48 modules at two different irradiance conditions have been computed and presented in Table 3. Performances of the other strings are identical. Overall efficiency is also equal to efficiency of solar module x efficiency of inverter.

Table 3. Overall efficiency evaluation

Parameter	Measured Value	Measured Value
Solar Irradiance (W/m^2)	**578**	1088
Module Temp (^{0}C)	62.2	71.2
PV Input Power (W)	319642.092	601679.232
Active Power L_{RN}(W)	15900	23020
Active Power L_{YN}(W)	15880	22820
Active Power L_{BN}(W)	15960	22880
Total (W)	47740	68720
Overall Efficiency (%)	14.93	11.42
Based on module and inverter efficiency (%)	0.169 x 0.8785 14.84	0.1157 x 0.99 11.45

6.4 Harmonics measurements

Solar inverter output should be harmonics free and should not cause increase in grid harmonic content when tied to local grid. Measurements are made to assess the value of harmonic injection by the inverter. Such measurements are complex in grid tied systems as a snap shot measurement at the point of common coupling (PCC) will not confirm whether the harmonics injection is from the inverter, from the local load or the one present in the supply grid. The harmonic content in the grid voltage without any load connected was measured first, using the power analyzer and was found that it is 2.3%. Then the harmonic content with load of the institution connected to the grid was measured and found that it was 2.7% and 8.6% in voltage and current respectively. Subsequently the capacitor banks were included in the circuit and it measured the harmonic content as 3.16% and 10.86 % in voltage and current respectively. When the inverter is tied with grid, with all the system connected, the harmonic content in the voltage and current was 3.79 and 4.64 respectively. These measurements indicated that the rise in voltage harmonics injected from the inverter is negligibly small and the overall value is within the limits. Current harmonics is also found low with the inverter connected.

7. Experience on integrated operation

Integrated operation of the system was generally smooth and trouble-free apart from the few minor difficulties faced. The peak value of power generated was 140 kW. Energy upto a value of 906 kWh was generated on a sunny day, 18th of March 2020. The power generation profile of the plant on that day is given in Figure 3. Energy export is taking place on holidays when the load in the institution is low. All the operating data is stored in the inverter and it can be used for analysis and performance assessment. The plant is in operation since October 2019 and delivering power.

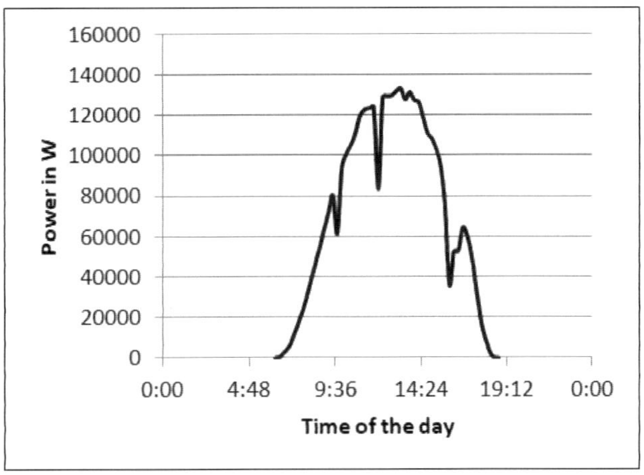

Figure 3. The Power Generation Profile of the Plant on 18th March 2020

8. Difficulties faced and remedial measures adopted

The Inverter is designed to deliver power at unity power factor always. During integrated operation of the system the power factor monitored at the substation feeder was oscillating with change in inverter output. This oscillation was high at low loads. The response time of Automatic Power Factor Control (APFC) unit and contactor based capacitor bank switching system was much higher than the time taken for output change of inverter due to irradiance variation. This was causing a delayed APFC correction and resultant power factor oscillation. The phenomenon is explained with the phasor diagram in Figure 4. The problem was analysed in detail and concluded that the power factor should be monitored at the load level before the inverter is linked to the distribution bus[5,6]. The issue can be resolved by connecting the inverter at the transformer end of the main distribution bus in the substation and the CT of APFC positioned on the same bus between the inverter node and the load feeder node, as shown in Figure 5.

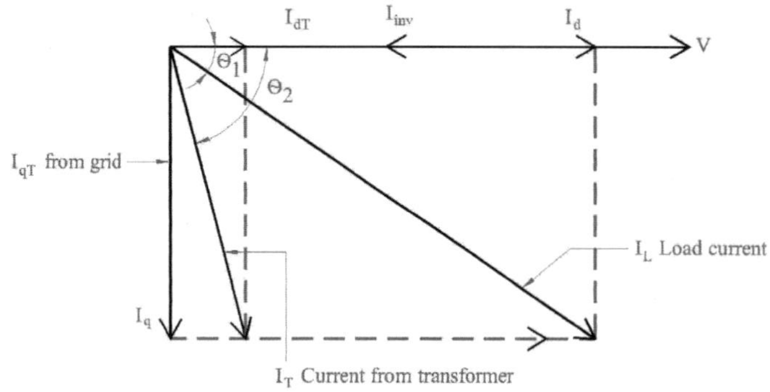

Figure 4. Phasor diagram of voltage and currents in the present scenario

V: grid voltage; I_L: load current; I_d: active component of I_L; I_q: reactive component of I_L; I_{dT} & I_{qT}: current components drawn from grid in the presence of PV generation; I_T: Line current drawn from grid in the presence of PV generation (it has a high power factor angle). θ_1 : Power factor angle without solar generation. θ_2 : Power factor angle with solar generation.

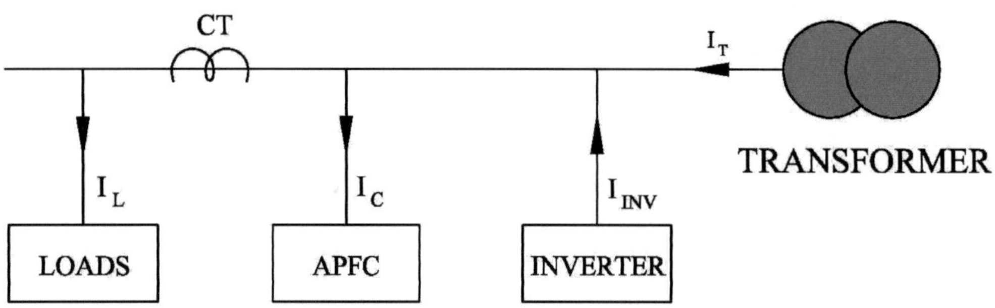

Figure 5. Required connection at substation

The operation of plant along with DG supply during grid failure was giving problems. When the grid fails, the solar plant will trip automatically. As per the time settings, within 45 seconds the DG will start automatically and will be connected to the MSB and feed the loads. After 3 minutes of steady operation of DG, the inverter should switch on and solar plant should be connected to the MSB. During this condition, reverse power flow if any, from the solar plant to the alternator of DG set, the RPR will send signal to the MCCB, trip the solar power plant and isolate it from MSB. The generator will continue to operate and would take the complete load alone. However in most of the cases, the solar plant was not getting started on DG power. This was analysed and found that the frequency fluctuations in the DG supply is not allowing synchronisation of inverters to MSB and continue the operation of inverters with DG supply. This issue was overcome by changing the allowable frequency band of the inverter in tune with the DG power frequency and by implementing better control on DG engine speed. The operation of the system was smooth when the load is higher than the power generated by the solar plant. At low load conditions, the solar plant was getting tripped on RPR signal and it was as per design. The MCCB was not of auto reset type and because of that even after grid power resumes, the solar plant was not getting connected to the substation bus. Sharing of load between DG and solar plant was desirable at low loads. This was causing concern as operator intervention is required to resume solar plant operation and a delay in it was leading to loss of revenue. On analysis, it was found that this problem can be solved only by hardware modification in the system by introducing an additional control. The additional control needed is the monitoring of load on DG and restricting it around 20% of the rated capacity by controlling the inverter power output. The inverter manufacturer confirmed the feasibility of this scheme and implementation is being carried out.

The overall efficiency of the plant was found lower than the specified value. This is because of the lower efficiency of solar module and also due to higher temperature coefficient of power of the module. Margin available for the input DC power is high and it is possible to add additional PV modules to system. Figure 6 shows the solar panel array at the roof top. Figure 7 shows the photo of inverter mounted on wall and the switch board positioned at substation near MSB.

Figure 6. Solar array at roof top

Figure 7. Wall mounted inverter and switch board near MSB

9. Conclusion

PV array efficiency is good as per generally expected commercial standards, but is lower than the specified value. This is due to high temperature coefficient of power of the modules. The issue is being taken up with module manufacturer. Overall performance of the Inverter is very good and the efficiency is excellent. Though the installed plant injects little harmonics, it is negligibly small and very well within the limits. Present loading of inverters is only upto 70%, as the rated capacities of PV array and inverter are equal. The actual required inverter capacity is less than kWp rating of the PV array. The scope exists for addition of PV array of about 40 kWp without overloading the present inverters and cable. Grid tied solar power plant is well-suited for academic institutions and the suitability was demonstrated. All the problems faced during integrated operation were addressed and solutions are being implemented.

10. References

[1] Planning and installing Photo voltaic systems , A guide for installers architects an engineers. Erathscan Routeldge, Taylor and Francis Group, Third edition 2013.
[2] Jain, Sachin, and Vivek Agarwal. "A single-stage grid connected inverter topology for solar PV systems with maximum power point tracking." *Power Electronics, IEEE Transactions on* 22, no. 5 (2007): 1928-1940.

[3] R. Teodorescu , P. Rodriguez, Marco Liserre. "Grid converters for photovoltaic and wind power systems", IEEE Press, 2011.
[4] Blaabjerg, Frede, Florin Iov, TamasKerekes, and Remus Teodorescu. "Trends in power electronics and control of renewable energy systems." In *Power Electronics and Motion Control Conference (EPE/PEMC), 2010 14th International*, pp. K-1. IEEE, 2010.
[5] https://www.gses.com.au/wp-content/uploads/2016/03/GSES_powerfactor-110316.pdf
[6] Abhishek Kumar Sinha, Vaibhav Shukla and Mukul Chankaya ; Performance Analysis of Grid connected Solar Farm by Power Factor Control ; J C T A, International Science Press, 9(41), 2016, pp. 207-217.

Acknowledgement
Authors express their sincere thanks to Shri Johnson Parayannilathu, Member board of directors, DTET, Rev. Fr. Dr. George Thanathuparambil, Director, VJCET and Msgr.Dr. Cheriyan Kanjirakompil, Manager, VJCET for their keen interest and encouragement during the course of the execution of this project. The support rendered by faculty members, EEE department of VJCET are commendable. Authors acknowledge the contributions of non- teaching staff members of EEE department of VJCET especially Shri. Dhinu Paul and Shri. Sabu Valaimattom for their untiring efforts for the successful completion of this project. Authors express their heartfelt gratitude to Prof. Dr Sasi K Kottayil, Amrita Vishwa Vidyapeetham, Coimbatore for his deep involvement in the measurement of plant parameters.

Design of Piezo-Resistive Type Acoustic Vector Sensor using Graphene for Underwater Applications

Smitha Prabhu[1], Amrita B Pai[1], Gurumukh Singh Arora[1], Kusshal M R[1], Veera Pandin[2], Goutham M A[3]

[1]Department of Electronics and communication, NMIT, Bangalore, India
[2]Center of NanoScience and Engineering, Indian Institute of Science, Bangalore, India
[3]Department of Electronics and communication, AIT,Chikamagalur,India

E-mail: *smitha.prabhu@nmit.ac.in*

Abstract. In recent years as the level of radiated noise of submarine drastically got decreased, the requirement of underwater acoustic sensor has become essential for the detection of low frequency acoustic signal. The latest advancement in the field of underwater sensors is the Micro Electro Mechanical Systems (MEMS) sensor. The major advantage of this sensor being determination of directionality along with the measurement of pressure of an acoustic signal. The biological inspiration behind the design of the sensor was derived from fish lateral line sensing system, with piezoresistive transduction principle to obtain the objective of miniaturization and low frequency signal detection. In this paper, we are reporting the design, simulation of an underwater two-dimensional MEMS acoustic vector sensor which exhibits better sensitivity, flexibility when compared to the traditional acoustic vector sensor. In this work, thin film of piezoresistve material, Reduced Graphene Oxide (Rgo) is taken as pressure sensing element instead of traditional polysilicon material and its performance in terms of sensitivity and directivity is analysed. MEMS acoustic vector sensors using Reduced Graphene Oxide and polysilicon were simulated on flexible kapton and silicon substrate respectively using COMSOL Multiphysics 5.5 vesrion. The simulation results indicates that designed vector sensor based on MEMS technology and piezoresistive effect of rGO on flexible kapton substrate is feasible and possesses better sensitivity of -149.47 dB when compared to the sensitivity of -171.37 dB of polysilicon based traditional acoustic vector sensor. The simulated value of eigen frequency of the sensor is found as around 47Hz, which almost matches with the theoretical value of resonance frequency of the sensor as per the design considerations of the sensor and hence capable of detecting low frequency acoustic signals.

1. Introduction

Underwater acoustic plays a vital role in underwater operations such as detection of the location of the submarines, ships, target detection, and in navigation. Sound wave in an elastic medium is known as acoustic wave. In water, acoustic wave is due to the molecular vibrations in the medium that travels at the speed of light. A sound wave propagating in water consists of alternating compressions and rarefactions of the water molecules. The rarefaction and compression of water molecules exerts restoring force. This restoring force that resists the motion is labeled as pressure. The variation in pressure is detected by the hydrophone, which converts them to the voltage proportional to the pressure. At any point in acoustic field the vector sensor will be able to measure both acoustic particle velocity and acoustic pressure. This vector hydrophone provides the direction and magnitude of pressure [1]. The vector sensor made up of an array of hydrophones, has drawbacks such as inconsistency and larger volume. A four-beam structure with an acoustic cylinder at the centre forms the basic structure of the vector sensor. An acoustic wave, on striking the cylinder leads to the distortion of the beam structure [2]. The shape of the support frame to which the four beams are attached plays an important role in improving the sensitivity of the sensor.

The largest centre of deflection is seen in the circular diaphragm having lowest stress on its edges when compared to the other diaphragms as in square diaphragm, and this helps in the distribution of stress on cantilever beams [3]. In recent times the level of noise transmitted from submarines has become lower and hence making SONARs inefficient to locate and track them. A single acoustic sensor was developed by China based on bionic, piezoresistive and MEMS technology [4]. The sensor is based on the piezoresistive transduction principle. The benefit of using piezoresistive transduction principle is that it detects the low-frequency signal even to the extent of zero hertz and it exhibits two-dimensional directivity.

The biological inspiration in the design of the MEMS vector hydrophone comes from the sensing system of the fish as indicate in Figure 1, which goes from head to tail, resembling an array, having sensing organs called stitches, spaced at intervals along the nerve fibre. Each stitch consists of many neuromasts as shown in Figure 2 and each neuromast in turn contains several mechanosensory hair cells. Stereocilia is the apical portion of the hair cell is. When the fish swims in water, stereocilia vibrates and thereby, acts as sensor for flow noise. This turbulence causes the movement of hair cell, intern causes the change in synapses connected to the main nerve fibre. It is observed that the lateral line of fish is sensitive to the motion of fluid of low frequency. The entire structure of the hydrophone resembles that of a neuromast. The hair cell is similar to that of the piezoresistors and the stereocilia is similar to the cylindrical rod [5].

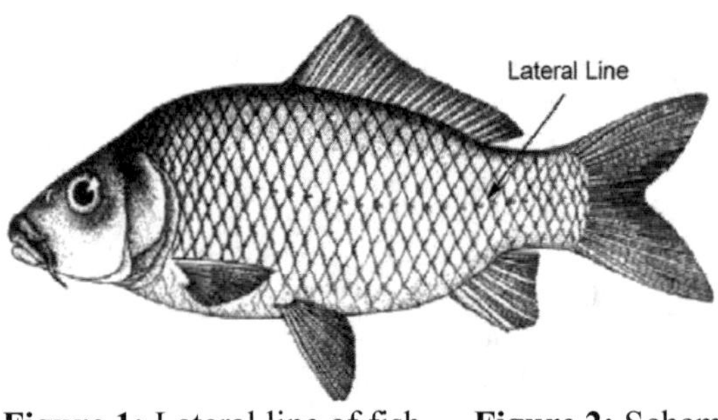

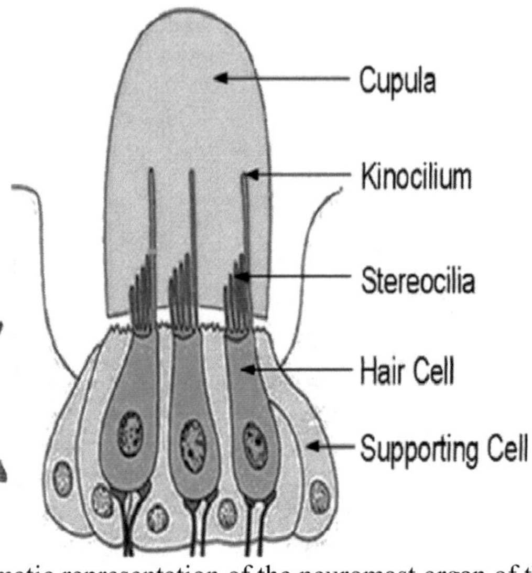

Figure 1: Lateral line of fish **Figure 2:** Schematic representation of the neuromast organ of the fish

On a conventional approach, Mengran Liu et al. [6] have used silicon as a substrate for the structure. Silicon as a substrate is not flexible and needs to undergo additional chemical processes to make it flexible which increases the fabrication cost. This process is used to decrease the thickness of the silicon substrate. The sensing element used in traditional piezoresistive pressure sensor is polysilicon and the fabrication process of it requires clean room facilities which requires several days. Many researchers have worked on silicon to improve the sensitivity of the model. Wen Dong Zhang et al [7] have designed polysilicon on an insulator and found3.44 mv/psi as its pressure sensitivity

In the present-day graphene has taken lead in Micro Electronic Mechanical Systems (MEMS) technology, especially in the domain of pressure sensors, due to their chemical, electronic, mechanical, and physical properties. One of the carbon alltropes, Graphene is utilized to develop a pressure sensor [8]. Graphene is made up of monolayer of carbon atoms, which gives them extreme thinness. The

extreme thinness of graphene layers, causes the deflection of the membrane. Graphene is a unique material with excellent mechanical flexibility and high crystallographic quality. It also provides a linear relationship with strain and resistivity. Extraction of pure graphene without any defects is highly complex as per Kaihua Cao et al [9]. Most of the techniques used in the synthesis of graphene do not yield a good amount of graphene. Reduced Graphene Oxide (rGO) is a material whose properties are similar to graphene, when compared to other members of the graphene family. Reduced Graphene Oxide can be easily synthesized in bulk quantities. It is highly isotropic and provides high conductivity [10]. Reduced Graphene Oxide is obtained by reducing Graphene Oxide (GO).
Based on Bionic, piezoresistive principle and MEMS technology an effort is made in this work to design and simulate an underwater bi-directional vector sensor. The proposed design makes use of Kapton as a substrate which is highly flexible in nature and Reduced Graphene Oxide as the piezoresistive sensing element. The device shows two-dimensional directivity, hence overcomes the short comes of scalar hydrophone. Combining MEMS technology and piezoresistive effect of reduced graphene oxide, along with its unique electronic properties, a low frequency sensor can be developed with a simple, low cost and scalable process.

2. Design & Mathematical Modelling

The structure of the bi-directional acoustic vector sensor mainly consists of four horizontal cantilever beams, a rod, support frame and 8 piezoresistors as shown in Figure 3.

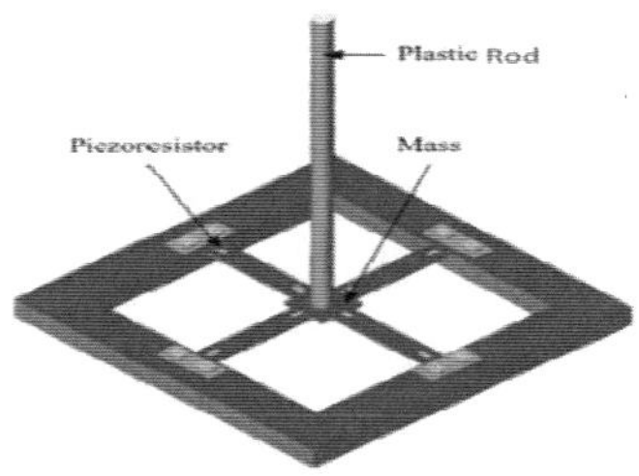

Figure 3: Structure of Acoustic Sensor

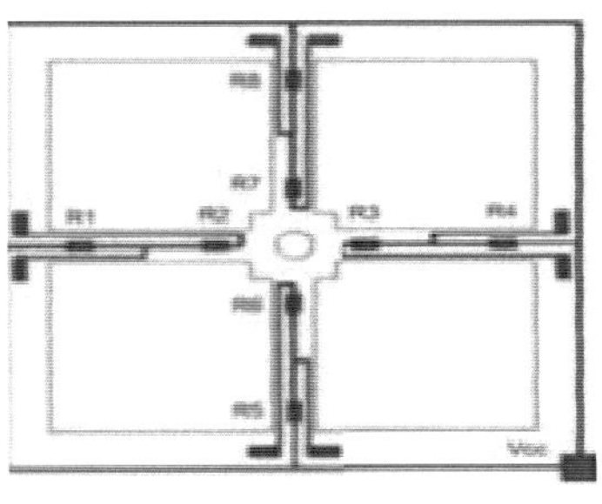

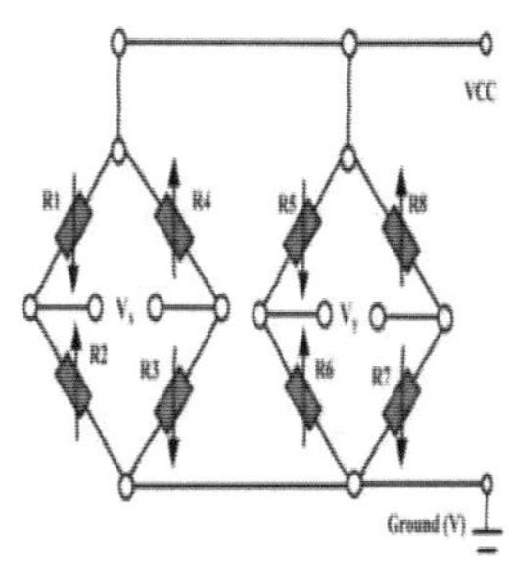

Figure 4a: 4 beam microstructure with eight piezoresistors

Figure 4b: Piezoresistors connec wheatstone bridge Configuration

Figure 4: Two Wheatstone bridge with eight piezoresistors

Piezoresistors R1, R2, R3, R4 are placed on the beam, which is situated along axial direction & R5, R6, R7, R8 are along radial direction beam as shown in Figure4a. Using these piezoresistors two wheatstone bridges are formed, one along axial and the other along radial direction respectively as shown in Figure4b.

2.1 *Working Principle of the Sensor*

When the pressure wave displaces the rod in axial or radial direction, the cantilever beams of the structure experiences strain. Being sensing elements, these piezoresistors sense the induced stress and hence cause a corresponding variation in the electrical resistance of the same. This change in the resistance of the resistors connected in a Wheatstone network on the sensing beams generates an electrical output voltage. As shown in Figure 4b, Wheatstone bridge is biased by using supply Vcc and for the equilibrium condition of the bridge, i,e when $R_1R_3 = R_2R_4$ ----- (1), output voltage Vx=0

If the pressure wave is incident along axial direction, then it changes the resistance of axial beam piezoresitors then the output voltage of the axial beam Wheatstone bridge is given by

$$V_x = \frac{(R_1+\Delta R_1)(R_3-\Delta R_3) - (R_2+\Delta R_2)(R_4-\Delta R_4)}{(R_1+\Delta R_1+R_2-\Delta R_2)(R_3+\Delta R_3+R_4-\Delta R_4)} V_{IN} \text{----- (2)}$$

2.2 *Sensor Design Criteria*

The dimensions of the sensor depend on Eigen frequency and the stress imparted to the structure. The main objective here is to build a low-frequency acoustic vector sensor. Hence identifying the operational frequency range of sensor is very essential.

Eigen frequency is given by $f = \frac{1}{2\pi}\sqrt{\frac{k}{m}}$... (3) $f = \frac{1}{2\pi}\sqrt{\frac{2Ewt^3}{mLH^2}\left(\frac{a^2}{L^2}+\frac{a}{L}+\frac{1}{3}\right)}$ ... (4)

where 'L' is length of the cantilever beam, 'H' is height of the rod, 'a' is the half-width of the central mass,' w 'is the width of the cantilever beam,' t' is the thickness of the cantilever beam, 'E' is the elastic modulus of the substrate and 'R' is the radius of the rod.

To finalise the dimensions of the sensor, the parametric analysis of all above mentioned parameters are carried out. Few results of analysis are shown in Figure5 and Figure6.

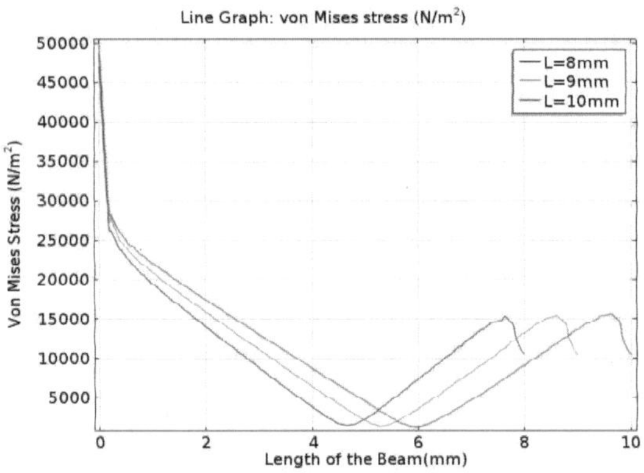

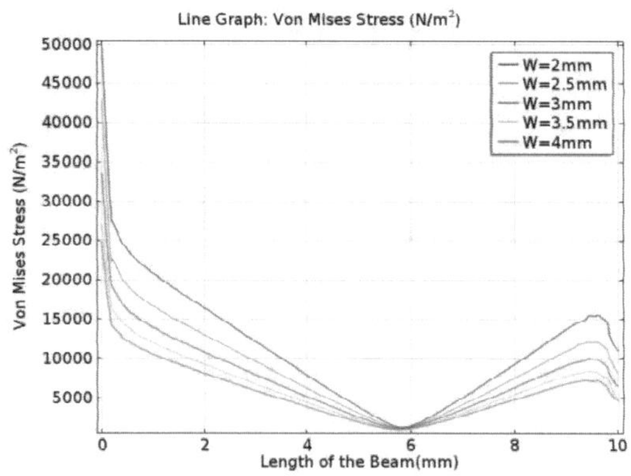

Figure 5: Parametric analysis of length of the beam **Figure 6:** Parametric analysis of width of the beam

It can be inferred from Figure5 and Figure6 that, as the width of the beam decreases and length of the beam increases the stress across the horizontal cantilever beam increases. Similarly, the larger radius of rod will also results in more stress and the size of the central mass exhibits similar impact on stress. Similarly, various dimensions of the structure are considered and its effect on the Eigen frequency and stress are analysed. The dimensions of the device are finalised after parametric analysis and are indicated in table 1.

Table 1. Parameter dimensions of sensor

Parameter	**Dimension (mm)**
Length of the cantilever beam (L)	10
Width of the cantilever beam(W)	2
Thickness of the cantilever beam(t)	0.175
Central mass Square(S)	6
Length of the rod(L_b)	25
Radius of the rod(R)	3

With these dimensions, the theoretical value of Eigen frequency of the structure was found to be around 41.06 Hz hence capable of detecting very low frequency acoustic signals.
Another design criterion for this sensor is, density of the material used for the rod. The sensing principle of this sensor follows acoustic theory of cylinder. This theory indicates that the density of the surrounding medium and the cylindrical rod should match, so as to have a synchronous vibration, as per equation$v = \frac{2\rho_o}{\rho+\bar{\rho}} v_o$...... (5), else the MEMS hydrophone cannot memorize the acoustic information. Hence while simulating, the material used for rod is taken as nylon whose density is almost equal to that of density of water.

3. Modelling Using Comsol Multiphysics

The geometry of the sensor is designed and simulated using COMSOL Multiphysics. Solid Mechanics physics and piezoresistive domain currents are chosen as the environment for the purpose of analysis. The first step is to create the geometry, followed by defining suitable materials for the same. The materials defined for rod, piezoresistors and connector strip are nylon, rGO and aluminium respectively. Kapton is the material used for the central block and the microstructure with density of 1330 kg/m3, Young's modulus of 3.1G Pa and Poisson's ratio of 0.34. The piezoresistive property of Reduced Graphene Oxide plays a important role in improving the sensitivity of the MEMS underwater acoustic sensor. The electrical resistance of the piezoresistor varies in accordance with the applied stress. When the sensing beams are stretched or compressed, the piezo resistors placed on these sensing beams experience a change in its length, area and resistivity. The resistivity of the piezo resistors depend on its piezoresistive coupling matrix. Piezoresistive coupling matrix of a material is a 6*6 matrix that defines the relationship between the stress components and its corresponding resistivity [11]. There exist only three non-zero independent stress components (π11, π12, π44) due to the crystalline nature of graphene, as given below:

$$\pi = \begin{bmatrix} \pi 11 & \pi 12 & \pi 12 & 0 & 0 & 0 \\ \pi 12 & \pi 11 & \pi 12 & 0 & 0 & 0 \\ \pi 12 & \pi 12 & \pi 11 & 0 & 0 & 0 \\ 0 & 0 & 0 & \pi 44 & 0 & 0 \\ 0 & 0 & 0 & 0 & \pi 44 & 0 \\ 0 & 0 & 0 & 0 & 0 & \pi 44 \end{bmatrix}.$$

On simplification, the change in the resistance of a piezoresistor is given by equation $\frac{\Delta R}{\mathrm{R}} = \pi_l\ \sigma_l + \pi_t\ \sigma_t$----- (6) where the transverse and longitudinal piezoresistive coefficients are given by π_t and π_l respectively. The longitudinal and transverse stresses on the surface of the piezoresistors are given by

$$\pi = \begin{bmatrix} 66.17e^{-11} & -30.65e^{-11} & -30.65e^{-11} & 0 & 0 & 0 \\ -30.65e^{-11} & 66.17e^{-11} & -30.65e^{-11} & 0 & 0 & 0 \\ -30.65e^{-11} & -30.65e^{-11} & 66.17e^{-11} & 0 & 0 & 0 \\ 0 & 0 & 0 & 90.35e^{-11} & 0 & 0 \\ 0 & 0 & 0 & 0 & 90.35e^{-11} & 0 \\ 0 & 0 & 0 & 0 & 0 & 90.35e^{-11} \end{bmatrix}.$$

After defining piezoresistive, structural and electrical properties of the material, the support frame was fixed by applying fixed constraint domain condition and boundary load pressure or force was applied on rod. The deformation of the structure for the applied boundary load of 1N is as shown in Figure7. Due to the deformation, stress gets generated on the beam and it is found to be maximum at beam central block interface and at the support frame as shown in Figure 8 and act as a optimal position for the placement of piezoresistors.

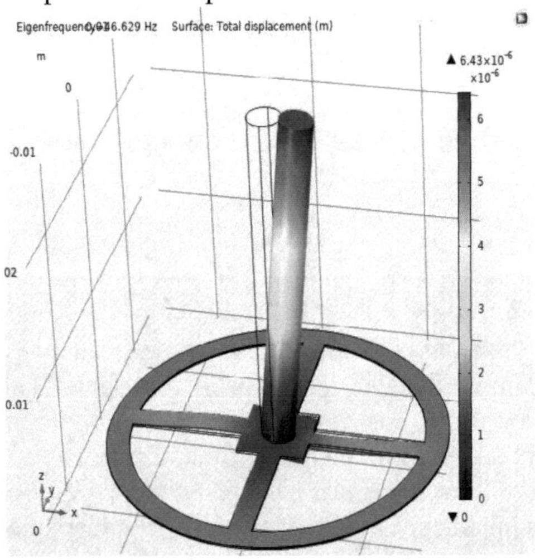

Figure 7: Displacement of the rod for 1N force

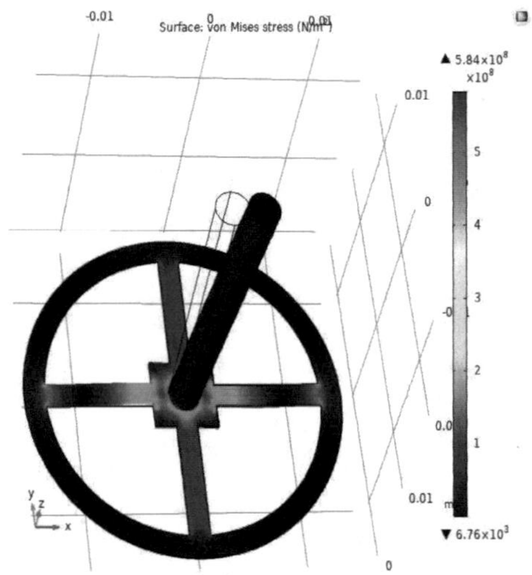

Figure 8: Surface stress distribution plot

4. Simulation Results

Performance of the sensor was analysed based on the shape of the support frame, substrate and piezoresistive materials used.

4.1 Shape of the Structure

The shape of the structure is one of the imperative factors to improve sensitivity of the sensor. Contemporarily most of the underwater acoustic sensors use square based structures. Circular structure design and square structural design of the sensor is compared and analysed interms of stress distribution as shown in the Figure 8 and Figure 9 respectively.

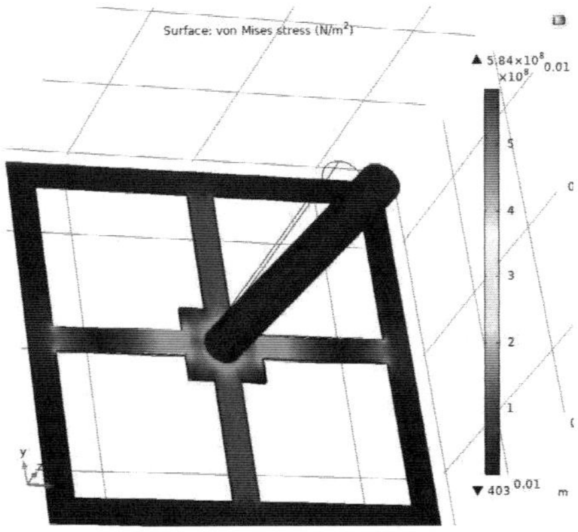

Figure 9: Surface stress distribution plot of square structure for 1N force

For precise comparison of stress distribution of both structures, stress versus arc length graph was plotted and they are as shown in Figure 10 and Figure11 respectively.

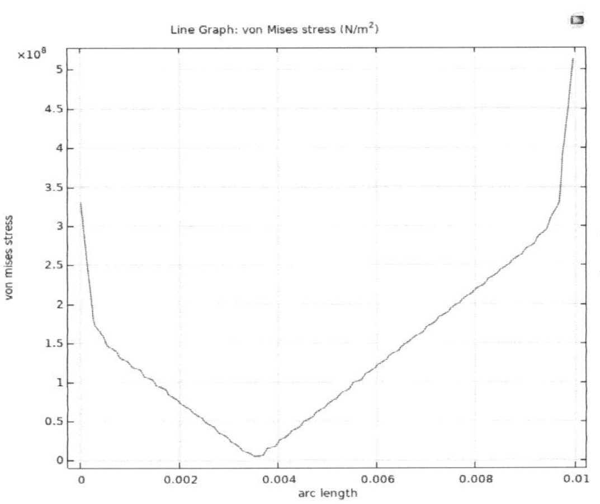

Figure10: Von Mises stress distribution on the single cantilever beam of circular structure when a force of 1N is applied

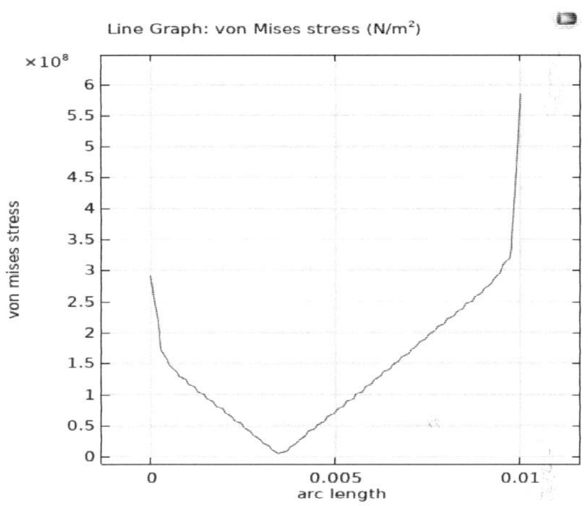

Figure11: Von Mises stress distribution on the single cantilever beam of square structure when a force of 1N is applied

On comparing the graphs in Figure 10 and Figure 11, it is observed that amount of stress at the fixed ends of the cantilever beam is more for circular-shaped sensor than the square-shaped sensor. At the end of the beam which is connected to the central mass, von mises stress for square shape is below $3*10^8 N/m^2$ when compared to $3.4*10^8 N/m^2$ of circular shape structure. These results indicate how circular structure design is better suited for piezoresistive effect based acoustic sensor.

4.2 Substrate Material Consideration

Kapton as a substrate has numerous advantages when compared to conventional silicon based substrates which includes low manufacturing cost, high flexibility and higher temperature tolerance. Kapton is naturally flexible in nature whereas silicon needs to undergo additional chemical processes to make it flexible.

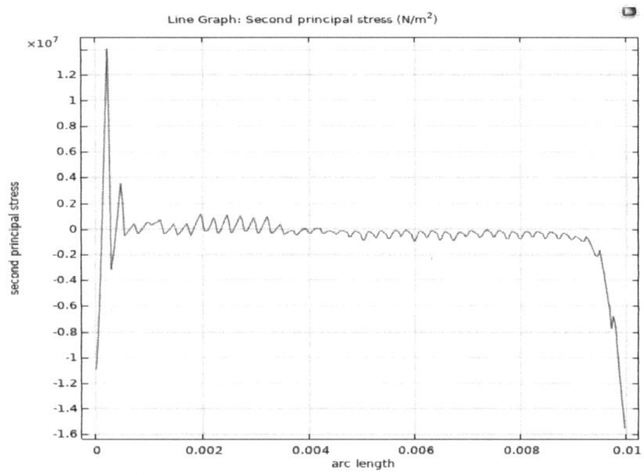

Figure 12: 2nd principle stress distribution plot of n-type silicon based single cantilever beam when a force of 1N is applied

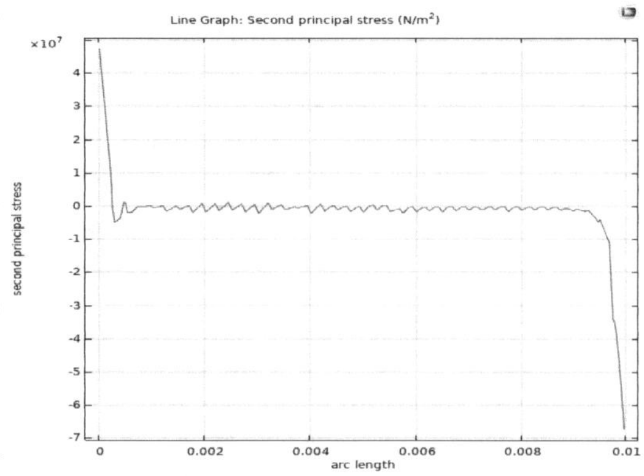

Figure 13: 2nd principle Stress distribution plot of Kapton based single cantilever beam when a force of 1N is applied

From, Figure 12 and Figure 13, it is observed that Kapton possesses a higher stress value when compared to n-type silicon, both at fixed end of the beam and at the interface of beam and central mass. Maximum stress variation inturn aids in improving the sensitivity of the device.

4.3 Reduced Graphene Oxide as a Piezoresistor

Piezoresistive sensing materials plays a vital role in defining the efficiency of the pressure sensor. In this paper, reduced graphene oxide is introduced and taken as a sensing material and paralleled against the conventionally used silicon. The sensor is simulated for various pressure levels (1Pa to 5Pa) for both rGO and polysilicon as sensing materials.

4.3.1 When various pressure levels are applied along the Positive X-axis direction on the nylon rod

For the applied pressure along axial direction, the beams which are along axial direction will be subjected to deformation due to tensile and compressive stresses on the beams which results in the variation of resistance of the corresponding beam.

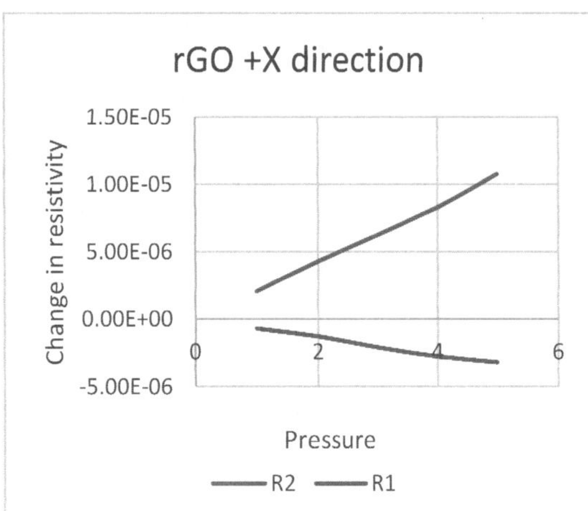

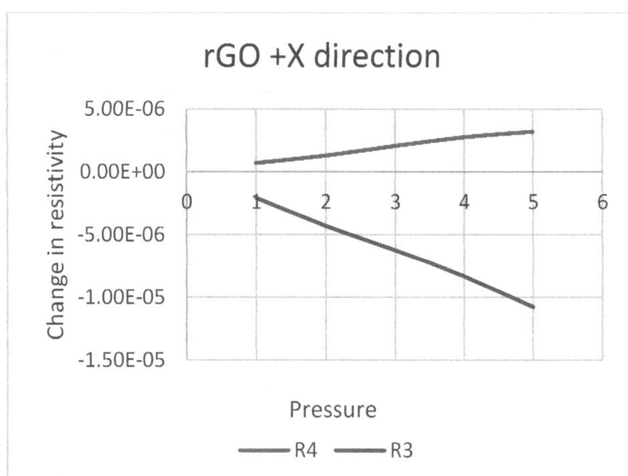

Figure 14: Change in resistivity of R1, R2 &R3, R4 piezoresistors

The pressure of 1Pa to 5Pa is applied in positive x-axis direction. Resistor R1&R3 experiences compressive stress, hence the change in resistivity keeps on decreasing for R1&R3 (Negative) and resistors R2&R4 experiences tensile stress and hence the change in resistivity value is increasing (Positive) for resistors R2&R4 as shown in Figure 14. The change in resistivity is in the order of 10^{-7} ohm-m to 10^{-6} ohm-m for R1, R4 and R2, R3 respectively. Chnage in resistivity of radial axis resistors are found to be negligible in the order of 10^{-10}ohm-m. But for polysilicon piezoresistive material, change in resistivity was found to be in the order of 10^{-9} ohm-m to 10^{-8} ohm-m for R1, R4 and R2, R3 respectively. Higher the change in resistivity more is the change in output voltage across Wheatstone network which leads to higher sensitivity. Hence, reduced graphene oxide is chosen as the piezoresistive material.

5. Directivity and Sensitivity

For the applied axial direction force on the rod, the resistivity of only the resistors placed on axial beam will show significant changes and whatstone bridge formed on the axial beam produces electrical output voltage, thereby indicating the directionality as x axis. For radial force, the wheatstone bridge formed on radial beam gives the output voltage indicating the direction as Y axis.
The sensitivity of the sensor depends on the output voltage of the Wheatstone network. The output voltage is observed for both polysilicon and rGO as piezoresistors by simulation. The output voltage observed on x axis Wheatstone bridge, when pressure wave acts on +X axis direction for polysilicon as piezoresistors is found to be 2.7 mV and for rGO as piezoresistors, it is 33.6 mV. Hence, rGO based sensor could achieve a sensitivity of -149.47dB when compared to -171.37dB of a polysilicon based sensor.

6. Conclusion

We have proposed design and simulation of the bi-directional Reduced Graphene Oxide based acoustic vector sensor which is suitable for underwater applications. It is observed that the rGO's graphene like properties are suitable for detection of the sound in underwater environment, when it is deposited on a uniform thickness flexible Kapton substrate. Due to its piezoresistive transduction principle and miniaturized design parameters, this sensor is best suited for detection of low frequency acoustic signals. This design has additional advantages which include simple structure, small volume and simpler fabrication process by eliminating the need of clean room, and low cost. The simulated value of natural frequency of the sensor is found to be around 47 Hz and it almost a match with the theoretical value of 41.06Hz. Its sensitivity is found to be around -149.47 dB. This sensor can detect horizontally moving underwater acoustic waves and also achieve low frequency acoustic wave detection with enhanced sensitivity.

REFERENCES

[1]. Hu Zhang, H J Chen, W Z Wang., 2014. An underwater acoustic vector sensor with high sensitivity and broad band Sensors & Transducers. Vol. 170, Issue 5.

[2]. Peng Wang_,Guo-jun Zhang ,Chen-yang Xue , Ji-jun Xiong.,2017. Engineering application of MEMS vector hydrophone and self-adapting root-MUSIC algorithm., *IEEE DOI: 10.1109/TRANSDUCERS.2011.5969568*

[3]. Shivaleela Melennavar, Ajaykumar C Kategeri., 2016.Theoretical Analysis of the Design of Different Shape Diaphragm for Piezoresistive Pressure Sensor, DOI: 10.17577/IJERTV6IS030408

[4]. Shang Chen, Chenyang Xue, Binzhen Zhang, Bin Xie and Hui Qiao.,2007.A Novel MEMS Based Piezoresistive Vector Hydrophone for Low Frequency Detection. *Proceedings of the IEEE,* **1839-1844**.

[5]. ChenyangXue, Shang Chen,Wendong Zhang,Binzhen Zhang,Guojun Zhang, Hui Qiao., (2007). Design,fabrication,and preliminary characterization of a novel MEMS bionic vector hydrophone. Microelectronics Journal 38, **1021-1026**.

[6]. Mengran Liu, Guojun Zhang, Zeming Jian, Hong Liu, Xiaopeng Song, Wendong Zhang., 2014. Design of Array MEMS Vector Vibration Sensor in the Location of Pipeline Internal Inspector. *TELKOMNIKA Indonesian Journal of Electrical Engineering*,**Vol. 12, No. 9, pp. 6651 ~ 6657.**

[7]. Wen Dong Zhang , Ling Gang Guan, Guo Jun Zhang.,2009. Research of DOA Estimation Based on Single MEMS Vector Hydrophone,Article, *Sensors*,9,**6823-6834.**

[8]. Nagarjuna Nella,V Gaddam,M.M Nayak, K.Rajanna, T.Srinivas, 2015. Highly flexible and sensitive graphene-silver nanocomposite strain sensor. *Proceedings of the IEEE.*

[9]. Kaihua Cao,Xiaodong Ye,Xinhua Guo, 2017. Design and Simulation of Two-Dimensional Graphene-Based Acoustic Sensor Arrays. *Proceedings of the IEEE*

[10]. Raluca Tarcana, Otto Todor-Boera, Ioan Petrovaia, Cosmin Leordeand, Simion, IoanAstileana Botiza,2015. Reduced graphene oxide today. **DOI: 10.1039/C9TC04916A**

[11]. Guo Jun Zhang,Lin Xian Liu,Wen Dong Zhang,Chen Yang Xue,.2014. Design of a monolithic integrated three-dimensional MEMS bionic vector hydrophone. Technical Paper *,Springer-Verlag Berlin Heidelberg*

Implementation of Sobel filter using CUDA

Akshita Akasapu[1], V Sailaja[2], G Prasad[3]

[1&2]Department of EEE, Amrita School of Engineering, Bengaluru, Amrita Vishwa Vidyapeetham, India.
[3]Special Projects and Photo processing division, National Remote sensing center, Indian Space research organization, Hyderabad, India.

[1]akshitaakasapu@gmail.com, [2]v_sailaja@blr.amrita.edu, [3]prasad_g@nrsc.gov.in

Abstract. The Sobel filter is an edge detection filter that results in highlighting and focusing the edges. It is used for extracting the image gradient and contours in image processing applications. The data sets we are using in the paper are PGM and TIFF images with different rows and column pixel dimensions. The sobel filter operations used are implemented in C and CUDA programming language and the time taken for the operations in the serial and parallel platforms respectively are analyzed.

Keywords. Image Processing, Edge Detection, Sobel operator, C programming language, CUDA programming language, Parallel Performance, Serial Performance, PGM file, TIFF file.

1. Introduction

Image processing is one of the most powerful tools used in fields with requirements of transmission of data especially in telecommunication, security systems, and remote sensing. The manipulation of the images using various algorithms applied to the input to get the desired output is the aim of image processing [1].The quality of the image under extreme undesirable and unfavourable conditions can be improved using Digital Image Processing(DIP) in several ways by adjusting the image properties using edge detection, noise reduction, etc [2].Acquisition, enhancement, restoration, morphological processing, segmentation, representation &description, object recognition are the key stages used in image processing. The main components of DIP are importing, analysis and manipulation and then output. Where scanning and digital photography are used for importing. Analysis and manipulation accomplished using various software applications. Image processing has extensive applications in many areas, including astrophysics, medicine, robotics, and remote sensing by satellites [3][10].

The process of portioning the image into constituent parts or objects using groups of pixels that are homogeneous concerning to some criterion is called Segmentation. The pixel collection in each group must be different and adjacent groups must be heterogeneous. Segmentation algorithms are mainly based and focused on region orientation rather than pixel-oriented. Segmentation finally results in the splitting of images into similar and connected areas. It mainly focuses on the division of images into meaningful regions.

Satellite Sensors collect data in different spectral bands and hence this data is represented in many dimensions. Due to the multiple dimensions, redundancy of data arises. To remove this and to emphasize and isolate the required data from the image, edge detection is applied.

Data is transmitted from satellites to the research centres on the ground with an objective to monitor the Earth's surface and atmosphere. This data is of huge volume and it is very difficult to store, retrieve, and analyze. Edge Detection is the general term for a technique to reduce the dimensionality of the image and isolate the required data from the huge data by emphasizing the edges. Using mathematical horizontal and vertical projection, along with gradient operation, the required data is isolated with the required dimensions from the original huge volume data set. It is therefore often the case that the data acquired from the reduced dimension data set will allow the user to spot patterns, obligate and requisite data, way more easily than without performing the Edge detection operation using Sobel Filter Operator.

2. Related work

Image processing is becoming most challenging in the fields like astrophysics, medicine, etc. Different segmentation algorithms can be used for the detection of images [10].One of the main steps in image processing is the segmentation of the image. The partitioning of the image is called segmentation. The method which helps in separating the foreground and the background of an image for accurate analysis is called image segmentation. Techniques involved in image segmentation are edge base segmentation, Fuzzy based segmentation, Threshold based segmentation and Region based segmentation [9]. Image segmentation can be used in the medical aspect to avoid problems such as noise and signal distribution during processing. But research has been conducted to analyze the algorithm in a wider range of applications as well [11].

Edge detection is a foundation tool in image processing mainly focusing on feature detection and extraction. This mainly aims towards the identification of the pixels in an image where there are sudden disruptions in the pixel values or any sharp differences in the image luminance. Different techniques have been developed and analyzed for the identification of the edges, one of the techniques is to find the edges to the respective image and then identify the image. Depending on the type of edges, several edge detection operators have been developed. Edge orientation, image noise and edge structure are the variables effected in the selection of edge detection operators. The majority of different edge detection methods may be grouped into Gradient and Laplacian [7]. But many times, the algorithms can get overloaded on the machine through series performance and hence, this would result in delayed execution of the algorithm and intensive computational problems when high resolution inputs are used. An effective solution to resolve this problem is by parallel performance.

The graphics processing unit (GPU) provides a constructive solution for the intensive computing of the image during image processing and provides high performance in real time applications. GPU features parallel programming techniques that ensure performance by highest standards and reduce the execution time of accelerating an extensive calculation algorithm [6].

3. Edge detection technique

The sobel operator is a derivative mask and is used for edge detection. Sobel operator is used to detect the horizontal and vertical edges in an image. Figure 1 explains the working of the filter.

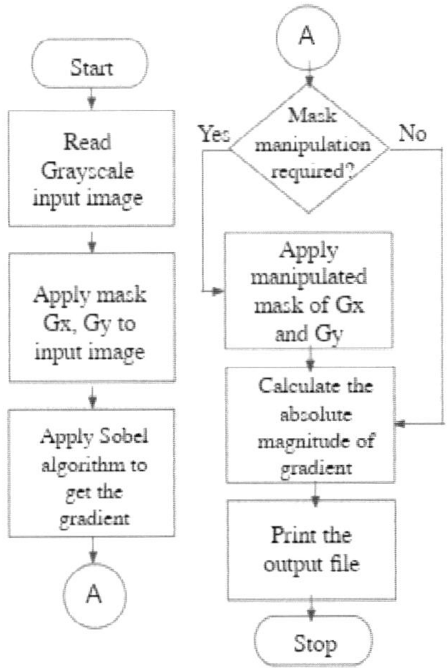

Figure 1 Flowchart of Sobel Filter

3.1. Horizontal and Vertical Gradient Components

+1	0	-1
+2	0	-2
+1	0	-1

Gx

Figure 2 Horizontal sobel operator

-1	-2	-1
0	0	0
+1	+2	+1

Gy

Figure 3 Vertical sobel operator.

The horizontal gradient component (Gx) in figure 2 and Vertical gradient component (Gy) in figure 3 are used to find the edges oriented to the respective operations. The operators can be applied separately to find the respective direction of edges or can be made into a single direction by finding the absolute magnitude gradient. The absolute magnitude is given by equation 1[7].

$$|G| = \sqrt{Gx^2 + Gy^2} \qquad (1)[7]$$

The angle of orientation of the edge is formulated by equation (2)[7].

$$\theta = \arctan(Gy/Gx) \qquad (2)[7]$$

3.2. Pseudo Convolution Operator

Two gradient components are computed and added in a single pass to obtain the absolute magnitude using the pseudo-convolution operator shown in Figure 3.

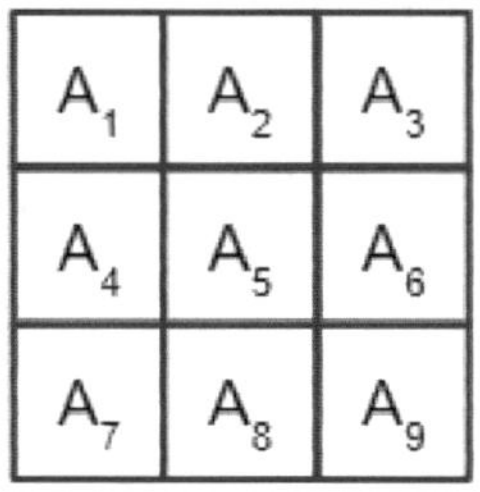

Figure 4 Pseudo-convolution kernels

Quick computation of approximate gradient magnitude is obtained from Pseudo-convolution kernel is shown in figure 4. The approximate magnitude is given by equation (3)[8].

$$|G| = |(A_1 + 2 \times A_4 + A_7) - (A_3 + 2 \times A_6 + A_9)| + |(A_7 + 2 \times A_8 + A_9) - (A_1 + 2 \times A_2 + A_3)| \quad (3)[8]$$

4. Implementation details

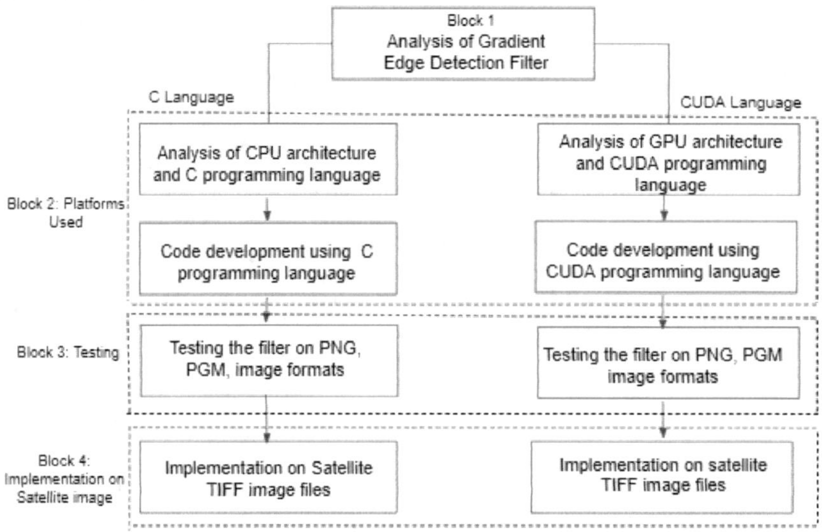

Figure 5 Implementation details

Figure 5 gives a brief description of how the implementation is carried forward and the various software and file formats used.

4.1. Block 1- Gradient Edge Detection Filter

In order to increase the clarity of an image, we have to find all the edges. Edges can be identified by applying algorithms called edge detection filters. Once all the edges have been identified, they can be applied to the image for making it sharper. In this paper, we have implemented this Edge Detection Filter on a serial platform on a Central Processing Unit (CPU) using C Programming Language and Parallel Platform on a GPU using the CUDA programming language.

4.2. Block 2- Platforms Used

4.2.1. Serial Processing: It is a method where a storage device moves through all the locations from where it aims to read or write and gathers information and then the data is retrieved from the device.

IOP Conf. Series: Materials Science and Engineering 1045 (2021) 012016 doi:10.1088/1757-899X/1045/1/012016

4.2.2. Parallel Processing: Process of evenly assigning computer processes between computer processors or cores. This requires a computer with multiple CPUs or a single CPU or GPU equipped with multiple cores. An operating system capable of supporting parallel processing, or software is written specifically to process instructions in parallel are essential.

4.3. Block 3- Test Implementation
The Sobel operator is applied to PGN files chosen as test datasets.

4.4. Block 4- Implementation on Satellite image
The data from the satellite arrives in the form of TIFF images. The obtained TIFF image is executed using the sobel edge detection operator.

5. Results

5.1. Serial processing using static approach
The C program has been tested with multiple inputs. Sobel operator is applied to the figure 6, using Serial Processing C code and the time taken for the execution from figure 8 is noted. The figure 7 shows the output image after application of the filter.

Figure 6 Input of serial processing static approach

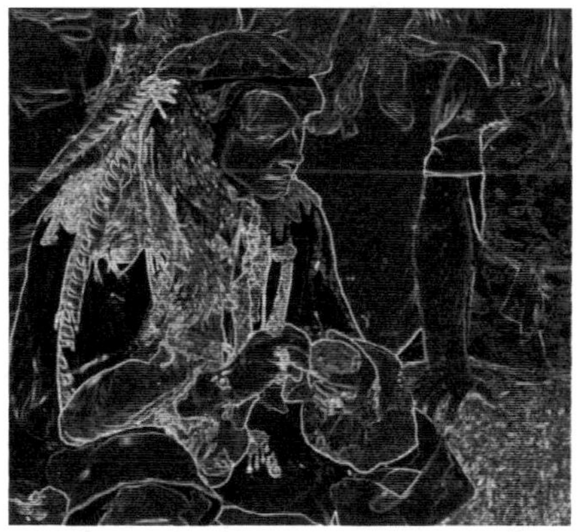

Figure 7 Output after edge detection

Figure 8 Time taken for execution

5.2. Serial processing multi thread approach
The C program has been tested with multiple inputs. Sobel operator is applied to the figure 9 using Serial Processing C code and the time taken for the execution from figure 11 is noted. The figure 10 shows the output image after application of the filter.

IOP Conf. Series: Materials Science and Engineering 1045 (2021) 012016 doi:10.1088/1757-899X/1045/1/012016

Figure 9 Input of serial processing multi thread approach

Figure 10 Output after edge detection for 5 threads

```
akshita@Akshita:~/akshita/Project/Cthread$ gcc -o thread sobel.c -lpthread -lm
akshita@Akshita:~/akshita/Project/Cthread$ ./thread imgin.pgm imgout.pgm 5
 Time taken to execute the filter operation is: 0.212515 seconds
```

Figure 11 Time taken for 5 thread execution

5.3. Parallel processing

The CUDA program has been tested with multiple TIFF inputs. Sobel Operator is applied to the figure 12 and figure 15 using the parallel processing CUDA code and figure 13 and figure 16 is the output image obtained. Then the time taken for execution from figure 14 and figure 17 is noted.

Figure 12 Input of parallel processing

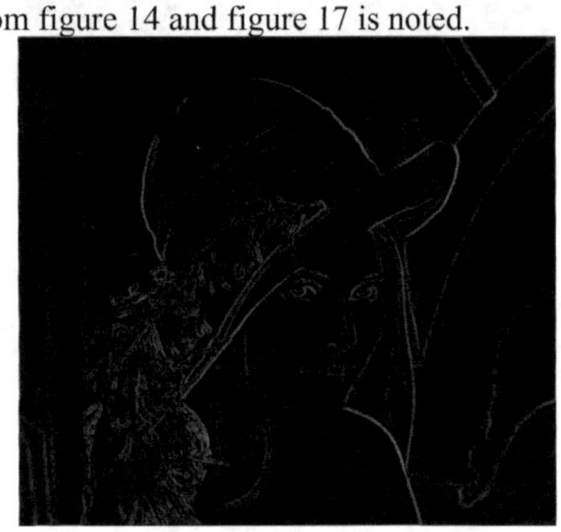

Figure 13 Output after edge detection

```
Time Required for GPU :  0 Seconds , 334 Milliseconds
```

Figure 14 Time taken for execution

Figure 15 Input of parallel processing satellite image

Figure 16 Output after edge detection

Time Required for GPU : 0 Seconds , 568 Milliseconds.

Figure 17 Time taken for execution

6. Observations

The paper is focused on CUDA and its parallel programming platform and the time gained is reviewed. The time gained is calculated using an edge detection filter algorithm. A well-known algorithm SOBEL for edge detection has been used as a part of the paper. A dataset of images is tested using both serial and parallel processing algorithms and the results are noted and observed. The results are further divided based on the image pixels used and the device used.

Table 1 Serial processing using static approach

	Input image name	Size (pixels)	Time (Seconds)
1	Lena.pgm	1024X1024	0.119251
2	Man.pgm	1024X1024	0.115749
3	House.pgm	1024X1024	0.114897

From table1, the results acquired after implementing the sobel edge detection filter, the following are the conclusions:

- On comparison with the multithreading approach with the constant pixel size input image, the static approach takes 0.014398 seconds more time for the operation.
- As the complexity of the input image increases, the number of edges increase and hence the time taken to find edges using technique decreases.
- It is observed that the time taken (in seconds) are in the order,
  LENA.PGM > MAN.PGM >HOUSE.PGM

  Since the number of edges are in the order,

  LENA.PGM < MAN.PGM <HOUSE.PGM
- Hence, for images with more edges, it is comparatively easier to detect the edges and emphasize the image and hence the time taken is less.

IOP Conf. Series: Materials Science and Engineering 1045 (2021) 012016 doi:10.1088/1757-899X/1045/1/012016

Table 2 Serial Processing Using Multi threading Approach

Input image name	Size (pixels)	Threads	Time (Seconds)
		1	0.028458
		2	0.053174
Lena.pgm	512x512	3	0.043227
		4	0.1025
		5	0.050856

Table 3 Serial Processing Using Multi threading Approach

Input image name	Size (pixels)	Threads	Time (Seconds)
		1	0.101351
		2	0.183654
Man.pgm	1024x1024	3	0.159045
		4	0.219699
		5	0.212515

From table 2 and table 3, the results acquired after implementing the sobel edge detection filter, the following are the conclusions:

- On comparing the two input images of different pixel sizes, we can observe that, for smaller pixel images, the time difference for multiple threads is constant (i.e. almost negligible).
- For the image with a larger pixel size, it is observed that as the number of threads increase, more strips are to be computed. Hence, the time taken also increases uniformly.
- Hence, as the image resolution and pixels increase, the time taken for the operation also increases uniformly along with the increase in the number of threads.

Table 4 Parallel Processing

	Input image name	Size (pixels)	Time (Seconds)
1	Lena.tiff	512X512	0.334
2	City.tiff	577X419	0.568
3	River.tiff	256X256	0.203
4	Satellite.tiff	571X579	0.601

From table 4, the results acquired after implementing the sobel edge detection filter, the following are the conclusions:

- From the obtained data, it is observed that, for an 8 X 8, i.e. 64 threads CUDA code, as the pixel's increase, as shown below,

RIVER.TIFF < LENA.TIFF < CITY.TIFF < SATELLITE.TIFF

IOP Conf. Series: Materials Science and Engineering 1045 (2021) 012016 doi:10.1088/1757-899X/1045/1/012016

The time taken (in seconds) also increases.

- It is observed that the time taken to compute the filter operation on the same Lena data set of 512X512 pixels on a GPU is greater than on CPU.

7. Conclusion

- CPU has a lesser number but more powerful cores compared to GPU.

- For smaller pixel images, the CPU gives and faster outputs compared to the GPU, but the GPU gives better results.

- For larger scale and high resolution data sets, GPU is preferred due to the capability of powerful parallel instruction processing techniques.

8. References

[1] Jain A Namdev and M Chawla 2016 *Parallel edge detection by SOBEL algorithm using CUDA C* (Bhopal)
[2] R C Gonzalez and R E Woods 2009 *Digital Image Processing* (New Delhi, India).
[3] https://www.britannica.com/technology/image-processing
[4] Vincent O R and Folorunso O 2009 *A descriptive algorithm for sobel image edge detection* (California: Informing Science Institute) Vol. **40** pp. 97-107.
[5] Gupta, Samta, Susmita Ghosh Mazumdar and M Tech Student 2013 *Sobel Edge Detection Algorithm.*
[6] H B Fredj, M Ltaif, A Ammar and C Souani 2017 *Parallel implementation of Sobel filter using CUDA* (Hammamet) 2017, pp. 209-212.
[7] Vairalkar, Manoj K, and S U Nimbhorkar 2012 *Edge detection of images using Sobel operator.*
[8] https://homepages.inf.ed.ac.uk/rbf/HIPR2/sobel.htm
[9] J N Chandra, B S Supraja and V Bhavana 2017 *A Survey on Advanced Segmentation Techniques in Image Processing Applications* (Coimbatore) pp. 1-5.
[10] J N Chandra, V Bhavana and H K Krishnappa 2018 *Brain Tumor Detection Using Threshold and Watershed Segmentation Techniques with Isotropic and Anisotropic Filters* (Chennai) pp. 0372-0377.
[11] P Tyagi, T Singh, R Nayar and S Kumar 2018 *Performance comparison and analysis of medical image segmentation techniques* (Bangalore) pp. 1-6.
[12] Sanders, Jason, and Edward Kandrot 2011 *CUDA by Example: An Introduction to General-Purpose GPU Programming* (Upper Saddle River, NJ Addison-Wesley).

[13] T Naqash and I Shafi 2011 *Edge sharpening in grayscale images using modified Sobel technique* (Karachi) pp. 132-136.

A bibliometric analysis on path planning methods for mobile robots published in scientific citation index-expanded indexed journals between 2000-2020

Aisha Muhammad[1,3], Mohammed AH Ali and Ibrahim Haruna Shanono[2,4]
[1]Faculty of Manufacturing Engineering, Universiti Malaysia Pahang, Malaysia
[2]Faculty of Electrical Engineering, Universiti Malaysia Pahang, Pahang, Malaysia
[3]Department of Mechatronics Engineering, Bayero University Kano
[4]Department of Electrical Engineering, Bayero University Kano

*Email: ayshermuhd@gmail.com

Abstract. Robots are currently replacing humans in different tasks in various sectors. Among the vital features desirable in autonomous robots is the capability of navigating safely through a given environment. Robot navigation is a process designed with the ability of avoiding any hitches or obstacles while aiming at a specific predefined position. Many studies have been proposed to find solutions to robot path-planning problems. This paper presents a bibliometric analysis of the relevant publications published from 2000 to 2020. The results show that 5385 articles were published in 1128 journals, hence indicating publication diversity. There is a steady rise in the yearly publication output, reflecting an increase in global research interest in the topic. In general, this research provides useful insight into path-planning research so that researchers in this area can better recognize the relevant research study topics and search for the appropriate research partners.

1. Introduction

At first, the manufacturing sectors were the major areas utilizing the application of autonomous mobile robots. However, recently, it is frequently use in other sectors: Medical, Military, mining, agriculture and so on. For the mobile robot to model its environment, localize its position, detect and avoid using a certain path planning technique, it is required to be provided with intelligent information. Choosing the proper path planning method is necessary in order for the robot to navigate in a clustered environment from a given initial position to the target without colliding with obstacles. Robot navigation can be classified into: global and local navigation. In global navigation, the positions of the elements are defined based on a reference axis while moving towards the target. In the later, the dynamic constraints of the environment are identified the relations among elements positions established.

1.1. Robot path planning

Robot path planning is the process of finding an enhanced collision-free path from a start to a predefined goal point through a certain given cluttered real world environment within the shortest possible time [1], [2]. Siegwart and Nourbakhsh [3] identified path planning as defining a trajectory through a map with which a robotic can reach a known goal point from its starting place while avoiding obstacles. The four integral path planning components that made up the navigation problem are: Perception, localization, motion control and path planning [4]. Figure 1 below shows the four integral parts of robot navigation problem. Localization as well as path planning are necessary components in navigating of mobile robots.

1.2. Bibliometric analysis

Bibliometric analyses, defined by Pritchard [5], is the application of mathematical and statistical methods to books and other means of communication. Bibliometric analysis has actually been used to systematically evaluate the progress of a details scientific field [6]. It has been used to review the distribution patterns of authors, journals, institutions, key phrases, topics, development patterns, as well as predict future research directions based on the search engine result. Bibliometric analysis has ended up being an important tool to measure the scientific progress in numerous fields [7].
There has been little effort made on the systematically review in path planning studies and in order to show the growth of the path planning, this article aims to analyse and visualize the existing path planning articles published with details such as the publication chronological distributions, the most appropriate scientific journals, and the popular key word phrases. Using " robot path planning " as the major key phrase, we determined over 14000 articles prior to being categorized into 5385 main related articles. All these are taken mainly from the Web of Science Core Collection. With the selected 5285 articles, we did an analysis by creating the connection between the abstract, title, publication, citation, research study area, geographical allocation and also the keyword phrases used. Through bibliometric co-citation, cartography analysis, and content analysis of the literature, this article fills the research gap on robot path planning review

The main aim of this survey paper is to perform a bibliometric analysis aiming to answers the following questions

- How are path planning articles clustered?
- What are future path planning research questions that provide opportunities to further the role of robot path planning?
- Which channels (journals, articles, and countries) are the most influential in Path planning?

describes the Bibliometric methods and results respectively. Finally, the last section concludes the survey.

2. Methods and Data sets

In bibliometric analysis, we aim to show affiliations among articles and also research subjects by examining how frequently an article is cited and co-cited by other articles. The article is the standard unit of evaluation [8], [9]. A fundamental presumption in bibliometric co-citation analysis is that published articles in academic journals develop their research study on articles already published. Bibliometric co-citation meta-analysis can help to disclose underlying commonalities among research study [10].
The ISI Web of Science (WOS) consists of different articles with their bibliographic references, institutions as well as authors [11]. Articles for all years (since1900) are available [12]. It is an item stemmed from "Thomson Reuters Institute of Scientific Information" (ISI) with more than 12,000 journals of different titles and multiple disciplinary [13]. The database gives effective searching choices by enabling various alternatives to filter the search results page [14]. In enhancement to the browsing alternatives, the WOS is additionally able to arrange the journals/articles based on particular specifications such as time added, publication date, citation, significance, and usage count. WOS database allows particular omission by article, author, institutions, countries and many more by fine-tuning the search result.
In order to avoid overlap between the data sources, Scopus and Google Scholar were left out in this study. The uncertainty of Google Scholar in carrying out research analysis and low data quality were some other reasons for its exclusion. Some database (like IEEE Explore, Springer and so on) only index their published articles while WOS combines all the high impact articles and journals from all of the databases. Also, WOS is used in this study due to its popularity in performing a bibliometric analysis of articles in scientific fields [15]. WOS database declares it has more quality than both Scopus and Google Scholar [16].

Topic search was "path planning". The search time span was set from the year 2000 to the year 2020. The retrieval time was 2020.03.28. Topic search suggests search from titles, abstracts, and keywords of the articles. A total of over 14304 articles prior to being categorized into 5385 main related articles were retrieved for bibliometric analysis. In this study, the articles publishing trend were first illustrated. The evaluation was performed based on productivity and research focus. Visual results were further provided using the VOSviewer tool [17] due to its availability and great features used in bibliometric analysis.

3. Result and discussion

3.1. Publication trend
Figure 1 shows the year-wise frequency of publications and the trend in the number of publications from 2000 to 2020. It is observed that just a few articles were published between 2000 and 2003. After 2004, the variety of publications began to rise slightly, and also there were more than 500 publications in 2007. Ever since, the number of articles each year has been increasing, which shows that path planning study has received more interest with publication in 2019 having the highest possible number of citations 917 with the year 2020 still counting.

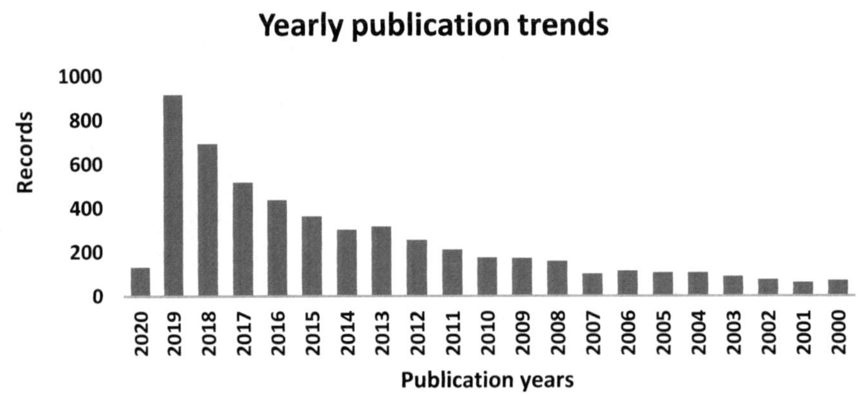

Figure 1. Publication trends

3.2. Publishing journals impact
Relative to the source journals, these 5385 publications were released in 1128 journals. Amongst all the 1128 journals, about 270 journals only released one article in the research study time. Table 1 shows the leading 20 most efficient journals, making up 34.746% of the overall journals. IEEE Robotics and Automation Letters was one of the most efficient journal with 196 write-ups, representing 3.64% of the publication Journal of Intelligent Robotic Systems ranked the 2nd with 171 articles, accounting for 3.179%. The third most productive journal was Robotics and Autonomous Systems, with 157 articles (2.916%).

3.3. Research focus
The keywords frequently used by the published articles in the journals were analyzed. Using the keyword, previously published as well as current articles in the journal can be traced and analyze to identify research gaps. Articles theme with authors keyword have been provided by WOS since 1990 [18], [19]. To measure the co-occurrence link within the path planning keyword phrases, 200 key phrases were drawn out. Nevertheless, the number is also large to classify them into understandable research clusters. Hence, the key phrases were manually further refined to eliminate those words that appeared twice as a result spelling changes, and repetition.181 key phrases were then picked. Fig 2 demonstrate

how the key phrases are interlinked with each other in the path planning research study domain. 4 clusters were developed, with the biggest cluster having 65 items while the smallest having 16.

Table 1. Publishing Journals

Source Titles	records	% of 5385
IEEE Robotics and Automation Letters	196	3.64
Journal of Intelligent Robotic Systems	171	3.175
Robotics and Autonomous Systems	157	2.916
International Journal of Advanced Robotic Systems	144	2.674
International Journal of Advanced Manufacturing Technology	133	2.47
IEEE Access	121	2.247
Robotica	121	2.247
International Journal of Robotics Research	113	2.098
IEEE Transactions on Robotics	108	2.006
Sensors	93	1.727
Advanced Robotics	63	1.17
Autonomous Robots	61	1.133
Robotics and Computer Integrated Manufacturing	61	1.133
Applied Sciences Base	53	0.984
Lecture Notes in Computer Science	49	0.91
International Journal of Robotics Automation	48	0.891
Mathematical Problems in Engineering	46	0.854
Computer Aided Design	45	0.836
IEEE Transactions On Automation Science and Engineering	45	0.836
Applied Soft Computing	43	0.799

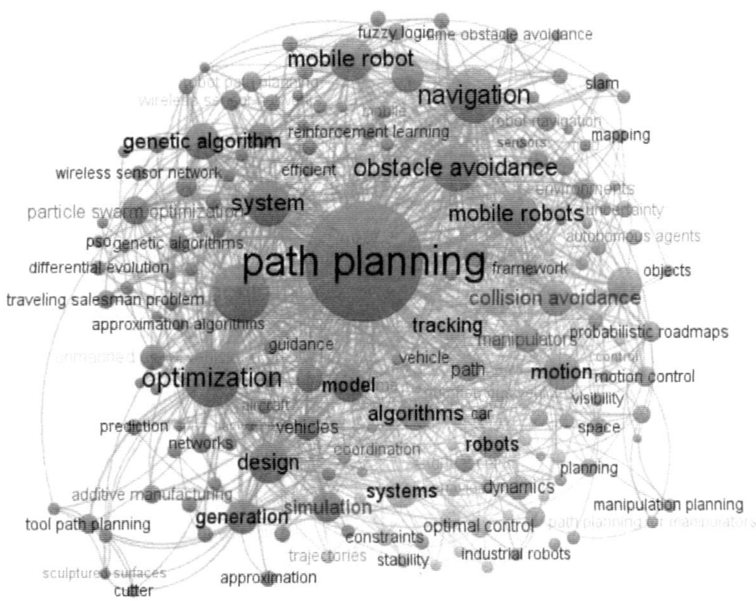

Figure 2: Authors keyword network

Cluster # 1 is the one in red color. As seen, it contains one big circles at the center, about three mediums once and then some small scattered ones. The size of a circle indicates the overall occurrence as well as link strength of the key words in the path planning research area. The leading research subject indicated by the big circle is the Path planning. Path planning is the process of finding a Collison-free path from a start to a predefine goal point through a certain given cluttered real world environment. Path planning plays a major rule in the navigation of autonomous vehicles. It is used in several applications such as automation, games, artificial intelligent etc. Other cluster members are algorithm, Optimization, differential evolution, robotics, genetic algorithm etc. Most of which are path planning algorithms and have proven to be effective in finding an optimal path for a robotic system.

Cluster #2 is the one with the green color, and it has 42 items. The most prominent research topic is motion and collision avoidance with 192 and 180 occurrences respectively. Collision avoidance is one of the key features required by autonomous. The next in the cluster is algorithm with 134 links and 169 occurrences. They are all effective path planning features that are often used in path planning research. Cluster #3 is the one with a Deep-Blue color and contain 37 items in total. The most prominent path planning research keyword is navigation. It appears to be the most significant research area in the entire clusters, as it has the biggest circle size and is densely connected to the other major clusters as shown in figure 3. The remaining phrases are mobile robot, robot navigation etc.

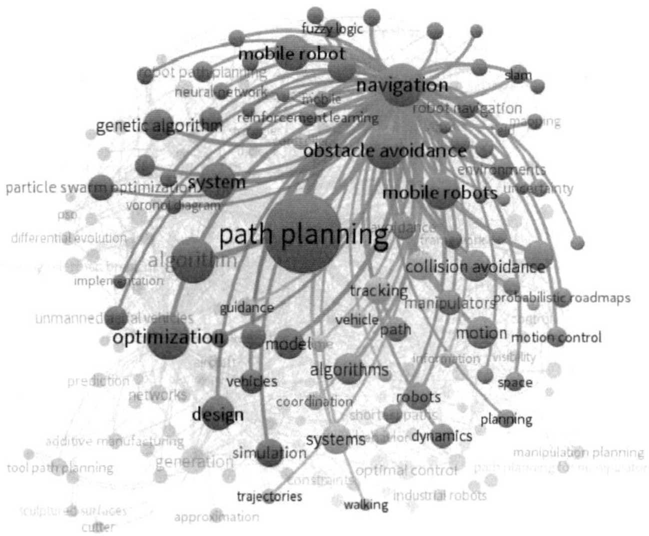

Figure 3. cluster #3 keyword network

Cluster #4 has 35 keywords, and its denoted by the Yellow color figure 4. The three dominant research phrases in the cluster are system, tracking, optimal control, constraints, trajectory coordination are some of the major trending research optimizers in the cluster.

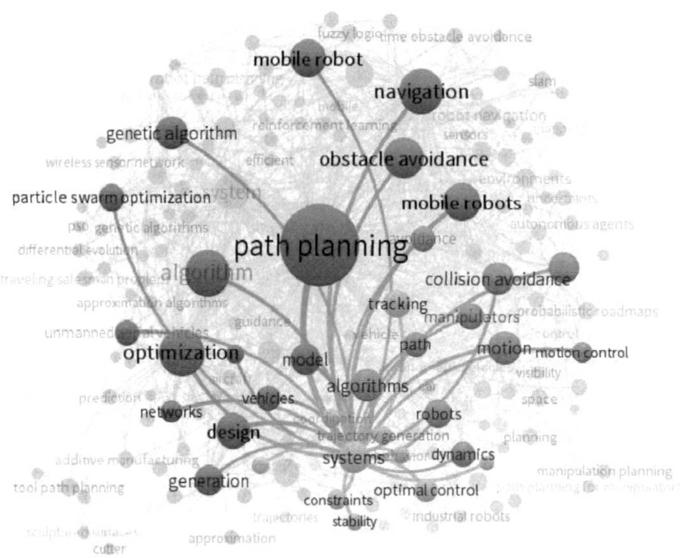

Figure 4. cluster #4 keyword network

4. CONCLUSION

Path planning has actually been practiced worldwide as such a lot more academic efforts have actually been made in this field. In order to summarize the research progression and also suggest future directions of path planning research, this study carries out a bibliometric analysis on the relevant publications published during 2000 to 2020. Result show that 5385 articles were published. The yearly publication rises rapidly after the year 2007, reflecting global interest in the topic. These publications were published in 1128 journals, indicating the publication diversity. IEEE Robotics and Automation Letters, Journal of Intelligent Robotic Systems, Robotics and Autonomous Systems ranked the leading three journals in regards to total publications. keywords search phrases was carried out to track the top topics and research trend. The hot topics includes: path planning, robot navigation, optimization and obstacle avoidance. With the rise of obstacle avoidance and optimization more research will focus concentrate on such an area in the near future. Generally, this research provides useful insights as well as future research study directions on path planning research study so that researchers in this area can better recognize their research study topics and also look for appropriate research partners.

Acknowledgments
The authors would like to thank Universiti Malaysia Pahang (UMP) for providing the research grant (PGRS1903182) and Ministry of High Education (MOHE) - under research grants: RDU180313 and the facilities.

References

[1] Amrit, K. J. Dheerendr, M. Sangheeta. Path Planning by a Robot is Static Environment using Hybridization of Particle Swarm Optimization and A* Algorithm. *International journal of computer Architecture and Mobility*, vol. 4, no, 2319-9229.

[2] Yuan-Qing Q, De-Bao S, Ning L, Yi-Gang C 2004 Path planning for mobile robot using the particle swarm optimization with mutation operator. In Third *International Conference on Machine learning and Cybernetics*, Shanghai.

[3] Roland S, Illah RN, Davide S. 2004 Introduction to autonomous mobile robots. MIT-Press.

[4] Thi T. M, Cosmin C, Duc TT, Robin DK. 2016 Heuristic approaches in robot path planning: A survey. *Robotics and Autonomous Systems* 2016; 86: 13-28.

[5] Pritchard A. 1969 Statistical bibliography or bibliometric*? Journal of Documentation* **25(4):** 348-349.

[6] Michael H, Simon S, Ken F.2009 The quantitative crunch: The impact of bibliometric research quality assessment exercises on academic development at small conferences. *Campus-Wide Information Systems* **26(3)**: 149-167.

[7] Mao G, Liu X, Du H, Zuo J, Wang L. *Way forward for alternative energy research: a bibliometric analysis during 1994-2013.* Renew sustain energy review 2015; 48: 276-286.

[8] Koskinen J, Isohanni M, Paajala H, Jääskeläinen E, Nieminen P, Koponen H, Tienari P, Miettunen J. 2008 How to use bibliometric methods in evaluation of scientific research? An example from Finnish schizophrenia research.. Nord. J. Psychiatry; 62: 136–143.

[9] McKerlich R., Ives C., McGreal R.. Comparing bibliometric statistics obtained from the web of sciences and scopus. *Int. Rev. Res. Open Distance Learn* 2013a;14:90–103.

[10] Aisha M, Mohammed AHA, Ibrahim HS. 2020 ANSYS –A Bibliometric study. In 10th *International Conference of Materials Processing and Characterization*. India,

[11] Mongeon P., Paul-Hus A. 2016 The journal coverage of web of science and scopus: a comparative analysis. *Scientometrics* ; 106: 213–228.

[12] Mingers J., Leydesdorff L. 2015 A review of theory and practice in scientometrics *Eur. J. Oper. Res*; 246: 1–19.

[13] Dehdarirad T., Villarroya A., 2015 Barrios M. Research on women in science and higher education: a bibliometric analysis. *Scientometrics* 103: 795–812.
[14] Fahimnia B., Sarkis J., Davarzani H. 2015 Green supply chain management: a review and bibliometric analysis. *Int. J. Prod. Econ* ; **162:** 101–114.
[15] McKerlich R., Ives C., McGreal R. *Comparing keywords plus of WOS and author keywords: a case study of patient adherence research.* Int. Rev. Res. Open Distance Learn 2013b; 14: 90–103.
[16] Chadegani AA., Salehi H, Yunus MM, Farhadi H, Fooladi M, Farhadi ., Ebrahim NA. A comparison between two main academic literature collections: web of science and scopus databases. *Asian Soc. Sci* 2013; 9: 18–26.
[17] Van Eck NJ, Waltman L. Software survey: VOSviewer, a computer program for bibliometric mapping. *Scientometrics* 2010; 84: 523–538.
[18] Sun J, Wang MH., Ho YS. *A historical review and bibliometric analysis of research on estuary pollution.* Mar. Pollution Bulletin 2012;64: 13-21.
[19] Wu X, Chen X, Zhan F B, Hong S. *Global research trends in landslides during 1991 – 2014: a bibliometric analysis.* Landslides 2015; 12: 215–1226.

IOP Conf. Series: Materials Science and Engineering 1045 (2021) 012018 doi:10.1088/1757-899X/1045/1/012018

Thermal analysis of a Thermoelectric Generator (TEG) using FEM technique

Prajwal K T[1,3] and Pritam Bhat[2]

[1] Department of Electrical Engineering, M S Ramaiah University of Applied Sciences, Bangalore, India

[2] Department of Mechanical and Manufacturing Engineering, M S Ramaiah University of Applied Sciences, Bangalore, India

[3] E-mail: ktprajwal@gmail.com

Abstract. Finite Element Method (FEM) is one of the numerical technique based on minimization of energy, developed to solve complex engineering problems by converting governing differential equations of the physical phenomenon into algebraic equations that can be easily solved by a computer. Finite Element Analysis (FEA) is a computer application that uses FEM technique to analyze the physics of given problem based on the user inputs provided such as material data, geometry, and other pre requisites to study the behavior and structural strength in case of structural mechanics. A device working on principle of Seebeck-Thomson effect is Thermoelectric Generator (TEG) that converts the difference of heat energy into electrical energy. Thermal analysis of a TEG is a necessary step in design engineering as the electricity generated by the device is dependent on the difference in heat flux (temperature difference) between the source and sink (two ends) of TEG. In this paper, Finite Element Analysis of Thermoelectric Generator is carried out. The model is constructed, and thermal analysis is carried out in this paper.

1. Introduction

Amid growing environmental concerns regarding the clean power generation, TEG has set its footprint as a promising alternative technology to generate electricity from available thermal energy. TEG is a device that works using thermoelectric materials that have a property by virtue of which, heat flux (temperature difference) is converted into electrical energy due to Seebeck-Thomson effect [1]. Due to the advancement in the availability of thermoelectric (TE) materials, TEG finds wide applications in waste heat recovery units/plants as they can readily convert low grade heat into electricity. [2] The other advantages of TEG are, there are no mechanical parts thus free of wear, tear, noise and vibrations without leaving any carbon footprint on the environment. Some specific domain such as wearable read application in other energy fields. [3]

A thermoelectric material subjected to differential heat flux at their ends, generate electricity while connected in a circuit known as thermoelectric module. A thermoelectric effect is produced due to presence of two dissimilar thermoelectric materials with a p-type (positively charged) and an n-type (negatively charged) semiconductor joined at their ends. Whenever there is a temperature difference between the two materials, it so happens that charges (electric current) flow in the circuit. By placing a resistive load in the circuit, voltage is generated. Generally, magnitude of electric current generated is directly related to the applied differential heat flux (temperature difference). [4].

Thermoelectric modules are used in power generating plants to recover waste heat rejected into the atmosphere. Due to very high temperatures that exist in exhaust of chimneys, thermoelectric devices are subjected to enormous stress and strains arising due to thermal load on the system. Thus,

IOP Conf. Series: Materials Science and Engineering 1045 (2021) 012018 doi:10.1088/1757-899X/1045/1/012018

a FEA analysis of TEG can help in predicting the induced thermal stresses and performance leading to optimized design suitable to the given application. The geometry of the TEG and the selection of right thermoelectric materials has a great effect on its efficiency as it has to be designed to work under harsh conditions that exist in power plants [3].

In this study a numerical model of a typical TEG is modelled in commercial FEA package ANSYS APDL to study the thermal characteristics and the behavior of thermoelectric materials under a temperature difference of around 3000 C. The thermal stresses and other parameters are evaluated, and the obtained results can serve a purpose in design and development of an efficient TEG specifically for waste heat recovery from an IC engine exhaust.

2. Methods and course of the study

A typical TEG is shown in Figure 1. It consists of heat source, heat sink, P and N- type junctions that are connected to resistive load. T_h and T_c are the heat source and heat sink temperatures respectively. As the heat load is absorbed at the hot end and resistive load Rl is applied, current I will flow through the circuit due to Seebeck effect. The Thomson effect is also introduced due to joule heating at the two ends [3].

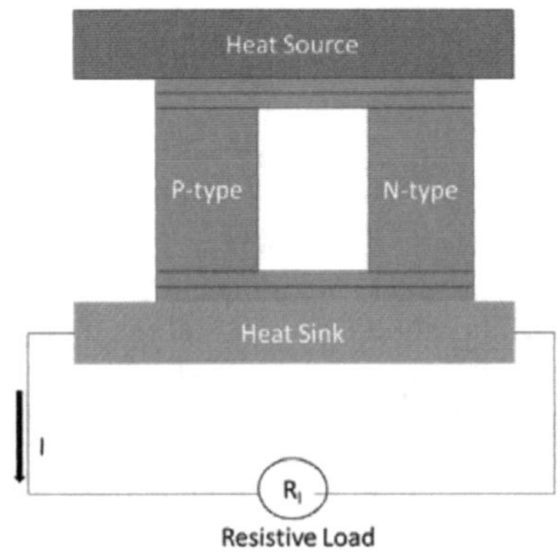

Figure 1.Schematic of the thermoelectric generator considered for FEA analysis

The TEG can be modelled using two approaches namely numerical model and thermal resistance model [2]. Numerical model can include the thermoelectric-fluid coupling effects, temperature dependent material properties. Thermal resistance model accounts for temperature dependent material properties as well as thermal contact resistance between different materials and can heavily safe the computing time as required in numerical models.

In this work, we have adopted a numerical modelling technique with few assumptions to perform the thermal analysis of TEG. The assumptions made are realistic and do not affect the solution of the problem. Several energy balance equations are required to numerically model the TEG. Equations used in FEM solver are detailed as follows [5-6].

The equation of heat flow **q** is

$$\rho C \frac{\partial T}{\partial t} + \nabla \,.\, \boldsymbol{q} = \dot{q} \tag{1}$$

Where ρ is the density, C is specific heat, t is the time, $\dot{q}$ is the heat generation rate per unit volume.

Electric charge equation is given by

$$\nabla \,.\left(\boldsymbol{J} + \frac{\partial \boldsymbol{D}}{\partial t}\right) = 0 \tag{2}$$

Where J is the electric current density, D is the electric flux density vector

Thermoelectric constitutive equations are as follows

$$\boldsymbol{q} = \boldsymbol{\pi} \,.\, \boldsymbol{J} - \boldsymbol{\kappa} \,.\, \nabla T \tag{3}$$

$$\boldsymbol{J} = \sigma .\, (\boldsymbol{E} - S \,.\, \nabla T) \tag{4}$$

And the dielectric medium constitutive equation is

$$\boldsymbol{D} = \epsilon \, . \boldsymbol{E} \tag{5}$$

Where π is Peltier coefficient, ϵ is the dielectric permittivity, κ is the thermal conductivity.

The scalar electric potential φ is introduced to obtain the electric field for irrotational electric field and time constant magnetic field.

$$\boldsymbol{E} = -\nabla . \varphi \tag{6}$$

Combining all the above equations, thermoelectricity equations are obtained as follows.

$$\rho C \frac{\partial T}{\partial t} + \nabla . (\boldsymbol{\pi} . \boldsymbol{J}) - \nabla . (\boldsymbol{\kappa} . \nabla T) = \dot{q} \tag{7}$$

$$\nabla \left(\epsilon . \nabla \frac{\partial \varphi}{\partial t} \right) + \nabla . (\sigma . S \nabla T) + \nabla . (\sigma . \nabla \varphi) = 0 \tag{8}$$

The above coupled thermoelectric equations are solved in FEM solver to obtain the heat absorbed and the electric current generated by the TEG. The thermal and electrical effects due to Seebeck and Peltier along with Thomson effects (Joule heating at source/sink) are considered in these equations. The output power and TEG efficiency is respectively given by

$$P = I^2 R_l \tag{9}$$

$$\eta_{TEG} = \frac{P}{Q_{in}} \tag{10}$$

Following are the assumptions made in the FEM solution to save the computing time [7].

- Radiation and convection heat losses are neglected
- Internal thermal contact resistance between the materials are ignored

2.1. Designing of TEG using fem technique

I. Construction

There are 3 layers in the top and 2 layers in the bottom, between the P, N junction layers.
Layer 1: Material 1: The topmost layer is the aluminum layer.
Layer 2: Material 2: Aluminum Ceramic
Layer 3: Material 3: Copper
Material 4: N-Type
Material 5: P Type
Layer 4: Copper
Layer 5: Aluminum Ceramic

The steps followed in the ANSYS APDL is as follows:

1. Select the preferences as thermal and electric.
2. Preprocessor - choose element type - add/edit/delete - add solid 279, circuit124
3. Material props - material models - add thermal conductivity, and electrical resistivity for all 5 materials and Seebeck coefficient for material 4 and 5.

Table 1. Geometric and Material property values used in FEA Model

Material Definitions and geometry	Al	Al_2O_3	Cu	N-type BiTe	P-type BiTe
Electrical Resistivity(ohm m)	0.25	1*10e18	0.18	9.8	13.8
Thermal Conductivity(W/mk)	200	37.2	398	1.78	1.22
Seebeck Coefficient	--	--	--	16.8	16.8
Thickness	2.5	2.5	2.5	2.5	2.5
Length	10	1	0.5	20	20
Width	50	40	40	25	15

*All dimensions are in mm.

4. Model the TEG model as per the dimensions in Table 1[8].
Modelling: create - using areas the given model, extrude them to a depth of 2.5
Booleans - glue- volumes.
Thus, the modelling of the TEG is complete. The model of the TEG is obtained as shown in Figure 2.

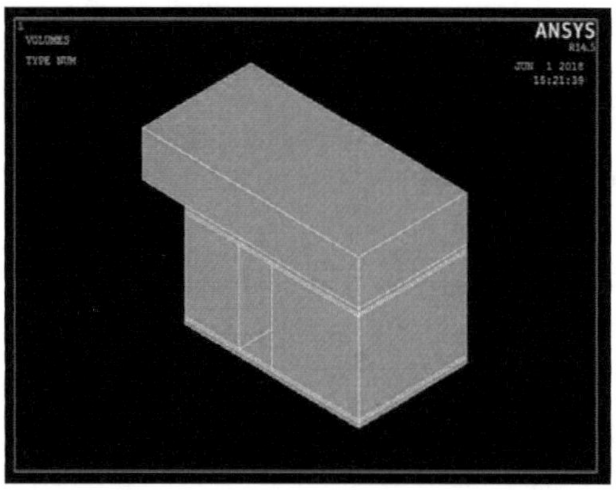

Figure 2. The geometric model of thermoelectric generator in ANSYS APDL

5. Meshing - mesh tool -
Set element attributes at volumes. Assign the material properties to different materials. Mesh the given model. Figure 3. Shows the tetrahedral elements used in meshing of TEG. The grid independence test was conducted in order to minimize error in results due to mesh size.

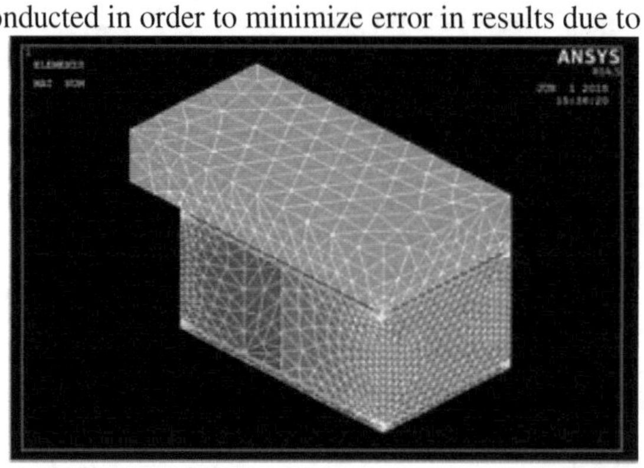

Figure 3. Meshed model of thermoelectric generator in ANSYS

6. Solution - loads - define loads - apply - thermal - temperature - on areas
The analysis of the thermal distribution is carried out by assigning the temperature of 327oC to the top face of aluminum plate and 27oC to the other side of the cold ceramic plate made up of AL2O3.
7. Solution - solve - current ls - ok
8. General postproc - plot results - nodal solution - nodal temperature. The results are obtained.

2.2. Selection of proper element type and boundary conditions in ANSYS

SOLID 279, CIRCU124 are the two element types used in the TEG model simulation in ANSYS. SOLID279 an element type that is used in ANSYS for analysis purposes. It is quadratic in nature and can be used for thermal analysis purposes. The element is subjected to spatial orientation in nature and can be represented by many nodes. Each node represents a temperature nodal equation.
The element is anisotropic in nature. The input to these elements can be heat. The heat can be represented by the convection and the radiation techniques. The temperature load can be represented by nodal or surface loads.
The solution production related by the element are in binary methods:

- Nodal temperatures comprised in the complete nodal solution
- Use SOLID279 Covered Thermal Solid to model heat conduction in layered thick shells or solids.

CIRCU124:
This element is used for electrical analysis. Using this method, the electrical output from the system can be analyzed. Using this there is an option of getting upto 6 nodes in direction as the element contains 6 nodes in different direction.
The three degrees of freedom that can be associated with the element is:

- VOLT (voltage)
- CURR (current)
- EMF (potential drop)

3. Research Results

The result analysis is carried out using the ANSYS APDL solution module. The TEG module is evaluated for the temperature distribution of the hot and cold side of the model. The solution can be obtained by first meshing the module. The model can be meshed using many types of meshes. In this triangular meshes is used.

Figure 3 illustrates the meshed model of the TEG. The meshing of the model is carried out to divide the model into several smaller pieces of finite elements to solve the equations and obtain the solution.

IOP Conf. Series: Materials Science and Engineering 1045 (2021) 012018 doi:10.1088/1757-899X/1045/1/012018

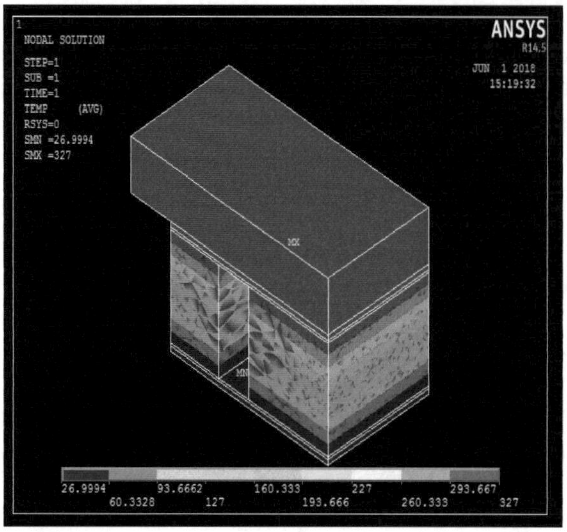

Figure 4. Temperature distribution of TEG in ANSYS APDL

The meshing can be done in done in many methods. Simple and efficient method of meshing is by using fine mesh.

Figure 4 shows temperature distribution of the generator. From the figure it can be seen that the hot side of the generator is experiencing highest temperature while cold side of the generator settles to lowest temperature. The hot side temperature ranges up to 327°C and cold side of the temperature is about 27°C.

This approves by the temperature loads allocated previously while modelling.

4. Conclusions

Using FEM, we have modeled thermoelectric generator and studied the thermal analysis of the same. FEM technique has proved to be the best way to analyze engineering structures. The FEA technique can be effectively leveraged to predict the performance of thermoelectric generators that helps to design and improve the deployment and potential of TEG in real life applications.

References

[1] Van Herwaarden, A. W., and P. M. Sarro. "Thermal sensors based on the Seebeck effect." *Sensors and Actuators* 10.3-4 (1986): pp. 321-346.

[2] Luo, Ding, Ruochen Wang, and Wei Yu. "Comparison and parametric study of two theoretical modeling approaches based on an air-to-water thermoelectric generator system." *Journal of Power Sources* 439 (2019): 227069.

[3] Shittu, Samson, et al. "Advancements in thermoelectric generators for enhanced hybrid photovoltaic system performance." *Renewable and Sustainable Energy Reviews 109 (2019)*: pp. 24-54.

[4] Ramadass, Yogesh K., and Anantha P. Chandrakasan. "A battery-less thermoelectric energy harvesting interface circuit with 35 mV startup voltage." *IEEE Journal of Solid-State Circuits* 46.1 (2010): 333-341.

[5] Antonova, Elena E., and David C. Looman. "Finite elements for thermoelectric device analysis in ANSYS." ICT 2005. *24th International Conference on Thermoelectrics, 2005*. IEEE, 2005.

[6] Zhang, Mengjun, et al. "Influence of Thomson effect on the thermoelectric generator." *International Journal of Heat and Mass Transfer* 137 (2019): pp. 1183-1190.

[7] Fan, Shifa, and Yuanwen Gao. "Numerical analysis on the segmented annular thermoelectric generator for waste heat recovery." *Energy* 183 (2019):pp 35-47.

[8] Zulkifli, Muhammad Nubli, et al. "Finite element analysis on the thermoelectric generator for the waste heat recovery of solar application." *AIP Conference Proceedings*. Vol. 1877. No. 1. AIP Publishing LLC, 2017.

Tuning and experimental assessment of second-order generalized integrator –frequency locked loop grid synchronization for single-phase grid assisted system

Bhavik Brahmbhatt[*1] **and Dr. Hina Chandwani**[2]

[1]Electronics and communication Department, Government Engineering Collage, Modasa, Gujarat, India
[2]Electrical and Electronics Department, The Maharaja Sayajirao University of Baroda,Vadodara, Gujarat, India

*email-bhavik0072009@gmail.com

Abstract. The phase-locked loop (PLL) is an essential part of the grid-tied system to synchronize control of converter with grid voltage, particularly affects the converter stability as well as performance under weak grid conditions. During abnormality in a grid, its bandwidth ought to be a11dequately brought down to achieve appropriate disturbance rejection capability with a compromise to slower detection speed. The researchers have done intended work in advanced PLLs to improve phase-angle detection speed by modifying the pre/in-loop filtering stage. A most concerning challenge with the PLLs is the means by which to additionally provide superior dynamic performance and reduced settling time without bargaining stability of system along with the capability of disturbance elimination To overcome this challenge, this paper describes the second-order generalized integrator (SOGI-Frequency Look loop (FLL) which offers filtering capability like band-pass filter, low-pass filter, and notch filter for adaptive frequency tuning as well as an orthogonal signal generation for the grid-tied photovoltaic inverter. The impact of control parameters on their dynamic performance in SOGI-FLL is tabulated from a step response of frequency estimation by taking a frequncy shift. Furthermore, Low-cost DSP based STM32F407VGT the microcontroller is employed to implement a SOGI-FLL to test under adverse grid conditions using Waijung Block-set of SIMULINK/MATLAB. The experimental results of SOGI-FLL have proven superior dynamic performance over type-2 PLLs by choosing the appropriate value of the control parameter of SOGI-FLL.

Keywords: Phase detection, phase-locked loop (PLL), grid synchronization,STM32F407VG microcontroller,WAIJUNGBlock-set

1. Introduction

The phase-lock loop is employed to synchronize its internal voltage controlled oscillator with a grid voltage signal in order to extract phase-angle which, further, utilized by different control system blocks of the photovoltaic converter. The phase-lock loop can estimate angular frequencyω_g, phase angle θ_g, and frequency f_g for the control block of grid assisted converter from grid voltage and also ensure the protection of converter against an islanded mode. The zero-crossing detection method was first applied as synchronizing technique for grid assisted photovoltaic system (PV),which uses comparators for finding polarity changes in the grid voltage signal. During abnormalities in grid voltage or the case of weak grids, it had a drawback of multiple zero-crossings detections, further which cause issues for extracting phase-angle ofgrid voltage. Stationary frame PLLs are suitable as well as efficient in single-phase grid assisted PV converter due to the only one voltage signal to synchronize as well as increase in the speed of synchronization in comparison with other methods. Classical PLL systems consist of a multiplier phase detector (PD) that yields an error signal ε_{pd} for the low-pass filter (i.e. loop filter) (LF); which is further created $\omega_e$for the phase-angle extraction. The Phase detector is having an inherent drawback of producing a (2ω) ripple components in ε_{pd} that further proliferates through the Loop Filter [1,2,4].It can be minimized by adding another low-pass filter in a loop that encounters this 2ω term at the cost of reduction in bandwidth and/or phase-margin, slower transient response, and decrease in overall speed of synchronization [1,4-7]. Synchronous frame (DQ) PLLs can be adopted to compute frequency and phase angle as well as transform the grid voltage into DC signals. It can be controlled by a simple PI controller with zero steady-state errors in output while PI regulators cannot be used in the stationary frame. However, DQ PLL requires two signals i.e. a direct signal and an orthogonal signal (quadrature signal) for the computation of frequency and phase angle[10-11]. Therefore, an orthogonal signal is derived from direct signal i.e. sensed grid voltage for the transformation from the stationary to the DQ frame by using different methods such as phase delay filters, differentiation of the input signal, Inverse Park's transform, Hilbert transform, [6-7,14],and the second-order generalized integrator (SOGI)[11-12]. In a phase delays filter, generally two low -pass filters with the 0.707 gain and 45° phase are cascaded and their output further divided with 0.5 to create signal with unity gain and 90° phase[1,4,5].But they can cause inaccuracies to arise as a result of slow varying characteristics of the grid within tolerable range whereas differentiation of the input voltage signal method is having an issue of noise amplification during the generation of the orthogonal signal [19, 20]. This paper describes a second Order Generalized Integrator (SOGI)-Frequency Locked Loop(FLL) method for generating the orthogonal component of a 1Φ and phase-angle exctraction for DQ PLL. By appropriate choice of control parameter values, SOGI-FLL offer band-pass, low pass, band stop filtering along with estimation of phase angle, frequency, amplitude as well as offset parameters. Therefore, it can also eliminate the 2ω ripple without using LPFs; subsequently increase inherent synchronization speed as well as bandwidth as compared with convention PLL. The SOGI-FLL method can be use for a phase-angle detection in the single phase as well as three phase grid tied inverter, active filters in power system. It can also used as stationatary-frame current controller and fundamental component extraction from any voltage or current signal.This paper describes how these additional feedback with appropriate selection of control parameter improve synchronization speeds during grid abnormality, increase robustness against input noise and disturbances, and estimate accurate and non-distorted values phase-angle for the control system.Reference [1-6] implemented SOGI-FLL in single-phase grid assisted PV system, but none of the above have focused the impact analysis of parameter k and γ and their effects in transient and steady-state response SOGI-FLL.In this paper,the Imapct analysis of control parameter is tabulated from step reponse of frequency estimation by taking a frequency shift. paper elaborate an analysis of second order generalized integrator and additional structure extensions that allow estimation of the parameters of a periodic signal.

2. Structure of Second-order generalized integrator (SOGI)

The structure of SOGI, as shown in Figure 1, and their transfer functions in equation (1) & (2) are indicating that two imaginary complex conjugated poles placed at $\pm j\omega_0$ that behaved like resonator oscillating at angular frequency ω_0. This feature offers infinite gain at resonator oscillating at angular frequency, ω_0 as observed from the bode plot of SOGI in figure 2 that can be useful in the implementation of voltage controlled oscillator (VCO) block in PLL.

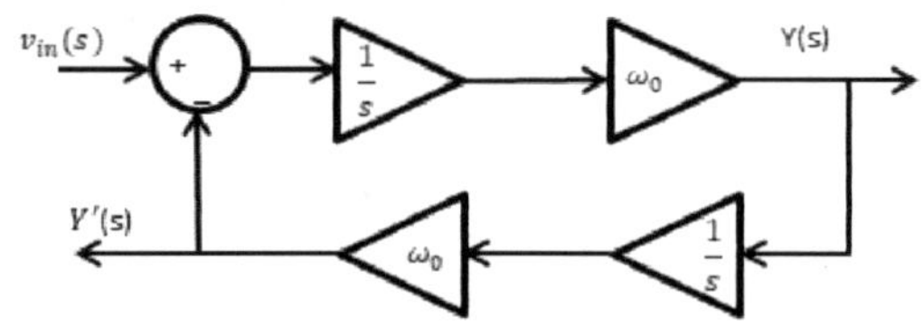

Figure 1. Basic structure of Second-order generalized integrator[11]

$$\frac{Y(s)}{v_{in}(s)} = \frac{\omega_0/_s}{1+\frac{\omega_0\omega_0}{s\ s}} = \frac{s\ \omega_0}{s^2+\ \omega_0{}^2} \tag{1}$$

$$\frac{Y'(s)}{v_{in}(s)} = \frac{Y'(s)}{Y(s)} * \frac{Y(s)}{v_{in}(s)} = \frac{\omega_0}{s} * \frac{s\ \omega_0}{s^2+\ \omega_0{}^2} = \frac{\omega_0{}^2}{s^2+\ \omega_0{}^2} \tag{2}$$

$$y(t) = L^{-1}[Y(s)] = L^{-1}\left[\frac{s\ \omega_0}{s^2+\ \omega_0{}^2} * v_{in}(s)\right] = L^{-1}\left[\frac{s\ \omega_0}{s^2+\ \omega_0{}^2} * \frac{\omega_0}{s^2+\ \omega_0{}^2}\right] = \frac{1}{2}[t\ \omega_0 \sin\omega_0 t] \tag{3}$$

$$y'(t) = L^{-1}[Y'(s)] = \frac{1}{2}[\sin\omega_0 t\ \ t\ \omega_0 \cos\omega_0 t] \tag{4}$$

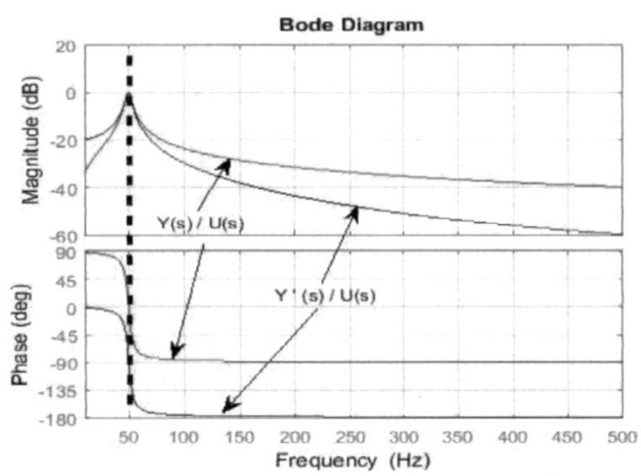

Figure 2. Bode diagram of transfer function (Y(s) / Vin (s))and (Y'(s) / Vin (s))of SOGI

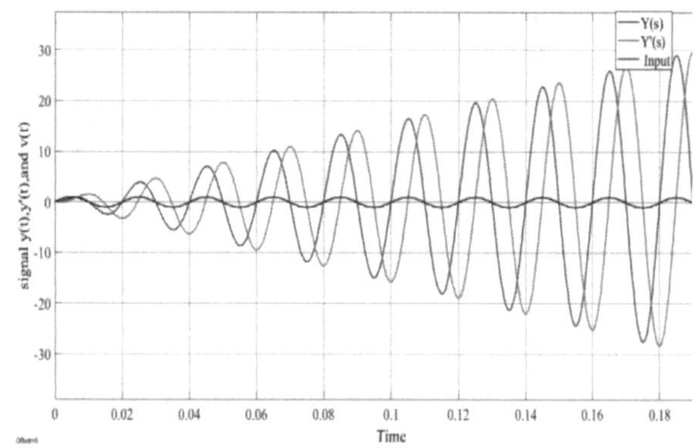

Figure 3. Step response of SOGI

Figure 3 shows the step time response of a SOGI structure as per eq.(3) & eq.(4) with $\omega_0 = 2\pi.50$ rad/s, respectively, that increase the amplitude of output signals which cause unstable system, when a unitary step is applied as input.To prevent the system to be unstable, input to the SOGI structure is modified as weighted k of (where k is the gain parameter of SOGI) the difference of input signal which is nothing but grid voltage$v_{in}(s)$and unity feedback of output signal Y(s) to the input as shown in Figure 4.

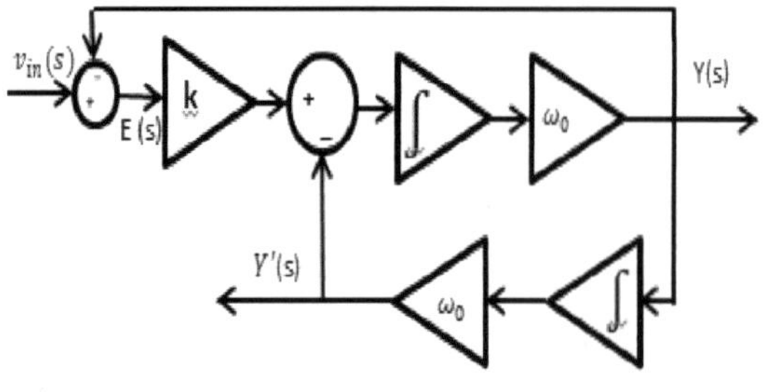

Figure 4. Block diagram of SOGI-OSG

As a result, the transfer functions for $D_v(s)$ and $Q_v(s)$ can be re-written as:

$$D_v(s) = \frac{Y(s)}{v_{in}(s)} = \frac{k\, s\, \omega_0}{s^2 + k\omega_0 s + \omega_0^2} \tag{5}$$

$$Q_v(s) = \frac{Y'(s)}{v_{in}(s)} = \frac{k\, s\, \omega_0}{s^2 + k\omega_0 s + \omega_0^2} \tag{6}$$

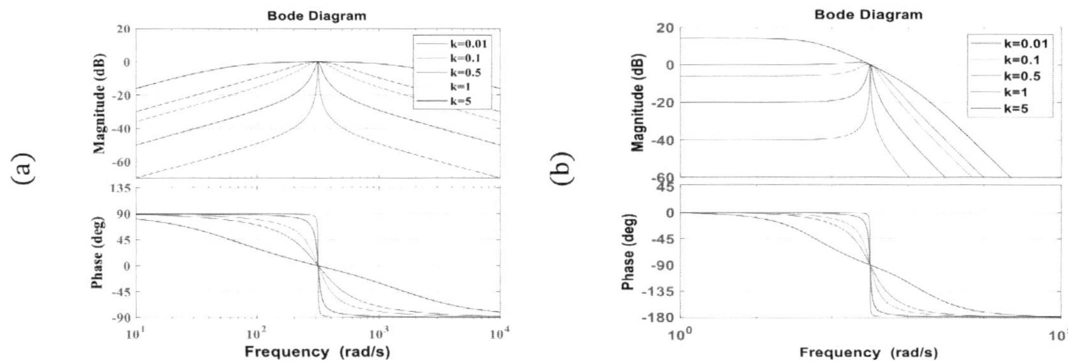

Figure 5 . Bode Diagram of (a) $\boldsymbol{D_v(s)}$, and (b) $\boldsymbol{Q_v(s)}$with different value of k

Bode Diagrams describe band-pass compatible nature of the output *Y*(*s*) (Figure 5) and nature of the output *Y'*(*s*) as low-pass compatible, while phase responses of curves at frequency 50Hz are observed at 0^0 and 90^0respectively which indicates that Y(s) is having90^0 phase lead to *Y'*(*s*) (Figure. 5). Moreover, The magnitude responses of Bode are maintaining 0 dB at the 50Hz fundamental frequency and also attenuating amplitude at 5th and 7th harmonics frequency i.e.250Hz and 350Hz.It is seen from Figure 5 that the transfer function Y(s) and *Y'*(*s*) are able to extract only fundamental components of grid voltage while eliminating harmonics components of grid voltage as second –order band-pass filter.

The SOGI-QSG is employed in SOGI-PLL to generate in-quadrature signal from single phase grid voltage for phase-angle extraction as depicited in Figure 6. The SOGI-PLL locks the phase angle of its internal oscillator to the input grid voltage signal as to when input frequency was detected while detected frequency allowed to remain properly tune SOGI-QSG.

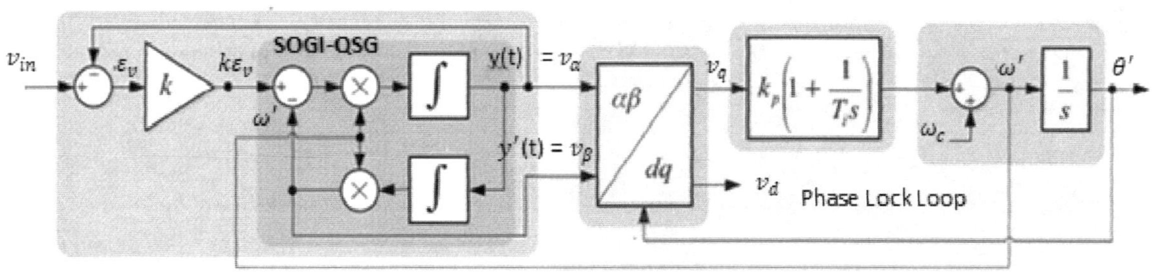

Figure 6. Basic diagram SOGI-PLL[14]

2.1 Frequency locked loop

However, SOGI has an inherent resonate character that can be used as voltage-controlled oscillator which emphasizes to design simple and robust single feedback control loop for an auto-adapting center frequency of SOGI resonator as per the input grid frequency. Frequency Locked Loop, FLL is nothing but the simple and robust extension of SOGI structure as shown in Figure 7. In order to make auto-tunable SOGI-QSG, it should pay attention to the voltage error signal $E_v\,(s)$, which is nothing but the difference between input $v_{in}\,(s)$ and output Y(s) and behave as notch filter with zero gain and180°phase-angle jump at center frequency as observed from a bode diagram in Figure 8, is described in term of transfer function by

$$E_v(s) = \frac{\varepsilon_{v(s)}}{v_{in(s)}} = \frac{s^2+\omega_0{}^2}{s^2+k\omega_0 s+\omega_0{}^2} \tag{7}$$

The transfer function of $E_v\ (s)$ and $Q_v\ (s)$ gives worthy information for auto-tunable frequency control system by taking common bode diagram of transfer function $E_v\ (s)$ and $Q_v\ (s)$, as depicted in Figure 8.

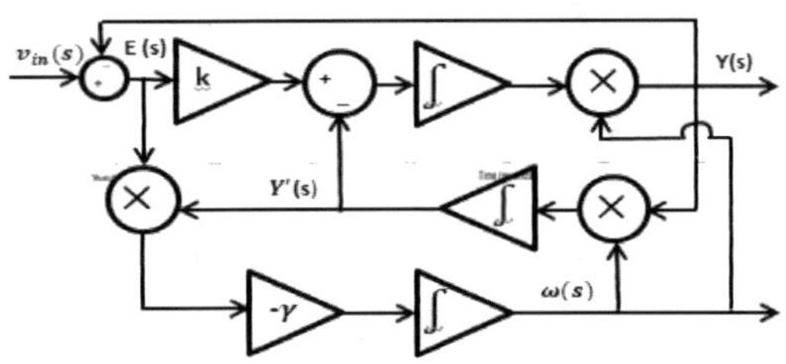

Figure 7. Basic structure of SOGI based Frequency lock loop

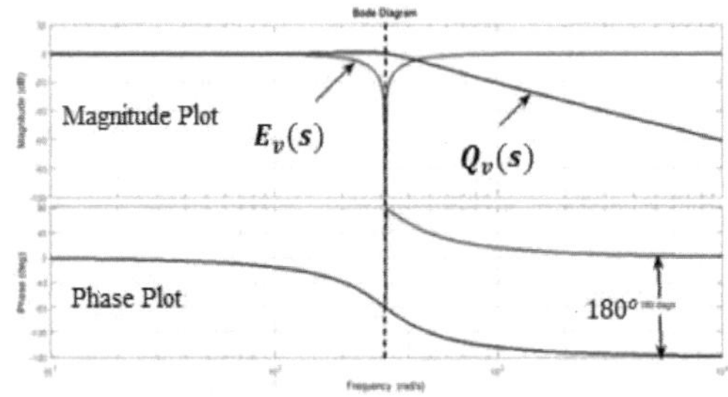

Figure 8. Bode Diagram of the $\boldsymbol{E_v(s)}$ and $\boldsymbol{Q_v(s)}$in an SOGI- QSG

The Bode diagram of transfer function $E_v\ (s)$ and $Q_v\ (s)$reveals that the phase of signals $E_v\ (s)$ and $Q_v\ (s)$ are in a phase when input frequency (ω) lower than SOGI resonance frequency (ω') i.e. $\omega < \omega'$ and out of a phase (180° phase difference) when $\omega > \omega'$, as indicated in Figure 8. Hence, a frequency error variable ε_f is derived from product of $E_v\ (s)$ and $Q_v\ (s)$, which remain positive when input frequency (ω) is lower than SOGI resonance frequency (ω'), zero when $\omega = \omega'$, and remain negative $\omega > \omega'$ in the SOGI-FLL.Moreover, the frequency locking loop can be designed by using frequency error variable, ε_f and a negative value of frequency loop controller gain,$-\gamma$ as shown in Figure 7. The frequency loop controller gain $(-\gamma\ )$ is used to achieve DC component of frequency error variable ε_f equal to zero by changing SOGI resonance frequency,ω', until equal to the input frequency, ω.A feed-forward variable, ω_c i.e. nominal value of grid frequency is provided in frequency locking loop to speed up the initial synchronization process. The SOGI-OSG and Frequency locking loop combined a structure diagram known as SOGI-FLL for single-phase grid synchronization system, as shown in Figure 7. The transient and steady-state behavior of the SOGI –FLL mainly depends on a suitable value chosen for control parameters γ and k in order to obtain desired response in the estimation of the frequency and amplitude of input signal.From the Figure 7, the state-space equations are described as:

$$x = \begin{bmatrix}\dot{x}_1\\ \dot{x}_2\end{bmatrix}=\text{Ax+Bu}=\begin{bmatrix}-k\omega' & -\omega'^2\\ 1 & 0\end{bmatrix}\begin{bmatrix}x_1\\ x_2\end{bmatrix}+\begin{bmatrix}k\omega'\\ 0\end{bmatrix}\text{u} \tag{8}$$

$$\text{y} =\begin{bmatrix}y\\ y'\end{bmatrix}=\text{Cx}=\begin{bmatrix}1 & 0\\ 0 & \omega'\end{bmatrix}\begin{bmatrix}x_1\\ x_2\end{bmatrix} \tag{9}$$

$$\dot{\omega}' = -\gamma\ x_2\omega'\ (u - x_1) \tag{10}$$

where x = $[x_1\quad x_2]^T$ and y = $[y\quad y']^T$ are the state variables and output vectors respectively.
The voltage error signal from state equations in eq.(8) and Figure 7can be written as

$$\tilde{\varepsilon}_v = (u - \tilde{x}_1) = \frac{1}{k\ \omega'}(\dot{\tilde{x}}_1 + \omega'^2\tilde{x}_2) \tag{11}$$

The steady–state frequency error signal$\tilde{\varepsilon}_f$ can be derived as:

$$\tilde{\varepsilon}_f = \omega'\tilde{x}_2\tilde{\varepsilon}_v = \frac{\tilde{x}_2{}^2}{k}(\omega^2 - \omega'^2) \tag{12}$$

The small-signal performance of FLL can be obtained by simplifying $(\omega^2 - \omega'^2) \approx \omega'(\omega' - \omega)$ taken into the consideration of steady –state condition i.e. $\omega = \omega'$

$$\dot{\omega}' = -\gamma\ \tilde{\varepsilon}_f = -\gamma\ \frac{\tilde{x}_2{}^2}{k}(\omega^2 - \omega'^2) \quad \approx -2\gamma\ \frac{\tilde{x}_2{}^2}{k}\omega'(\omega' - \omega) \approx -\frac{\gamma V^2}{k\ \omega'}(\omega' - \omega) \tag{13}$$

A value of γ should be normalized by using feedback and described as:

$$\gamma = \frac{k\omega}{V^2}\Gamma \tag{14}$$

3. Comparative Analysis

Reference [1]-[6] implemented SOGI-FLL in single-phase grid assisted PV system, but none of the above have focused the impact analysis of parameter k and γ and their effects in transient and steady-state response SOGI-FLL. The dynamic performance of frequency estimation in SOGI-FLL is examined by experiencing a -5 Hz frequency step change (i.e. frequency jump (50Hz to 45Hz) and phase angle shift (0^0 to45^0) at 200msec with feed-forward taking initial frequency value i.e. $2\pi * 50$.

Table 1. IMPACT ANALYSIS OF SOGI-FLL

Parameter	Progressive change	Transient response	Steady-state response	Filtering	Settling Time
k	Increasing	Good	Good	Good	Reduce
	Decreasing	Poor	Poor	Poor	Increase
γ	Increasing	Poor	moderate	No effect	Reduce
	Decreasing	Good	Good	No effect	Increase

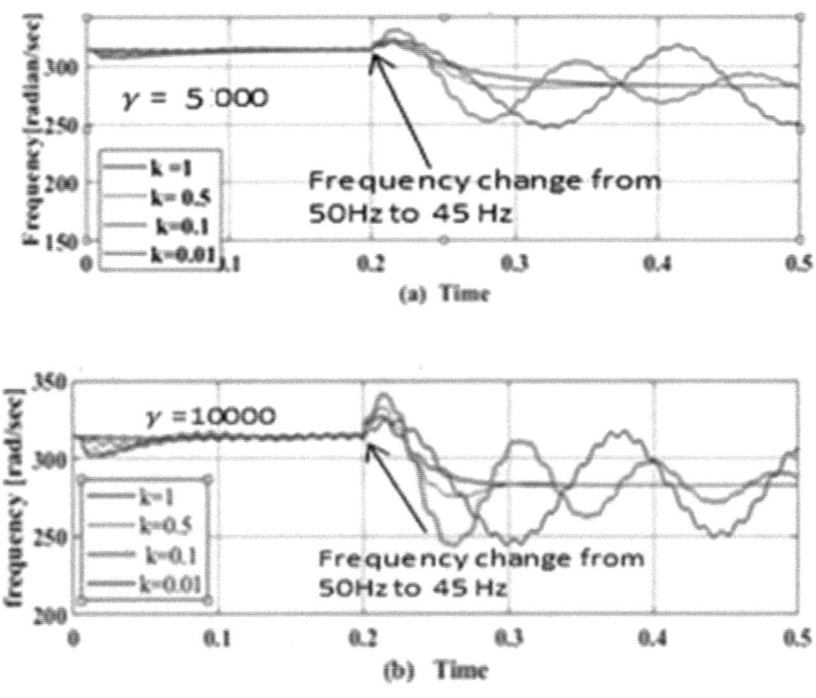

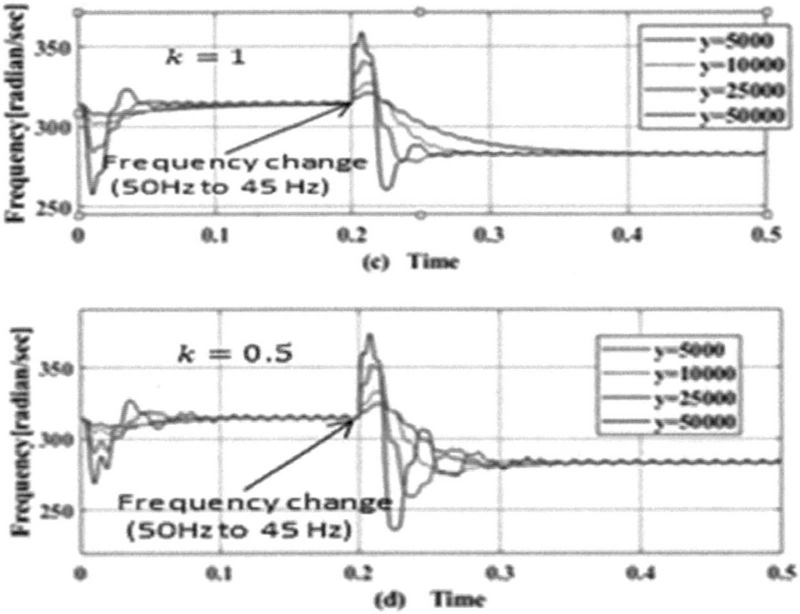

Figure 9.Step response of Frequency Estimation (a) different value of k ,constant value γ=5000, (b) different value of k ,constant value γ=10000, (c) different value of γ ,constant value k=1, and (d) different value of γ ,constant value k=0.5

In Figure 9 (a) & (b), the dynamic response of frequency estimation is observed by taking a progressive value of k and constant value of γ, in the Figure 9 (a) & (b), and progressive value of γ and constant value of k , in Figure 9 (c) & (d). Impact analysis of parameters k and γ in SOGI-FLL have been tabulated from the Figure 9 (a)-(d) in table-1.The para-meter k is used to amplify signal (ε_v) , which affects the transient response and also bandwidth and of SOGI-FLL. The selection of the gain k is compromising between good signal filtering and dynamic response of system (Figure.5 & Figure 9). The selection of γ value is a trade-off between precision of frequency estimation and dynamics of SOGI-FLL. Narrower bandwidth increases the rise time (t_r= 0.35/ BW), further, degrades successive value stabilization of frequency estimation, however, adversely improves the other parameters.

4. Experimental Results

The SOGI-FLL is implemented and tested by using a downscaled STM32F407VGT6 microcontroller and waijung block-set environment in Simulink/Matlab. The discrete- SOGI-FLL model is converted into c code, that further, compiled and dumped into microcontroller by KEIL-IDE (KEIL 4 and above, ST-utility driver) and physical interface USB cable. The STM32F407VGT6 is a low cost, 32-bit microcontroller having 2 Digital analog converters(12bit-DAC) with 168 MHz crystal frequency, and supported by waijung environment (i.e. model-based programming) in the SIMULINK/MATLAB.The experimental results are verified with simulation results to present the effectiveness of SOGI-FLL using model-based programming. The convenient parameter value of SOGI-FLL are chosen as k = 0.5, $\gamma = -5000$,sampling time t_s = 50 usec, and grid voltage frequency f_g = 50 Hz. The DC-offset is generated due to ADC conversion block and other discrete blocks cause uncertainty in phase-detection in SOGI – FLL.However, many researchers presented many improved structures of SOGI-FLL. The SOGI-FLL used along with first-order high-pass filter (i.e known as DC blocker) with cut-off frequency 1-5 Hz to eliminate DC offset without moving toward a complex modification in structure.

For the validation, SOGI-FLL is tested under various cases individually in the experimental set-up as follow:

Test 1: It is conducted on the grid voltage affected by voltage sag of 0.6 p.u,

Test 2: It is conducted on the grid voltage experiences a -5 Hz frequency step change (i.e. frequency jump (50Hz to 45Hz) and phase angle shift (0^0 to 45^0) .

Test 3: It is conducted on the grid voltage affected by harmonics.

In Figure 10, experimental results are obtained to measure settling time of SOGI-FLL on the grid voltage affected by voltage sag of the magnitude 0.6 p.u at t=200 msec .These results are only visible in the Digital oscilloscope have a frequency above 70 MHz and change must be applied after 200msec.

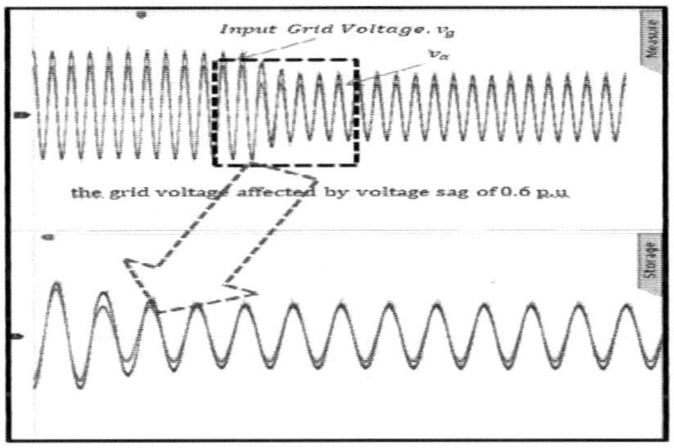

Figure 10. Experimental results: output signal v_α of SOGI-FLL with input grid voltage characterized simultaneously dc offset, voltage sag of 0.6 p.u

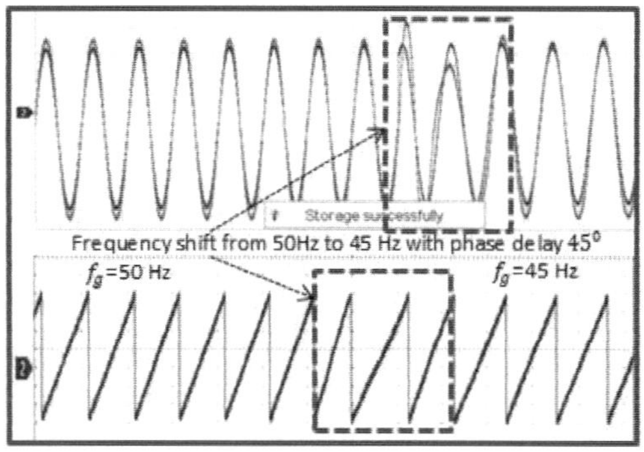

Figure 11. Experimental results: output signal v_α of SOGI-FLL with input grid voltage encountered with frequency jump (from 50Hz to 45Hz) and phase angle shift (0^0 to 45^0), and also phase-angle detected by SOGI-FLL

Figure 10 shows that v_α/v' of SOGI-FLL settled down before the third cycle i.e 50 msec when grid voltage affected by sag of 0.6 p.u. As shown in the Figure 11, the dynamic response of SOGI-FLL is evaluated by performing experimental test-2 i.e. grid voltage experiences a -5 Hz frequency step change (frequency jump from 50Hz to 45Hz) and phase angle shift (0^0 to 45^0) at t =500msec. The phase angle shift (from 0° to 45°) is intentionally made only to observe frequency change occurs at t =500msec, as shown in figure 18. The frequency and phase- angle detected by SOGI-FLL are settled down before the third cycle observe from the time instant at which the frequency shift made in grid voltage.

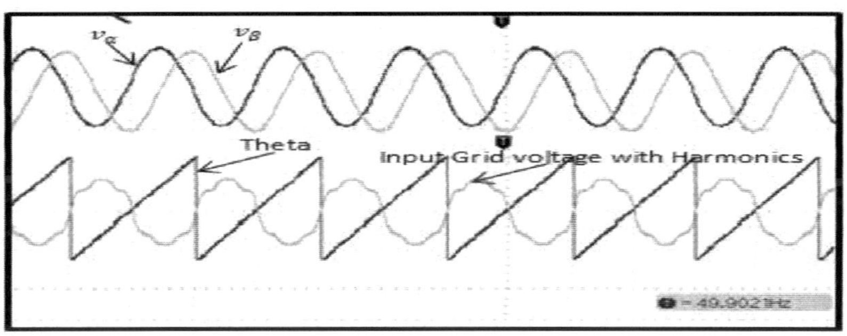

Figure 12. Experimental results: output signal $v_\alpha \& v_\beta$ of SOGI-FLL with input grid voltage effected by harmonics component, and phase angle shift (0^0 to45^0), and also phase-angle detected by SOGI-FLL

Due to the unavailability of AC grid simulator, this test is conducted by generating sin wave inside SIMULINK/MATLAB and frequency controlled by externally through the ADC pin of a microcontroller. As shown in Figure 12, the dynamic response of SOGI-FLL is evaluated by performing experimental test-3 i.e. grid voltage affected by 3rd, 5th and 7th order harmonics with proportional amplitudes of 35%, 15%, and 8%, respectively, with respect to fundamental grid voltage.The SOGI-FLL is confirmed as immunity to distorted grid voltage as presented in Figure 12. Figure 12 shows that the detected phase-angle free from the harmonics distortion present in the supply voltage as well as the 100Hz frequency component in phase-angle.Besides, SOGI-QSG is experimentally proven as band-pass filter through output signal of SOGI-FLL i.e. v_α/v' and v_β/qv'also free from the effect of harmonics distortion as shown in figure 19. The SOGI-FLL works in the same fashion even in the situation when the grid voltage is affected by multiple abnormalities.

5. Conclusions

In this paper, attention was paid to estimate the frequency and further phase angle detection by tan-arc method for changing frequency and voltage of single-phase utility. The additional structure in a second-order generalized integrator (SOGI) provides a disturbance rejection capability along with selective harmonics filtering ability, informal to design and implement on wijung-block set of SIMULINK/MATLAB environment to perform at desired bandwidth.The main challenging in a design point view is to identify the operative conditions, and further estimate suitable control parameters for SOGI-FLL. The selection of appropriate control parameters is a trade-off between the dynamic response, filtering capability and the desired accuracy in detection of frequency and phase angle especially during a non-ideal grid conditions for single-phase grid-tied inverter. The second challenging part is to create the synchronization process adaptive as well as a fast dynamic response to frequency changes that may happen during the power balancing in the grid system. The practical design constrains of a non-linear system FLL was discussed in this paper. Experimental results have justified that rejection of dc offset achieve through a high-pass filter with cut-off frequency1-5Hz in place of complex DC offset rejection PLL method. It is also proven that appropriate selection of control parameters provides a desire settling time, improves bandwidth, good filtering capability for the lower order harmonics, accurately estimate frequency and a superior response during grid abnormality.

6. References

[1]. F. Blaabjerg, R. Teodorescu, M. Liserre, and A. V. Timbus, "Overview of control and grid synchronization for distributed power generation systems," *IEEE Trans. Ind. Electron.,* **vol. 53, no. 5,** pp. 1398–1409,Oct. 2006.

[2] S. Kouro, J. I. Leon, D. Vinnikov, and L. G. Franquelo, "Grid-connected photovoltaic systems: An overview of recent research and emerging PV converter technology," *IEEE Ind. Electron. Mag.,* **vol. 9, no. 1,** pp. 47–61, Mar. 2015.

[3] S. Golestan, J. M. Guerrero, and J. C. Vasquez, "Three-phase PLLs: Areview of recent advances," *IEEE Trans. Power Electron.,* **vol. 32, no. 3**,pp. 1894–1907, Mar. 2017.

[4] S. Golestan, J. M. Guerrero, and J. C. Vasquez, "Single-phase PLLs:A review of recent advances," *IEEE Trans. Power Electron.*, **vol. 32,no. 12**, pp. 9013–9030, Dec. 2017.

[5] M. Karimi-Ghartemani, Enhanced phase-locked loop structures forand energy applications. *John Wiley & Sons*, 2014.

[6] X. He, H. Geng, and G. Yang, "A generalized design framework ofnotch filter based frequency-locked loop for three-phase grid voltage,"*IEEE Trans. Ind Electron*.,**vol. 11, no. 99**, pp. 1–1, 2017.

[7] B. Brahmbhatt and H. Chandwani, "Single phase transformerless photovoltaic inverter with reactive power control," in 2016 *IEEE 1st International Conference on Power Electronics, Intelligent Control and Energy Systems (ICPEICES),* Delhi, India, 2016, pp. 1–5, doi: 10.1109/ICPEICES.2016.7853435.

[8]. R. M. S. Filho, P. F. Seixas, P. C. Cortizo, L. A. B. Torres, and A. F.Souza, "Comparison of three single-phase PLL algorithms for UPSapplications," *IEEE Trans. Ind. Electron*., **vol. 55, no. 8,** pp. 2923–2932,Aug. 2008.

[9]. Thacker, D. Boroyevich, R. Burgos, and F. Wang, "Phase-lockedloop noise reduction via phase detector implementation for single-phasesystems," *IEEE Trans. Ind. Electron*., **vol. 58, no. 6,** pp. 2482–2490,Jun. 2011.

[10] M. Karimi-Ghartemani and M. R. Iravani, "A method for synchronization of power electronic converters in polluted and variable-frequency environments," *IEEE Trans. Power Syst*., **vol. 19, no. 3**, pp. 1263–1270, Aug. 2004.

[11] K.Możdżyński, K. Rafał, MałgorzataB.Rafał, "Application of the second order generalized integrator in digital control systems"*Archives Of Electrical Engineering*,**vol. 63(3)**, pp. 423-437 (2014).

[12] P. Rodr´ıguez, L.Luna, R. Teodorescu, F. Blaabjerg, "A Stationary Reference Frame Grid Synchronization System for Three-Phase Grid-Connected Power Converters Under Adverse Grid Conditions"*IEEE TRANSACTIONS ON INDUSTRIAL ELECTRONICS*,pp.1-6(2016).

[13] R. Teodorescu, M. Liserre, and P. Rodríguez, Grid Converters for Photovoltaic and Wind Power Systems: Teodorescu/Grid Converters for Photovoltaic and Wind Power Systems. Chichester, UK: *John Wiley & Sons, Ltd, 2011.*

IOP Conf. Series: Materials Science and Engineering 1045 (2021) 012020 doi:10.1088/1757-899X/1045/1/012020

A review on training and leadership development: its effectiveness for enhancing employee performance in Indian construction industry

Pramod Kumar Misra[1,2] and Jitendra Mohanty[2]

[1]*Head – Talent Management, G R Infraprojects Ltd, India*
[2]*School of Management, KIIT (Deemed to be University), India*

E-mail : pmisraphd@gmail.com

Abstract : The second largest contributor to Indian GDP is the construction industry. It is the largest employing industry having huge economic potential which was valued over $126 billion in 2016 and it continues on a steady growth path. It is expected that the value of the real estate and construction market will increase seven-fold by 2028. Growth is likely to be underpinned by stable government support for infrastructure development, as well as through expanded private sector involvement. However, the industry suffers from inefficiencies that result in time and cost overrun in their projects. Some of the potential causes of inefficiencies originate from the underdeveloped labour market and poor organizational structures that operate in the industry. An exhaustive list of factors found to drive the industry to its inefficient operation resulting invariably to failures. If we put all these factors in one basket, they reduce to one single word; "lack of training of workforce and poor management", whether it refers to the skilled, semi-skilled or unskilled labour force and/or the poor project management from the organizational side of the project. The present study extensively reviews problems of construction company that originate from the labour market.

1. INTRODUCTION

1.1. Construction Industry: An Overview of the Its Growth and Problems

Infrastructure and industrial development of a country primarily depends upon its robust construction industry. Historically, the largest employing industry, construction industry, is reckoned as oldest and second largest contributor to gross domestic product of India next to agriculture industry. After 1947, construction activities in India were not only limited to buildings and basic transportation infrastructure but also extended foundations for modern architectural, engineering, and construction services in order to fulfil the necessity for accelerated real-estate, industrial and infrastructure development. Government owned the construction industry gradually increased its share and contributions by private companies [1].

With increasing population density of India with 1.2 billion current population, its estimated that by 2030 around 0.59 billion would be residing in cities of India. In order to meet the both requirement for urban areas and expectations of citizens, existing urban areas needs to be rejuvenated, as well as, new inclusive smart cities needs to be created. Thus, it would be responsible for rapid development in housing, commercial, industrial, and infrastructure sectors [2].

Several researchers reported that, 70% population of India would be in working-age by 2025, whereas, significant proportion of population density would have been retired in other developed nations. This large demographic dividend might become advantageous for India over other developed nations. Literature revealed adverse effects unemployability due to lack of skill if not handled properly that it will take no time to transform demographic dividend into demographic disaster [3].

Few reports suggest that, Infrastructure development plans in India are faced with the major challenge of lack of sufficient skilled workers (plumbers, machine operators etc.) and supervisors, that are responsible for sluggish nature and enhanced overall cost of construction projects.

Nation's economic growth and social development is mainly dependent upon its skills and knowledge of demographic dividend. The development of bigger skilled human capital leads to productive, innovative and competitive economy. Generally, the performance is distinguished from knowledge by any individual's skill competency level. Rapid growth of Indian construction industry is mainly due to rapid growth of housing sector leading to rapid growth of economy of India. If skill and efficient project management handled properly, the current introduction of green building and techniques in housing sector can lead into be an average 6.2% sustainable growth till 2027. Such growth rate is essential to fulfil the demands of several Indian government's flagship project initiatives like smart cities. In order to fulfil the resident's significant needs, as well as, to explore the possibility to improve his life quality, smart cities play important role [4].

Apart this, in all rural areas Government of India working on "housing for all" mission (a flagship project initiative named as "Pradhan Mantri Awaas Yojana-Gramin"). During 2015-22, this aims to provide 20 million affordable houses across the nation. It includes the financial assistance to be provided to 10 million households for "pucca house - solid and permanent house", during 2016-19 [5]. The other major contributor to the growth of the industry is infrastructure development including basic transportation infrastructure.

1.2. Construction Inefficiencies in the construction industry and the reasons there of

The available evidence on the performance of the industry is not encouraging even though the construction sector gaining prominence in India recently with the opening up of Indian markets and the commencement of several mega developmental construction projects [6]. World Economic Forum reported in the infrastructure area, India possesses 87th ranks in the global competitiveness index [7]. The "Ministry of Statics and Programme Implementation" reported that 40% of infrastructure projects, costing around 4.45 million dollars, get delayed usually 1 - 252 months [8]. A fall short of investments targets in infrastructure sectors other than in telecom during the eleventh five-year plan has been reported in the Conference of Industry Secretaries Report [9]. Apart this, such alarming issue is also evident in countries beyond the Indian territories [10]. Hence, the construction sector faces problems and challenges globally, especially in under-developed and developing nations, which need massive technological and organizational paradigm shift towards contemporary approaches. Also, reports have emphasized on rescue of industry and help to adjust under way in the national economies and to develop the capacity and the ability to adapt to the future development.

De Saram et. al. argued that lack of effectiveness, which causes the failure of anticipated quality, productivity, and safety levels along with cost over-runs and delay, poor claim settlement ratio, and continued conflicts on many primary projects, that leads to hinderance in national development [11]. Several industry experts reported that performance enhancement of the construction industries of developing nations can be done by prediction of relations between infrastructure development, construction industry development and the broader role of national development. Also, they reported about several countries those emphasized improvements needed for the construction industries' performance enhancement at different stages of socio-economic growth. In many countries industries' spontaneous developments is administered by several dedicated agencies. Such similar steps are the dire need of the time [12].

1.3. Training and skill development in construction industry in India

As discussed, construction sector, India's second-largest employing industry, spontaneously looks for skilled employees at all levels, from supervisors, plumbers, masons, electricians etc. to site managers, architects, and so on. They experience lack of skilled labour and supervisors that not only restricts the progress of work and leads to delay in meeting deadlines i.e. timely output. There's a considerable gap between the skill-set that the employees in the industry today possess and the need for business. For a specific construction project, significant skill-set differences identified in employee's knowledge and his skill to undertake that job-role. Construction industries significantly face issues of unskilled or semi-skilled candidate with poor knowledge in relevant job-role available during talent acquisition with for key activities for example masonry, brickwork, carpentry, so on. It's well reported on the success stories of industries and the key-role of enhanced-capability and learned and skilled human-capital to establish them as global leader [13].

In order to ensure the employee's performance in order to achieve desired outcome, effective training sessions needed to refine their knowledge, skills and abilities that leads them highly motivated and committed for the organization. Through spontaneous investment and planning, significant industries prioritize the capacity building, refine skills and core competency of human capital, shape their employees motivated and committed for any varied, uncertain and competitive future, thus maintain the high employee's performance. Through different training activities employee becomes aware of key interest of the organization, that motivates employees to give their best and makes them committed towards organizational goal which leads the employee to exhibit high performance, failure to which, the existing and fresh employees remains out-of-date and a fall short to achieve the target becomes usual due to lack of adequate skill. Apart this, proper training also enhances the employee retention. Thus, training is elixir for the organization that rebuilds employee's understanding on their responsibilities [14].

The training and development programs primarily focused on enhancement of the employees' performance. The employee's present performance and the performance desired by organization is bridged by effective training sessions. Mentoring, coaching, individual participation and group participation are different training delivery methods. Organizational performance can be improved by the active participation of employee in team-work through enhancement of the team performance.

Training programs helps to think beyond in order to make best possible use of human capital in favor of gaining competitive advantage through employees' individual development. Hence, industries must plan employee training activities that would be useful to develop employee's skill and core competency as per organizational need [15].

Effective trainings refine and rebuild the creative and critical thinking ability which leads to enhance the employee's in-time productive suitable decision-making capability. Thus, it enhances the effectiveness of customer engagement, timely claim settlement and customer satisfaction. It is all about accomplishment of organizational goals that appears to essentially need improvement of the skills. It also effectively reduces the employee's anxiety or frustration originated due to the work-stress and work-life imbalance [16].

Mainly, the government's impetus along with urgent industrial requirement, accelerated skill development in India. Dr. M. Singh, former PM of India, expressed concern of central government over lack of skilled workers "As our economy booms and as our industry grows, I hear a pressing complaint about an imminent shortage of skilled employees. As a country endowed with huge human resources, we cannot let this be a constraint" [17]. There was no institutional mechanism available to impart skill enhancement earlier, as well as, master craftsmen used to train traditionally to the construction workers. Furthermore, earlier system not only fails to impart skill in quantitative terms but also fails to impart of new technologies and work methods those are concluded in various research and development programs. Apart this, for all working life women workers they remain under unskilled labor as they are not trained in any trade

Till the date the exact details of human capital volume in construction sector and the skill diversity has not been well documented. A research predicted that about 79% workers out of 31 million construction workers are unskilled. These unskilled workers in construction industry comprises of 21 million seasonal workers along with regular workers. Apart this, 23-27% of the workforce in construction industry are women. The National Skill Development Coordination Board (NSDCB), National Council on Skill Development and National Skill Development Corporation (NSDC) are three-tier structure along with a national policy on Skill Development was set up in order to institutionalize skill development at various levels in the country in 2009 [18].

1.4. The relationship between training and employee's performance: The role of human resource management

Several researchers argued that best practices of human resource management have strong positive impact on organizational performance. Guest emphasized that qualitative nature of the capability, core-competency and skills of workers influenced by human resource management practice such as training and development programs which enhances on-job employee performance leading to organizational performance [19]. In order to fulfill the goals, employees in the organization have to deliver the best which necessitates needs-identification and development of core-competence in the deprived employees through training. Harrison [20] reported the enhancement of the organizational performance and identified the role of training and learning for enhanced overall employee performance leading to fulfil the corporate goals. Conversely, implementation of training programs leads to an effective employee performance improvement through covering performance issues by bridging performance gap. In order to enhance performance of the employee and their certain skills and core-competency trough implementation of a relevant training intervention that would be useful to bridge the organizational standard and the actual performance. Apart from this, Identification of non-performing workers along with their assessment of core-competency level, skill-set, industry specific requirements can be exercised through training activity in an organization. Lack of inspiration to use their competencies, diffident on capabilities, work-life imbalance are primary reasons for decline trend in employee's performance. Through a structured training intervention which takes the care of all above said aspects, the highly motivated employee exhibit desired performance to meet firm expectations, as well as, resolves issues in organization.

Swart et al. argued that qualitative training activity not only highly motivates of employee and address their requirement but also enhance employee's performance. Competitively archive organization's target, development of core-competencies, as well as, efficient skill of employees can be done through training [21]. Enhanced employees' performance leads to organization's success along with comparatively higher return. This is mainly due to shaping the behavior of employees through training activity to exhibit enhanced job-skill of employee and strong motivation to participate in developmental activity. Apart from this, waste minimization along with best possible utilization of resources of the organization are achieved through workers those are well-trained. Ohabunwa highlighted the success of the organization primarily depends on authorizing the employees with full confidence, as well as, discharging the assigned responsibility which ultimately depends on well-trained employees [22].

Farooq & Khan reported that in the organizations for creating good working environment, the managers are putting their best effort through enhancing the capability of the employees. The goals of the organization are being achieved through capacity building, for which effective training activities are being conducted by managers to refine the employee's desired skills, core-competency and abilities. Thus, Managements are striving hard to enhance the performance of employee, as well as, to create firm's global positive image. In order to avoid mistakes and job errors at a particular job, the employees need to have commendable authority over desired skills and core-competencies through various effective training activities along with the advancement in technology. Through proper training and development programs employee development is one among the crucial functions of human resource management [23].

Employee development, also known as employee's capability building, lead to enhanced job satisfaction leads to enable for higher return along with higher productivity, Thus, enables to meet the standard performance level as of whole organization [24]. Armstrong defined the training as "Training is systematic development of the knowledge, skills and attitudes required by an individual to perform adequately a given task or job" [25]. Furthermore, E B Flippo also emphasized, "Training is the act of increasing knowledge and skills of an employee for doing a particular job" [26]. The process of enhancing the employee's skills, aptitudes and abilities of the to perform specific jobs referred as 'training'. It helps to refine the skills of existing employee's as well as develop new skills needed. Aswathappa has argued that "Successful candidates placed on the jobs need training to perform their duties effectively" [27].

Expected benefits can be observed when a structured training activity is systematically conducted. Assessment of need of the program, structured programs design, systematic implementation, assessment of the out-come of this training and development program are the various stages of training system. From this detailed study skill areas that need refinement or in-depth knowledge will be identified along with best fit training-mode and training-module.

1.5. The Leadership Style of Employee Performance

The organizational goal and success primarily depend upon the role of leadership style which is a crucial factor to deal issues encountered related to planning, management, human resources and leadership. Thus, skills development in leadership quality of higher-management of the construction industry becomes a major and key aspect for companies' performance enhancement. Efficient communication, project planning, project management, project performance, organizational commitment and job satisfaction are the various requirement that raised the leadership demand in developing countries especially in Indian construction industry. Also, it is argued that effective performance of organization primarily depends on high-quality work where employees' satisfaction in job plays a key role. Fulfillment of commitment to organization and high level of job-satisfaction is greatly influenced by the employee's training program. The detrimental commitment towards organization goal and job satisfaction are found to be outcome of the ineffective administration and management in significant construction industry [28].

For proper communication, the strong leadership is a key attribute needed to share project related information among the members of the team. Team communication depends on manager's leadership ability and enhancement in the effectiveness of the team observed through appropriate exchange of project data. Also, for maintaining overall organization productivity in complex competitive environment, the team work, as well as, appropriate exchange of project data and experience plays significant role. It is also found that in a project environment cohesiveness, collaboration of the team and employee's relationship are primarily influenced by management's leadership policy [29].

Success and growth of organization depends on rehabilitation of employee's performance for which training is key and an imperative tool. Organization's employees and employers both are the beneficiary of this training program. In practice a trained employee exhibits more productivity with high efficiency. An exhaustive training and development can enhance the quality of the current employees in the organization. Awareness of significance of the job role, strong motivation and inspiration among the workers, enhanced productivity, exchange of job role relevant information as well as industrial experience to deal the adverse issues can be achieved through training program. Highly committed and motivated employee, High level of morale and job-satisfaction, technological adoptability along with innovative practices of strategies and products, higher performance that result in higher financial gains along with reduced loss due to human error, as well as, reduced employee turnover are the common advantages of employee's training.

Several researchers emphasized the role of job-satisfaction over organization success. Also, reported that satisfied employees exhibited enhanced performance along with higher employee retention as compared to unhappy employees of the organization. Furthermore, higher turnover of the employees indicates significant presence of unhappy employees, those exhibit poor performance. Thus, management find the satisfied and happy employees as easy to motivate and these employees contribute for higher organization's performance [30]. The employee could be only satisfied when they themselves feel competent enough to perform their jobs, and this can be achieved via suitable training activity. Thus, improvising the performances and the motivational level of the workforce, top executives need to create better working environment where training practices plays key role. Leonard-Barton stated the training as a systematic process of constant learning that an organization needs to enhance knowledge of its employees as a source of gaining competitive advantage over its competitors [31].

Employee performance relating to Human Resource management of the firm / organization/ company refers to all the aspects those directly or indirectly influence employee's performance related to a specific job role. It has significant role in identification of poor performers and help them to refine their skill and core-competency. Followings are the five different elements that constitute performance.

- Planning stage: Defining objective, methodology, task outlining, accomplishment schedule
- Monitoring stage: Objectives are fixed and spontaneous performances are measured in real time and share the feedback on accomplishment status to the appropriate authorities.
- Rating stage: deficiencies in performance become evident and ratings are provided to summarize the performance of the employee that becomes useful for performance comparison among various employees over a period of time. This ensures certain standards are met by the employees and address the issues appropriately in the existing standards.
- Developing stage: The main objective is to uplift the poorly performing employees of the organization over a period of time and refine their skill and core-competency.
- Rewarding stage: Identify organization's best performers and their outstanding behavior at the end of each cycle along with some motivational reward

Thus, organizational development influenced by excellent leadership capacities of the managers along with suitable leadership style., employee's performance, job-satisfaction, and commitment towards organizational goal and its efficient accomplishment should be taken into consideration for organizational development. This research work aims to study the issues in Indian construction industry and analyze the significance of leadership qualities. This will be useful for supervisors in Indian construction industry through refining the leadership skills along with the proper implementation guideline.

2. OVERVIEW OF LITERATURE

Considering the capital formation, participation to GDP and employment, construction sector appears as significant in Indian economy crosslinked to several other sectors. Taking into account objectives of national economic, the whole industry should be considered for the development of construction industry. Development of construction industry as per International Council for Research and Innovation in Building and Construction Task Group 29 (CIB TG29) refers to a deliberate and managed process for construction industry's effectiveness and capacity enhancement aiming to fulfil demand of building and civil engineering products for the growth of national economy, as well as, to provide a sustained development favoring social and economic objective. construction sector faces many problems and challenges irrespective of its geo-political positioning. Developed nations like Hong Kong, UK, Singapore, USA and Canada were compelled to focus for drastic developments in the construction industries to address the persistent issues in this sector which was highlighted by Kumaraswamy. These nations have established different councils, high-level boards or bureau for development of industry that will undertake the controlling, monitoring and planning through commissioned high-profile study along with high-powered reports. The major issues of construction industries along with the initiatives launched to overcome and address these issues for industry development by these nations explained through lucid examples [32].

2.1. Training and Development of Employees

In order to ensure the gain in the global competitive advantage, the organizations are bearing the expenditure towards training and development that seems to appear as an indispensable strategic tool for enhancement of effective performance of individual employee along with overall performance of organization. However, for any organization to achieve its stated goals and objective in this competitive world, adequate and relevant training and development of staff cannot be over emphasized. It is expected that identification of need for employee's training and structured training programs design would be useful for the optimal utilization of the organizational workforce along with fulfilment of the goal of the organization. The improvement in the employee's performance occurs through training and development on current jobs and future assignments that transfer the relevant skills, knowledge and competence for their performance enhancement [33, 34].

Several authors emphasized on the contribution of training towards the competitive and challenging business format in this competitive world. It improves the employee's work-life balance along with organization's substantial development through fulfillment of the need for smooth functioning of work at edge. Especially at the managerial level in the organization, job role is not only limited to deliver the physical skills, but also more emphasis given for knowledge, behavior, attitudes and values along with specific skills through qualitative as well as quantitative developmental advancement process. Furthermore, unlike training, development is a continuous process having no specific areas and objectives. Hence, possible exploration of role, importance and advantages of training along with its positive developmental impact towards the organizational growth has to be studied in every organization. Process of identification of their responsibility for the excellence of organizational performance as well as employee skills known as quality of work life. Constructive development for optimization of quality of work life of the employees in such organizational motives referred by training. Furthermore, such training and development programs become useful for enhancement of the behavior and attitude of employee towards the job, as well as, uplift their morale [35, 36].

It is found that higher quality of the workforce has significant contribution towards organizational success, for which organizations have to come across many restrictions for a sustained quality of the work force. Such restrictions faced by the organization include: retention of the current workforce; inspiration of existing employees through different techniques; intellectual, self-motivated and passionate talent recruitment; maintaining the organizational status in the competitive market, as well as, attraction of the qualitative workforce towards the organization. In order to gain competitive advantages along with for business survival and becoming a successful pillar in the market, it is found training as useful tool. Achievement the objectives of organization depends on the ability of the workforce enhancement for which training seems to appear as key parameter. Hence, behind the success of business goals, more emphasis given towards contribution of good training programs. Thus, the organization's operation becomes dynamic nature through significant training activity which is essential to enable consistent results delivery through improved quality of work life [37].

Efficient and optimal performance of employees play important role to fulfil organizational goals and objectives for which adequate training of their employees is not negotiable for organizations. Training and development of employees seems to appear as a strategic determination for enhancing effectiveness of the organization, as well as, performance efficiency through the learning of the core-competency, attitude, capability and skills at ease [38].

Improvement in performance of the organization along with gaining competitive edge happens mainly due to enhanced workers skill, core-competency and knowledge for which training is essential. Acceptance towards new techniques and technologies, enhanced competences of employees, innovation, invention, are the outcomes of training and development. Organizations must conduct the need identification for training and development followed by selection of suitable techniques and strategize the implementation plan. Towards last step, it is important to evaluate the outcomes [39].

The enhancement of motivation and retention of the employee significantly depend on career-enhancing skills development of employees. Well-trained and knowledgeful employees are valuable asset of organization, whereas this enhances their efficiency and effectiveness in discharging their duties. Desired organizational objectives achieved along with the positive impact on organization behavior through experience of learning during training. Thus, for the development of the employees, training and development programs are the basic structural and functional foundations those are significant for enlightening the employees through different situations which enhances the capability, performances, efficiency and excellence of the employee. Employee's personal and professional skills, knowledge, and abilities developed within the framework of Training and Development. Development of contented & productive employees along with efficient, adaptable and profitable organization are the outcomes of proper execution of training programs, hence imparts knowledge to the employees related to various organizational issues [40].

In the environmental uncertainty along with the fast pace changing world, business organizations understand its limitation while addressing the potential issues and challenges. Furthermore, in in order to remain competitive in the market, firms should invest in training programs to make their employees competent enough to take effective decision in time through various uncertainties. Decreasing the unintentional job rotation of the workers; building and maintaining capabilities, both on individual and organizational level; long-term commitment towards its workers; enhances the retention capacity of a talented workforce; participation in the process of organizational change; increases the employee's motivational level are the various significant outcomes of the effective training in the firm. These contributions provide a gain in global competitive advantage and enhancement of productivity of the organization through enhancement of employee performance [41].

2.2. Issues Relating to Project Management

Indian construction industry is experiencing unprecedented pace of projects. But, many of these projects in private as well as public sector get delayed, face time overruns, are not completed as per quality specifications or fail in effective stakeholder management. Identified reasons are different aspects of PM that can be managed by including PM professionals in the team [42]. Construction industry has a unique 6 character that government is the client for major construction and engineering projects including infrastructure projects and the industry comprises Small and medium enterprises (SMEs), which may work on projects independently or as sub-contractors for large organizations. Client's influence is one of the most significant factors found to be affecting time performance on Indian projects and effective project management can be achieved on a project if it is 'Client driven' [43, 44].

M. Desai and R. Bhatt conducted a field survey considering 50 participants including 13 architects, 17 contractors and 20 developers. They reported 59 causes for delayed residential construction projects under 9 major groups belonging from construction industry in India [45]. Salunkhe and Patil reported on the delay factors those influence the project success, as well as, methodology to eliminate or minimize such delay in projects in order to improve the project performance. After conducting exhaustive literature survey and interview, they reported the seven categories construction delay factors directly influencing project success as key parameters [46]. Zarina et.al. developed a conceptual framework, where Project Management Action, Project Related Factors, Human Factors, Project Procedures, and External Issues are five variables for project success. Their research findings are useful for evaluating the performance of organizational project management. Here, attempts have been made for identification of the extent of the relationship between project performance and CSFs [47]. D. F. Ofori conducted survey on project management practices of Ghanaian organizations following an exploratory approach. They reported on the critical success factors for projects in Ghana, as well as, identification and assessment of the quality of project in management practices [48]. A. W. Shaikh et.al. reported on causes of delay and key factors causing it along with their identification [49].

Furthermore, regardless of the advancement of the construction companies the leadership issues in the Indian construction industry in recent times are well known. Employee's individual achievements along with fulfillment organizational goal must be ensured through effective organizational leadership traits. Lack of essential leadership styles to fulfil the desired organizational goals has disrupted the growth of the industry due to the inconsistency in leadership. The organizational goals accomplishment proportionately linked to rate of employee's job satisfaction. Not only well-being of the employees but also organizations suffer from adverse impact of toxic, destructive, or tyrannical leadership, when management occupied with poor and ineffective leadership quality [50].

Leaders face major challenge to manage the present diverse workforce to fulfil the goals of organization along with employee job satisfaction which appears as significant management issues for the current construction market. In order to enhance the organizational operational performance, it is essential to handle the workforce through suitable leadership strategy. Several researchers reported that leadership styles seem to have distinct control over job satisfaction of the employee. Also, study relating job satisfaction, organizational commitment and the relationship between leadership styles is insignificant. After exhaustive study of literature, the present study evaluates leadership traits based on most crucial and extensive aspects.

3. CONCLUSIONS

In this paper an attempt has been made to review the various problems of construction company that originate from the labour market. It is also found that the training and leadership development of employees has effectively improve the Performance. The present review proposes to undertake an evaluation of Indian construction company and study extensively its problems that originate from the labour market on the one hand and at its organizational front – project management in all its dimensions.

REFERENCES

[1] Agarwal M 2013 *The opportunity and challenge of India's infrastructure* (Mumbai: Price water house Coopers) p 10.
[2] Dept. for Promotion of Industry and Internal Trades 2014 *Make in India Construction* (New Delhi: Invest India).
[3] DMG Events India 2015 *A report on Construction market in India* (Mumbai: DMG Events India) p 13.
[4] India Smart Cities Challenge 2016 *What is a smart city?* (Government of India, New Delhi: Ministry of Housing and Urban Affairs).
[5] Ministry of Housing and Urban Poverty Alleviation 2015 *Pradhan Mantri Awas Yojana* (New Delhi: Ministry of Housing and Urban Affairs).
[6] Iyer K C and Jha K N 2006 *J. of Construction Engineering and Management* **132** 871.
[7] Schwab K 2013 *The global competitiveness report 2013–2014* (Geneva: World Economic Forum) p 569.
[8] http://www.mospi.nic.in/
[9] Ministry of Personnel, Public Grievances and Pensions 2012 *Grievance analysis and systemic reforms recommendation* (Government of India Delhi: Ministry of Personnel, Public Grievances and Pensions) p 259.
[10] Kumaraswamy M M 2006 *J. of Construction in Developing Countries* **11** 73.
[11] De Saram D D, Rahman M M and Kumaraswamy M M 2001 *Proc. of the FEISCA Conf. on Engineering Management for Accelerated Development in South & South-East Asia* (Colombo: Institution of Engineers).
[12] Ofori G 2015 *J. of Construction in Developing Countries* **20** 115.
[13] McKinsey Quarterly 2006 *An executive take on the top business trends: a McKinsey Global Survey*.
[14] Singh J, Mangal M and Cheng J C 2017 *Proc. of 25th Annual Conf. of the Int. Group for Lean Construction* Doi: 10.24928/2017/0321
[15] Jie S 2005 *J. of Management Development* **24** 656.
[16] Hollenbeck J R, De Rue D S and Guzzo R 2004 *Human Resource Management* **43** 353.
[17] Government of India 2011 *Reports and Publications* (Government of India, New Delhi: Ministry of Statistics and Programme Implementation).
[18] NSDC 2017 *Executive Summary for the state of Maharastra (2012-17, 2017-22)* Available at: https://nsdcindia.org/sites/default/files/files/maha-executive-summary.pdf
[19] Guest D 1997 *Int. J. of Human Resource Management* **8** 263.
[20] Harrison R 2000 *Employee Development* (Silver Lakes, Pretoria: Beekman Publishing).
[21] Swart J, Mann C, Brown S and Price A 2005 *Human Resource Development: Strategy and Tactics* (Oxford: Elsevier Butterworth-Heinemann Publications).
[22] Ohuabunwa M S I 1999 *The challenges of Globalization to the Nigerian Industrial Sector* Nigerian Tribune Report December 14 pp 20-21.
[23] Farooq M and Khan M A 2011 *Far East J. of Psychology and Business* **5** 23.
[24] Jia Fang L, Tjosvold D and Shi K 2010 *J. of Applied Social Psychology* **40** 101.
[25] Armstrong M 2001 *A Handbook of Human Resource Management Practice* 8th ed (NY: Kogan Page) p 981.
[26] Flippo E B 1984 *Personnel Management* 6th ed (New York: McGraw Hill).
[27] Aswathappa K 2000 *Human resource and Personnel Management* (New Delhi: TMH Publishing).
[28] Shah M N, Raitani V, Oza A and Gupta K 2017 *Journal of Construction Management* **XXXII** 30.
[29] Tabassi A A, Ramli M and Dashti K F 2013 *Strategic Leadership Review* **3** 48.
[30] Pani B, Chandrasekhar P and Singh S 2018 *Mater. Today Proc.* **5** 27845.
[31] Leonard-Barton D 1992 *MIT Sloan Management Review* **34** 23.
[32] World Bank 1984 *The construction industry: Issues and strategies in developing countries* (Washington DC: World Bank).
[33] India Brand Equity Foundation (IBEF) 2017 *Real Estate* https://www.ibef.org/download/Real-Estate-March-2017.pdf
[34] *Construction in India, Key Trends and Opportunities to 2021* Available at: https://prn.to/2pdEASP
[35] McKinsey May 2016 *Organization: Agility and Organization Design Organization Practice* Available at: https://mck.co/3bJp6gH
[36] Wienerberger India 2017 *Major Challenges Facing the Construction Industry in India in 2018* Available at: https://gosmartbricks.com/?p=5054
[37] Rowland J 2018 *India's construction industry to grow strongly* Available at: http://bit.ly/2vEv3eo
[38] Nebhwani L 2018 *Construction challenges in India* Available at: http://bit.ly/321NkhG
[39] Prasad J 2017 *Int. J. of Engineering Research and Application* **7** 18.

[40] Pani B, Chandrasekhar P and Singh S 2019 *Bull Mater Sci* **42** 217.
[41] Pani B, Chandrasekhar P and Singh S 2018 *IOP Conf. Ser.: Mater. Sci. Eng.* **455** 12068.
[42] Desai M C and Desale S V 2013 *Int. J. of Latest Trends in Engineering and Technology* **2** 115.
[43] Pani B, Chandrasekhar P and Singh S 2019 *Polym. Compos.* **40** 1433.
[44] Doloi H 2012 *Construction Management and Economics* **30** 869.
[45] Desai M and Bhatt R 2013 *Int. J. of Engineering Trends and Technology* **4** 762.
[46] Salunkhe A and Patil R S 2014 *Int. J. of Latest Trends in Engineering and Technology* **3** 256.
[47] Alias Z, Zawawi E M A, Yusof K and Aris NM 2014 *Procedia - Social and Behavioral Sciences* **153** 61.
[48] Ofori D F 2013 *Int. J. of Business and Management* **8** 14.
[49] Shaikh A W, Muree M R and Soomro A S 2010 *Sindh University Research Journal (Science Series) SURJ* **42** 11.
[50] Chukwura F A 2017 *The impact of elected leadership styles and behaviors on employee motivation and job satisfaction* Doctoral dissertation 10262875.

Investigating the fast cooling process of silicene by MD simulation

Vo Van On, Nguyen Thanh Hung, Dang Thi Khanh Huyen, Le Thi Phuong Trinh

Group of Computational Physics and Simulation of Advanced Materials – Institute of Applied Technology – Thu Dau Mot University

E-mail: onvv@tdmu.edu.vn

Abstract. This paper presents the results of two-dimensional silicene cooling studies by MD simulation with a sample of 6400 atoms. Silicene after melting to 3500K, it is cooled at a rate of 10^{13}K /s to a temperature of 300K. Investigation of the dependence of energy on temperature shows a jump in the average total energy of molten silicene at the temperature T = 1772K. Investigations on the radial distribution function g(r), coordinate number distribution, ring distribution, and angular distribution all show that the freezing temperature of silicene is about 1772K. When cooled to 300K, silicene is in crystalline form, but the ratio of the defects is quite high and approximately 37.5%.
Keywords: silicene, structure , and properties of silicene, molecular dynamics simulation of silicene

1. Introduction

Following the discovery of graphene, a new 2-dimensional form of carbon (C) by Russian physicists Andrei Geim and Konstantin Sergeevich Novoselov, the scientists started investigating related two-dimensional forms of materials to Si 2D, Ge 2D, ... Owing to the promise of new technologies, they are very desirable. While three-dimensional silicon has long been used in physics for important applications in electronic engineering, two-dimensional silicon content has not been extensively studied. The presence of 2-dimensional silicon theoretically predicted in 1994 [1], and in 2007, Guzman-Verri et al. rediscovered and named silicene [2]. Compared to its two-dimensional carbon counterpart graphene[3,4], silicene has a similar but buckled honeycomb structure, which means that the atoms are not strictly in sp2 hybridized states. Silicene exhibits properties similar to those of graphene according to theoretical investigations [5, 6]. In addition, the puckered structure implemented in highly symmetrical configurations can be exploited in some electronic applications [7]. Similar to graphene, silicene also has many potential applications in photoelectric and photoelectronic equipment [8], solar cell production [9], and hydrogen storage [10], for example. One advantage of silicene over other 2D materials is that it blends easily with nanomaterials based on silicone. Polycrystalline silicone experiments have also recently been conducted by the authors [11], suggesting great potential for the use of silicene. Bocchetti et al. [12] simulated the melting of silicene with the original and improved versions of the Tersoff voltage parameter (called ARK) for the silicon atom using the Monte Carlo method. The results show that the temperatures of the phase transition are 3600K and 1750K, respectively. Using a reactive force field, Berdiyorov et al. [13] simulated the effect of defects by MD simulation on the heat stability of silicene and found that pure silicene remained stable up to 1500K. Tjun Kit Min et al. [14] also investigated the melting process of silicene via MD simulation and obtained the phase transition temperature with an optimized 1500K Stillinger-Weber potential. Vo Van Hoang et al. have also carried out some studies on two-dimensional silicene materials and at around 1775K obtained the phase transition temperature from molten to solid silicone [15]; amorphous silicene (a-silicene) approximately 1350K [16], approximately 1950K for crystallization transition temperature and approximately 1350K for glass transition temperature with tetra-silicene [17], approximately 1620 K for P-silicene [18]. We also studied fast-melting silicene by MD simulation at a rate of 10^{13} K/s and obtained a phase transition temperature of about 1772K [19]. How does the heating rate affect the phase transition temperature in silicene from a liquid-state to a solid-state? This research presents the results of the silicene cooling process at a rate of 10^{13} K/s by the MD simulation. The paper is organized as follows: calculations and simulation results are presented in Part 2; the conclusion is presented in Part 3.

2. Calculations and simulation results

The simulation process is designed as follows: the model includes 6400 Si atoms, which are cooled according to the NPT mode and periodic boundary. First, the model after being melted to 3500K, it was relaxed in 5.10^4 simulation steps and the results obtained show that such a model is stable because the volume and temperature are stable. Then, the Si liquid was cooled from 3500K down to 300K with a rate of 10^{13}K/s, including 320 thousand simulation steps. This cooling rate is relatively fast and we expect to also obtain crystalline Si. The potential of Stillinger-Weber (SW) type is used in the simulation process, it consists of the 2-particle term and 3-particle term with a dimensionless parameter λ controls the relative strength between them as follows [20],

$$U = \sum_i \sum_{j>i} U_2(r_{ij}) + \lambda \sum_i \sum_{j \neq i} \sum_{k>i} U_3(r_{ij}, r_{ik}, \theta_{ijk}) \tag{1}$$

The first term consists of a repulsion force at a very short distance and an attractive force at a short distance,

$$U_2(r) = A\varepsilon[B(\frac{\sigma}{r})^p - (\frac{\sigma}{r})^q]\exp(\frac{\sigma}{r - a\sigma}) \tag{2}$$

The second term is the sum of all triples of three-form interaction,

$$U_3(r_{ij}, r_{ik}, \theta_{ijk}) = \varepsilon[\cos\theta_{ijk} - cos\theta_0]^2 \exp(\frac{\gamma\sigma}{r_{ij} - a\sigma})\exp(\frac{\gamma\sigma}{r_{ik} - a\sigma}) \tag{3}$$

LAMMPS software is used to run the MD simulations, ISAACS software to calculate the ring and angular distribution, the VMD software to display the atomic configuration image.

2.1 Investigate the correlation between the average total energy per atom and the temperature

Figure 1 shows the dependence of total energy on temperature when cooling silicene in molecular dynamic simulation at rate 10^{13}K/s.

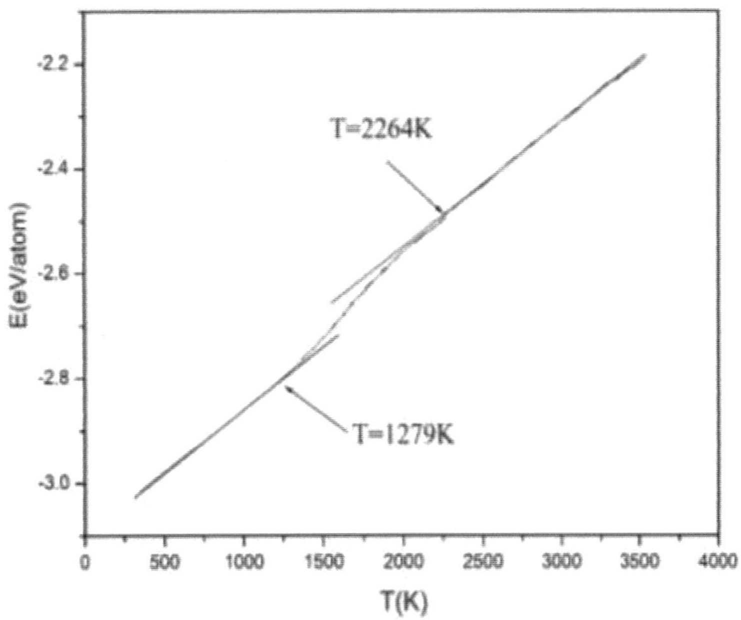

Figure 1. Temperature dependence of total energy when melting 2-dimensional Silicene model

From Figure 1, we can see that at a temperature of 2264K, the curve deviates from the tangent. At 1279K, the curve also deviates from the tangent, the phase transition temperature is the average of these two temperatures, to be 1772K. The temperature of this phase transition is only slightly different from that of the silicene phase transition when cooled at a rate of 10^{10} K/s by the team of Vo Van Hoang to be 1775K [9].

2.2 Radial distribution function

The graph in Figure 2 shows the radial distribution function (RDF) at different temperatures during silicene cooling.

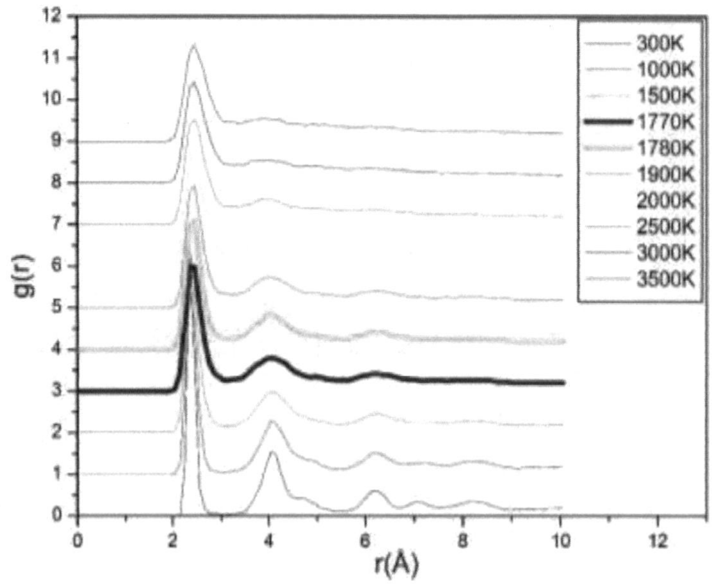

Figure 2. Temperature dependence of radial distribution functions when cooling silicene model, bold lines are for T = 1700K and T = 1780K which are closed to T = 1772K

From the graph, we can see that in the temperature range from 3500K to below 1780K appeared another atomic distribution peak. At a temperature of 1770K (close to the phase transition temperature), there appear many peaks in the g(r) function, indicating the appearance of crystal formation. This result reinforces the above results showing that the phase transition temperature is about 1772K. From the transition temperature onward, the peaks form more and more clearly. At 300K, the plot has discrete pointed peaks, indicating that crystalline silicene has formed in the model.

2.3 Coordination number distribution function

The graphs in Figure 3 show the distribution of coordination number at different temperatures of 3500K, 1770K, and 300K.

IOP Conf. Series: Materials Science and Engineering 1045 (2021) 012021 doi:10.1088/1757-899X/1045/1/012021

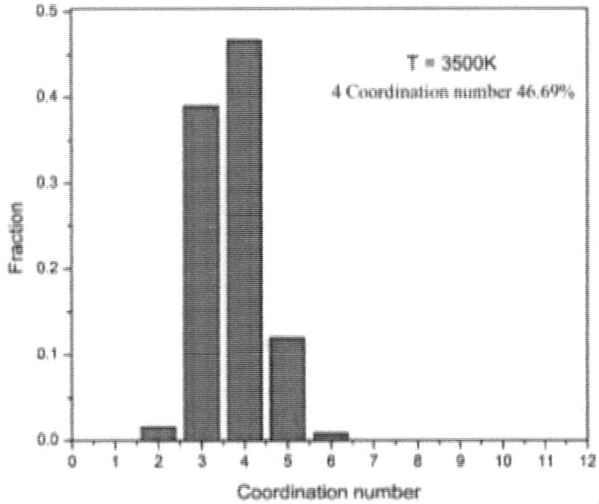

a)

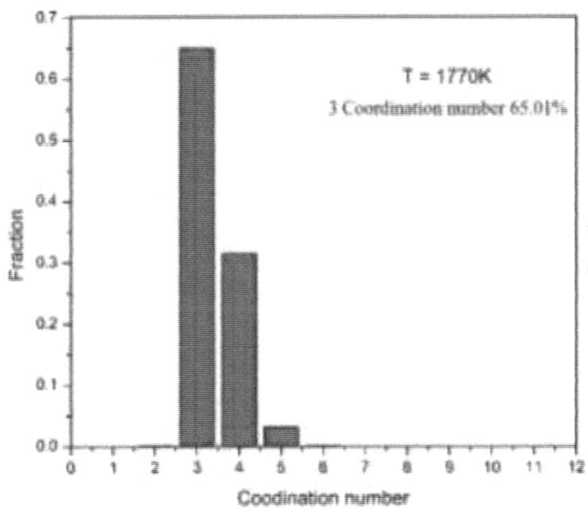

b)

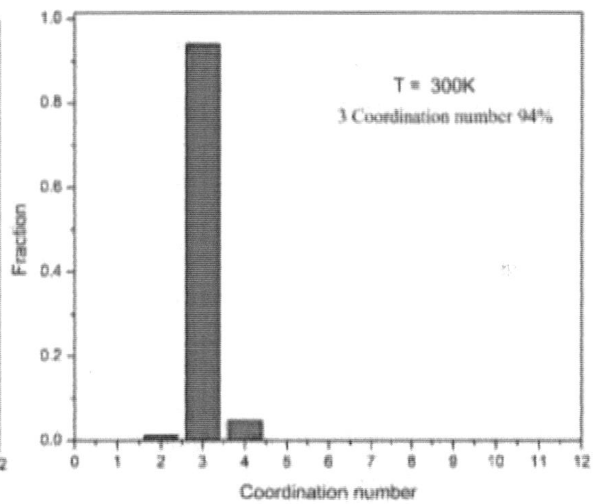

c)

Figure 3. Coordination number distribution (a) 3500K, (b) 1770K (c) 300K

At temperature T = 3500K, there are no coordinate numbers greater than 8. The coordinate numbers 2, 3, 4, 5, 6, 7 have an uneven distribution. The coordinate number percentages of 3 and 4 are 38.92% and 46.69% respectively. Besides, the 2, 5, 6 coordinate number percentages are small proportions of 1.55%, 11.91%, and 0.812% respectively. At the temperature close to the transition temperature of T = 1770K, the coordination number of 4 tends to decrease and reaching 31.57%, on the contrary, the number coordination of 3 continues to increase and reaching 65.01%. At 300K, the number coordination of 4 decreases to 4.87%, the 3-coordinate increases continuously, and reaching 94%, the 2-coordinate is of small value, about 1.14%, the remaining coordinates disappear.

2.4 Ring distribution function

The graphs in Figure 4 show the ring distribution at different temperatures of 3500K; 1770K and 300K.

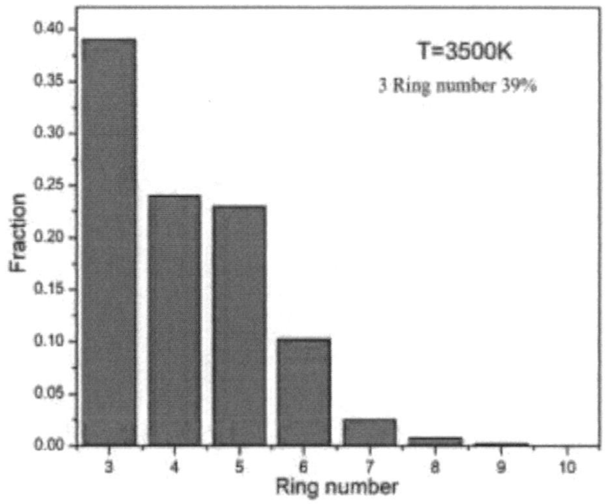

a)

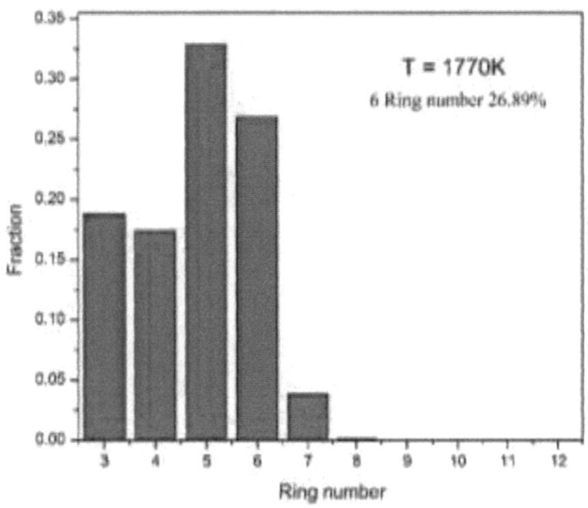

b)

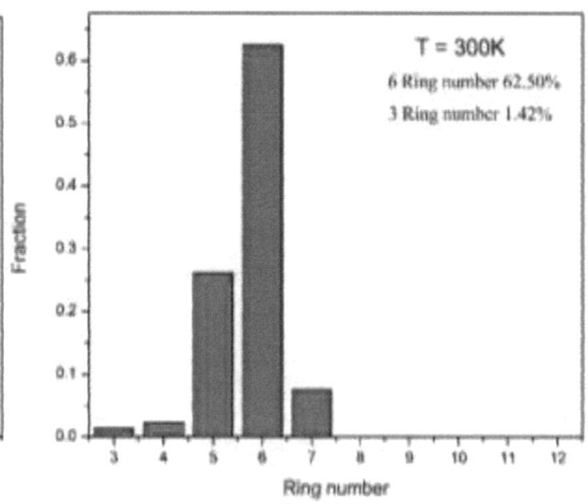

c)

Figure 4. Ring distribution (a) 3500 K, (b) 1770 K (c) 300 K

At temperature 3500K, the value of ring 3 dominates with a ratio of 39.04%. Percentages of rings 4, 5, 6 are 24.09%, 23.05%, 10.27% respectively. Percentages of rings 7, 8, 9, 12 are very small, insignificant. At temperature T = 1770K, ring 3 tends to decrease gradually and reaches 8.86%, ring 6 increases rapidly and reaches 26.89% at the phase transition temperature. At temperature T = 300K, rings of 3, 4, 5 continue to decrease, ring 6 increases and dominates, reaching the ratio of 62.50%.

2.5 Mean ring distribution

Figure 5 shows the change of the average ring distribution on the temperature. The graph is plotted for 20 different temperatures

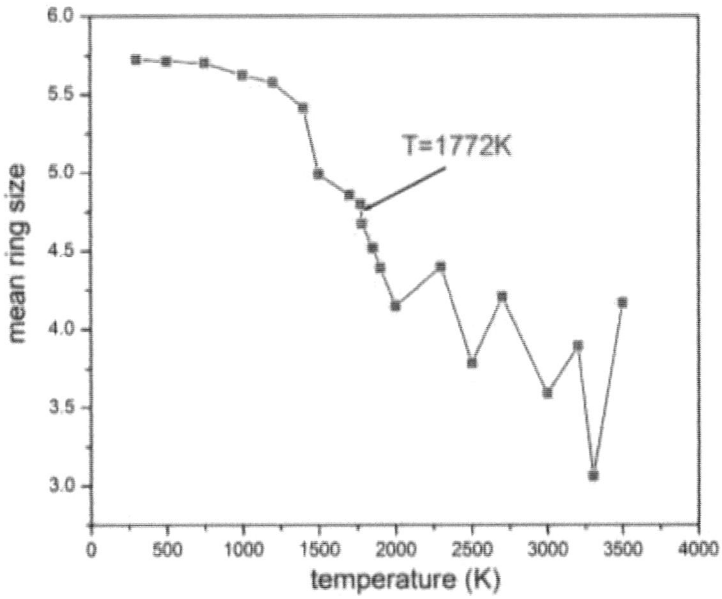

Figure 5. Temperature dependence of mean ring size

The mean ring size values distribute in the range of 4.15 before the phase transition temperature. At phase transition temperature, the mean ring size changes soared. After the transition temperature, the mean ring size did not change much, having a value in the range of 5.41. The mean ring size values tend to increase with decreasing temperature from 3500K to 300K. The mean ring size reaches 6 when the temperature is close to 300K.

2.6 The angle distribution function

The graphs in Figure 6 show the distribution of the bonding angle at different temperatures of 3500K; 1770K and 300K.

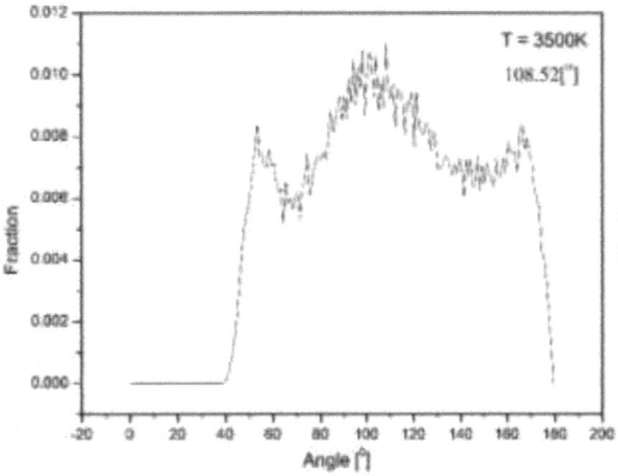

a)

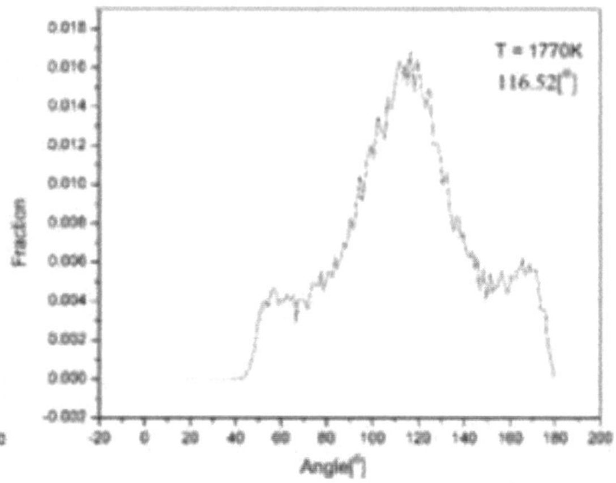

b)

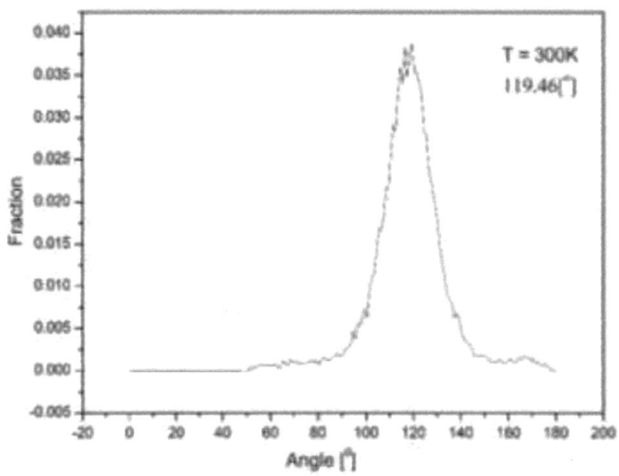

c)

Figure 6. Bond - angles distributions (a) 3500 K, (b) 1770 K (c) 300 K

At 3500K, the Si - Si - Si bond-angle distributes mainly in the 53°-168° range. The curve has a continuous variation. The ratio of Si-Si - Si angle value increases and decreases continuously, it concentrates mostly at the value 108.5°, with a ratio of 1.10%, but the peak is not formed. From the value of angle 121°, the angles tend to decrease steadily as the temperature decreases; the peak of the curve is more visible. At temperatures close to the phase transition temperature, 1770K, the peak of the curve is more pointed, reaching a maximum at 116.52° with a ratio of 1.68%. At temperature 300K, the value of the bonding angle narrows in the range 51.47° to 179.63°, reaching the highest ratio at the bonding angle of 119.46° with a ratio of 3.87%. The peak of the curve has a change of magnitude and percentage; it tends to shift to the right as the temperature decreases.

2.7 Interatomic distance distributions

The graphs in Figure 7 show the interatomic distance distributions at different temperatures of 3500K; 1770K, and 300K.

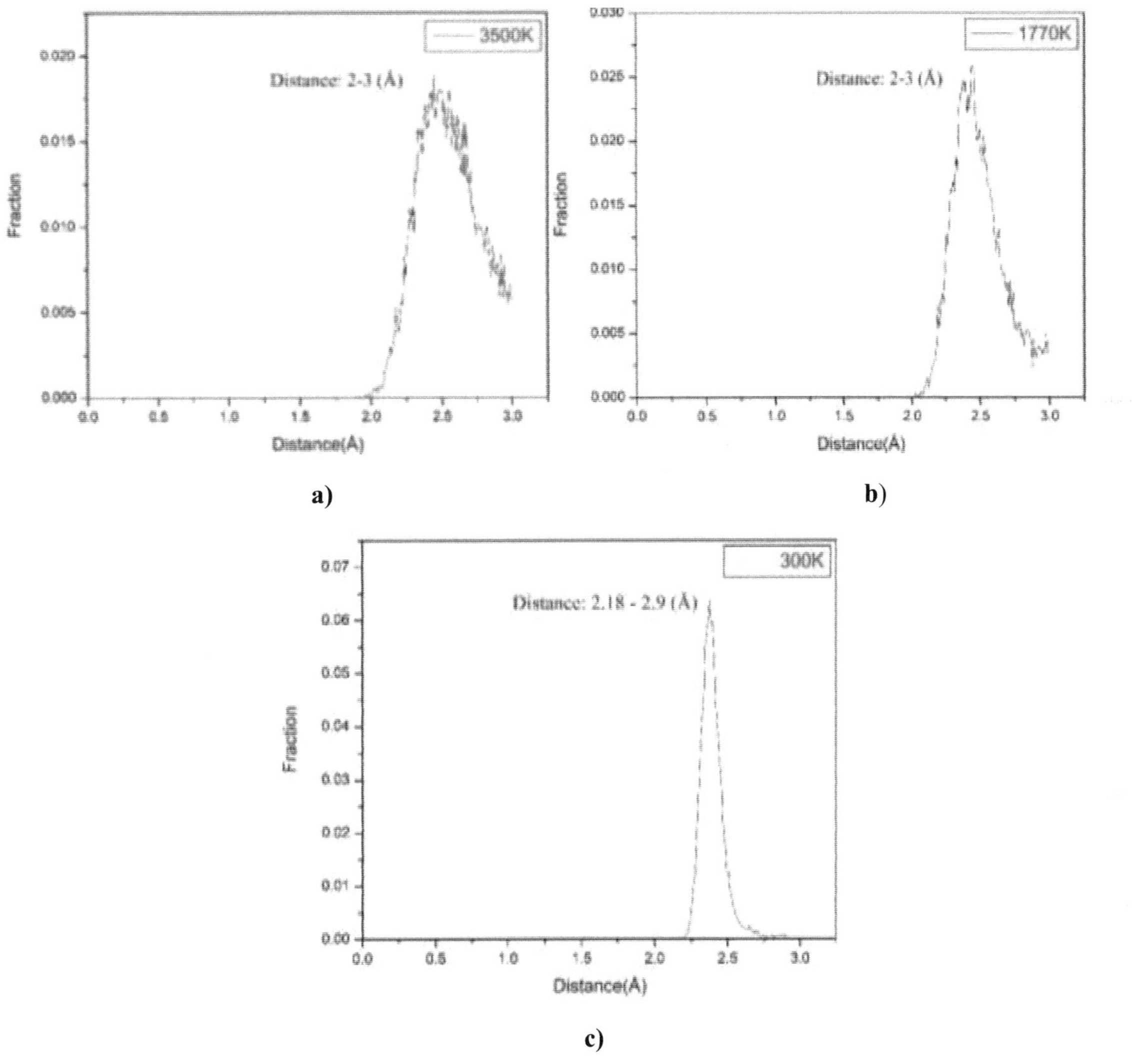

Figure 7. The interatomic distance distribution (a) 3500 K, (b) 1770 K (c) 300 K

The interatomic distance change at pre-phase transition temperatures from 3500K to 1770K is not clear. At 300K, the interatomic distance varies from 0.1- 0.2 (Å).

2.8 The distribution function of mean angle

The graph in Figure 8 shows the dependence of the mean bond-angle distribution on the temperature

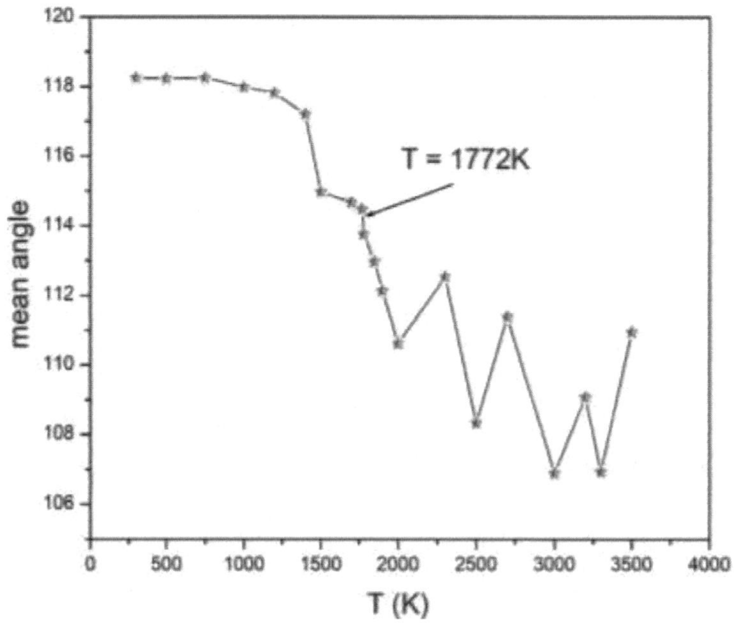

Figure 8. Temperature dependence of the mean bond angle distribution.

The values of the average bond angle ranged around 110.67° before the phase transition temperature. After the phase transition, this value is about 117.25°. When reducing the temperature from 3500K to 300K, the average bond angle tends to increase intermittently.

2.9 Visualization 2D of the crystalline silicene model

The visual image of the silicene model at a temperature of 300K is shown in Figure 9 below. It can be seen that at a cooling rate of 10^{13} K/s, silicene forms two-dimensional crystals with the majority of rings 6, with a ratio of 62.5%. Structural defects form rings 5, 7, 8, which have a pretty high ratio. The defects of the ring-merging type also appear such as merging 2 rings, 3 rings. These will significantly increase the electrical and thermal conductivity of silicene. This result is significantly different from the result obtained by the Vo Van Hoang group [15], they also cooled silicene from 3500K to 300K but the lattice structure defects ratio is less than 3%. This is because the cooling rate in the study of the Vo Van Hoang group is 2.10^{10} K/s which is much lower than 10^{13} K/s in our study.

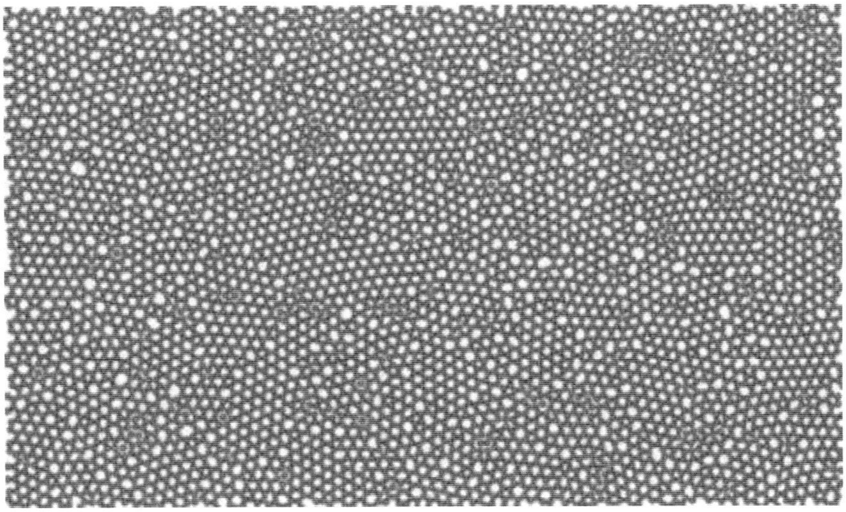

Figure 9. The defects of ring structure 5, 7, 8, and ring merging appear in the model at 300K.

3. Conclusion

The paper presents the results of the study of cooling silicene liquids at a rapid cooling rate of 10^{13} K/s. Simulation results show that the phase transition temperature from liquid to crystal is about 1772K. Investigations of the ring distribution, the coordination number distribution, the distribution of the angles, the distance distribution also showed phase transition at this temperature. As the temperature decreases, silicene gradually forms a crystal structure with the majority of Si atoms distributed in the lattice. The crystal has a nearly symmetrical structure, with lattice failure with a large percentage, about 37.5%. This failure of the lattice causes the electrical and thermal conductivity of silicene to increase significantly compared to perfect silicene.

4. References

[1] Takeda K and Shiraishi K , 1994 *Phys. Rev.*B, **50**, 14916;

[2] Guzman-Verri G G and Lew Yan Voon L C , 2007 *Phys. Rev.*B **76**, 075131;

[3] Zhang L and Wang X, 2014 *Phys. Chem. Chem. Phys.* **16**, 2981–8;

[4] Zhang L, Becton M and Wang X, 2015 *Mater. Sci. Eng.* A **620,** 367–74;

[5] Guzmán-Verri G G and Lew Yan Voon L C, 2007 *Phys. Rev. B* **76,** 075131;

[6] Cahangirov S, Topsakal M, Aktürk E, Şahin H and Ciraci S, 2009 *Phys. Rev. Lett.* **102,** 236804;

[7] Jose D and Datta A, 2012 *J. Phys. Chem.*C**116** 24639–48;

[8] Ming Hu, Xiaoliang Zhang, and Dimos Poulikakos , 2013 *Phys. Rev.* B **87**,195417;

[9] Bing Huang, H. J. Xiang, and Su-Huai Wei , 2013 *Phys. Rev. Lett.***111,** 145502;

[10] Deepthi Jose and Ayan Datta , 2011 *Phys. Chem. Chem. Phys.* **13**, 7304 –7311;

[11]. Ning Liu, Jiawang Hong, Ramana Pidaparti and XianqiaoWang, 2016 *2D Mater.* **3** 035008;

[12] V. Bocchetti, H. T. Diep, H. Enriquez, H. Oughaddou, and A. Kara , 2014 *Journal of Physics: Conference Series* **491**, 012008;

[13] G. R. Berdiyorov and F. M. Peeters, 2014 *RSC Adv.* **4**, 1133-1137;

[14] Min, T. K., Yoon, T. L., & Lim, T. L, 2018 *Materials Research Express* **5**(6), 065054;

[15] Van Hoang Vo, and Huynh Thi Cam Mi, 2014 *Journal of Physics D: Applied Physics* **47**, 495303;

[16] Van Hoang Vo, and Nguyen Truong Long, 2016 *Journal of Physics: Condensed Matter* **28,** 195401;

[17] Van HoangVo, Nguyen Hoang Giang, and To Quy Dong, 2019 *Computational Materials Science* **158**, 406-413;

[18] Hoang Vo Van, Nguyen Hoang Giang and To Quy Dong, 2020 *Philosophical Magazine,* 1-20;

[19] Vo Van On et al, 2020 Journal of Physics: Conference Serials (ID. PSM 6004- ICAPSM 2020) (accepted);

[20] Stillinger F H and Weber T A, 1985 *Phys. Rev.B* **31**, 5262.

Improved progressive collapse resistance of irregular reinforced concrete flat slab buildings under different corner column failures

Suyash Garg*, **Vinay Agrawal and Ravindra Nagar**

Civil Engineering Department, Malaviya National Institute of Technology, Jaipur, Rajasthan, India, 302017

* Corresponding author email: suyash.garg512@gmail.com

Abstract. As the flat slab building with irregular geometric configurations is commonly designed to advance urban development and for architectural demands, its behaviour to resist progressive collapse must be examined. The building's progressive collapse occurs when one or more vertical structural load-bearing elements such as columns are removed due to an extreme load caused either by natural or manmade hazards. The present analytical research examines the failure criteria and improvement in the progressive collapse resistance of five and ten-storey irregular R.C flat slab building by incorporating perimeter beams along with strengthened perimeter columns in the building. The progressive collapse study is conducted by removing different corner columns on the ground floor as per the GSA guidelines (2003) and assessing the spread of damage. Static analysis is performed using the structural analysis program ETABS 2016. For each case, the results have been taken in terms of demand capacity ratio (DCR) at critical sections and, chord rotation and joint displacement at locations of removal of columns. The results showed that the presence of strengthened columns provides sufficient stiffness and the perimeter beam provides load paths for the building's gravity loads, thus making it resistive to progressive collapse under column failure.

Keywords: Flat Slab; Perimeter Beam; Progressive Collapse; Linear Static Analysis; Demand Capacity Ratio; Vertical Displacement; Chord Rotation

1. Introduction

When the system collapses progressively, initial localized disruption is distributed to adjacent components that could cause the whole structural system to fail. The most effective way to limit the spread of localized destruction is to preserve the stability and ductility of the building structure. The ASCE 7-05 [1] provides design guidelines for enhancing the progressive collapse resistance of building structures but does not provide explicit rules for their execution. In two design guidelines released by the United States, General Service Administration (GSA) [2] and the Department of Defense (DoD) [3], recent design approaches to minimize the risk for progressive collapse in structures could be found. The guidelines of the GSA and DoD suggested the Alternate Path Method (APM) for progressive collapse analysis where a single column in the ground level is assumed to be unexpectedly missing and analyses are carried out to assess the ability of the weakened structure to pass through the missing column. The APM focuses on the vertical deflection of the building after the column has been suddenly removed. It is thus a threat-independent design-oriented approach for introducing additional redundancy into the system to prevent the spread of collapse.

IOP Conf. Series: Materials Science and Engineering 1045 (2021) 012022 doi:10.1088/1757-899X/1045/1/012022

Irregular flat slab structures are a common form of structural reinforced concrete constructions in buildings, which enable architects to view picturesque and perspective by using obtuse and acute angles in the plan and to take advantage of low floor heights. Flat slabs are more likely to collapse compared to conventional slab-beam-column structures, due to non-existence of beams that could contribute to redistributing the load previously held by lost columns. The collapse of South Korea Department store in 1995 [4], and the 16-story flat slab structure during construction in Boston, Massachusetts in 1971 [5], are the result of brittle column-slab connections [6-8].

In resisting progressive collapse of RC flat slab structures, Kokot et al. [9] studied the problem by instantaneously eliminating three different columns alternately, and in each scenario, the structural response is determined by linear and nonlinear dynamic time-history analyses. Liu et al. [10] numerically accessed an older reinforced concrete flat plate system against progressive collapse using a macro model incorporated with a punching failure criterion for slab-column connections. The results showed that the RC flat plate buildings are highly vulnerable to progressive collapse when an interior or exterior column is immediately removed without the use of either shear or integrity reinforcement. Qian and Li [11,12] analyzed the load resisting mechanism and quasi-static behaviour of reinforced concrete flat slab systems and evaluated the influence of drop panels on the efficiency of RC flat slabs in preventing progressive collapse. The research findings showed that flat slabs without drop panels were highly susceptible to progressive collapse.

The purpose of this paper is to evaluate the effectiveness of perimeter beams along with strengthened perimeter columns on the static response of 5 and 10-storey irregular flat slab building when exposed to four damage circumstances by the sudden removal of different corner columns at the ground level. The enhancement in the resistance of progressive collapse is evaluated by comparing the vertical displacements (Δ_{crl}) and chord rotation at the locations of removal of columns, and Demand Capacity Ratios (DCR) for the critical columns adjacent to the removed column.

Notations	
GSA	General Services Administration
APM	Alternate Path Method
UFC	Unified Facilities Criteria (DOD 2005)
B-N	Building without Perimeter Beam
B-PBSC	Building with Perimeter Beam and Strengthened Perimeter Column
ACC	Acute Corner Column
OCC	Obtuse Corner Column
RICC	Right Corner Column
RECC	Re-Entrant Corner Column
Δ_{crl}	Vertical Displacement under column removal location
θ	Chord Rotation at column removal location
DCR	Demand Capacity Ratio
Q_{UD}	Acting force determined in the member
Q_{CE}	Expected ultimate unfactored capacity of the member
DL	Dead Load applied to the structure
LL	Live Load applied to the structure

2. Description of Analytical Models

The 3-D models of 5 and 10-storey flat slab buildings having 4x4 bays with a storey height of 3.0 m and span of 8.0 m in the plan are constructed using the ETABS software [13]. The Figure 1 shows the 3-D view while the plan of the studied building is shown in Figure 2. The studied flat slab building is considered irregular as per the GSA [2] guidelines, having geometrical irregularities like Acute, Obtuse, Right and Re-entrant corners in the building plan, as shown in Figure 2. Two types of building models are studied, i.e. building without any perimeter beam (B-N) and building with perimeter beams and strengthened perimeter columns (B-PBSC). The flat slabs are modelled using the Equivalent Frame Method [14-16]. The thickness of the flat slab is considered as 200 mm. For each building, the interior and exterior flat slab beams are of size 1933 x 200 mm and 966 x 200 mm, respectively. The perimeter beams considered are of size 450 x 450 mm throughout its length. For each building, the column size is taken as 450 x 450 mm and the size of the strengthened perimeter column is considered as 750x750 mm. A live load of 3.0 kN/m^2 (office loading) and a superimposed dead load of 2.0 kN/m^2 (interior partition, mechanical and plumbing load) is applied to the floors. The grade of steel and concrete are considered as Fe500 and M25, respectively. The buildings are designed for gravity loads according to IS 456:2000 [17].

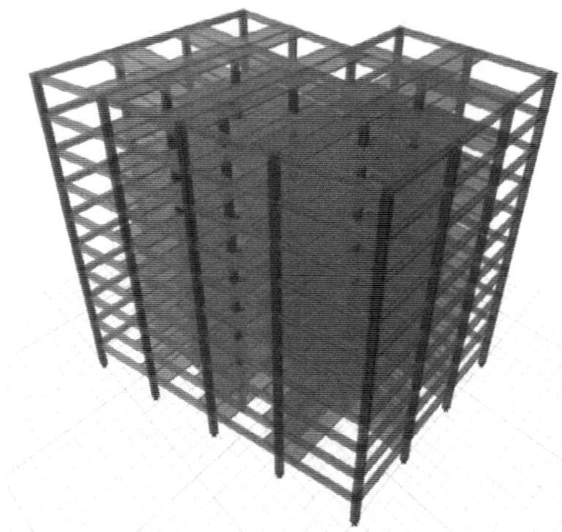

Figure 1. 3D view of the studied 10-storey building.

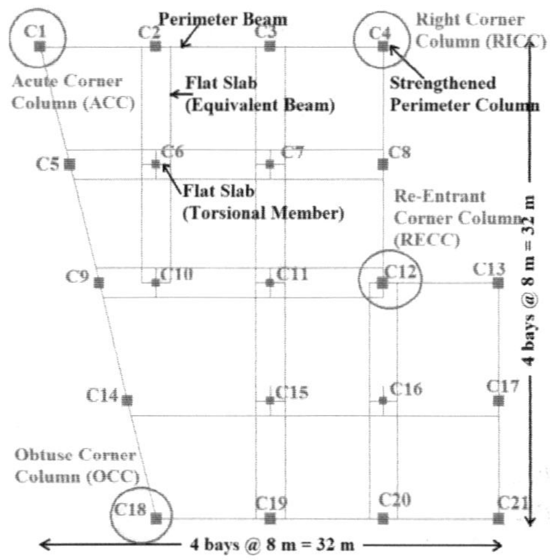

Figure 2. Plan of the studied 5 and 10-storey buildings.

3. Analysis Methodology

The four cases of removal of columns (ACC, RICC, RECC and OCC) at ground level are shown in Figure 2 and the deflected shapes of buildings in ACC, RICC, RECC and OCC are shown in Figures 3, 4, 5 and 6, respectively. The effect of perimeter beams together with the strengthened column on Demand Capacity Ratios for the columns and chord rotations at locations of removal of columns was evaluated for each case. With normal structures, the DCR should not exceed 2 and for irregular structures 1.5 or else they are considered as susceptible to progressive collapse as being severely damaged. GSA has defined DCR according to Equation 1. The calculation of the connection rotation is determined in a frame via the chord rotation. The chord rotation is compared to the permissible plastic rotation angle for this connection according to the GSA and DoD recommendations. The chord rotation as per DoD is calculated according to Eq. 2.

$$DCR = Q_{UD} / Q_{CE} \quad (1)$$

Where, Q_{UD} = Acting force computed in the component (shear, moment, axial force)
Q_{CE} = Expected ultimate, unfactored capacity of the component.

$$\theta = \Delta / L \quad (2)$$

Where, θ = chord rotation (plastic rotation angle) in radians,
Δ = vertical displacement, L = length of member.

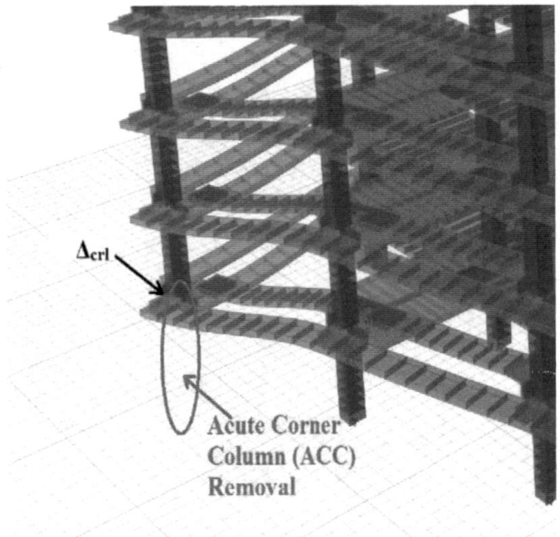

Figure 3. Deflected Shape of the building when ACC removed.

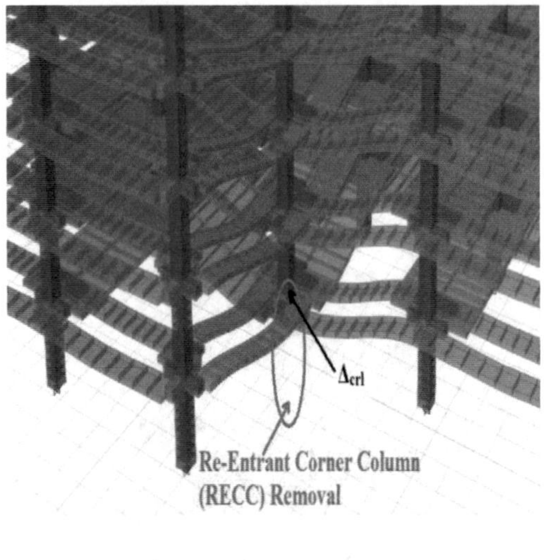

Figure 4. Deflected Shape of the building when RECC removed.

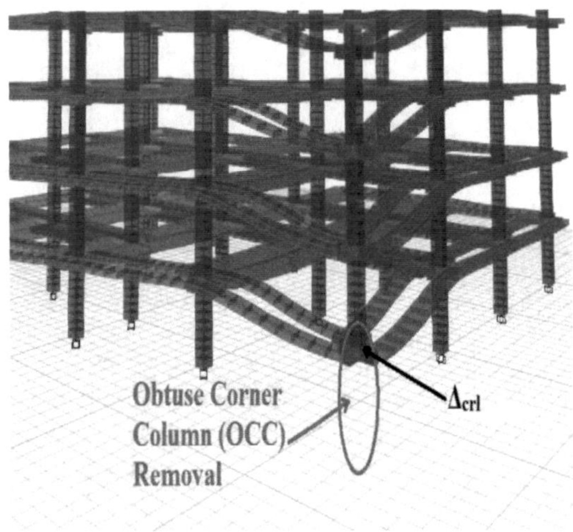

Figure 5. Deflected Shape of the building when OCC removed.

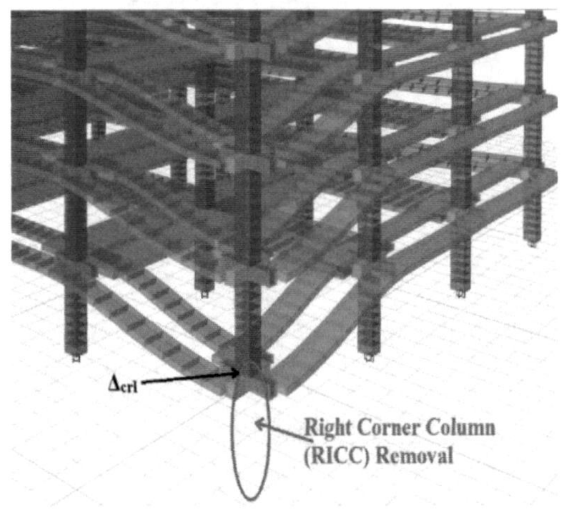

Figure 6. Deflected Shape of the building when RICC removed.

4. Results and Discussion

4.1. Impact of Perimeter Beams with Strengthened Columns on DCR of Columns adjacent to Removed Column

The responses of columns C2 and C5 are evaluated when Acute Corner Column ACC (C1) is removed in the ground floor. In the scenario of removal of Right Corner Column RICC (C4), the responses of columns C3 and C8 are evaluated. The responses of columns C8, C11, C13 and C16 are evaluated when Re-Entrant Corner Column RECC (C12) is removed in the ground floor. In the situation of removal of Obtuse Corner Column OCC (C18), the responses of columns C14 and C19 are evaluated.

For the irregular flat slab building without any retrofitting measure, under GSA factored loading 2(1.2DL+0.5L), in all four circumstances of removal of columns (ACC, RICC, RECC and OCC), the analysis results in Tables 1, 2, 3, 4 and 5 show the development of progressive collapse in the columns adjacent to the removed column as their DCR values exceed the limit 1.5 [2] for irregular buildings. The addition of perimeter beams along with the strengthening of perimeter columns has significantly reduced the DCR values for the axial force and moments of the critical columns for most of the column removal cases. This could be due to the reason that the strengthened columns provided extra stiffness and rigidity to the perimeter frame. The DCR values of critical columns of the 10-storey building are found marginally higher than that of the 5-storey building but the effect of perimeter beams together with strengthened columns on DCR is similar for both 5 and 10-storey buildings.

Table 1. DCR of C2 and C5 (axial force and moments) when C1 (ACC) removed.

	DCR (C2)		DCR (C5)	
	AXIAL	MOMENTS	AXIAL	MOMENTS
5-Storey				
B-N	1.86	1.85	1.81	1.82
B-PBSC	1.43	1.42	1.38	1.40
10-Storey				
B-N	1.88	1.87	1.83	1.83
B-PBSC	1.45	1.45	1.41	1.40

Table 2. DCR of C3 and C8 (axial force and moments) when C4 (RICC) removed.

	DCR (C3)		DCR (C8)	
	AXIAL	MOMENTS	AXIAL	MOMENTS
5-Storey				
B-N	1.72	1.75	1.76	1.78
B-PBSC	1.38	1.42	1.40	1.43
10-Storey				
B-N	1.77	1.81	1.78	1.82
B-PBSC	1.40	1.47	1.44	1.48

Table 3. DCR of C8 and C11 (axial force and moments) when C12 (RECC) removed.

	DCR (C8)		DCR (C11)	
	AXIAL	MOMENTS	AXIAL	MOMENTS
5-Storey				
B-N	1.62	1.65	1.66	1.68
B-PBSC	1.37	1.43	1.65	1.67
10-Storey				
B-N	1.72	1.74	1.68	1.72
B-PBSC	1.47	1.48	1.67	1.70

Table 4. DCR of C13 and C16 (axial force and moments) when C12 (RECC) removed.

	DCR (C13)		DCR (C16)	
	AXIAL	MOMENTS	AXIAL	MOMENTS
5-Storey				
B-N	1.61	1.64	1.65	1.68
B-PBSC	1.37	1.42	1.63	1.65
10-Storey				
B-N	1.71	1.73	1.71	1.75
B-PBSC	1.48	1.49	1.69	1.73

Table 5. DCR of C14 and C19 (axial force and moments) when C18 (OCC) removed.

	DCR (C14)		DCR (C19)	
	AXIAL	MOMENTS	AXIAL	MOMENTS
5-Storey				
B-N	1.96	1.92	1.89	1.90
B-PBSC	1.47	1.46	1.42	1.42
10-Storey				
B-N	1.98	1.95	1.92	1.92
B-PBSC	1.49	1.48	1.44	1.45

4.2. Impact of Perimeter Beams with Strengthened Columns on Vertical Displacement and Chord rotation at the location of removal of column.

For the building B-N, under GSA factored loading, the worst case of OCC removal gives the highest deflection of 763.2 mm, whereas the lowest displacement is in the case of the removal of RECC with 525.5 mm, as shown in Table 6. This may be explained by the fact that the OCC is tied to only two edge columns (comparatively far apart) whereas the RECC is tied to four columns. The influence of the addition of perimeter beams with strengthened columns on the restriction of Δ_{crl} is much greater in OCC, ACC and RICC removal than that in case of RECC removal. When buildings B-N and B-PBSC are compared, it is noticed that the inclusion of perimeter beams with strengthened columns has reduced Δ_{crl} by 61.19 % (ACC), 62.64 % (RICC) and 60.24 % (OCC) in comparison to 30.59 % (RECC). This can be attributed to the better stiffening of the perimeter of the irregular flat slab and connecting the enlarged span created by the removal of the column by the perimeter beams in the cases of ACC, RICC and OCC removals in comparison to RECC removal, where the two interior columns C11 and C16 are left untied. Additionally, removal of OCC showed higher Δ_{crl} when compared to that with the removals of ACC and RICC. This could be due to the larger area covered in the obtuse shape compared to smaller areas protected in the cases of ACC and RICC. Table 7 shows the chord rotation at different locations of removal of columns. Further, Figure 7 shows that the addition of perimeter beams with strengthened columns decreased the chord rotation and kept it below the 0.05 limit (for flat slab structures) defined in GSA and DoD.

Table 6. Vertical Displacement at the location of removal of column.

	Δ_{crl} (in mm)			
	C1 (ACC)	C4 (RICC)	C12 (RECC)	C18 (OCC)
5-Storey				
B-N	734.2	717.8	525.5	763.2
B-PBSC	284.9	268.1	364.7	303.4
10-Storey				
B-N	760.4	742.7	549.1	787.3
B-PBSC	304.2	287.8	384.4	325.6

Table 7. Chord Rotation at the location of removal of column.

	θ (in rad)			
	C1 (ACC)	C4 (RICC)	C12 (RECC)	C18 (OCC)
5-Storey				
B-N	0.091	0.089	0.065	0.095
B-PBSC	0.035	0.033	0.045	0.037
10-Storey				
B-N	0.095	0.092	0.068	0.098
B-PBSC	0.038	0.035	0.048	0.040

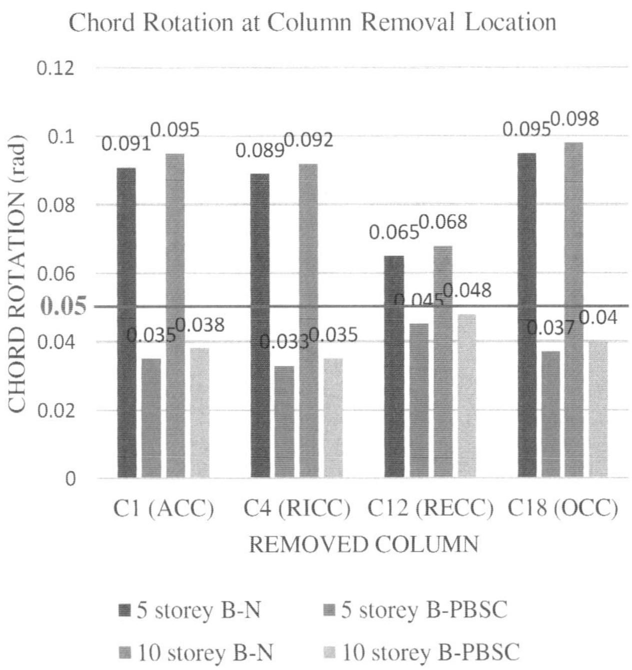

Figure 7. Chord rotation at different locations of removal of columns compared to the limit of 0.05 radians.

5. Conclusion

Adding the perimeter beams and the strengthened perimeter columns to the irregular flat slab building, performance enhancement is assessed in view of critical column DCR, chord rotation and vertical displacement at removed column joint using progressive collapse analysis following the GSA guidelines. The building underwent four circumstances of the removal of different corner columns at its ground level. From the findings of the cases reviewed, the conclusions drawn are as follows:

- In all four cases of removal of corner columns (ACC, RECC, OCC and RICC), the irregular flat slab building B-N demonstrated the evolution of progressive collapse as the DCR values in the critical columns and the chord rotation values in the removed column joint exceeded the 1.5 and 0.05 limits, respectively.
- Among all the four corner column removal scenarios, the effect of the addition of perimeter beams with strengthening perimeter columns to the irregular flat slab building is more significant in the ACC, OCC and RICC removal cases than in the RECC removal case.
- The addition of perimeter beams, in order to form moment frames throughout the perimeter of the building, bridges the enlarged span created by the removal of the column whereas the strengthened columns provide sufficient rigidity and strength to carry the extra dead weights.
- The incorporation of perimeter beams along with strengthened periphery columns, in all the column removal cases, is found to be very effective in preventing the progressive collapse of the irregular flat slab building considered in this study. It enhances all the performance indicators; DCR of the critical columns, joint displacement and subsequently the chord rotation at the locations of removal of columns.

Acknowledgements
The authors want to express their gratitude to the Malaviya National Institute of Technology, Jaipur for their support through the use of ETABS software.

References

[1] ASCE/SEI 7-2005 *Minimum design loads for buildings and other structures* (Virginia-ASCE).
[2] GSA 2003 *Progressive collapse analysis and design guidelines for new federal office buildings* and major modernizations projects (The US General Services Administration).
[3] UFC 4-023-03 2005 *Unified Facilities Criteria (UFC) Design of buildings to resist progressive collapse* (The USA; Department of Defense).
[4] Park TW 2012 Inspection of collapse cause of sampoong department store *Fore. Sci. Int.* **217** 119–126.
[5] King S and Delatte NJ 2004 The collapse of 2000 commonwealth avenue: punching shear case study *J. Perform. Constr. Facil.* **18** 54–61.
[6] Foraboschi P 1995 Modeling of collapse mechanisms of thin reinforced concrete shells *J. Struct. Eng.* **121** 15-27.
[7] Foraboschi P 2012 Shear strength computation of reinforced concrete beams strengthened with composite materials *Comp. Mech. Compu. Appl. Int. J.* **3** 227-252.
[8] Foraboschi P 2016 Structural layout that takes full advantage of the capabilities and opportunities afforded by two-way RC floors coupled with the selection of the best technique to avoid serviceability failures *Eng. Fail. Anal.* **70** 387–418.
[9] Kokot S, Anthoine A, Negro P and Solomos G 2012 Static and dynamic analysis of a reinforced concrete flat slab frame building for progressive collapse *Eng. Struct.* **40** 205–217.
[10] Liu J, Tian Y and Orton SL 2014 Robustness of older flat plate buildings against dynamic progressive collapse *Structural Congress Boston ASCE* 880-890.
[11] Qian K and Li B 2013 Experimental study of drop-panel effects on the response of reinforced concrete flat slabs after the loss of corner column *ACI Struct. J.* **110** 319–329.
[12] Qian K and Li B 2015 Resilience of flat slab structures in different phases of progressive collapse *ACI Struct. J.* **113** 537–548.
[13] User's guide ETABS 2016 *Integrated building design software* (Berkeley: Computers & Structures Inc.).
[14] Sen S and Singh Y 2015 Seismic performance of flat slab buildings *Adv. In Struct. Eng.* **2** 897-907.
[15] Hwang SJ and Moehle JP 2000 Models for laterally loaded slab-column frames *ACI Struct. J.* **97** 345-353.
[16] Cano MT and Klingner RE 1988 Comparison of analysis procedure for two-way slabs *ACI Struct. J.* **85** 597-608.
[17] IS 456 2000 *Plain and Reinforced Concrete - Code of Practice* (New Delhi: Bureau of Indian Standards).

IOP Conf. Series: Materials Science and Engineering 1045 (2021) 012023 doi:10.1088/1757-899X/1045/1/012023

Smart aerial monitoring system to prevent human wildlife intervention

M Sri Harsha[1], P Sai Teja[2], Punyasloka Mahapatra[3] and R Anand[4]

Department of Electrical and Electronics Engineering, Amrita School of Engineering, Bengaluru, Amrita Vishwa Vidyapeetham, India
[1]merugu625@gmail.com, [2]pakalapati0527@gmail.com, [3]shlokmahapatra@outlook.com, [4]r_anand@blr.amrita.edu

Abstract- In India, agriculture is one of the crucial factors for inspecting India's economy. Proper supervision of *this sector leads to an exceptional yield. India, being a densely populated country, h*as encountered frequent and often fatal conflicts due to the share of land and other natural resources among humans and the elephants, in specific. Recent statistics have exhibited that approximately 28000 elephants are distributed over 3% over India's geographical area. *Human* encroachment continues to grow around elephant habitats which results in colossal damage to the crop yield and human life. Technologies like AI and ML are prominent in unraveling the distinct challenges of the modern society. This paper focuses on the design and implementation of a IoT and ML enabled smart drone to intervene the association between humans and elephants to protect both human life as well as the crop yield.

Keywords— Internet of Things (IoT), Machine Learning (ML), Unmanned Aerial Vehicle (UAV), Image Processing, TensorFlow

1. INTRODUCTION

Over the past few years, Unmanned Aerial Vehicles (UAVs) became the epicenter in implementing the functions of various businesses and government organizations and have deeply penetrated through sectors which were stagnant or lagging behind in solving the problems of the society. Unmanned aircraft systems, proved efficient for military applications (greater than 80%), in recent years are being integrated with smart technologies for environment protection and forested areas, agricultural sector, etc. areas. They are interfaced with smart technologies like Image Processing, IoT, Data Analytics, Computer Vision to tackle nearly everything from disease control to vacuuming up ocean waste and crop management. Use of drones in multiple domains increased exponentially as more and more industries and businesses started realizing scope and potential it has in becoming the driving technology of the future globally. Agriculture is an essential domain where it provides basic employment opportunities and self-sustainability and hence it is the backbone of Indian economy. Technology has a crucial role to play to satisfy the demands of the farmers for a good harvest. Drone advocates have cited precision agriculture by employing GPS and Big Data to gather statistics, automate redundant processes to increase the crop yield while resolving water and food crisis in various geographical locations. [1] UAV's armed with GPS and high precision digital and thermal cameras have therefore become a prime point of research all around the globe. The GPS, inertial measuring techniques, and other sensors helped in the autonomous navigation of UAV's. [2] Today's situations demand surveillance to maintain the integrity at a particular habitat and ensure the safety and freedom of its people. An aerial monitoring system is most reliable technology in agricultural sector in terms of surveillance and assistance.

Statistics explicate that crop- raiding has been the most widespread form of human–elephant conflict and results in devastating economic losses for farmers, loss of human lives and the killing or capture of elephants. With an estimate between 24000 and 32000, 60% of Asia's elephants live in India. In 2001, 150 people were killed in elephant attacks in India. This was down from a peak of around 300 people a year in the early 1990s. Between the period 1999 to 2004, at least 150 people and 200 elephants have been killed in the state of Assam with rice beer being the sole attraction and the reason for the devastation. According to WWF (World Wildlife Fund), 900 people get killed annually due to elephant conflicts, and over 100 elephants end up getting electrocuted. Crops over 330 sq. km is destroyed every year amounting to billions of dollars. The population of elephants in India has decreased approximately by 10% over the last 5 years. Indian environment ministry stated that between April 2014 and May 2017, 1,144 people were killed in elephant conflicts. Illegal poaching, habitat destruction for farming lands, and change in global climatic conditions stand as the major factors contributing to this issue. Human Elephant Conflict (HEC) has become a challenging task for wildlife conservationists all over the world especially in the Asia and Africa. [3] Most attacks are a consequence of the clearance of the forests for human settlements. Agriculture is disrupting traditional elephant migration routes leading to violent raids of crops. Small and light weight drones have large advantages over satellites and other aerial photography techniques, with the spatial analysis of regions without the interference of clouds. Drones' flight at lower altitudes leads to higher spatial resolution of images than that of the satellites. [4] Smart Technologies like IoT (Internet of Things) and Image Processing are independently acclaimed for their applications in smart security, pattern recognition etc. but their combination so far is non-existent. [5]
This paper concentrates on a real-time smart drone system to construct a barrier between the farmer's and the elephants to protect his assets for a profitable harvest. An Aerial Monitoring Drone equipped with Machine Learning, Image Processing and IoT (Internet of Things) is employed to design a monitoring system to alert the farmers and to execute the counter action to prevent the elephants' conflicts.
This paper is divided into eight sections. Section 1 explicates the statistics of human-elephant conflicts and their reasons across various countries. Section 2 elucidates the design specifications of the UAV or the drone. Section 3 expounds the literature survey and Section 4 illustrates a brief idea of the proposed system methodology to intervene the conflicts. Section 5 illuminates the construction of the custom classifier model for refining and analyzing the captured images through Image Processing and Machine Learning algorithms. Section 6 describes a IoT-Enabled alert system to generate real-time user alerts and Section-7 exemplifies a concise description of the counter-action implementation to confront the scenarios of conflicts arising all over the world. Section-8 delineates the conclusion and the future scope of the system discussed in this paper followed by the references section.

2. UAV SYSTEM DESIGN AND SPECIFICATIONS

The proposed smart sprayer system has been illustrated in Figure 1. It is controlled by the Pixhawk Flight Controller which has a 32-bit ARM Cortex M4 core with FPU. The quadcopter embodies four BLDC (Brushless DC) motors, four ESCs (Electronic Speed Controllers) and four propellers correspondingly. [6]
When the UAV is armed, the flight controller operates in correspondence to the transmitter and the receiver. Receiver receives the signal from the transmitter and the corresponding signal is sent to the raspberry Pi for controlling the servo motor. This servo action is used to expel the chili powder from the container.

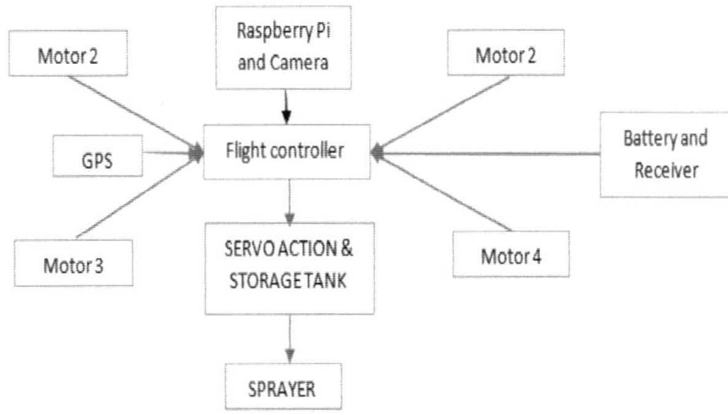

Figure 1 Block Diagram of the UAV with Sprayer Action

The forces: thrust and lift and the rotary motions: roll, pitch and yaw are accountable for the stable flight of a quadcopter. The thrust force can be altered by varying the speeds of the four motors by means of the remote controller.

Technical details of the distinct modules of the drone have been described in Table 1. The weight of the completely built structure is approximately equal to 1000 gms. The calculations and the weight approximations of the payload of the drone system are exemplified in Table 2.

The most crucial detail to be considered while designing the UAV is that the thrust to weight ratio must be greater than or equal to 2:1. [7] The total average weight approximated from Table 2. equal to 1000 gms.
The total thrust to hover the quadcopter is calculated below.

By multiplying the average weight with 2, the thrust
required is 1000*2=2000 gms.
The result is bumped by a factor of 20% to prevent the complexity of hovering through extreme climatic conditions.
2000*(20/100) =400 gms,
Hence, the total thrust required is 2000+400=2400 gms.

On the whole, 2500 gms of thrust is required for hovering the drone or the UAV. Since there are 4 motors employed in the drone, the average thrust on each motor becomes 2500/4= 625 gms.
Table 3. illustrates the variation of the thrust with the variation in the propeller size. The thrust produced by 1045 propellers will be approximately equal to 800*4 = 3200 gms which is much greater than 2500 gms. Therefore, 1045 propellers are equipped to the drone to produce sufficient thrust and lift to implement the necessary action.

Pixhawk 2.4.8 with its in-built accelerometer, gyroscope and barometer stabilizes the flight consequently. FS-iA6B, 6 channel, 2,4 GHz transmitter and receiver are deployed to hover the quadcopter. The radio controller is used to govern the transmission and reception of commands to the drone. It is calibrated through Mission Planner software on a PC. Electronic Speed Controllers (ESCs) of 30 A each are maneuvered to control and regulate the PWM signals monitor the motor speeds based on the instructions of Pixhawk. The sprayer payload is basically a compact canister of 200 gm to spray the chili powder on the elephant. The servo motor receives the control from the flight controller and subsequently expels the powder from the canister.

Table 1. Technical Delineation of components of the UAV

Component Name	Specifications
Brushless DC motor	1000 kV
Pixhawk flight controller	2.4.8 version
Transmitter and receiver	2.4 GHz (10 channel)
Frame (Glass fibre and polyamide nylon)	Length: Width: 450mm: 450mm Height:55mm
Electronic Speed Controllers(ESC)	30 A
Propellers	1045R,1045 Diameter: 10 inch Pitch: 4.5 inch
3s Li-Po Battery	11.1(nominal) 12.6V(max) 2200 mAh 30C
Flight Time	10 min (avg)

Table 2. Weight estimation of the payload of the UAV

Component	Weight In Gms	Quantity	Final Weight In Gms
Flight Controller	40	1	40
Propellers	7.5	4	30
Receiver	7	1	7
Motors	30	4	120
Esc	12	4	48
Battery	190	1	190
Frame	250	1	250
GPS Module	33	1	33
Raspberry Pi	42	1	42
Sprayer(Loaded)	200	1	200
Miscellaneous	40		40
Total			**1000 (Approx)**

Table 3. The effect of propeller size on thrust.

Propeller Size	Thrust Produced (Approx.)
10-Inch 4.5 Pitch (1045)	800gms
9-Inch 4.5 Pitch (0945)	475gms
8-Inch 4.5 Pitch (0845)	475gms

It is utmost significant to design the UAV to determine the accurate hardware components for the construction of the UAV depending on the application and its executable functions to fulfill the purpose with high efficiency.

3. LITERATURE SURVEY

Authors in [8] have discussed the numerous techniques for wildlife monitoring and the framework for automatic detection and recognition of individuals in different patterned species like tigers, zebras and jaguars. This paper provides the fundamental method of implementing the system of image detection in a real-time environment using various image processing and machine learning techniques. [9] delineates the attributes of the aerial surveillance system and advancements that were used in constructing it which include the Firmware ArduCopter 3.2, the 915 MHz Telemetry Radio Transmitter, the Raspberry Pi B+ Microcontroller with Wireless Adapter, the LiPo battery, distinct softwares some of which are Mission Planner and APM Planner. This paper expounds a method of quick and reliable supervision at a challenging cost making it convenient to use it widely at private, institutional and organizational level. [10] expounds the optimization of load on the UAV, adopting and implementing IoT into the drone system and an efficient drone design for handling a real-time crisis. The authors in [11] have proposed a crisis management system for real-time emergency notification of recipients using mobile phones and smart wearables. They have described a mechanism for easy and efficient notification to the users within the range of danger in the occurrence of a disaster through SMS or push notifications to the mobile phone or the smart device. [12] portrays a Region Proposal Network (RPN) that shares full-image convolutional features with the detection network, thus enabling nearly cost-free region proposals. Convolution Neural Networks (CNNs) are one of the most crucial aspects for the application of object detection and classification. The authors in [13] and [14] have propounded a comparative study of the performance of distinct CNNs in detecting and classifying a real-time object. Superiority of the performance of Faster-RCNN over YOLO, in terms of precision, has been explained. [15] sets forth a strategy of setting up a visual based surveillance method in the real time conditions enables a significant use to provide the disaster priorly and therefore reduce the human elephant conflict. The attributes such as hue, saturation, color contours and texture are exploited for the detection of elephant in this method. The authors in [16] have discussed a Bolt cloud enabled Internet of Things (IoT) system to visualize and control the basic pollutants like oxides of Carbon, Sulphur and Nitrogen, and Particulate matter, in closed environment of university layouts in real time. Bolt based IoT system provides pollution level information directly to smart device in real time. The proposed system in this paper expounds an efficient mechanism of interfacing the sensor, MQ135, over the Bolt IoT module which has an in-built Wi-Fi module (ESP8266) that is setup using Bolt application that shows the list of all the connected devices and then using a password authentication and Wi-Fi name to transmit the alerts to the user whenever the sensor senses an increase in threshold of the quantity being measured by the sensor.[17] puts forward a productive and a highly optimized notification and security user alerting system. It describes the collaboration of numerous devices in accomplishing a stable and interconnected system. Employing a dynamic and a precise IoT based system is utmost essential in generating user alerts and interfacing various hardware modules such as sensors. [18] describes the distinct factors and parameters that need to be taken into account to model a highly optimized and efficient network of interconnected devices. [19] provides a brief review of the various machine learning algorithms frequently in use. The author highlighted the merits and the demerits of the algorithms from their application perspective. The algorithms were evaluated in terms of their performance, learning rate etc.

4. PROPOSED SYSTEM METHODOLOGY

Drones are capable of reaching to the most remote areas with highest accuracy with no manpower and with minimal amount of effort, time and energy. This is one of the massive reasons for their prominence in evident sectors like military, agriculture, technology and commercial.

The implementation details for the proposed system for the application of intervention between the humans and the elephants predominantly consists of 3 modules:

i. *Custom Classifier for Elephant-Detection*
ii. *User Alert System*
iii. *Counter-Action Implementation*

Figure 2. expounds the brief principle of implementation of the proposed system by interfacing smart technologies to the Unmanned Aerial Vehicle (UAV). This method uses the basic idea of interfacing IoT and image processing concepts in machine learning to construct a framework for intervening the real-time conflicts of humans and elephants.

The same principle elucidated in Figure 2. can be extended to avoid conflicts between any other animal.

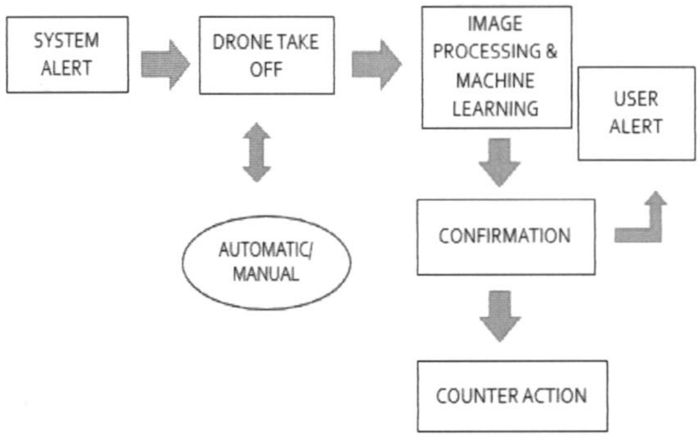

Figure 2 Block Diagram of the Proposed System

A sensor facilitates the requirement to capture the motion of any object in its vicinity. One such sensor is an ultrasonic sensor which estimates the motion of an object with the application of ultrasonic signals can be employed in the fields of the farmers which is the region of interest but with a lower range. Although the LIDAR sensor is quite expensive, it has a higher range and can be used for higher and complex applications.

The sensors are positioned all around the field to capture and monitor the movements of the wild elephants. Upon detection, the drone flight is initiated. The proposed system focuses on the manual flight of the drone. A Raspberry Pi is mounted on the drone for satisfying the requirement of image processing. The Raspberry Pi Camera Board, which is available in 2 Mega pixels and 8 Mega Pixel variants is mounted to the CSI port of the raspberry Pi. Here 8MP camera is suggested as it provides high resolution images and video streams provide accurate results in processing. This camera board captures the image of the wild elephant. Image Processing and ML algorithms, discussed in the further sections, are embedded in the Raspberry Pi as a software and execute a continuous and rigorous comparison to generate an alert to the user on successful detection of an elephant. The Bolt-IoT module, with in-built Wi-Fi/GSM chip and a cloud platform, is brought into action for transmitting the alert to the famer over a server. A servo motor is interfaced with the Raspberry Pi to accomplish the servo action for expelling the chili powder on the elephant from the container. The chili powder, a natural remedy, acts as a natural repellant for eluding the conflicts between humans and wild elephants.

5. CUSTOM CLASSIFIER MODEL

A large percentage of the human brain dedicates itself to visual process. Images grab and process the attention of a person at an alarming speed. One of Machine Learning's most successful paradigm is supervised learning which lets us build a generalization model by the continuous processing and rigorous training of images of the desired object.

The main agenda of digital image processing is to extract and predict certain important information and augment its visual quality making it more understandable by human analysts or for autonomous machine perception. For constructing a remote sensing application, vital algorithms and procedures must be developed to process the data to generate the estimated output. Images acquired from the sensors at different times are emphasized to facilitate the extraction of significant information. Further, the images are preprocessed, segmented and then classified to produce an inference graph. [20]

Distinct computer algorithms and object-detection techniques are utilized to accomplish and extract the results from the digital images. Object detection technology has seen a rapid adoption rate in various and diverse industries. A single deep neural network can be designed to automatically solve all the problems reducing the human intervention. One of the major problems is the large amount of time spent in the training of the images to achieve precise results. Ultramodern object detection classifiers

mostly rely on region proposal algorithms for conjecture of object locations. Methodologies like *Fast R-CNN* reduced the run time of these detection classifiers, depicting the region proposal computation as a bottleneck.
The custom classifier model proposed makes use of several machine learning algorithms to achieve the required results. Generating and designing the custom classifier model comprises of the crucial steps of image labelling, setting up the platform, generating the training data by selecting a specific pre-trained model and extracting the inference graph to detect the elephant.
Image labelling is the fundamental step to gain insights into the distinct entities in an image. It is critical to determine the type of annotation tool prior to labelling the images for precise results and high efficiency of the model. LabelImg is the annotation tool used in the proposed system to label the numerous images in the elephant dataset. *Qt* is the graphical interface employed and annotations are saved as *XML* files in *VOC* format. Each and every image is assigned a unique keyword, metadata, for the application of training the images. This metadata is fed to various Machine Learning algorithms. The obtained dataset is divided into training and testing data. The quality of the training determines the quality of the final model. Fig. 3. represents the sample labelling of an elephant image among the cluster of images collected for the dataset.

Figure 3 Sample Labelling of an image of an elephant

TensorFlowGPU and *TensorFlowCPU* along with tensorboard are preinstalled to setup an environment possible for the application of image processing. *CUDA* and *CuDNN* need to be installed to support *TensorflowGPU* which is more efficient than the CPU version in terms of the processing and the analysis time. Several additional packages of Python need to be encapsulated to support the modules of the Machine Learning algorithms.
Region proposal methods and region-based convolution networks are the recent advances in object detection. *Region-CNN (R-CNN)* is one of the ultramodern CNN-based deep learning approaches. In Region Proposal Networks, layer judges the anchors that belong to background of foreground by soft max, then it uses method of bounding box regression to modify anchors for obtaining accurate proposals. [21] Region proposals are comprehensively being implemented in several applications of image processing such as computer vision, robotics, pattern recognition and segmentation. These region proposals are superior to the earlier state-of-the-art algorithms in terms of the processing speed, accuracy and computational complexity. The combined algorithm of CNN depends on two critical phases: (1) the use of the region proposal algorithm in generating the most efficient set of region proposals, and (2) process the raw image data into the convolutional neural network. [22]
The object-detection-system of the proposed system in this paper brings *Fast-RCNN* into play for high accuracy and efficiency of detecting an elephant. It structures on two modules. The first module is a convolutional neural network that proposes regions. The second module is the detector that makes use of the proposed regions. The entire network of *F-RCNN* applies these two modules. *Fast R-CNN* network is mainly used to classify the objects and refine the boundaries of these particular regions. Both the modules of Faster RCNN devote the parameters of the convolution layer obtained through the process of feature extraction, making the detection tasks process at a relatively greater speed. Proposal feature maps are utilized by the classifier layer to calculate the required category, and applies bounding box regression again to obtain the exact location of the detection box. Various boxes are scrutinized by a classifier and the existence of objects is detected by the regressor. The CNN recognizes the object class

and the bounding box. After the images iterate through the convoluted areas, the feature map for the image is developed.
The location of each unique image is passed through a sliding window where all the features are extracted by the network for detection. The main function of the RPN network is to check the existence of an object. The bounding box will direct this object to the detector for further detection. [23] Fast *Region-based CNNs* make use of GPUs, whereas the region proposal methods used in the research are processed on the CPU, indicating that the CPU takes longer time for running the training. The inference time is directly proportional to the resolution of the feature map. Thus, the inference time for generating object proposals increase with the number of feature map locations. [24] An obvious way to accelerate proposal computation is to reimplement it for the GPU. The training of data utmost essential in determining the quality and the accuracy of the ultimate custom classifier model. Usually, Faster R-CNN is highly accurate whereas the SSD and F-RCN are quite faster. Faster R-CNN using Inception Resnet feature extractor with 300 proposals gives the highest accuracy at a speed of 1 FPS. *PYTHONPATH* is utilized to configure the environment variables. *Protocol Buffers (Protobufs)*, methods of serializing structured data, are compiled so that TensorFlow can configure the model and the training parameters. The initial subset of data to rigorously train a model through various technologies to produce sophisticated results is the training set. The set of observations necessary for evaluating the performance and generalizing the classifier model based on a performance metric is the testing data. *TFRECORD* is employed to generate the testing data, the testing labels, the training data and the training labels. Configure a label map and edit with the training configuration file and finally pipeline the training model. The label map is created and the model is pipelined to improve the accuracy of the model and achieve a successful algorithm. The training data is continuously executed to minimize the error on each iteration for a productive and highly accurate model. The characteristics of the entire training process can be viewed on Tensorboard as explicated in Figure 4. and final inference graph is generated and exported to the main application.

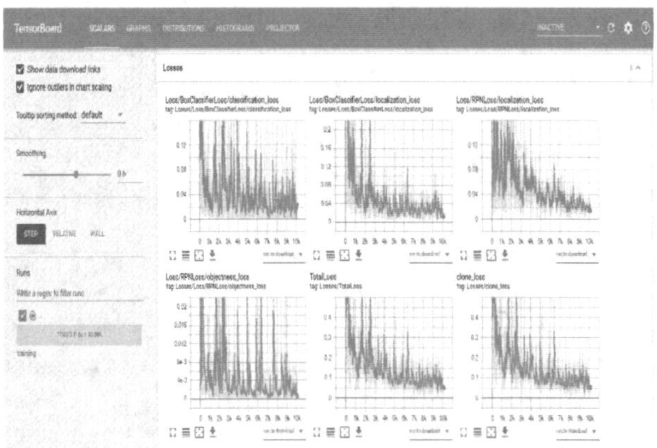

Figure 4 Depiction of the whole training process on Tensorboard

Figure 5 Sets of Images elucidating the detection of an elephant using the custom classifier model in a video stream

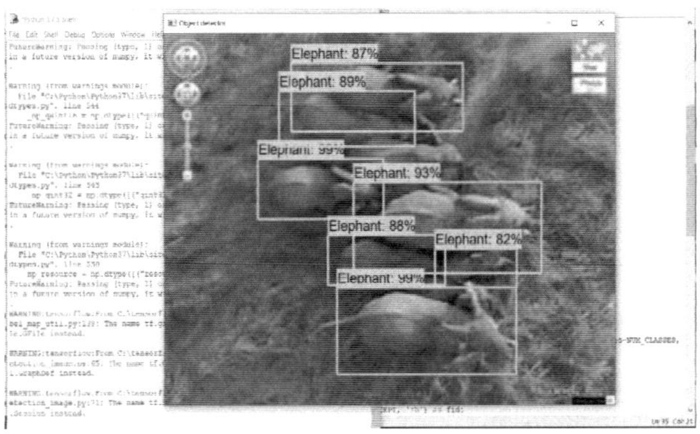

Figure 6 Detection of a herd of elephants in a particular region of interest using the custom classifier model

The generated final inference graph is embedded into a Raspberry Pi equipped with a Camera Module V2 for capturing and processing the image for elephant detection through the custom classifier model. Fig. 5. depicts the output images as a result of the object detection of an elephant by the custom classifier model in a video stream comprising of numerous frames. Fig. 6. illustrates the detection of a herd of elephants which is of utmost importance for a real-time application custom classifier object- detection model.

6. USER ALERT SYSTEM

Technology is evolving rapidly every day in today's era with all the smart physical devices connecting to each other on a network for exchanging info and increasing the productivity. Internet of Things (IoT), a system of interrelated computing devices and physical machines, are provided with unique identifiers and the ability to transfer data over a network of machine to machine communication(M2M). Devices, connectivity and platforms are the inherent qualities of the scheme. To build any IoT application the most critical issue is to select the right combination of sensors and also in choosing networks and communication modules. [25] IoT has distinct applications in the fields ranging from smart grids, wearables to industrial internet. IoT is commonly termed as a network of networks. Due to this, it can accomplish significant number of tasks with utmost efficiency and accuracy. Internet of things (IoT) has gained a lot of importance as the infrastructure required for heavy use of internet is almost available easily and transmission of data over internet has become more efficient by setting up record speeds of transmission. The next upcoming years are expected to see an exponential growth in this smart technology for securing and interconnecting the devices with cloud. IoT environments encompass three most crucial requirements: Confidentiality, Integrity and Availability. This makes sure that the IoT analytics systems do not inundate with spurious traffic, or that the communication channels or the Network Element are not intentionally jammed. [26]

Bolt, a cloud-based platform, with an in-built Wi-Fi/GSM chip and prominent for an integrated cloud-based operating platform supports a secured connection of devices and sensors to the Internet. With the application of Bolt-Cloud, monitoring and visualizing data, and tracking machine learning algorithms turns out to be more efficient and accurate. It provides the proficiency to control the connected devices and collect the data safely and securely.

Figure 7. depicts the pin schematic diagram of the Bolt Wi-Fi module with ESP8266-12E and Table 4. expound the technical specifications of the distinct parameters in the module.

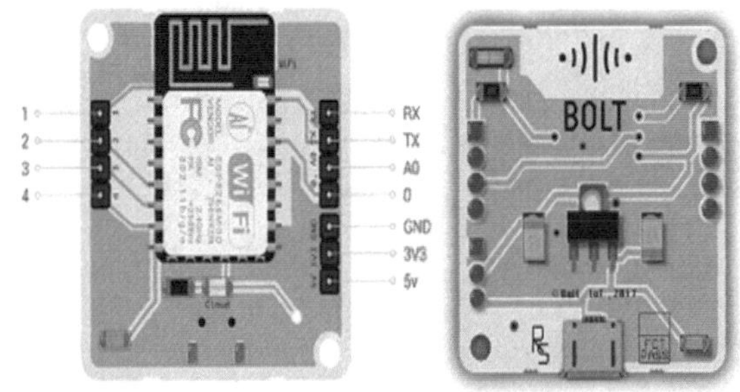

Figure 7 Bolt ESP8266-Based IoT Platform Module

Table 4. Technical Specifications of the Module

Parameters	Details
Connectivity and Processing Module	ESP8266 with custom firmware
MCU	32-bit RISC CPU: Tensilica Xtensa LX106
Power	5 V/1 A DC via Micro-USB port or 5 V and GND pins
Operating Voltage	3.3 V
CPU Clock Frequency	80 MHz
MCU Internal Memory	64 KB of instruction RAM; 96KB of data RAM
MCU External Memory	4 MB Flash memory [QSPI]
GPIO pins	5 Digital pins [3.3V logic]
ADC	1 pin 10 bit ADC [0-1V input]
PWM	All 5 Digital pins capable of PWM [Software PWM]
Connectivity	
WiFi	802.11 b/g/n Automatic AP mode if not connected to WiFi WEP/WPA/WPA2 authentication
UART	8-N-1 3.3V TTL UART [using TX, RX, GND pins] [9600 baudrate]
Cloud	Default: Bolt Cloud
Boot Time	Less Than 1 second

Bolt-IoT plays a key role in generating user alerts for implementing counter action for intervening the human-wildlife conflicts. Nevertheless, it can be used for interfacing the position sensors for determining the position of the elephant in the considered region of interest. The proposed system brings this phenomenon of Bolt-IoT into action to generate the user alerts by integrating the Bolt-IoT module with distinct delivery channels like SMS, E-Mail, Telegram. Alerts are typically delivered through a notification system and the most common application of the service is machine-to-person communication. The ability of the alert notification systems to be integrable makes them functional with

third party applications like other educational mobile applications, alarm instruments, IT systems, cloud services, access control systems, weather forecasting systems, and other domestic sensors, manufacturing systems, mining hazardous materials detection systems, and motion and thermal cameras. Each cloud communication platform has a unique API key (Application Programming Interface) and a device id to interface the Bolt device.
The user alert is generated on the successful detection of an elephant by the custom classifier object detection model. This generated user alert enables the user to initiate the flight of the drone and execute the necessary counter action for tackling the problem. Bolt-Cloud, being one of the most significant advantages of the Bolt-IoT module, enables storage of immense amounts of data and transmitting real-time user alerts accordingly.
Fig. 8. expounds the block diagram of the proposed user alert system in this paper. In this paper, a SMS alert system is discussed which focuses on the transmission of real-time SMSs by interfacing a cloud communications platform with Bolt-IoT module for transmitting the user alerts to the user.
Twilio, a cloud communications platform as a service, allows users to efficiently interface the bolt device to programmatically make and receive calls, text messages and perform other communication functions using web service APIs. The Bolt device is authorized with the Twilio Account SID, the AUTH Token, the API Key and the Bolt Device Id.
Further, we need to interface the Bolt IoT Device and Twilio for sending SMS Alerts on the To_Number configured in the Twilio dashboard by implementing code in the UBUNTU environment in VMware Workstation 12 Player.

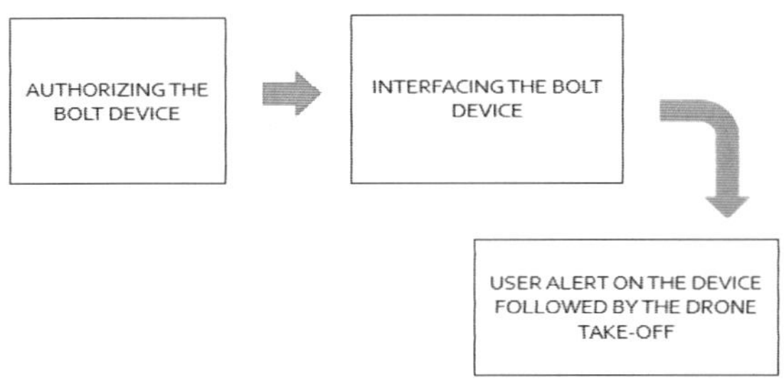

Figure 8 Block Diagram of the Proposed User Alert System

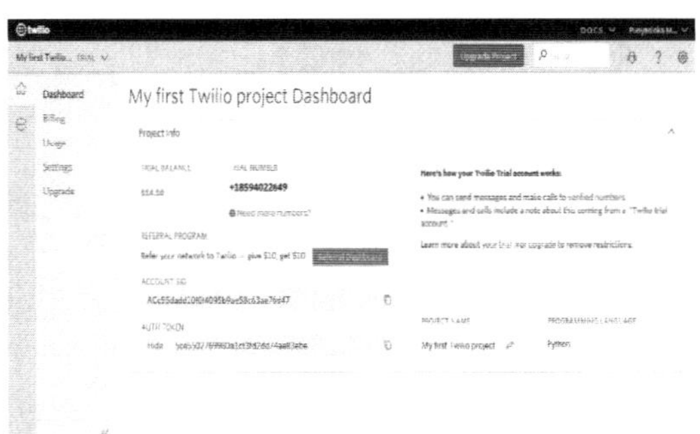

Figure 9 Twilio Dashboard for configuring SMS Alert System

Fig. 9. exemplifies the dashboard of Twilio for configuring the SMS alert system by interfacing the Bolt-IoT device with the implementaion of the code taking into consideration the ACCOUNT SID, the FROM_NUMBER, the TO_NUMBER, the API Keys and the threshold to decide the condition for generating the SMS alerts to the user.

The proposed SMS alert system in this paper can be extended for various other prominent delivery channels such as that of E-MAIL, Telegram which generate the alerts in the same technique based on a threshold.

7. COUNTER-ACTION IMPLEMENTATION

Successful detection of the elephants by the custom classifier model for object detection and an efficient SMS alert system was implemented in the previous sections of this paper.
The implementation of the image processing, machine learning algorithms and the user alert system finally leads to the implementation of the counter-action to resolve and intervene the conflicts between the elephants and the humans.
Extensive research and intensive study on animal behaviors explicate that elephants are extremely sensitive to sound, smell and touch. The elephants being sensitive to sound, are terrified of the sound made by the bees. Hence, a 'Honey Bee Buzzer" is practically being implemented by the Alipurduar division of the North East Frontier Railway (NFR) to avoid elephant deaths and to keep elephants off the railway tracks. This device generates the sound of buzzing bees at two rail gates in the division.
This paper proposes an innovative idea based on elephants' characteristic of being sensitive to smell. Chili powder, known for its pungent smell, is filled in a tank that is armed with a motor that operates the opening and closing of the nozzle of the tank when an appropriate signal is given to it. A servo motor which uses positive feedback system to control motion and the final position of the shaft is deployed for the mechanical servo action of expelling the powder from the canister. Fig. 10. illustrates the SG-90 servo motor for actuating the nozzle of the canister. The servo, being an automatic device that uses error-sensing negative feedback to correct the action of a mechanism and having the ability to generate highly precise angles, is used for the counter implementation as proposed in this paper. The chili powder is inserted into a canister prototype which is equipped with the servo for expelling the chili powder when it receives the appropriate signal from the Raspberry Pi. This setup is mounted to the drone and the servo motor is connected to the Raspberry Pi. When the drone identifies and confirms the presence of an elephant, based on the SMS alert generated after the successful results from the custom classifier elephant detection model, the chili powder is expelled from the canister through the servo action. The elephants are terrified of the smell of the chili powder and retreat in the same path.
This is how the conflicts between humans and elephants can be avoided in distinct parts of the world.

Figure 10 SG90 servo motor for actuating canister nozzle

8. CONCLUSION AND FUTURE SCOPE

Studies have unveiled that the expansion of human settlements and agricultural fields across Asia and Africa has resulted in widespread loss of animals' habitat and the extinction of their species relative to their historical size and overall range. These endangered animals either destroy human lives and their vegetation or end up getting killed or injured by the people to ensure safety in distinct geographical locations of the world. Strict measures need to enforced to maintain ecological stability in an environment.

The technology of drones, image processing, and cloud solutions has given us an edge in gathering the requirements for materializing this idea of reducing human-wild animal intervention. The monitoring system discussed in this paper deals with the association of the smart technologies like Image Processing, Machine Learning and Internet of Things (IoT) to generate real-time alerts and provide a constructive servo action to counteract the conflicts to save both the animals and human property. Through rigorous assessments of setbacks and by multiple iterations of testing the sole ideology of this project is achieved to the fullest possible form. In this process, the math and science behind each technology and tool is well understood and brought in a better insight into what one can do with knowledge over such superior and versatile development mediums. Through this project and this application there arises a hope where such advanced equipment can be a boon for protecting wildlife and conserving them amid a rapid increase in human activity. Though this application is a product ready, there is a need for research and development in some areas that help to achieve greater speeds and accuracy while minimizing the cost involved to productize it. With ever-changing technology, new attributes and attachments can be added to the prototype which helps in serving the purpose of the idea.

References

[1] Frank Veroustraete 2015 *The Rise of the Drones in Agriculture* (EC Agriculture) 325-327.

[2] Urbahs Aleksandrs and Jonaite Ieva 2013 *Features of the use of unmanned aerial vehicles for agriculture applications* (Aviation) 170-175.

[3] Mohamed Fazil and Mohamed Firdhous 2018 *IoT-enabled smart elephant detection system for combating human elephant conflict* (3rd International Conference on Information Technology Research)

[4] Yiqing Guo, Xiuping Jia, David Paull, Junpeng Zhang, Adnan Farooq, Xiaolin Chen and Md Nazrul Islam 2019 *A drone based sensing sytem to support satellite image analysis for rice farm mapping* (IGARSS) 978-1-5386-9154-0/19

[5] Ayush Kapoor, Suchetha I Bhat, Sushila Shidnal and Akshay Mehra 2016 *Implementation of IoT (internet of things) and image processing in smart agriculture* (International Conference on Computational Systems and Information Systems for Sustainable Solutions) 978-1-5090-1022-6/16

[6] Manoj Vihari M, Dr Usha Rani Nelakuditi and Purna Teja M 2018 *IoT based unmanned aerial vehicle system for agriculture applications* (ICSSIT) 978-1-5386-5873-4/18

[7] Dhanalakshmi K S, Karthika S, Saravanakumar P and Jacinth Emril A 2015 *Implementation of highly automated uav with gps for product delivery using labview* (International Journal of Advances in Electronics and Computer Science) ISSN: 2393-2835 **2** Issue-11 pp:39-42.

[8] Gullal Singh Cheema and Saket Anand 2020 *Automatic detection and recognition of individuals in patterned species*" 2005.02905

[9] Zainab Zaheer, Atiya Usmani, Ekram Khan and Mohammed Qadeer A 2016 *Aerial surveillance system using uav* 978-1-4673-8975-4/16/2016 IEEE

[10] Sanjana, P and Prathilothamai M 2020 *Drone design for first aid kit delivery in emergency situation.* 2020 (6th International Conference on Advanced Computing and Communication Systems) 10.1109/icaccs48705.2020.9074487

[11] Mohammed Ghazal, Samr Ali, Marah Al Halabi, Nada Ali and Yasmina Al Khalil 2016 *Smart mobile-based emergency management and notification system* (4th International Conference on Future Internet of Things and Cloud Workshops)

[12] Shaoqing Ren, Kaiming, He Ross and Girshick Jian Sun 2016 *Faster r-cnn: towards real-time object detection with region proposal networks* (IEEE Transactions on Pattern Analysis and Machine Intelligence) 10.1109/TPAMI.2016.2577031,

[13] Murali, A, Nair B B and Rao, S N 2018 *Comparative study of different cnns for vehicle classification* (IEEE International Conference on Computational Intelligence and Computing Research) 10.1109/iccic.2018.8782419

[14] Gokul, R, Nirmal, A, Bharath, K, Pranesh, M and Karthika R 2020 *A comparative study between state-of-the-art object detectors for traffic light detection* (International Conference on Emerging Trends in Information Technology and Engineering) 10.1109/ic-etite47903.2020.449

[15] Gowtham Ramesh, Senthilkumar Mathi, Sini Raj Pulari and Vidya Krishnamoorthy 2017 *An automated vision-based method to detect elephants for mitigation of human-elephant conflicts* 10.1109/ICACCI.2017.8126187

[16] Pallavi Asthana and Sumita Mishra 2018 *IoT enabled real time bolt based indoor air quality monitoring system* (International Conference on Computational and Characterization Techniques in Engineering and Sciences, Integral University, Lucknow, India)

[17] Brundha S M, Lakshmi P and Santhanalakshmi S 2017 *Home automation in client-server approach with user notification along with efficient security alerting system* (International Conference On Smart Technologies For Smart Nation)

[18] Sankaran S 2016 *Modeling the performance of IoT networks* (IEEE International Conference on Advanced Networks and Telecommunications Systems) 10.1109/ants.2016.7947807

[19] Susmita Ray 2019 *A quick review of machine learning algorithms* (International Conference on Machine Learning, Big Data, Cloud and Parallel Computing)

[20] Leila Fonseca M G, Laércio Namikawa M and Emiliano Castejon F 2009 *Digital image processing in remote sensing* (Brazilian Symposium on Computer Graphics and Image Processing)

[21] Xueqiang Song, Peng Jiang and He Zhu 2019 *Research on unmanned vessel surface object detection based on fusion of ssd and faster-rcnn* (Chinese Automation Congress)10.1109/CAC48633.2019.8997431

[22] Peilei Dong and Wenmin Wang 2016 *Better region proposals for pedestrian detection with r-cnn* (IEEE, VCIP, Chengdu, China) 978-1-5090-5316-2/16/c

[23] Deepa R, Tamilselvan E, Abrar E S, Shrinivas and Sampath 2019 *Comparison of yolo, ssd, faster rcnn for real time tennis ball tracking for action decision networks* (IEEE) 978-1-7281-3250-1/19

[24] Lars Sommer, Nicole Schmidt, Arne Schumann and J¨urgen Beyerer 2018 *Search area reduction fast-rcnn for fast vehicle detection in large aerial imagery* (ICIP) 978-1-4799-7061-2/18

[25] Lekhaa1 T R, Rajeshwari S, Aiswarya Sequeira J and Akshayaa S 2019 *Intelligent shopping cart using bolt esp8266 based on internet of things* (5th International Conference on Advanced Computing and Communication Systems) 978-1-5386-9533-3/19

[26] Daniel Minoli, Kazem Sohraby and Jacob Kouns 2017 *Iot security (IoTsec) considerations, requirements, and architectures* 978-1-5090-6196-9/17

Production of high strength bioliquid from municipal solid waste (MSW) using mixed culture

N Zainol, N S M Sidek and A Ya'acob

College of Engineering, Universiti Malaysia Pahang, Lebuhraya Tun Razak, 26300 Kuantan, Pahang, Malaysia

Abstract. Landfilling is one of the most commonly used methods of municipal solid waste (MSW) disposal. It is necessary because there are residues in all waste management processes that cannot be reused or recovered further and are essentially landfilled. For the degradation of MSW, thermal, mechanical and biological pretreatment techniques are used. MSW could be treated and converted into liquid biomass (bioliquid) by using enzymes. This research focused on the production of high strength bioliquid in MSW. The experiments were conducted in three stages namely: acclimatization process, preliminary study and factorial analysis to determine the production of high strength bioliquid in MSW. The selected factors were ratio of AMC and substrates (1:2 and 1:5), temperature (room temperature and 37 ^{o}C), pH (adjusted pH and no pH adjustment), PET plastics size (1cm^2 and 4cm^2) and Hydraulic Retention Time (3 hours and 9 hours). Design Expert software (Version 7) was used to construct an experimental table where all the factors were randomized. Two-level factorial analysis (TLFA) was used to analyze the most contributing factor and interaction between the factors. The results showed that the most significant factor was temperature, while the interaction factors were between temperature and HRT. The best conditions of COD increment were determined at AMC/S ratio 1:5, adjusted pH, PET plastics size of 1cm x 1cm, at room temperature and HRT of 3 hours. The experimental and targeted final increments of COD were 2.19 and 1.99 respectively. Since the error was less than 10%, it is acceptable, and the objectives of this research were successfully achieved.

KEYWORDS: municipal solid waste (MSW); bioliquid; enzyme; factorial analysis; design expert software

1. Introduction

In Malaysia, solid waste generation is estimated at about 26 million tons in 2007. According to Dong Energy, 'REnescience' handles unsorted MSW and industrial waste similar to MSW, using enzymes to convert biodegradable waste material into liquid biomass (bioliquid) (Lending, 2016). Hence, the solid and recyclable fractions of the waste are able to separate the biodegradable material. The bioliquid is suitable as feedstock to produce biogas and other chemicals production such as lactic acid and butanol which can be used in the chemical industry or as transportation fuel.

Bioliquid is liquid substrate made from biomass for energy purposes. Bioliquid is usually produced from waste such as municipal solid waste or agricultural waste. The bioliquid is highly suitable as a feedstock for the production of biogas, offering a high biogas yield that makes it possible to generate energy efficiently. Biogas production produced digestate as a by-product, which contains phosphorous that can act as fertilizer. In power plants that generate electricity and heat, solid fuels are ideal for replacing fossil fuels. High bioliquid intensity means high levels of chemical oxygen demand (COD), which implies a higher amount of organic oxidizable content in the substrate. Therefore, the production of the high strength bioliquid is very important as a substrate for production of high biogas yield.

There are a few factors give contribution to the production of high strength bioliquid such as temperature, oxygen availability, water content, pH, particle size and density (Farquhar and Rovers, 1973). Two-Level Factorial Analysis (TLFA) is employed to determine the most affecting factor to produce the high COD bioliquid. According to Box and Draper (1987), the most important aspect of TLFA is the design of experiments. This research focused on the production of high COD bioliquid in MSW. The purpose of TLFA is to identify the variables that have large effects on the bioliquid production.

2. Materials and Methods

2.1. Sample Collection and Analysis

The samples were organic waste and soil mixed culture (SMC). The samples were collected at Sungai Ikan, Kuala Terengganu. The SMC was collected at the bottom level of the site that was in contact with the organic waste at the site.

2.2 Acclimatization of Mixed Culture

This process was conducted to adapt the soil mixed culture to the biodegradation process. The organic waste and PET plastic were mixed with water at volume ratio of 1:3 (known as substrate). Then the substrate was mixed with the SMC and the COD concentration was determined. Initially one liter SMC was added into acclimatization bioreactor. Then, 500 mL substrate was added daily into the bioreactor for 14 days. The reactor was agitated manually to make sure the SMC was well mixed with substrate. The working volume for acclimatization bioreactor was eight liters. COD concentration was measured daily within this acclimatization process. After the acclimatization process, the SMC was knowns as acclimatized mixed culture (AMC).

2.3. Sample analysis using DR/2400 HACH Spectrophotometer

. 2 mL of filtered sample was poured into a chemical oxygen demand (COD) reagent vial. Then, the vial was inverted gently several times to mix. The vial was left in the COD digestion reactor for 2 hours. After 2 hours, the vials were put in tube rack to be cooled to room temperature. COD of samples was determined by using HACH Spectrophotometer.

2.4. Experimental Setup: Factorial Analysis

There were five selected factors in this study as shown in Table 1. The factors were ratio of AMC/S ratio, pH, temperature, PET plastics size and Hydraulic Retention Time (HRT). Experimental design set-up was performed by Design Expert software. Analysis factors were performed in this study to identify the design factors that have large effects for the experiment. The experiment was conducted based on Table 2. Next, experimental data were analyzed to determine the most contributing factor and interaction between factors by using Design Expert software.

Table 1. List of selected factors and their ranges.

Factors	Low Level	High Level
AMC/S ratio	1:2	1:5
pH	Initial	Adjusted (7)
Temperature	Room Temperature (27 °C)	37 °C
Specimen Size	1cm^2	4cm^2
HRT	3 hours	9 hours

3. Result and Discussion

3.1. Screening of factors affecting on the COD increment

Two-level factorial analysis was used to determine the effective effect on COD increment from MSW. Table 2 shows 16 runs of experiments and the result of COD increment. The most contributing factor and interaction between the factors that have effect on COD increment were analyzed using ANOVA, at more than 95% of confidence level. The ranges for response of COD increment (fold) from 0.31 to 2.02. Table 3 shows the percentage contribution of each factor on COD increment which is temperature (Factor C) has the highest percentage with value 37.75%, followed by HRT (Factor E), AMC/S Ratio (Factor A), pH (Factor B) and lastly PET Plastic Size (Factor D) with the least value of 0.034%.

Table 2. Experimental setup for factorial analysis.

		Factor 1	Factor 2	Factor 3	Factor 4	Factor 5	Response 1	Response 2	Response 3
Std	Run	A: Ratio of AMC and organic waste	B: pH	C: Temperature (°C)	D: Specimen Size (cm$^{2)}$	E: HRT (hour)	Initial COD (%)	Final COD (%)	Fold (%)
1	2	1:2	No pH adjustment	Room Temperature	1	9.00	82630	71560	0.87
2	11	1:5	No pH adjustment	Room Temperature	1	3.00	191860	47630	0.25
3	16	1:2	Adjusted pH	Room Temperature	1	3.00	81860	40660	0.5
4	1	1:5	Adjusted pH	Room Temperature	1	9.00	81860	81200	0.99
5	13	1:2	No pH adjustment	37.00	1	3.00	74830	84330	1.13
6	5	1:5	No pH adjustment	37.00	1	9.00	89060	99160	1.11
7	10	1:2	Adjusted pH	37.00	1	9.00	56800	67330	1.19
8	8	1:5	Adjusted pH	37.00	1	3.00	55730	112560	2.02
9	9	1:2	No pH adjustment	Room Temperature	4	3.00	201530	61530	0.31
10	4	1:5	No pH adjustment	Room Temperature	4	9.00	58300	85200	1.46

11	6	1:2	Adjusted pH	Room Temperature	4	9.00	70700	72600	1.03
12	15	1:5	Adjusted pH	Room Temperature	4	3.00	75760	42200	0.56
13	12	1:2	No pH adjustment	37.00	4	9.00	59060	80100	1.36
14	7	1:5	No pH adjustment	37.00	4	3.00	68930	93200	1.35
15	14	1:2	Adjusted pH	37.00	4	3.00	70230	77000	1.1
16	3	1:5	Adjusted pH	37.00	4	9.00	56400	57260	1.02

Table 3. Contribution of factors in bioliquid production.

Term	% contribution
A: AMC/S Ratio	3.28
B: pH	0.66
C: Temperature	37.75
D: PET Plastic Size	0.034
E: HRT	6.66

3.2. Analysis of variance (ANOVA) for COD increment

Table 4 shows the ANOVA summary for COD increment to estimate the coefficient of the model, check the significance of each parameter and indicate the interaction strength of each parameter. The ANOVA study revealed an effective model with a *p*-value below 0.05. The main factors, AMC/S ratio (Factor A), pH (Factor B), temperature (Factor C) and HRT (Factor E) are significant model terms. For interactions between factors, pH and PET Plastic Size (Factor BD), pH and HRT (Factor BE) and temperature and HRT (Factor CE) are significant model terms. Meanwhile, the model terms of PET Plastic Size (Factor D) and AMC/S ratio and pH (Factor AE) are not significant with *p*-value greater than 0.1. The determination coefficient (R^2) and adjusted determination coefficient (R^2 adj) were 0.9985 and 0.9926 respectively (Table 5). Olmez (2009) indicates that at least 0.80 should be a good match for a model R^2, indicating that the regression models clarified the process well. The final empirical model (actual factors) was determined according to Equation 1 to express the COD increment as a function of independent variables.

$$COD\ increment\ (Fold) = 1.02 + 0.079A + 0.036B + 0.27C + 8.125E - 003D + 0.11E + 0.017AB - 0.063AE - 0.13BD - 0.11BE - 0.086CD - 0.23CE + 0.081DE \quad (1)$$

Where A is AMC/S Ratio, B is pH, C is temperature, D is PET plastics size and E is HR

Table 4. ANOVA of COD increment.

ANOVA for selected factorial model						
Source	Sum of Square	Df	Mean Square	F value	P-value Prob > F	
Model	3.07	12	0.26	168.06	0.0007	Significant
A-AMC/S ratio	0.10	1	0.10	66.19	0.0039	
B-pH	0.02	1	0.02	13.33	0.0355	

C-Temperature	1.16	1	1.16	762.36	0.0001
D-PET Plastic Size	1.056E-03	1	1.056E-03	0.69	0.4661
E-HRT	0.20	1	0.20	134.45	0.0014
AB	4.556E-03	1	4.556E-03	2.99	0.1821
AE	0.064	1	0.064	41.86	0.0075
BD	0.28	1	0.28	182.71	0.0009
BE	0.18	1	0.18	120.00	0.0016
CD	0.12	1	0.12	77.03	0.0031
CE	0.83	1	0.83	546.75	0.0002
DE	0.10	1	0.10	68.29	0.0037

Table 5. Values for coefficient of determination (R-Squared.

Std. Dev.	0.039	R-Squared	0.9985
Mean	1.02	Adj R-Squared	0.9926
C. V. %	3.84	Pred R-Squared	0.9577
PRESS	0.13	Adeq Precision	49.714

3.3. Analysis of main effect and interaction effect between factors on COD increment

Figure 1 illustrated the main effects and interaction effects of the factors on the COD increment. Main effects show two main factors that contribute to COD increment. Factor C (Temperature) has shown the highest effects followed by Factor E (HRT). In Table 4, Factor C does give the highest contribution among other factors. Based on Table 2, for Factor C, room temperature (27 ^{o}C) was set as low level and 37 ^{o}C was set as high level. Factor A, B, C, D and E gave a positive effect toward the COD increment. It means the response is increasing when the factor is increasing. To obtain higher COD increment, all these positive effect factors should be increasing too. This means that to increase the COD increment the temperature should be set at high level (37 ^{o}C). It demonstrates that PET Plastic Size and HRT (Factor DE) had a positive influence on the COD increment for interaction effects. The other interaction gave negative effect on the response. The negative effect arises when the factor is not proportional to the response value. Therefore, as the negative impact factor grows, the COD increment decreases.

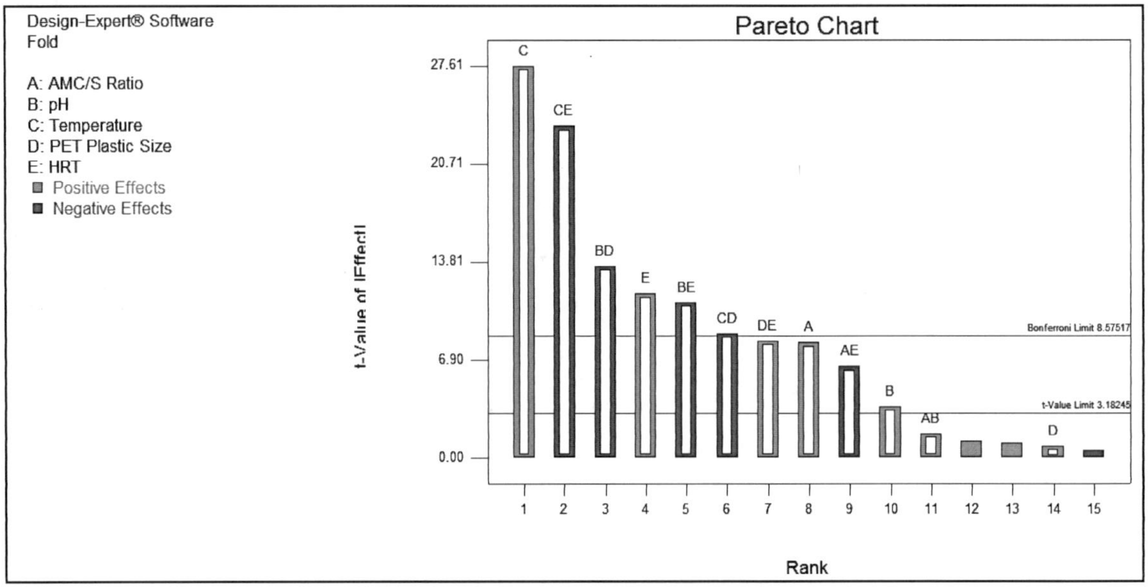

Figure 1. Pareto Chart for factorial analysis on COD increment.

3.4. Effect of independent processing parameters on COD increment

Figure 2 shows the effect of independent variables on the COD increment. Factor C gave the highest contribution to COD increment. Figure 2a shows the slope which indicates the COD increment value in MSW increase as the temperature increase. COD increment was higher when temperature at 37 °C with value 1.19, meanwhile, it decreased to 0.65 when temperature was at room temperature (27 °C). It showed that higher COD increment was achieved by using temperature at 37 °C. During the biological process, low temperatures decreased the rates of microbial growth and substrate usage (Kim et al., 2006; Stuckey, 2010). Beside, a study by Kashyap et al. (2003) claimed that low temperatures may also result in an exhaustion of cell energy, leakage of intracellular substances and complete lysis. This high difference of COD increment gives the temperature is the major independent factor that affects COD production. From Figure 2b, the COD increment increase as the HRT increase. At HRT is lower (3 hours), the COD increment is 0.14 lower than HRT higher (9 hours) which is 1.16. It also showed that higher COD increment achieved at HRT higher (9 hours). Figure 2(c) showed that the slope slightly increases when the ratio increases from 1:2 to 1:5. Thus it can conclude from Figure 2(c) where there is no significant difference in ratio to the COD increment. The benefits of acclimatized culture are easy to adapt for various toxic wastes, simplicity of operation and maintenance. Compared to pure culture, acclimatized mixed culture has a better advantage where the substrate must be sterile and more factors affect or retard its growth and results (Swaminathan and Ramanujam., 1999). It was proved that the amount of microbe does not give a big difference in COD increment.

IOP Conf. Series: Materials Science and Engineering 1045 (2021) 012024 doi:10.1088/1757-899X/1045/1/012024

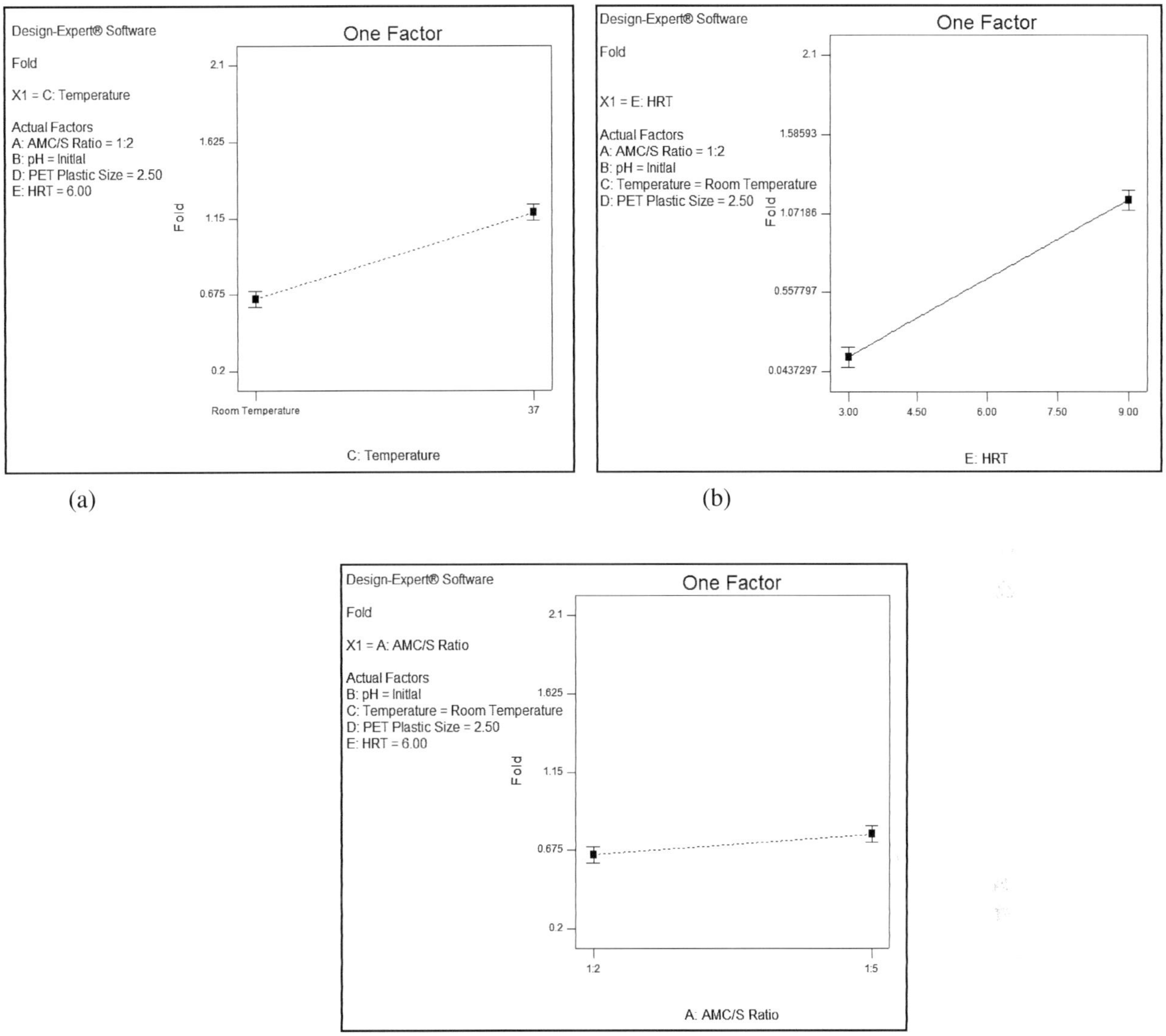

Figure 2. Most effective independent factors in COD increment.

3.5. Interaction effects between factors on COD increment

Data analysis from Design Expert software indicate the interaction gives positive effect with the highest contribution is interaction between Factor D (PET Plastic Size) and Factor E (HRT) with a percentage of 3.38%. Figure 3(a) shows the interaction between Factor D and E. The COD increment was highest when HRT at 9 hours and PET Plastic Size at 4 cm^2 with value of 1.47. For both HRT which are at 3 hours and 9 hours, when PET Plastics Size increase, the COD increment also increased. The COD increment was 0.85 when the PET Plastic Size is 1 cm^2 at the HRT of 9 hours. The COD increment increased till 1.46 at the same HRT but different in PET Plastic Size at 4 cm^2. For both factors, the COD increment was increasing when PET Plastic Size and HRT were increasing. Figure 3(b) shows the interaction between Factor C (temperature) and E (HRT). For Factor C and E, room temperature (27 ^oC) was considered as low level, and 37 ^oC were as high level. The COD increment was highest when HRT at 9 hours and temperature at 37 ^oC with value of 1.24. The COD increment slightly increased at from room temperature to 37 °C at same HRT which is at 9 hours. On the other hand, the COD increment was lowest when HRT and temperature at 3 hours and room temperature respectively with value of 0.14. However, the COD increment increased till 1.13 as temperature increasing to 37 °C. For both factors, the COD increment was increasing when temperature and HRT were increasing. According to Luostarinen et al. (2007), high temperature result in higher conversion rates of organic matter in the anaerobic process. Meanwhile, as a result of slow hydrolysis of volatile solids at a given HRT, more organic matter typically stays undegraded at low temperatures (Seghezzo, 2004).

Figure 3(c) shows the interaction between Factor B (pH) and D (PET Plastic Size). From the graph, it shows a big difference in COD increment at no pH adjustment with different PET Plastic Size. For PET Plastic Size of 1 cm^2 and 4 cm^2, the COD increment was at 0.42 and 0.87 respectively. Meanwhile, for the adjusted pH, it shows there is a small difference in COD increment. For PET Plastic Size of 1 cm^2 and 4 cm^2, the COD increment was at 0.72 and 0.65 respectively. It shows that at adjusted pH, the PET Plastic Size did not give a big effect on COD increment but slightly different between 1 cm^2 and 4 cm^2.

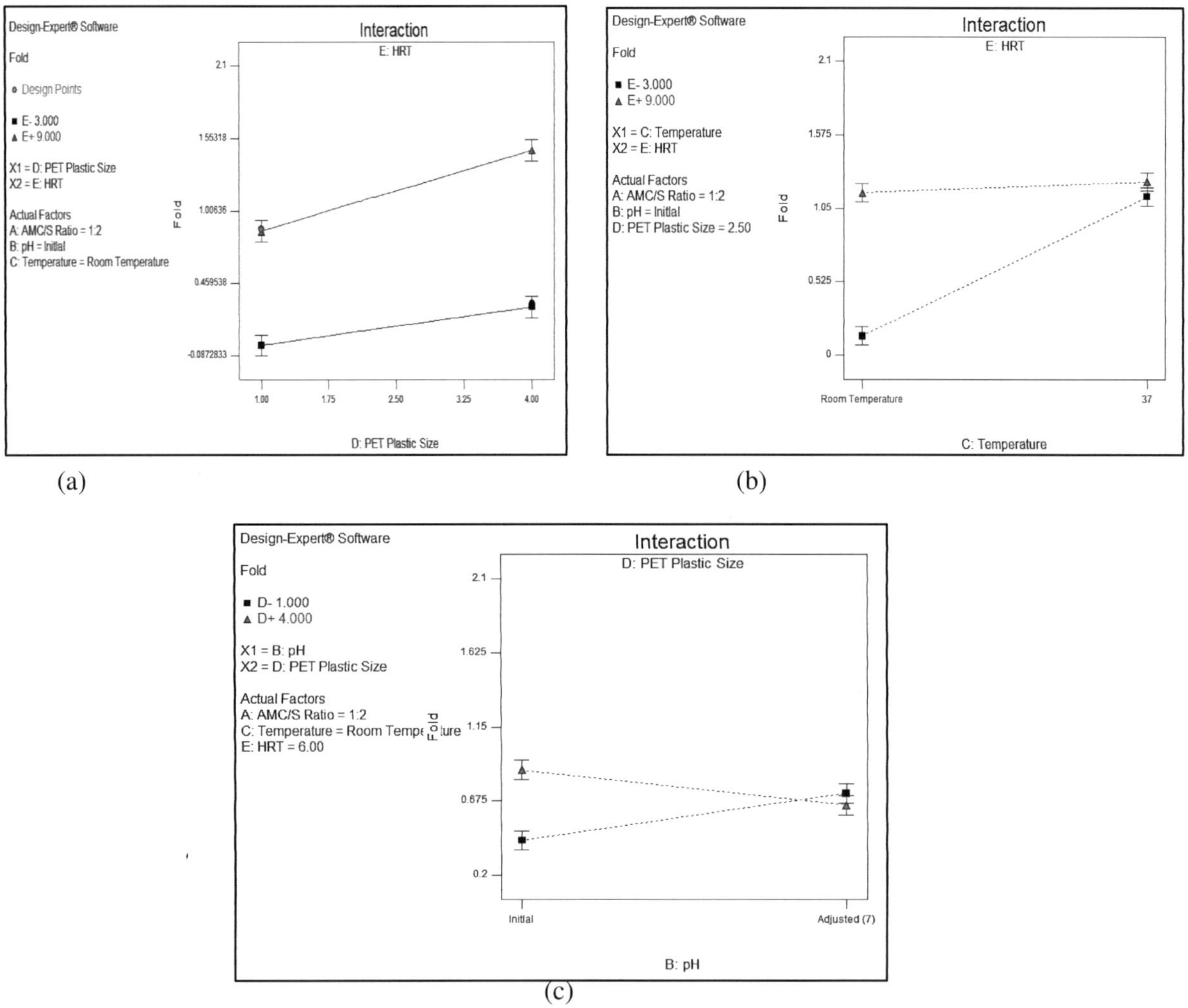

(a) (b) (c)

Figure 3. Most effective interaction factors in COD increment

3.6. Validation experiment

The COD increment values were predicted using the suggested best conditions from Design Expert Software. Table 6 shows the criteria for best conditions determination.

Table 6. Best conditions criteria.

Criteria	Goal	Value
AMC/S Ratio	In range	1:2 – 1:5
pH	In range	No pH adjustment – Adjusted pH
Temperature	In range	Room Temperature – 37°C
PET Plastics Size	In range	1 cm^2 – 4 cm^2
HRT	In range	3 hour – 9 hours
Fold	Maximize	-

Table 7 shows the suggested best conditions and their predicted COD increment. Triplicate experiments were conduct to validate the predicted and experimental COD increment. The error from this experiment was 9.9%, 20.18% and 35.74% for Run 1, 2 and 3 respectively, calculated by using Equation 2.

$$\text{Error (\%)} = 100\% \times \left|\frac{\text{predicted} - \text{experimental}}{\text{predicted}}\right| \quad (2)$$

Table 7. Validation experiment data.

Factor 1 A: AMC/S Ratio	Factor 2 B: pH	Factor 3 C: Temperature	Factor 4 D: PET Plastics Size	Factor 5 E: HRT	Predicted COD Increment	Experiment COD Increment (%)		
1:5	Adjusted pH	37 °C	1 cm x 1 cm	3 hours	1.99	Run 1	Run 2	Run 3
		Error				2.19	1.59	1.28

4. Conclusion

The purpose of this research was to establish the most important factors in COD increment. Design Expert software (Version 7) was used to construct the experimental design and analyzed experimental data through TLFA to analyze the most contributing factors and interaction factors. Five selected factors were AMC/S ratio, pH, temperature, PET plastics size and Hydraulic Retention Time (HRT). The contribution factor rank of COD increment were temperature > HRT > AMC/S ratio > pH > PET Plastic Size. This comes to the conclusion only temperature and HRT were the significant factors. The targeted final increment of COD 1.99 was achieved from Design Expert software. The final increment of COD value from experimental data was 2.19. Since the error was below than 10%, it is acceptable. Thus, the objective of this research was successfully achieved.

Acknowledgements

The authors thank Universiti Malaysia Pahang for financial assistance through research grant with RDU1901183 (FRGS/1/2019/TK10/UMP/02/17).

References

[1] Box G E P and Draper N R 1987 Empirical Model Building and Response Surfaces John Wiley & Sons, New York

[2] Farquhar G J and Rovers F A 1973 Gas production during refuse decomposition *Water Air Soil Pollut.* **2** 483 https://doi.org/10.1007/BF00585092

[3] Kashyap D R, Dadhich K S and Sharma S K 2003 Biomethanation under psychrophilic conditions: a review *Bioresour. Technol.* **87** 147–153

[4] Kim J K, Oh B R, Chun Y N and Kim S W 2006 Effects of temperature and hydraulic retention time on anaerobic digestion of food waste *J. Biosci. Bioeng* **102** 328– 332

[5] Lending U F 2016 REnescience – DONG *Energy Enzymatic treatment of waste – Value from waste* All-Energy Conference Glasgow 5 May 2016

[6] Luostarinen S, Sanders W, Kujawa-Roeleveld K and Zeeman G 2007 Effect of temperature on anaerobic treatment of black water in UASB-septic tank systems *Bioresource Technology* **98**(5) 980-986

[7] Seghezzo L 2004 *Anaerobic Treatment of Domestic Wastewater in Subtropical Regions* Ph. D. Dissertation Wageningen: Wageningen University.

[8] Stuckey D C 2010 Treatment of municipal solid waste leachate using a submerged anaerobic membrane bioreactor at mesophilic and psychrophilic temperatures: analysis of recalcitrants in the permeate using GC-MS *Water Res* **44** 671–680

[9] Swaminathan G and Ramanujam T K 1999 Effect of substrate concentration on biodegradation of 2,4-dichlorophenol using modified rotating biological contactors *Bioprocess Eng.* **21** 169 173

IOP Conf. Series: Materials Science and Engineering 1045 (2021) 012025 doi:10.1088/1757-899X/1045/1/012025

Cosmic acceleration in LRS bianchi type I space- time with bulk viscosity in general relativity

Sanjay Sharma[1], Laxmi Poonia[2*]

[1-2]Department of Mathematics & Statistics, Manipal University Jaipur Rajasthan -303007, India

[*1]**E-mail:** *laxmi.poonia@jaipur.manipal.edu*

Abstract: In present paper our intention to construct a locally rotational symmetric bianchi type I space time under framework of massless scalar field have flat potential. In existence of bulk viscosity the developed model provides inflationary solutions. The System of nonlinear Einstein field equations are solved by using suitable condition $V(\varphi) = K$ (constant) and the bulk viscosity coefficient is inversely proportional to scalar of expansion i.e. $\xi\theta = \alpha$ yields the negative deceleration parameter, which favorable to cosmic accelerating universe in current scenario. The model isotropize at particular cases. The Proper volume for model increase with time represents eternal inflation of present universe. The physical and structural aspects of model are discussed in significantly manner.

Keywords: Bianchi Type I space time, Bulk viscosity, Eternal inflation, Flat Potential

1. Introduction

In recent years, Einstein theory [1] has become subject of interest for its attainment in explaining the mystery of early universe. Bianchi models of cosmology provide a mathematical framework to study the realistic picture of present universe. Bianchi Type I space time becomes remarkable tool to understand the origin of early evolution of universe more general than standard model. It also provides the essential description of large scale behavior of physical universe in a simple manner in modern cosmology. In general relativity it is curious to study the cosmological problem to developed mathematical model that exactly predict the result of astronomical facts. Many cosmological problems are needed to be investigated by cosmologist to understand the cosmic structure formation after big-bang.

Inflation means extremely quick expansion of earliest universe by a factor of 10^{78} with volume driven by negative pressure with vacuum energy density. Guth [2] first introduced the basic idea of inflation while faced the problem why we monopole do not exist today. He investigated the false vacuum with positive energy leads an expansion of space exponentially. Inflationary scenario of universe provides a potential solution to cosmological problem like Horizon problem, Isotropy, Homogeneity and Magnetic Monopole. Inflationary phenomenon for homogenous and isotropy space time (FRW model) is investigated by many authors [3-5]. Rothman and Ellis [6] have resulted out solution of isotropic problem if we deal with anisotropic space time metric that isotropize in special case. Some Bianchi Type I inflationary model for flat potential in general relativity is constructed by Bali and Jain[7] and Bali[8]. Bali and Singh [9] provide LRS Poonia Bianchi Type I space time metric for stiff fluid with variable bulk viscosity. Bali and [10] studied Bianchi Type VI_0 cosmological model to study inflationary scenario of universe and observe late time acceleration. Naidu et.al [11] constructed inflationary model in Kantowski –Sachs space with constant deceleration parameter. Many authors [12-15] discussed the cosmological phenomenon under consideration of bulk viscosity and study the effect on evolution. Khalatnikov and Belinskii [16] studied FRW model by taking bulk viscosity as function of energy density. Reddy [17] have developed spatially, anisotropic and homogenous Bianchi Type V space time to study inflation and structure of early universe.

In current paper, we have presented a LRS Bianchi Type I inflationary space time with bulk viscosity in presence of flat potential to discuss inflationary scenario of universe and cosmic acceleration. The spatial volume increased with time which represent accelerating phase of universe. Model isotropize at special case. This study will provide some sufficient fact to study the astrophysical phenomenon. The negative deceleration parameter yield de-sitter universe. Geometrical and physical behavior of the model are discussed.

2. The metric and fields equations

We consider LRS Bianchi Type I metric can be written in the form

$$ds^2 = -dt^2 + A^2dx^2 + B^2(dy^2 + dz^2) \tag{1}$$

where A and B are metric coefficients and function of parameter t only

The gravitational field coupled minimally with scalar field $V(\varphi)$is given by

$$L = \int \left(R - \frac{1}{2}\, g^{ij}\varphi_{,i}\varphi_{,j} - V(\varphi)\right)\sqrt{-g}\ dx^4 \tag{2}$$

The Einstein's field equation $(8\pi G = c = 1)$ for massless less scalar field φ is given by

$$R_{ij} - \frac{1}{2}Rg_{ij} = -T_{ij} \tag{3}$$

The energy momentum tensor T_{ij} in presence of viscosity for scalar field is given by

$$T_{ij} = \varphi_{,i}\varphi_{,j} - \left(\frac{1}{2}\varphi_{,l}\varphi^{,l} + V(\varphi)\right)g_{ij} - \xi\theta\left(g_{ij} + u_iu_j\right) \tag{4}$$

with

$$\frac{1}{\sqrt{-g}}\partial_{,i}\left(\sqrt{-g}\,\varphi_{,i}\right) = -\frac{dV(\varphi)}{d\varphi} \tag{5}$$

where V is the scalar potential, φ is Higgs field, ξ is coefficient of viscosity and θ is the expansion scalar

The commoving derivative is considered as $u^i = (0,0,0,1)$

The system of field equations (3) for metric (1) become

$$\frac{{B_4}^2}{B^2} + 2\frac{B_{44}}{B} = \frac{1}{2}{\varphi_4}^2 - K - \alpha \tag{6}$$

$$\frac{A_{44}}{A} + \frac{B_{44}}{B} + \frac{A_4\, B_4}{AB} = \frac{1}{2}{\varphi_4}^2 - K - \alpha \tag{7}$$

$$\frac{{B_4}^2}{B^2} + 2\frac{A_4 B_4}{AB} = -\left[\frac{1}{2}{\varphi_4}^2 + K + 2\alpha\right] \tag{8}$$

For inflationary solution, we have assumed $V(\varphi) = K,\ \xi\theta = \alpha$

We consider $\xi\theta = \alpha$ (constant) from Brevik et.al.[18] because it have significant role to connect with existence of LR cosmology using FRW metric

From equation (5)

$$\varphi_{44} + \left(\frac{A_4}{A} + 2\frac{B_4}{B}\right)\varphi_4 = 0 \tag{9}$$

on solving provide

$$\varphi_4 = \frac{\mu}{AB^2} \tag{10}$$

where μ is constant of integration

From equation (6) and (7) we have

$$\frac{A_4 B_4}{AB} + \frac{A_{44}}{A} - \frac{{B_4}^2}{B^2} - \frac{B_{44}}{B} = 0 \tag{11}$$

which gives

$$\left(\frac{A_4}{A} - \frac{B_4}{B}\right)_4 + \left(\frac{A_4}{A} - \frac{B_4}{B}\right)\left(\frac{A_4}{A} + 2\frac{B_4}{B}\right) = 0 \tag{12}$$

on integration we get

$$\left(\frac{A_4}{A} - \frac{B_4}{B}\right) = \frac{\mu}{AB^2} \tag{13}$$

now from equation (6), (7) and (8) we have

$$\left(\frac{A_4}{A} + 2\frac{B_4}{B}\right) = \frac{\lambda^2}{A^2B^4} - K - \frac{1}{2}\alpha \tag{14}$$

Let assume $AB^2 = \eta$ in equation (13) and provide eq. (17)

$$\left(\frac{A_4}{A} - \frac{B_4}{B}\right) = \frac{\mu}{\eta} \tag{15}$$

Equation (14) become

$$\left(\frac{{A_4}^2}{A^2} + 2\frac{{B_4}^2}{B^2}\right) = \frac{\lambda^2}{\eta^2} - K - \frac{1}{2}\alpha \tag{16}$$

$$\left(\frac{A_4}{A} + 2\frac{B_4}{B}\right) = \frac{\eta_4}{\eta} \tag{17}$$

IOP Conf. Series: Materials Science and Engineering 1045 (2021) 012025 doi:10.1088/1757-899X/1045/1/012025

Equations (15) and (17) gives relation

$$3\frac{A_4}{A} = \frac{\eta_4}{\eta} + 2\frac{\mu}{\eta} \tag{18}$$

and

$$3\frac{B_4}{B} = \frac{\eta_4}{\eta} - \frac{\mu}{\eta} \tag{19}$$

using equations (18), (19) in (16) we obtained

$$\left(\frac{\eta_4}{\eta}\right)_4 + \frac{1}{9}\left(\frac{\eta_4}{\eta} + 2\frac{\mu}{\eta}\right) + \frac{2}{9}\left(\frac{\eta_4}{\eta} - \frac{\mu}{\eta}\right)^2 = \frac{\lambda^2}{\eta^2} - K - \frac{1}{2}\alpha \tag{20}$$

on solving gives

$$\eta\eta_{44} - \frac{2}{3}\eta_4{}^2 + K\eta^2 + \frac{1}{2}\alpha\eta^2 + \frac{2}{3}\mu^2 - \lambda^2 = 0 \tag{21}$$

Let use the transformation $\eta_4 = f(\eta)$

so that $\eta_{44} = f\frac{df}{d\eta}$

Equation (21) become

$$2f\frac{df}{d\eta} - \frac{4}{3}\frac{f^2}{\eta} = (2K+\alpha)\eta + \frac{2\lambda^2 - \frac{4}{3}\mu^2}{\eta} \tag{22}$$

Provide $\eta_4{}^2 = \frac{3}{2}\eta^2(2K+\alpha) - \frac{3}{4}\left(2\lambda^2 - \frac{4}{3}\mu^2\right) + D\eta^{\frac{4}{3}}$ (23)

where D is the constant of integration.

Since equation (6),(7),(8) and (9) have four unknown A,B, φ and $K = V(\varphi)$

For find solution take $D = 0$, which gives

$$\frac{d\eta}{dt} = \sqrt{\frac{3}{2}\eta^2(2K+\alpha) - \frac{3}{4}\left(2\lambda^2 - \frac{4}{3}\mu^2\right)} \tag{24}$$

leads to

$$\frac{d\eta}{\sqrt{\frac{3}{2}(2K+\alpha)(\eta^2+\beta^2)}} = dt \tag{25}$$

where

$$\beta^2 = \frac{\mu^2 - \frac{3}{2}\lambda^2}{\frac{3}{2}(2K+\alpha)} \tag{26}$$

gives $\eta = \beta sinh\sqrt{\frac{3}{2}(2K+\alpha)}\,(t+t_0)$

where t_0 is constant of integration (27)

IOP Conf. Series: Materials Science and Engineering 1045 (2021) 012025 doi:10.1088/1757-899X/1045/1/012025

and

$$\eta_4 = \beta\sqrt{\frac{3}{2}(2K+\alpha)}cosh\sqrt{\frac{3}{2}(2K+\alpha)}\,(t+t_0) \tag{28}$$

From equation (18), (27) and (28) we get

$$3\frac{A_4}{A} = \frac{\eta_4}{\eta} + 2\frac{\mu}{\eta} = \frac{\beta\sqrt{\frac{3}{2}(2K+\alpha)}cosh\sqrt{\frac{3}{2}(2K+\alpha)}\,(t+t_0)+2\mu}{\beta sinh\sqrt{\frac{3}{2}(2K+\alpha)}\,(t+t_0)} \tag{29}$$

$$3\frac{B_4}{B} = \frac{\eta_4}{\eta} - \frac{\mu}{\eta} = \frac{\beta\sqrt{\frac{3}{2}(2K+\alpha)}cosh\sqrt{\frac{3}{2}(2K+\alpha)}\,(t+t_0)-\mu}{\beta sinh\sqrt{\frac{3}{2}(2K+\alpha)}\,(t+t_0)} \tag{30}$$

on integrating we get

$$A = \tau^{\frac{1}{3}}sinh^{\frac{1}{3}}\left[\sqrt{\frac{3}{2}(2K+\alpha)}\,(t+t_0)\right]tanh^{\overline{3\beta\sqrt{\frac{3}{2}(2K+\alpha)}}^{\,2\mu}}\left[\frac{1}{2}\sqrt{\frac{3}{2}(2K+\alpha)}\,(t+t_0)\right] \tag{31}$$

From equation (19),(27) and (28) we get

$$B = \gamma^{\frac{1}{3}}sinh^{\frac{1}{3}}\left[\sqrt{\frac{3}{2}(2K+\alpha)}\,(t+t_0)\right]tanh^{\overline{3\beta\sqrt{\frac{3}{2}(2K+\alpha)}}^{\,-\mu}}\left[\frac{1}{2}\sqrt{\frac{3}{2}(2K+\alpha)}\,(t+t_0)\right] \tag{32}$$

where τ and γ are the constant of integration

From equation (10)

$$\varphi_4 = \frac{\lambda}{AB^2} = \frac{\lambda}{\beta sinh\sqrt{\frac{3}{2}(2K+\alpha)}\,(t+t_0)} \tag{33}$$

Using the transformation $t + t_0 = T$, $\tau^{\frac{1}{3}}x = X$, $\gamma^{\frac{1}{3}}y = Y$ *and* $\gamma^{\frac{1}{3}}z = Z$ we get the metric in the form

$$ds^2 = -dT^2 + sinh^{\frac{2}{3}}\left[\sqrt{\frac{3}{2}(2K+\alpha)}\,T\right]tanh^{\overline{3\beta\sqrt{\frac{3}{2}(2K+\alpha)}}^{\,4\mu}}\left[\frac{1}{2}\sqrt{\frac{3}{2}(2K+\alpha)}\,(t+t_0)\right]dX^2$$
$$+\ sinh^{\frac{2}{3}}\left[\sqrt{\frac{3}{2}(2K+\alpha)}\,T\right]tanh^{\overline{3\beta\sqrt{\frac{3}{2}(2K+\alpha)}}^{\,-2\mu}}\left[\frac{1}{2}\sqrt{\frac{3}{2}(2K+\alpha)}\,T\right](dY^2+dZ^2)$$

(34)

IOP Conf. Series: Materials Science and Engineering 1045 (2021) 012025 doi:10.1088/1757-899X/1045/1/012025

3. Physical and geometrical features of model

Scalar of the expansion for model is given by

$$\theta = \frac{A_4}{A} + 2\frac{B_4}{B} = \frac{\eta_4}{\eta} = \sqrt{\frac{3}{2}(2K+\alpha)}\coth\left[\sqrt{\frac{3}{2}(2K+\alpha)}\,T\right] \tag{35}$$

Shear scalar for model is given as

$$\sigma = \frac{\mu}{\sqrt{3}\beta sinh\sqrt{\frac{3}{2}(2K+\alpha)}\,T} \tag{36}$$

The Hubble parameters in X, Y and Z direction is given as

$$H_1 = \frac{\beta\sqrt{\frac{3}{2}(2K+\alpha)}cosh\sqrt{\frac{3}{2}(2K+\alpha)}\,T + 2\mu}{3\beta sinh\sqrt{\frac{3}{2}(2K+\alpha)}\,T} \tag{37}$$

and $$H_2 = H_3 = \frac{\beta\sqrt{\frac{3}{2}(2K+\alpha)}cosh\sqrt{\frac{3}{2}(2K+\alpha)}\,T - \mu}{3\beta sinh\sqrt{\frac{3}{2}(2K+\alpha)}\,T} \tag{38}$$

Also $$\frac{\sigma}{\theta} = \frac{\mu}{\sqrt{3}\beta\sqrt{\frac{3}{2}(2K+\alpha)}cosh\sqrt{\frac{3}{2}(2K+\alpha)}\,T} \tag{39}$$

Higgs field for model is given by

$$\varphi = \int \frac{\lambda}{\beta sinh\sqrt{\frac{3}{2}(2K+\alpha)}\,T} + N \tag{40}$$

The Proper volume V is obtained as

$$V = \beta sinh\sqrt{\frac{3}{2}(2K+\alpha)}\,T \tag{41}$$

Deceleration Parameter is given as

$$q = -\left(1 + 3sech^2\left(\sqrt{\frac{3}{2}(2K+\alpha)}\,T\right)\right) \tag{42}$$

4. Conclusion and discussion

The proper volume for developed model increases with time. When T→ ∞ then the proper volume (R^3) tends to infinity, it represent inflationary scenario of universe in Bianchi Type I model contain massless scalar field with flat potential. The expansion (θ) tends to infinite when T tends to zero and θ→ finite term when T tends to infinity. The physical quantity $\frac{\mu}{\beta}$ measures the anisotropy in the model. The Shear (σ) scalar tends to zero when T tends to infinity. The rate of Higgs field decreases with time. The model (34) start with a big-bang at T=0 .The deceleration parameter (q) is negative represents the accelerating phase of universe. Since $\frac{\sigma}{\theta} \rightarrow 0$ for large value of T it shows that the model approaches isotropy. The Hubble parameter decreases with time.

References:

[1] Einstein, A. 1916 The Foundation of generalized theory of relativity *Annalen der physik* **49** pp.769-822.

[2] Guth, A.H.1980 Inflationary Universe: A Possible solution of the horizon and flatness problem *Phys. Lett.* B **91**pp. 99-102

[3] Linde, A.D. 1982 A New inflationary scenario: A Possible solution of horizon, flatness, homogeneity, isotropy and primordial monopole problems *Phys. Lett. B***108**pp. 389-393

[4] La, D. and Steinhardt, P.J. 1989 Extended Inflationary Cosmology *Physical Review Letters* 62, pp. 376-378.

[5] Burd, A.B. and Barrow, J.D. 1988 Inflationary Models with Exponential Potentials *Nucls. Physics* 308, pp.929-945.

[6] Rothman, T. and Ellis, G.F.R.1986 Can inflation occur in anisotropic cosmologies *Phys. Lett. B***180** pp. 19-24

[7] Bali, R., Jain, V.C. 200 2Bianchi Type I Inflationary Universe in General Relativity *Pramana*, **59** pp.1-7.

[8] Bali, R. 2011 Inflationary Scenario in Bianchi Type I Space-time *International Journal of Theoretical Physics* **50** pp. 3043-3048.

[9] Bali, R., Singh, S. 2014 LRS Bianchi Type I Stiff Fluid Inflationary Universe with Variable Bulk Viscosity *Canadian Journal of Physics* **92** pp. 365-369.

[10] Bali, R. and Poonia, L.2013 Bianchi Type VI_0 Inflationary Cosmological Model in General Relativity *International Journal of Modern Physics* **22** pp. 593-602.

[11] Reddy, D.R.K. and Naidu, R.L. 2008 A Higher Dimensional Inflationary Universes in General Relativity *International Journal of Theoretical Physics* **47** pp. 2339-2343.

[12] Zimdahl, W. 1996 Bulk Viscous Cosmology *Physical Review D* **53** pp-5483-5493.

[13] Saha, B. 2005 Bianchi Type I Universe with Viscous Fluid *Modern Physics Letters* **20** pp.2127-2143.

[14] Peebles, P.J.E. and Ratra, B.2003 The Cosmological Constant and Dark Energy *Review of Modern Physics* **75** pp.559-606.

[15] Sahni, V. and Starobinsky, A. 2000 The Case for a Positive Cosmological Lambda term *International Journal of Modern Physics D* **9** pp. 373-443.

[16] Belinskil, V.A. and Khalatnikov, I.M. 1975 Influence of Viscosity on the character of cosmological evolution *Journal of Experimental and Theoretical Physics* **42** pp.205-210.

[17] Reddy, D.R.K. 2009 Bianchi Type V Inflationary Universe in General Relativity *International Journal of Theoretical Physics* **48** pp. 2036-2040.

IOP Conf. Series: Materials Science and Engineering 1045 (2021) 012026 doi:10.1088/1757-899X/1045/1/012026

Cold forging of silicon carbide particulate (SiCp) reinforced aluminium metal matrix composite preforms

M Agrawal[1] and S Singh[2]

[1]School of Mechanical Engineering, KIIT University, Bhubaneswar-751024, Odisha (India) and Dept. of Mechanical Engg, Sarala Birla University, Ranchi- 835103 (Jharkhand), manish.tanishq@gmail.com
[2]School of Mechanical Engineering, KIIT University, Bhubaneswar-751024, Odisha, (India), ssingh@kiit.ac.in

Abstract. The present paper investigate the forging of Aluminium metal matrix composite with Silicon Carbide particles (SiCp) as reinforcements. The generic shaped preform specimens were forged under cold conditions (room temperature) using open-die sets. The Aluminium metal matrix composite material under consideration was fabricated via liquid stir casting manufacturing route, where the pre-heated Silicon Carbide particles were mixed with the molten Aluminium casting alloy. The preforms with standard dimensions were machined obtained from the cast. To analysis the deformation characteristics, two different generic shapes of the preform were considered, *i.e.* disc and rectangular shape preforms. The present paper presents both theoretical (based on *'Upper Bound'* approach) and experimental investigations, where theoretical analysis considers a modified interfacial frictional law and preform free barreling. It is expected that the present work will be useful to assess various deformation characteristics during forging of metal matrix composites.

Keywords. Silicon Carbide particulate (SiCp), Aluminium metal matrix composite, Open-die forging, Disc preform, Rectangular preform

1. Introduction

Metal matrix composites mainly consists of two components, *e.g.* metal matrix and reinforcements. The main function of reinforcements is to support the distributed load from the base metal matrix and enhance the overall strength and stiffness of the composite materials [1]. In recent past, it has been established by many researchers that these materials had successfully replaced conventional monolithic materials due to enhanced mechanical properties, *e.g.* tensile strength, compressive strength, yield strength, stiffness, hardness, elastic modulus and resistance to wear [2-9]. These hybrid materials find applications in various industries like automobile, defense, aircraft, aerospace, transportation, construction, marines, corrosion-resistant products, household equipment and electrical-electronics equipments [10-12]. Additionally, these composites are also being utilized to manufacture various tailor-made industrial products by suitably combining the matrix and reinforcements using different mechanical processing routes [13-15].

Many researchers had reported their investigations regarding processing of conventional monolithic materials till date but very few had researched on the forging applications of metal matrix composites [16-22]. Thus, the present research work was undertaken as an attempt to investigate this process to understand the complex interaction of various underlying deformation characteristics involved during the forging of SiCp reinforced Aluminium metal matrix composites (AMC). The chief objective was to manufacture successfully an industrial-grade metal matrix composite using Aluminium casting alloy and appropriate reinforcements and then further cold-forge it to fabricate various engineering components. It is expected that these components shall have superior mechanical and related properties as compared to components made from the monolithic materials.

2. Fabrication of SiCp Reinforced Aluminium Metal Matrix Composites

The Aluminium metal matrix composite under consideration was manufactured via liquid metal stir casting route using Aluminium casting alloy (LM6 Grade). The reason of choosing LM6 Aluminium alloy was due to inherent presence of the high percentage of silicon content (*i.e.* 10-13 wt%) [23]. It has been observed that though Aluminum has excellent mechanical properties but its strength is too low to be used directly to fabricate functional components. Hence, its properties can be modified by adding carbide particles in suitable proportions in order to enhance the overall application capabilities. In the present paper, the SiC particles was added primarily to enhance the overall strength and stiffness of the base metal, *i.e.* matrix material, so that engineering products with superior properties can be later fabricated via various manufacturing routes.

Though, there are numerous manufacturing routes available for the fabrication of metal matrix composites as reported by various researchers; the present authors utilized liquid metal stir casting manufacturing route to fabricate the SiCp reinforced AMC. The chief reason for this selection was the operational simplicity and lower processing cost associated with the process [24, 25]. Basically, it involved melting of LM6 Aluminium casting alloy in an electric resistance furnace having clay-graphite crucible. Magnesium in proportion of 5 wt% was added to the molten LM6 metal in the furnace with an objective to generate a strong bond between the base metal and reinforcements. It has been reported by researchers that Magnesium helps in decreasing the surface energy of the molten LM6 Aluminium alloy by reducing the wetting angle between the constituents and thus improve the overall flowability of the molten metal [26]. Pre-heated SiC power particles in a temperature range of 850 - 900°C having grain size less than +400μ was added to the molten metal. This mixture was briskly stirred using a custom-fitted mechanical impeller mounted on top of the furnace at a speed of about 500 rpm as shown in figure 1 [27, 28]. The optimum temperature around 720°C was maintained during the mixing process and after ensuring the complete mixing of the components, same was poured in a pre-fabricated silica mould. This mould was suitably fabricated using core and parting sands with requisite amounts of bentonite and charcoal powder having parting gate structure. The sealing of mould was done with utmost care to decrease the occurrence of any defects like mould erosion, air entrapment, run out and dross formation. The mixture was allowed to cool down to room temperature and finally the cast was obtained by breaking the core. All cast samples were solution heat treated at 500 ℃ for 60 minutes in an electric furnace according to ASM heat treatment standards to ensure uniform microstructure and mechanical properties [29].

Different groups of AMC samples having 5.0 wt%, 7.5 wt%, 10.0 wt% and 13.0 wt% of SiCp were initially fabricated to study the basic characterization of mechanical properties and influence of SiCp particles on the alloy properties. Basis the analysis of the characterization results, two groups of preform samples, *i.e.* AMCs with 5 wt% and 13 wt % SiCp were selected for the cold-forging experimentations. Finally, the standard preform specimens were fabricated by machining the cast to required dimensions followed by polishing of the surfaces to desired finish. Table 1 shows the chemical composition of LM6 Aluminium alloy and thermo-physical properties of the alloy and SiC powder particles along with the dimensions of the preform specimen fabricated.

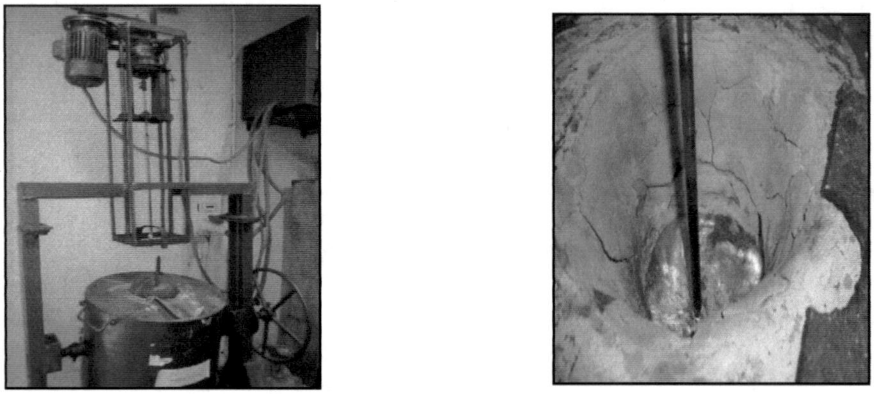

Figure 1. Electric resistance furnace with custom-made top-mounted mechanical impeller (stirrer)

Table 1. LM6 Aluminium alloy, SiCp and preform details

Chemical Composition of LM6		Thermo-Physical Properties		
Elements	**Percentage (%)**		**LM6**	**SiCp**
Si	10-13	**Density (gm/cm^3)**	2.66	3.2
Cu	0.1	**Average Particle Size (mesh)**	--------	400
Mg	0.1	**Thermal Conductivity (W/m-K)**	155	100
Fe	0.6	**Specific Heat (J/Kg-K)**	960	1300
Mn	0.5	**Dimensions of SiCp AMC Preforms**		
Ni	0.1			
Zn	0.1			
Pb	0.1	**Preform Shapes**	**Dimensions**	
Sb	0.05	**Cylindrical Preform**	D = 20 mm, H = 20 mm	
Ti	0.2	**Disc Preform**	D_0 = 15mm, H_0 = 10mm	
Al	Remaining	**Rectangular Preform**	L_0 = 20mm, B_0 =10mm, H_0 = 10mm	

3. Experimental Investigation & Characterization

The mechanical characterization of SiCp reinforced Aluminium metal matrix composite was performed using standard cylindrical specimens having 20 mm height and 20 mm diameter [30]. Investigations into mechanical properties, *e.g.* ultimate tensile strength, compressive strength, yield strength, hardness, impact toughness and ductility (*i.e.* percentage elongation) of SiCp reinforced AMC as fabricated above was performed using various preform samples. The results of these characterization of mechanical properties are shown in figures 2(a) - 2(c).

From figure 2(a), it is evident that as percentage of reinforcements increases in the composite, the hardness also increases and is maximum for 13.0 wt% SiCp. It is also observed that rate of increase in hardness is high initially and decreases later indicating that any further addition of SiCp shall not substantially enhance the hardness value of the material. Figure 2(b) shows the variation of compressive strength, ultimate tensile strength and yield strengths of the preforms and it can be observed that all three strength values increases with the addition of reinforcements in the matrix metal and highest values are observed for 13.0 wt% sample. The trend is also similar to figure 2(a), where it can be observed that the rate change of these mechanical properties becomes asymptote with x-axis after certain wt% of SiCp and this case it's found to be around 10-12 wt% SiCp. Graph of figure 2(c) clearly demonstrates that with increase in SiCp in the matrix metal though the impact toughness increases but the elongation (measure of the ductility) decreases substantially. Thus, for subsequent mechanical processing of SiCp AMCs into industrial components, there is a limiting value of SiCp which can be successfully added or mixed in the metal matrix for the fabrication of AMCs components. In the present research, this limiting value of SiCp is found to be 13.0%, which is also confirmed by the recent investigation [31].

To investigate the flowability of composite materials fabricated in the present research, the flow curve of the material has been established via a standard compression tests using cylindrical specimens. These preform specimens were compressed axially on a 100-ton hydraulic press at room temperature under several layers of thin teflon sheets and the corresponding load values were recorded for different height reductions. Further, the effective stress and effective strain were calculated from the experimental data of forging loads implied along with respective deformations recorded during the present experimentation, which is shown in figure 2(d). The stress-strain equation was assumed to be given by the power law and the general equation $\bar{\sigma} = a\bar{\varepsilon}^{b} MPa$ was used to draw this curve (where 'b' is strain hardening exponent and 'a' is the stress co-efficient or flow stress constant and found to be 269.31 MPa and 0.28 respectively for 13 .0 wt% SiCp).

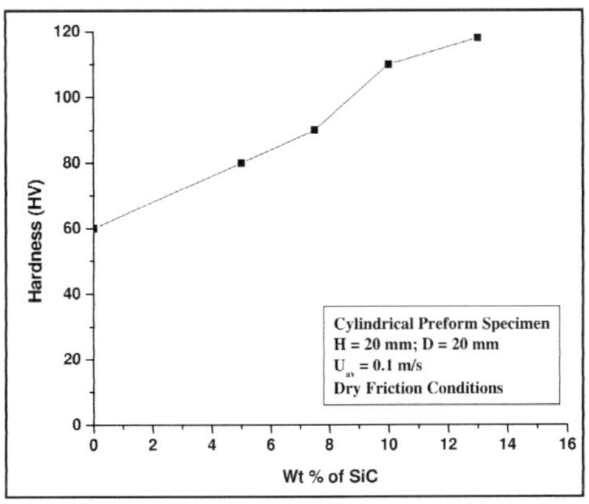

(a). Variation of hardness values

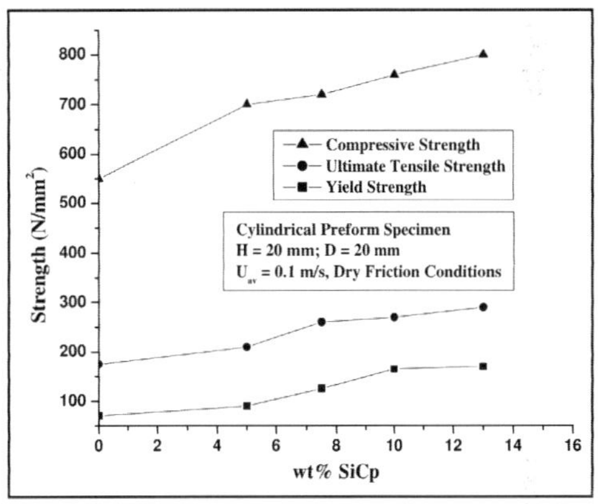

(b). Variation of compressive, ultimate tensile and yield strengths

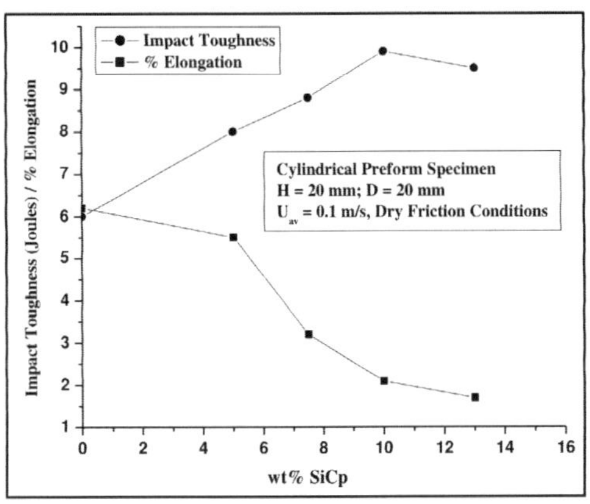

(c). Variation of impact toughness and % elongation

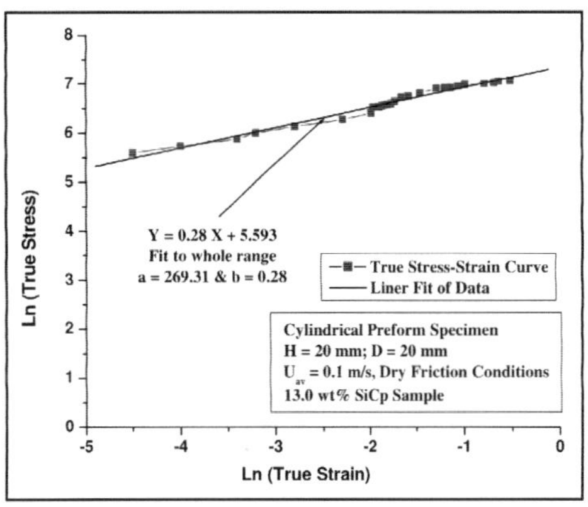

(d). Stress-Strain curve of SiCp AMC

Figure 2. Mechanical characterization of SiCp AMC preforms

The interfacial friction factor was estimated using ring compression test with annular ring-shaped specimens having standard dimensions. The specimen dimension ratio was 6:3:2 for outside diameter, inside diameter and height respectively. The experiments were performed under dry lubricated conditions (*i.e.* without any lubricant) and under lubricated conditions (with grease as lubricant). Figure 3 shows the SiCp reinforced Aluminium metal matrix composite specimens after ring-compression tests. The shear friction model $\tau_{friction} = mk$ was used to estimate the coefficient of friction, where *'k'* and 'm' are shear flow stress friction factor respectively. The curves were drawn for percentage decrease in the inner diameter along with preform height reduction and was mapped with the standard calibration curves for *'m'* values. Friction factor values of 0.33 & 0.15 were estimated for dry friction & lubricated conditions respectively.

SiCp AMC preform specimens were cold-forged using a 200 tons capacity hydraulic press till the onset of fracture. The upper die platen of the press was stationary and lower die platen moved upwards with preforms kept over its surface. The values of deformation (*i.e.* percentage height reduction) and corresponding forging load were recorded during the process. Various experimental runs were performed using different generic shapes preforms and interfacial frictional conditions, *i.e.* lubricated conditions using grease and dry-interfacial frictional conditions (without any lubricant). The process was carried till the onset of fracture to estimate the maximum forgeability of the preforms at room temperature. The disc and rectangular preforms before and after forging for 5 and 13 wt% of silicon carbide particulate reinforcements respectively are shown in figure 4. It can be observed that initiation of the cracks was observed at the equatorial bulged regions of the preform. Also, lower forgeability of the preforms was observed for higher wt% of SiCp for same magnitude of the deformation. To illustrate the distribution of SiCp reinforcements within the Aluminium metal matrix, the photomicrographs for preforms were obtained at 500X as shown in figure 5 [32]. It can be seen that 13 wt% preform is having higher density of silicon carbides particulate distribution within the base metal.

Figure 3. SiCp AMC specimens after ring-compression test

(a) Before Forging **(b)** After Forging

Figure 4. Aluminium metal matrix composite preforms before and after forging experimentation

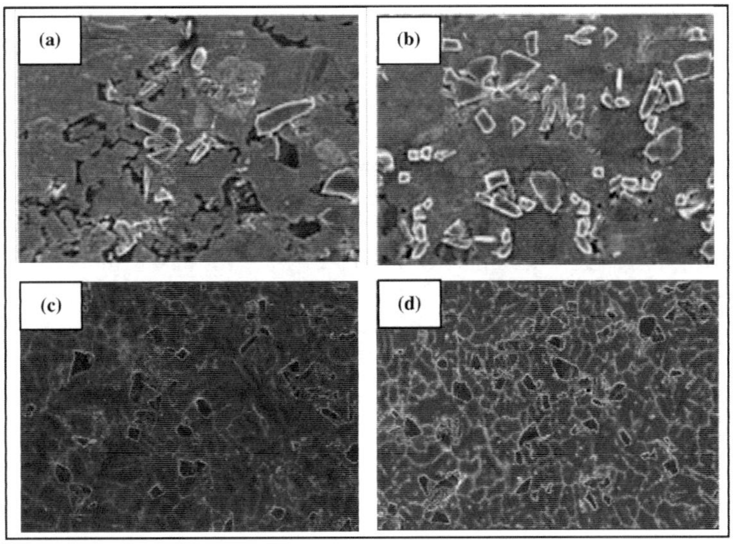

(a) 5% wt **(b)** 7.5 wt% **(c)** 10.0 wt% **(d)** 13.0 wt%

Figure 5: Digital Micrograph of various AMC samples with different wt% SiCp

4. *'Upper Bound'* Theoretical Analysis

The theoretical analysis during forging of SiCp reinforced AMC disc and rectangular preforms considered in the present research is based on the *'Upper Bound'* approach. The deformation of disc and rectangular preforms are considered under axi-symmetric and plane strain conditions respectively. To simply the mathematical formulations, few assumptions were made during the analysis, which are as follows:

- The die platens of the hydraulic press used during the forging process were considered to be incompressible, rigid and parallel
- The complete deformation of AMC preforms was assumed to be homogeneous and independent of the hydrostatic stress component and represented by Von-Mises yield criterion:

$$\sigma_o = \left[\sqrt{3J_2'}\right] \tag{1}$$

- The sticking interfacial friction was assumed to be a function of adhesion factor and the friction condition was assumed to be hybrid in nature (*i.e.* include both sticking and sliding friction components) [33, 34]. The modified interfacial frictional law for both the deformation cases, *i.e.* axisymmetric and plane strain conditions respectively were given as follows:

$$\tau = \mu\left\{P_{av} + \varphi_0\left[1 - \left(\frac{R_m - R}{nR_0}\right)\right]\right\} \textit{ (Disc preform)} \tag{2}$$

$$\tau = \mu\left\{P_{av} + \varphi_0\left[1 - \left(\frac{B_m - B}{nB_0}\right)\right]\right\} \textit{ (Rectangular preform)} \tag{3}$$

Where, 'R_m' and 'B_m' are the measures of the sticking zones for the axisymmetric and plane strain deformations respectively, which is mathematically approximated as follows [35]:

$$R_m = R_o - \frac{H_0}{2\mu}\ln\left(\frac{1}{\mu\sqrt{3}}\right) \tag{4}$$

$$B_m = B_0 - \frac{H_0}{2\mu}\ln\left(\frac{1}{\mu\sqrt{3}}\right) \tag{5}$$

- The *'Compatibility Conditions'* for both axisymmetric and plane strain deformations were approximated using *'Volume Constancy'* principle and respective equations are given as follows:

$$\dot{\varepsilon}_{rr}+\dot{\varepsilon}_{\theta\theta}+\dot{\varepsilon}_{zz}=0\,(Disc\ preform) \quad (6)$$

$$\dot{\varepsilon}_{xx}+\dot{\varepsilon}_{yy}+\dot{\varepsilon}_{zz}=0\,(Rectangular\ preform) \quad (7)$$

- The velocity discontinuities have been neglected in the present analysis along with the associated redundant energy dissipation assuming it to be very low
- The circumferential and lateral flow during deformation of the preforms has been neglected for axi-symmetric and plane-strain conditions respectively

The equations for various energy dissipations, *i.e.* internal energy dissipation *'W_i'*, frictional shear energy dissipation *'W_f'* and inertia energy dissipation *'W_a'* were formulated from the kinematically admissible velocity field and corresponding strain rates (Refer Appendix) satisfying the compatibility conditions and flow rule for both the above deformations conditions as mentioned below:

4.1 Axi-Symmetric Deformation (Disc Preform)

$$W_i=\frac{2}{\sqrt{3}}\sigma_0\int_V\sqrt{\frac{1}{2}\dot{\varepsilon}_{ij}\dot{\varepsilon}_{ij}}\,dV=\left[\frac{\pi\sqrt{3}\beta^2H_0\sigma_0R_0e^{-\beta/2}\left(1+\frac{\beta^2}{12}\right)U}{32\mathrm{Sinh}\left(\frac{\beta}{4}\right)^2}\right] \quad (8)$$

$$W_f=\int_{S_T}\tau|\Delta U|\,ds=\left[\frac{2\pi\mu\beta R_0^3e^{-\beta/2}\varphi_0U}{H_0\left(1-e^{-\beta/2}\right)}\right]\left[\left(\frac{P_{av}}{\varphi_0}\right)+\left(1+\frac{2}{3n}-\frac{R_m}{nR_0}\right)\right] \quad (9)$$

$$W_a=\int_V\rho\left(a_{ij}U_{ij}\right)dV=\left[\frac{\pi\beta\rho_0R_0^2U^3}{3\left(1-e^{-\beta/2}\right)}\right]\left\{\begin{array}{l}\left[\frac{\left(1+e^{-\beta/2}+\beta e^{-3\beta/2}\right)}{2}\right]+\left[\left(\frac{R_0^2}{3H_0^2}\right)\left(\frac{3H_0}{R_0}-2\right)\left(1+e^{-\beta/2}+e^{-\beta}\right)\right]\\+\left[\frac{\beta R_0\dot{U}\left(1+e^{-\beta/2}-3e^{-3\beta/2}\right)}{\left(1-e^{-\beta/2}\right)^3U^2}\right]\end{array}\right\} \quad (10)$$

4.2 Plane Strain Deformation (Rectangular Preform)

$$W_i=\left[\frac{\sqrt{3}\beta^2B_0^2\sigma_0U}{48\left(1-e^{-\beta/2}\right)}\right]\left[\left(\frac{4H_0^2}{B_0^2}\right)+\beta^2\left(1+\frac{8H_0}{\beta^2B_0}\right)\left(1-e^{-\beta/2}\right)\right] \quad (11)$$

$$W_f=\left[\frac{\mu\beta B_0^3e^{-\beta/2}\varphi_0U}{8H_0\left(1-e^{-\beta/2}\right)}\right]\left[\left(\frac{P_{av}}{\varphi_0}\right)+\left(1+\frac{1}{3n}-\frac{B_m}{nB_0}\right)\right] \quad (12)$$

$$W_a=\left[\frac{\beta^4\rho_0H_0^2U^3}{\left(1-e^{-\beta/2}\right)^2}\right]\left\{\left[\frac{\left(2+e^{-\beta/2}+e^{-\beta}\right)}{\beta^3}\right]+\left(1+e^{-\beta/2}\right)\left(\frac{2}{\beta^2}-\frac{B_0^2}{H_0^2}\right)+\left[\frac{B_0\dot{U}e^{-\beta}}{\beta U^2}\right]\right\} \quad (13)$$

The inertia effects, *i.e.* effect of die velocity on the magnitude of energy dissipations during forging process under consideration; a concept of inertia factor *'ξ'* has been introduced, which is defined as the ratio of inertia energy dissipated to the total energy supplied by die platen during plastic deformation. It is mathematically expressed as follows:

$$\xi(\%)=\left(\frac{W_a}{J^*}\right)100 \quad (14)$$

The average forging load for both the cases (*i.e.* disc and rectangular preforms) was computed separately by substituting the above energy dissipation equations in below equation [36]:

$$F_{av}=4J^*\left(U\right)^{-1}A_{av} \quad (15)$$

Where, '*J**' is the total energy dissipations (*i.e.* sum of internal, frictional and inertia energy dissipations) during plastic deformation.

5. Results & Discussions

Figure 6 depicts the experimental variation of effective stress and strain for disc and rectangular AMC preforms and it's clearly evident that at cold conditions (*i.e.* room temperature) under dry interfacial friction conditions, the maximum forgeability of the preforms is about 47-49% of the height reduction. The initiation of the cracks on the preforms vertical surfaces are observed at stress level of 280 - 290 MPa. Also, AMC rectangular preforms having higher SiCp percentage requires higher forging load to exhibit same amount of deformation. This is attributed to the constraint deformation conditions prevailing majorly due to the presence of sharp corners in rectangular shaped preforms. Figure 7 shows the variation of height reduction of preforms with average forging load. It can be observed that the experimental and theoretical results are in close agreement validating the present theoretical analysis. Additionally, it is found that deformation rate is exponentially higher after the average forging load attains a critical threshold value of 10-15 tons. This trend continues till the initiation of the cracks on the equatorial vertical surfaces of the preforms. Finally, the forgeability of disc preforms having 5 wt% SiCp was found be the highest.

To investigate the rate change of deformation, the variation of strain rate and forging load was investigated thoroughly and the plot is shown in figure 8. It can be seen that maximum strain rate of magnitude 24 (1000 mm / mm / sec) occurred at an average forging load of 44-45 tons. Also, the strain rates decreases slightly after attaining the highest value during the end of forging operations, which is due to the volume consolidation of reinforcement particles within the matrix metal due to the compressive forces. Thus, it can be concluded that effect of SiCp particulate on the magnitude of engineering stress, engineering strain and strain rates is predominant up to a certain deformation level, which in the present case was found to be around 40% height reduction irrespective of preform shapes.

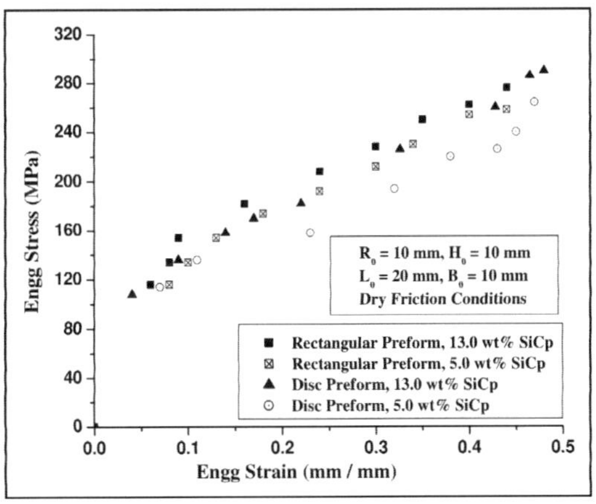

Figure 6. Engineering stress (MPa) vs Engineering strain (mm/mm)

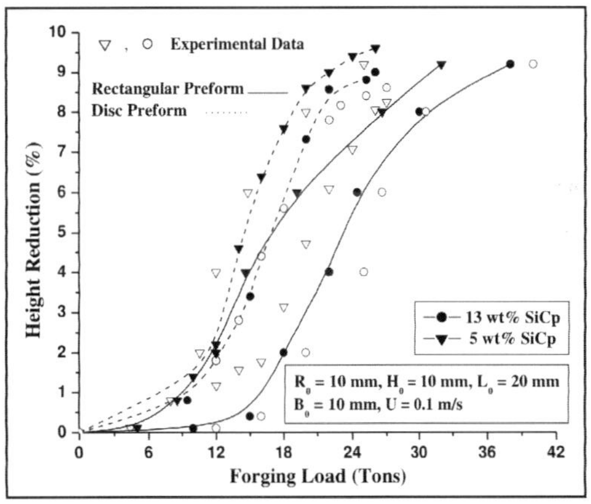

Figure 7. Preform height reduction (%) vs forging load (tons)

Figure 9 shows that plot of axial and radial strains with average forging load during deformation of disc and rectangular preforms under dry interfacial lubricated conditions. It is clearly observed that the values of both the axial and radial strains increases exponentially with the increase in the average forging load. Also, the magnitudes of axial strains are found to be higher than corresponding radial strains for same loads, which satisfies the normal flow rules. Additionally, this an estimate of the measure of Poisson's ratio for SiCp reinforced Aluminium metal matrix composite material under consideration can also be computed from the same.

As mentioned above, the investigation of inertia effects on the overall deformation of the AMC preforms is critical and hence a plot of total energy dissipation during deformation with forging load has been shown in figure 10. It can be seen that total energy dissipations are higher for rectangular preforms in comparison with disc preforms and it's also higher for high die accelerations and percentage of SiCp. Thus, it can be concluded that the energy requirements are higher for the deformation of rectangular preforms due to constraint deformation conditions. Additionally, to critically examine this effect, another plot of the 'Inertia Factor (ξ)' with die speed has been shown in figure 11. It can be seen that as die velocity and acceleration increases, the inertia factor also increases exponentially and at the end of the forging process the energy dissipation due to inertia forces can be as high as 30% of the total energy dissipated. Thus, it must be taken into account during the theoretically analysis of forging processes to arrive an accurate estimates of the forging loads and associated deformations.

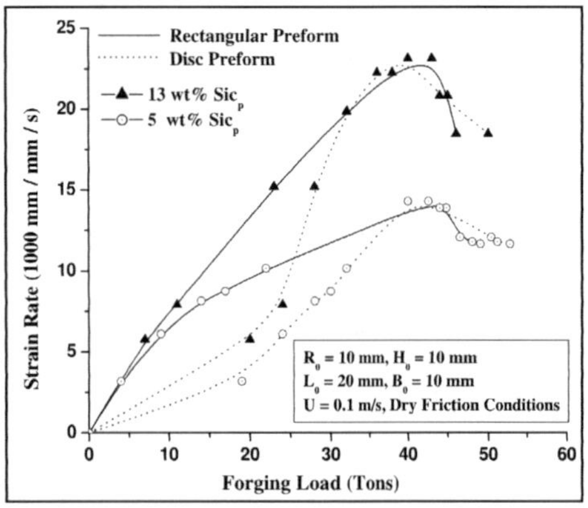

Figure 8. Strain rate (mm/mm sec) vs forging load (tons)

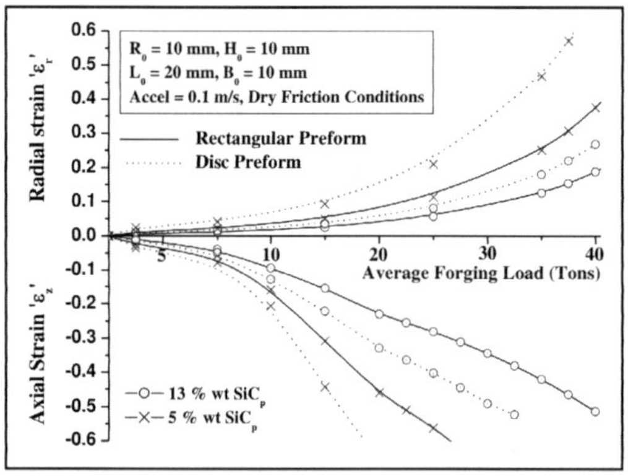

Figure 9. Radial & axial strains vs average forging load (tons)

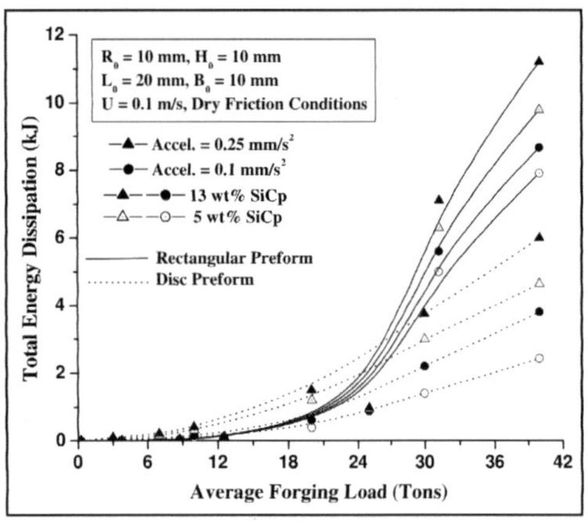

Figure 10. Total energy dissipation (kJ) vs forging load (tons)

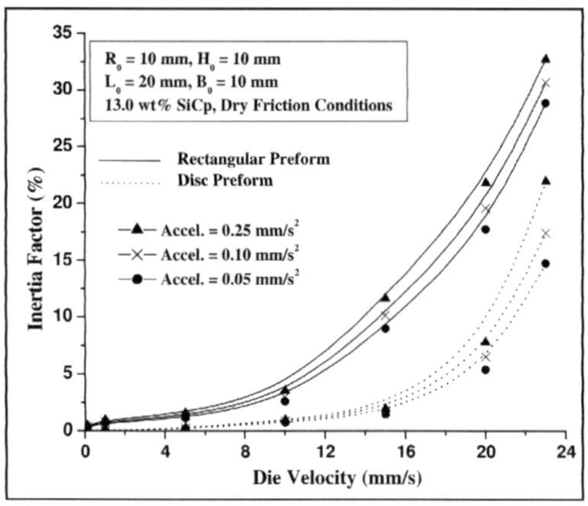

Figure 11. Inertia factor (%) vs die velocity (mm/s)

6. Conclusions

Based on the above investigations, some of the major conclusions are as follows:

- The maximum forgeability of the SiCp reinforced Aluminium metal matrix composite material under consideration was found to about 45-47% of the height reductions at cold conditions (*i.e.*

room temperature) under dry (without lubrication) interfacial frictional conditions. The initiation of the cracks on the free vertical surfaces of the preforms was found to be at maximum stress level of 280 - 290 MPa. The deformations in preforms having 5 wt% SiCp are found to higher indicating that additions of SiCp in matrix metal decreases the overall forgeability of preforms.

- The requirement of stress to induce same magnitude of strain and deformation was higher for preforms having high SiCp wt%. This was also found higher for the rectangular preforms due to high constraint deformations.
- The highest strain rate in order of 24 (1000 mm / mm / sec) was noticed during the open-die forging operations and the effect of particulate additions on engineering stress, engineering strain and strain rate was predominantly pronounced till 44-45 % of deformation, *i.e.* preform height reduction. It was observed that particulate particles in a metal matrix consolidate under compressive stress conditions especially during the end of forging operations.
- The energy requirements during open-die forging operations at cold conditions was found to increase with the addition of SiCp in the metal matrix and was additionally found to be higher for the higher die velocity and die accelerations. Also, the rectangular preforms required higher energy for deformation due to constrain deformations. This was also successfully portrayed using the plot of *'Inertia Factor'*.

It expected from the present research work that clear insights into the complex interaction of various deformation characteristics during cold open-die forging of SiCp reinforced Aluminium metal matrix composites has been established, which can further help successfully fabricate the engineering components.

References

[1] Cavaliere P and Evangelista E, 2006 Isothermal forging of metal matrix composites: recrystallization behavior by means of deformation efficiency *Comp. Sci. and Tech.* **66** pp 357-62

[2] Ismail O, Cocen U and Kazim O 2000 The effect of forging on the properties of particulate-SiC-reinforced Aluminium-alloy composites *Comp. Sci. and Tech.* **60** pp 411-19

[3] Badini C, Vecchia GML, Fino P and Valente T 2001 Forging of 2124/SiCp composite: preliminary studies of the effects on microstructure and strength *J. Mater. Process. Tech.* **116** pp 289-97

[4] Ma FC, Lu WJ, Qin JN, Zhang D and Ji B 2007 The effect of forging temperature on microstructure and mechanical properties of in situ TiC/Ti *Comp. Mater. and Design* **28** pp 1339-42

[5] Sulaiman S, Sayuti M and Samin R 2008 Mechanical properties of the as-cast quartz particulate reinforced LM6 alloy matrix composites *Proc. of 10th Int. Conf. Advances in Mater. and Process.Tech.* **201** pp 731-35

[6] Ceschini L, Minak G and Morri A 2009 Forging of AA618/20 vol% Al2O3p composite: effects on microstructure and tensile properties *Comp. Sci. and Tech.* **69** pp 1783-89

[7] Murashkevich AN, Lavitskaya AS, Alisienok OA and Zharskii IM 2009 Fabrication and properties of SiO2/TiO2 *Comp. Inorganic Mater.* **45(10)** pp 1146-52

[8] Ramesh B and Senthilvelan T 2010 Formability characteristics of Aluminium based composites - A review IACSIT *Int. J. Engg. and Tech.* **2(1)** pp 1-6

[9] Wu K, Deng K, Nie K, Wu Y, Wang X, Hu X and Zheng M 2010 Microstructure and mechanical properties of SiCp/AZ91 composite deformed through a combination of forging and extrusion process *Materials and Design* **31** pp 3929-32

[10] Kainer KU 2006 Basics of metal matrix composites. Metal matrix composites: custom-made materials for automotive and aerospace engineering *(Wiley-VCH Verlag Gmbh KGaA Weinheim)*

[11] Matejka V, Lu Y, Jiao L, Huang L, Martynkova GS and Tomasek V 2010 Effects of silicon carbide particle sizes on friction-wear properties of friction composites designed for car brake lining *applications Tribology Int.* **43** pp 144-51
[12] Mithun PS and Devaraj MR 2011 Development of Aluminum based composite material *Int. J. Appl. Engg. Res.* **6(1)** pp 121-30
[13] Gronostajski JZ, Marciniak H and Matuszak A 1996 Production of composites on the base of AlCu4 alloy chips *J.Mater. Process. Tech.* **60** pp 719-22
[14] Szczepanik S and Sleboda T 1996 The influence of the hot deformation and heat treatment on the properties of P/M Al-Cu composites *J. Mater. Process. Tech.* **60** pp 729-33
[15] Gronostajski JZ, Kaczmar JW, Marciniak H and Matuszak A 1998 Production of composites from Al and AlMg2 alloy chips *J.Mater. Process. Tech.* **77** pp 37-41
[16] Surappa MK 2003 Aluminum matrix composites: challenges and opportunities *Sadhana* **28(1-2)** pp 319-34
[17] Narayanaswamy R, Senthilkumar and Pandey KS 2008 Some features on hot forging of powder metallurgy sintered high strength 4% Titanium Carbide composite steel performs under different stress state conditions *Mater. and Design* **29** pp 1380-1400
[18] Singh S, Chand S, Roy S and Chandrasekhar P 2020 Influence of dispersoid content on compressibility, sinterability and mechanical behaviour of B4C/BN reinforced Al6061 metal matrix hybrid composites fabricated via mechanical alloying *Metals and Materials International* (https://doi.org/10.1007/s12540-020-00739-0)
[19] Jha AK and Kumar S 1994 Compatibility of sintered materials during cold forging *Int. J. Mater. Product Tech.* **9** pp 281-99
[20] Singh S, Jha AK and Kumar S 2001 Analysis of dynamic effects during high-speed forging of sintered preforms *J. Mater. Process. Tech.* **112(1)** pp 53-62
[21] Singh S, Jha AK and Kumar S 2007 Dynamic effects during sinter forging of axi-symmetric hollow disc preforms *Machine Tools and Manuf.* **47(7-8)** pp 1101-13
[22] Singh S, Jha AK and Kumar S 2007 High-speed sinter-forging of annular ring component Tamkang *J.Science and Engineering* (*Journal of Applied Science and Engineering - JASE*) **10(1)** pp 17-28
[23] Singh S, Jha AK and Kumar S 2007 Upper bound analysis and experimental investigations of dynamic effects during sinter-forging of irregular polygonal preforms *J. Mater. Process. Tech.* **194(1)** pp 134-44
[24] Narayanaswamy R, Ramesh T and Pandey KS 2008 Some aspects on cold forging of Aluminum-Iron powder metallurgy composite under triaxial stress state condition *Mater. and Design* **29** pp 891-903
[25] Narayanaswamy R, Ramesh T and Pandey KS 2008 Some aspects on cold forging of Aluminum-Alumina powder metallurgy composite under triaxial stress state condition *Mater. and Design.* **29** pp 1212-27
[26] Chandrasekhar P and Singh S 2011 Investigation of dynamic effects during cold upset-forging of sintered aluminum truncated conical preforms *J. Mater. Process. Tech.* **211** pp 1285-95
[27] Verma D, Chandrasekhar P, Singh S and Kar S 2013 Investigations into deformation characteristics during open-die forging of SiCp reinforced Aluminium metal matrix composites *J. Powder Tech.* **2013** pp 1-14.
[28] Singh S, Chand S and Chandrasekhar P 2018 Analysis of dynamic effects during rotary forging of cylindrical sintered preforms *Materials Today: Proc.* **5(9-3)** pp 19313-20
[29] Sutradhar G, Behera R, Dutta A, Das S, Majumdar K and Chatterjee D 2010 An experimental study on the effect of silicon carbide particulates (SiCp) on the mechanical properties like machinability and forgeability of stir-cast aluminum alloy metal matrix composites *Indian Foundry J.* **56(5)** pp 43-50
[30] Chung CY and Lau KC 1999 Mechanical characteristics of hipped SiC particulate-reinforced Aluminum alloy metal matrix composites *Intelligent Process. and Manuf. of Mater.* **2** pp 1023-28

[31] Singh S, Chand S and Chandrasekhar P 2018 Investigation of dynamic effects during cold upset-forging of Silicon Carbide particulate reinforced Aluminium metal matrix composite preforms *Materials Today*: *Proc.* **5(9-3)** pp 20201-09
[32] Chawla N, Williams JJ and Saha R 2002 Mechanical behavior and microstructure characterization of sinter-forged SiC particle reinforced aluminum matrix composites *J. Light Metals* **2** pp 215-27
[33] Male AT and Cockcroft MG 1965 A Method for the determination of the coefficient of friction of metals under condition of bulk plastic deformation *J. Inst. Metals* **93** pp 38-46
[34] Downey CL and Kuhn HA 1971 Deformation characteristics and plastic theory of sintered powder materials *Int. J. Powder Metal.* **7** pp 15-21
[35] Rooks AW 1974 The effect of die temperature on metal flow and die wear during high-speed hot forging *Proc. of 15th Int. MTDR Conf.* p 487
[36] Avitzur B 1968 *Metal Forming Processes and Analysis (McGraw Hill New York)*

Appendix

The boundary conditions, kinematically admissible velocity field and corresponding strain rate equations are given as:

(i). Axi-Symmetric Deformation (Disc Preform)

$$U_z = 0 \quad \text{at} \quad z = 0 \tag{A.1}$$

$$U_z = -U \quad \text{at} \quad z = H_0/2 \tag{A.2}$$

$$U_r = \frac{\beta e^{-\beta z/h} Ur}{2(1-e^{-\beta/2})h} \tag{A.3}$$

$$U_z = -\frac{(1-e^{-\beta z/h})U}{(1-e^{-\beta/2})} \tag{A.4}$$

$$U_\theta = 0 \tag{A.5}$$

$$\dot{\varepsilon}_{rr} = \frac{\partial U_r}{\partial r} = \frac{\beta e^{-\beta z/h} U}{2(1-e^{-\beta/2})h} \tag{A.6}$$

$$\dot{\varepsilon}_{\theta\theta} = \frac{U_r}{r} = \frac{\beta e^{-\beta z/h} U}{2(1-e^{-\beta/2})h} \tag{A.7}$$

$$\dot{\varepsilon}_{zz} = \frac{\partial U_r}{\partial z} = -\frac{\beta e^{-\beta z/h} U}{(1-e^{-\beta/2})h} \tag{A.8}$$

$$\dot{\varepsilon}_{rz} = \frac{1}{2}\left[\frac{\partial U_z}{\partial r} + \frac{\partial U_r}{\partial z}\right] = -\frac{\beta^2 e^{-\beta z/h} Ur}{4(1-e^{-\beta/2})h^2} \tag{A.9}$$

$$\dot{\varepsilon}_{r\theta} = \dot{\varepsilon}_{\theta z} = 0 \tag{A.10}$$

(ii). Plane Strain Deformation (Rectangular Preform)

$$U_z = U \quad \text{at} \quad z = 0 \tag{A.11}$$

$$U_z = 0 \quad \text{at} \quad z = H_0 \tag{A.12}$$

$$U_x = \left[\frac{\beta e^{-\beta z/h} Ux}{(1-e^{-\beta/2})h}\right] \tag{A.13}$$

$$U_z = -\left[\frac{\left(e^{-\beta/2} - e^{-\beta z/h}\right)U}{\left(1-e^{-\beta/2}\right)}\right] \tag{A.14}$$

$$U_y = 0 \tag{A.15}$$

$$\dot{\varepsilon}_{xx} = \left[\frac{\beta e^{-\beta z/h} Ux}{\left(1 - e^{-\beta}\right)h}\right] \tag{A.16}$$

$$\dot{\varepsilon}_{zz} = -\left[\frac{\beta e^{-\beta z/h} U}{\left(1 - e^{-\beta/2}\right)h}\right] \tag{A.17}$$

$$\dot{\varepsilon}_{yy} = 0 \tag{A.18}$$

$$\dot{\varepsilon}_{xz} = \frac{1}{2}\left(\frac{\partial U_x}{\partial z} + \frac{\partial U_z}{\partial x}\right) = -\left[\frac{\beta^2 e^{-\beta z/h} Ux}{2\left(1 - e^{-\beta/2}\right)h^2}\right] \tag{A.19}$$

$$\dot{\varepsilon}_{xy} = \dot{\varepsilon}_{yz} = 0 \tag{A.20}$$

Nomenclature

U_{ij}	velocity field
$\dot{\varepsilon}_{ij}$	strain rate field
a_{ij}	acceleration field
U	die velocity
U_{av}	average die velocity
ΔU	interfacial relative velocity
$\dot{U}$	die acceleration
p	die pressure
P_{av}	average pressure
F_{av}	average forging load
A_{av}	average cross sectional area
S	surface area
V	volume
σ_o	flow stress of SiCp AMC material
ρ	density of SiCp AMC preform
τ	frictional shear stress
$J_2^{/}$	second invariant of stress
D	diameter of cylindrical preform
H	height of cylindrical preform
R_o	radius of disc preform
R_m	sticking zone radius of disc preform
B_o	width of rectangular preform
B_m	sticking zone width of rectangular preform
L_0	length of solid rectangular preform
H_0	height of disc & rectangular preforms
W_i	internal energy dissipation
W_f	friction energy dissipation
W_a	inertia energy dissipation
J^*	external energy supplied
μ_{eff}	effective coefficient of friction
ϕ_o	specific cohesion factor
β	barreling factor
ζ	inertia factor
n	constant quantity

IOP Conf. Series: Materials Science and Engineering 1045 (2021) 012027 doi:10.1088/1757-899X/1045/1/012027

Comparison of reducing agent for reduced graphene oxide as cathode for zinc-ion hybrid capacitors

S Prasertkaew[1], T Dejthammathorn[2] and P Pattananuwat[2, 3, a]

[1]Petrochemistry and Polymer Science, Faculty of Science, Chulalongkorn University, Bangkok, Thailand
[2]Department of Materials Science, Faculty of Science, Chulalongkorn University, Bangkok, Thailand
[3]Research Unit of Advanced Materials for Energy Storage, Chulalongkorn University, Bangkok, Thailand

[a]Prasit.pat@chula.ac.th

Abstract. The comparative study of using the different reduced graphene oxide sources, obtained from thiourea and l-ascorbic acid as reducing agent, for zinc cathode capacitors is investigated. The chemical structure and the surface morphology of both reduced graphene oxide samples are examined by infrared spectrum and scanning electron microscopy, respectively. Serving as cathode for zinc-ion hybrid capacitors, the as-prepared reduced graphene from l-ascorbic acid reveals the better the specific discharge capacity (41.50 mAh g^{-1}) than that of the obtained from thiourea (40.78 mAh g^{-1}). The cycle stability of zinc-ion hybrid capacitors coin cell in 2M $ZnSO_4$ electrolyte is examined by multi-cycles of GCD at 0.1 A g^{-1}, demonstrating that rGO obtained from l-ascorbic acid can endow the retention up to 72.8% after 200 cycles, while rGO obtained from thiourea is remained at 63.1%.

1. Introduction

Nowadays, energy storage devices and electric vehicles have rapidly developed and set high demands for energy density and power density. Supercapacitors possess a fast charge-discharge rate, and high-power density but their low energy density limits their large-scale applications. In contrast, batteries reveal high-energy-density but low energy power, poor rate performance and inferior cycling stability [1]. To overcome these disadvantages of supercapacitors and batteries, the concept of hybrid capacitors has been developed consisting of a battery-type faradaic electrode as an energy source and a capacitive electrode as a power source. Zinc-ion hybrid capacitors (ZICs) have been increased attention in recent progress for many researchers due to high safety, low cost, abundant zinc resource, low redox potential of zinc and high volumetric capacity. In addition, the energy power of ZICs strongly depends on the Zn2+ storage behaviors in cathode materials [2]. Several cathode materials have been developed for ZICs, including conductive polymers and carbon materials.

Considerable much effort has been devoted to developing the carbon materials for ZICs such as activated carbon, carbon nanotubes, graphene and so on. Among them, graphene is a material having one atomic thickness materials that consist of sp2 hybridized carbon atoms arranged in a honeycomb lattice, becoming a popular topic in material science because of its amazing characteristics such as excellent structural, large surface area, mechanical, thermal, and electrical properties [3] with talented applications in the field of batteries, catalyst supports, nanoelectronics, composites and supercapacitors. Commonly, monolayer graphene sheets can be synthesized in large quantities by chemical reduction or thermal treatment of graphene oxide (GO). Graphite powder can be used as precursor to synthesize GO and reduced graphene oxide (rGO), respectively. After graphite powder oxidation under acid condition, the functional groups of oxygen were created between multi-layers of carbon sheet. The GO layers were mostly bonded with epoxy, hydroxyl groups. While carboxyl and carbonyl groups were bonded the rim of the layer [4]. These oxygens containing functional groups change the GO van der Waals interaction, making it hydrophilic and exfoliating in an aqueous solution. Thus, GO is greatly stable dispersion in other polar organic solvents and water.

One of the effective methods is exfoliation of GO by ultrasonication to obtain exfoliated GO before chemical reduction process [5]. However, when the functional groups of oxygen were removed to get the graphene sheets, its losses the ability to dispersed in water and agglomeration will be occurred. The different reducing agents were applied to synthesize rGO from GO via chemical reduction such as sodium borohydride, hydrazine, hydrazine monohydrate and hydrogen sulfide [6,7]. All these reducing agents are hazardous. In view of this, the present investigation has been focused on the use of thiourea and l-ascorbic acid as a reducing agent. Comparing to other reducing agents, thiourea and l-ascorbic acid are more environmental-friendly. rGO prepared by this method possesses promising electrochemical properties, which can be used as cathode for ZICs applications. Therefore, in this research, the development of graphene as a capacitive cathode for ZICs are proposed, expecting to possess a high power/energy density and long cycle life.

2. Experimental Section

2.1. Materials
Graphite powder was purchased from Sigma-Aldrich Co., Ltd. All other chemicals used were reagent-grade products.

2.2. Synthesis of graphene oxide
Graphene oxide (GO) used in this work was synthesized by oxidation of graphite powder through the modified Hummer's method [8].

2.3. Synthesis of reduced graphene oxide with thiourea and l-ascorbic acid
GO was dispersed in deionized water at 1 mg/ml by sonication for 30 min. Then, thiourea or l-ascorbic acid as a reducing agent was added into the dispersed GO solution and continually sonicated for 30 min. The suspension was hydrothermal at 120 °C for 3 h. The obtained product was washed several times with deionized water and dried in freeze dry for 24 h.

2.4. Structural and chemical characterization
The chemical structure and surface morphology of samples were characterized by Fourier transform infrared spectrometer (FTIR, Nicolet 6700, using KBr pellets in the 4000-500 cm^{-1} region) and scanning electron microscopy (SEM, Quanta 250), respectively.

2.5. Electrode assembly and electrochemical testing
The cathode was fabricated by mixing 80 wt.% of active material (rGO-thiourea or rGO-l-ascorbic acid), 10 wt.% of conductive carbon and 10 wt.% of polytetrafluoroethylene (PTFE) binder in ethanol solvent under stirring to form slurry. After that the slurry was pasted onto a graphene foil substrate. The as-prepare cathode samples were dried in air at 60°C for 24 h. Then, punched the cathode samples into a disk and assembled into CR2032 type coin cell with zinc foil as anode, glass fiber as separator, and 120μL 2M $ZnSO_4$ as electrolyte. The electrochemical properties of the electrode materials were evaluated by cyclic voltammetry (CV) and galvanostatic charge-discharge (GCD). CV measurements

were studied by applying scan rate at 5 mV s^{-1} and GCD method was used at 0.1 A g^{-1}. In addition, cycle stability was plot by specific discharge capacity at 0.1 A g^{-1} for 200 cycles.

3. Results and discussion

3.1. Morphology study

FT-IR spectra are used to elucidate the functional groups of GO, rGO-thiourea and rGO-l-ascorbic acid. As seen in figure 1, the FT-IR spectrum of GO clearly shows the absorption band at 1722 cm^{-1} assigning to the stretching vibration of C=O in the carboxylic groups. The vibrational bands at 3427 and 1619 cm^{-1} are ascribed to the stretching vibration of hydroxyl group (-OH). Obviously, it is clear that the reduction process via hydrothermal of GO with l-ascorbic acid can eliminate the oxygen function groups on GO. While the using of thiourea still remains the partial oxygen function groups as the present of the low intensity peak of 3427, 1722 and 1619 cm^{-1}, confirming the better efficiency of l-ascorbic acid as reducing agent for GO.

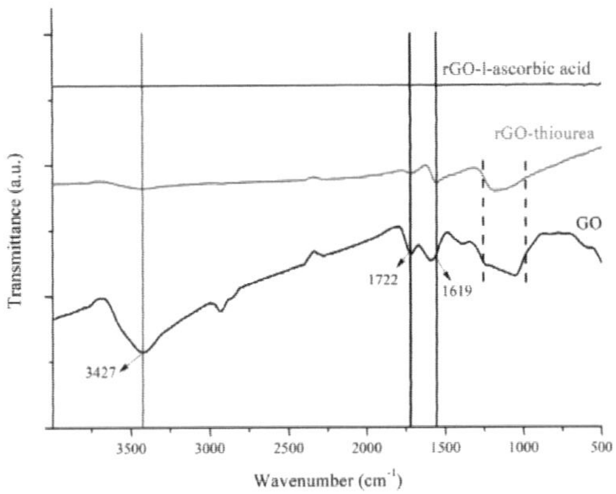

Figure 1. FT-IR spectra of GO, rGO-thiourea and rGO-l-ascorbic acid

The surface morphology and microstructure of the rGO-thiourea and rGO-l-ascorbic acid are characterized by SEM observation. As shown in figure 2a, rGO-thiourea shows the portion of 2D structure sheet coexisting with the twist sheet structure. In contrast, rGO-l-ascorbic acid reveals the sheet-to-sheet separation with 3D porous microstructure as seen in figure 2b, expecting to improve to facilitate ion diffusion and transport of electrolyte on surface [9].

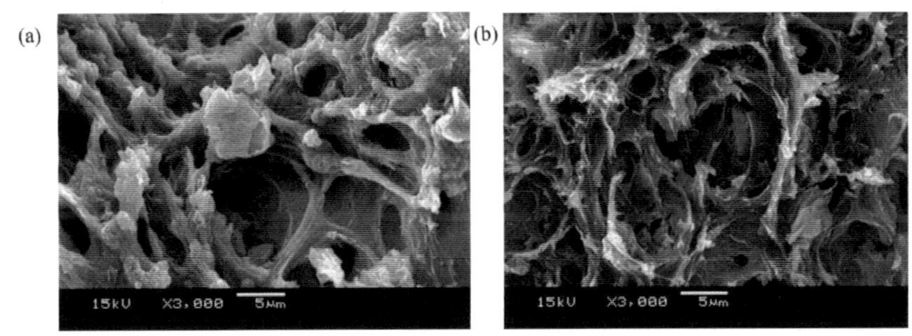

Figure 2. SEM image of (a) rGO-thiourea and (b) rGO-l-ascorbic acid.

IOP Conf. Series: Materials Science and Engineering 1045 (2021) 012027 doi:10.1088/1757-899X/1045/1/012027

3.2. Electrochemical properties

To evaluate the ZICs efficiency of rGO-thiourea and rGO-l-ascorbic acid, CV measurement is employed to characteristic of ZICs using CR2032 cell battery at 5 mV s^{-1} with a voltage window of 0.0 V to 1.6 V in 2M $ZnSO_4$. As shown figure 3a, CV curve of rGO-l-ascorbic acid sample obviously shows a little cliff of redox peak at 1.2 V, considering as the formation of Zn^{2+} insertion into the rGO-l-ascorbic acid structure on cathode. In fact, the oxidation peak at 1.1-1.2 V is ascribed to the Zn^{2+} insertion on cathode, accompanied by the reduction of Zn^{2+} to zinc on anode [10]. In contrast, as shown on figure 3b for CV of rGO-thiourea, the redox peak was unidentified in both of reduction and oxidation curves, indicated the lower adsorption/desorption of Zn/Zn^{2+} for rGO-thiourea. Notably, these CV results are corresponding to SEM results, confirming that 3D structure of rGO-l-ascorbic acid can facilitate more Zn^{2+} adsorption/desorption than that of 2D structure of rGO-thiourea.

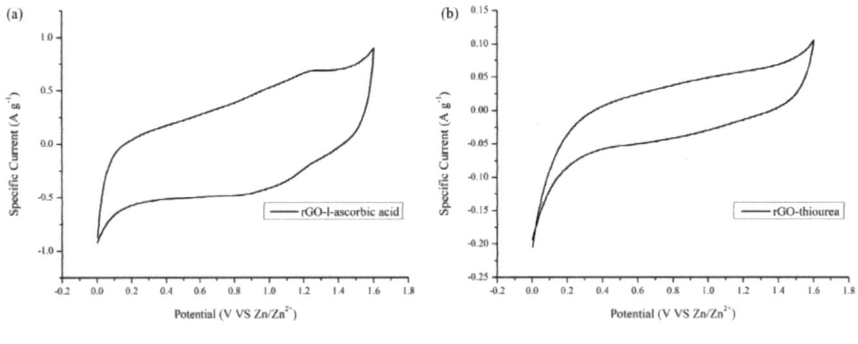

Figure 3. CV curve at scan rate 5 mV s^{-1} of (a) rGO-l-ascorbic acid and (b) rGO-thiourea.

The galvanostatic charge/discharge curves (GCD) of the rGO-l-ascorbic acid and rGO-thiourea are further investigated as shown in figure 4a. Both of rGO-l-ascorbic acid and rGO-thiourea can accommodate Zn^{2+} adsorption/desorption during charge/discharge process for ZICs, revealing the unsymmetrical triangular curves. rGO-l-ascorbic acid can achieve the discharge capacity of 41.5 mAh g^{-1}, while rGO-thiourea yields the discharge capacity of 40.78 mAh g^{-1}. The cycling stability of the rGO-l-ascorbic acid and rGO-thiourea are shown in figure 4b. The rGO-l-ascorbic acid can remain more than 72% of its initial capacity even the cycle number extended to 200 cycles at current density of 0.1 A g^{-1}, indicating the excellent cycling stability.

As a comparison, the capacity retention of the rGO-thiourea at 0.1 A g^{-1} is 63.1% after 200 cycles. Considerably, the structure stability of the rGO-l-ascorbic acid is superior to the rGO-thiourea. This might be because of the fact that 3D framework rGO-l-ascorbic acid structure can remain stability of structure during charge/discharge process, while 2D stacking graphene structure of the rGO-thiourea are collapsed and decrease space in the structure that interrupted Zn^{2+} adsorption/desorption process.

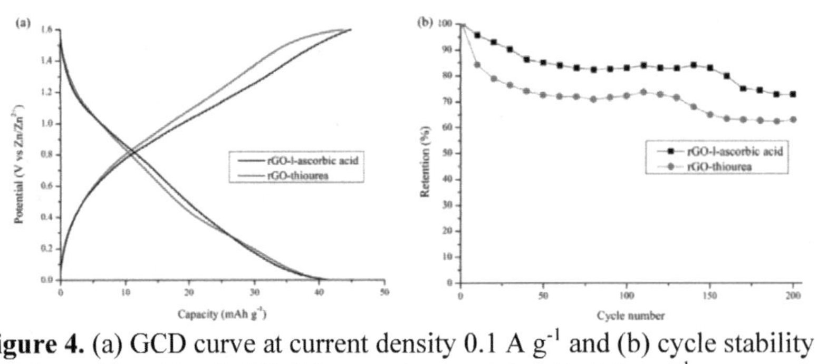

Figure 4. (a) GCD curve at current density 0.1 A g^{-1} and (b) cycle stability of rGO-l-ascorbic acid and rGO-thiourea at current density 0.1 A g^{-1}.

4. Conclusion

In conclusion, the present study demonstrates that the using of l-ascorbic acid exhibits an excellent reducing performance than that of the using thiourea. The obtained rGO-l-ascorbic reveals a high stability 3D framework of graphene structure, enhancing the cyclic stability (retention of 72.8% after 200 cycles). Reversible ion adsorption/desorption on rGO cathode and Zn^{2+} deposition/stripping on Zn anode enables the ZICs to repeatedly and rapidly store/deliver electrical energy.

References

[1] Goodenough J B 2014 *Energy Environ. Sci.* **7** 14.
[2] Zeng X, Hao J, Wang Z, Mao J and Guo Z 2019 *Energy Stor. Mater.* **20** 410
[3] Geim A K and Novoselov K S 2007 *Nat. Mater.* **6** 183
[4] He H, Klinowski J, Forster M and Lerf A 1998 *Chem. Phys. Lett.* **287** 53
[5] Lotya M, Hernandez Y, King P J, Smith R J, Nicolosi V, Karlsson L S, Blighe F M, De S, Wang Z, Mcgovern I T and et el. 2009 *J. Am. Chem. Soc.* **131** 3611
[6] Wang G, Shen X, Wang B, Yao J and Park J 2009 *Carbon* **47** 1359
[7] Si Y and Samalski E T 2008 *Nano Lett.* **8** 1679
[8] Vikrant S, Sonia G, Brindan T, Meenakshi S, Gaurav S, Manas R, Manav S, Niroj S, Kalpana B, Deepu P and et el. 2015 *Electrochi. Acta.* **160** 244
[9] Nanjundan A K and Jong B B 2015 *J. Nanotechnol.* **26** 1
[10] Dong L, Ma X, Li Y, Zhao L, Liu W, Cheng J, Xu C, Li B, Yang Q H and Kang F 2018 *Energy Stor. Mater.* **13** 96

Video quality evaluation and testing verification of H.264, HEVC, VVC and EVC video compression standards

Shreyanka Subbarayappa[1,2] and K R Rao[2]

[1]Electronics and Communication Engineering Department, Ramaiah University of Applied Sciences, Bangalore, KA 560054, India

[2]Electrical Engineering Department, University of Texas at Arlington, Arlington, TX 76019, U.S.A

Abstract. This paper is a comparative analysis with respect to video compression, coding efficiency with respect to performance and quality of the compressed video of Advanced Video Coding (H.264), High Efficiency Video Coding (HEVC), Versatile Video Coding (VVC) and MPEG-5 Essential Video Coding (EVC). The approach used for coding and comparative analysis is by keeping the video quality (QP) and group of pictures (GOP) constant for all the four codecs. Performance with respect to quality and coding efficiency is tabulated with respect to objective quality metrics analysis like (PSNR, M-SSIM, Bitrate, compression ratio) and subjective quality analysis. From both these techniques, it is observed that the VVC is superior to EVC, HEVC and H.264 in terms of compression. Therefore, more bitrate savings is observed for VVC which is 15% to 45% more when compared to EVC. Due to the complexity of the encoder block, VVC takes almost twice the time taken to encode EVC and HEVC taking the least time to encode the video bit streams. The conclusion of the quality of video keeping system parameters constant, showed VVC performing better than EVC, followed by HEVC and then H.264 for both objective and subjective analysis. Time taken to decode the encoded sequences is almost same with very less delta difference between the video codecs. The codec configurations with algorithmic enhancements is also explained in this paper.

1. Introduction

Innovations in the communication systems and technology are growing tremendously and the growth seen is unimaginable in the last forty years. In multimedia communication systems, technology has transformed from analog television to digital television in the video domain. Mobile phones are known as smart phones as they are used, not only to make voice calls, but also used to send emails, video calls, transfer data, GPS, taking pictures and so on. Due to the widespread user applications, compression on data becomes important to save system resources. Video has occupied 80% of all major traffic of data transfer in 2019 and is expected to cover 85% by 2021 [2]. Video is also continuously increasing in size and quality from standard-definition (SD) video to ultra-high definition (UHD – 4K, 8K and 12K) video. More data or size in video requires higher transmission bandwidth or more disk space to store, which is very expensive. This drives into the betterment in compression and hence a demand for a new video codec. Several algorithms are implemented to achieve compression of Image or Video with respect to the user's demand on the quality of output as well as for applications it is used for. These algorithms working together are classified in terms of codecs. These video codecs are developed to target specific application usages. Advances in video compression technology are used to reduce the utilization of the system resources like processing time, memory use, network bandwidth and battery life. This is possible by reducing the complexity of the video codecs without compromising on the output video quality. There are two separate streams in the next generation video coding development work. VVC driven by Joint Video Exploration team (JVET) and EVC driven by Moving Picture Expert Group (MPEG). These two codecs are the extended versions with respect to the advances in compression technology to HEVC.

Versatile Video Coding (VVC) standard which was finalized on July 6th, 2020 by the JVET team of ITU-T SG 16 WP 3 [11] and ISO/IEC JTC 1/SC 29/WG 11 [10] is due by October 2020. The latest software version of VVC available is VTM-8.0 [14]. This codec promises 40% more compression than its predecessor (HEVC) with having the same perceptual quality of video [6]. VVC supports lossless and subjectively lossless compression with resolutions varying from 4K to 16K along with 360 videos. It supports video contents of 8 to 16 bit depth with chroma formats ranging from 4:2:0 to 4:4:4. Some applications of VVC are high dynamic range (HDR) video, Multiview coding, still picture coding, panoramic formats. Complexity of the codec was observed to be 10 times more than that of HEVC, but that is waived off with the better quality of the video output with more compression achieved.

Essential Video Coding (EVC) standard which is under development by the MPEG team of ISO/IEC JTC 1/SC 29/WG 11 is due by end-2020. The latest software version of EVC available is ETM-4.0 [15]. The coding efficiency of EVC is almost like HEVC with slightly better video quality than HEVC. It is developed with some licensing conditions, i.e. royalty-free for the baseline profile and with IPR for the main profiles.

High Efficiency Video Coding (HEVC) also known as H.265 standard was developed by Joint Collaborative Team on Video Coding (JCT-VC) of ITU-T SG16 WP-3 and ISO/IEC JTC1/SC29/WG11 and published the first version in June 2013. The next version included applications like range extensions (RExt), Multiview extensions (MV-HEVC) and scalability extensions (SHVC) and published in 2015. The coming versions applications included extensions in 3D video and screen content coding (SCC) which were published in the 1st quarter of 2017. HEVC promised in offering 30%-50% better compression of video with respect to H.264 and having the same perceptual quality of the video [1]. It supported video contents up to 8K resolution with bit depth from 8 to 12 bits. The latest software version of HEVC is HM-16.0 [13].

Advanced Video coding (AVC) also known as H.264 was developed by Joint Video Team (JVT) of ISO/IEC MPEG & ITU-T VCEG (ISO/IEC JTC1/SC29/WG11 and ITU-T SG16 Q.6) and published in 2003. H.264 codec is the first codec to support HD videos and with a very large video compression possibility with good perceptual output video quality. In today's time, it is the most commonly used video format with 81% [19] of the video industries using it for distribution, compression and recording of the video content. With the later years of its initial development, some new features were added to the codec, i.e. fidelity range extensions, multiview video coding, scalable video coding, 3D-AVC and MFC stereoscopic coding. H.264 promised 50% better bitrate efficiency when compared to MPEG-2 part 2 [5]. It supported contents of bit depth 8 to 14 bits and chroma formats from 4:2:0 initially in baseline profile to 4:4:4 at the later stage for high profile. This codec is known for its broad application range like digital video compression at low-bitrate, internet streaming to broadcasting, and digital cameras for nearly lossless coding. The latest software version of AVC is JM-19.0 [12].

2. Structure of H.264, HEVC, VVC and EVC Codecs

2.1. H.264 structure [5]

The H.264 codec structure starts with a block for video frame to be divided into maximum 16x16 macroblocks. The first frame of the video undergoes an intra prediction of Transform coding which is 2 Dimensional DCT (2-D DCT) [7] [20] followed by scaling and quantization. These quantized transform coefficients are given to the entropy block where we select either CABAC or CAVLC as entropy coding techniques to get the bit-stream of the 1st frame [8]. Simultaneously we have the decoder block within the encoder block which decodes the present frame by performing inverse quantization and inverse transform and stores the frame for performing motion estimation, which is needed for the next frame. A deblocking filter is added at the decoder section within the encoder block to smooth the ringing effect produced by 2-D DCT in the output video. There are 9 intra prediction modes with 5 luma prediction modes and 4 chroma prediction modes. The encoder framework of H.264 is analyzed in Figure 1.

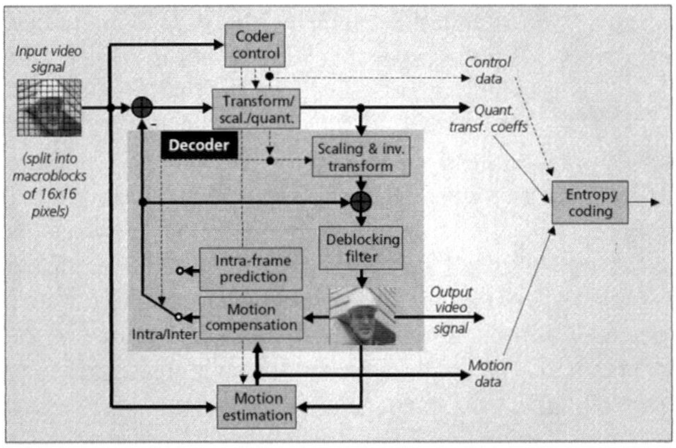

Figure 1. H.264 video coding encoder framework [5] [9]

2.2. HEVC Structure [3]

The HEVC codec is a hybrid motion compensated codec with the block structure splitting in the Quadtree format. Coding Tree Unit (CTU) for each block ranges up to 64X64. The codec constitutes total 35 intra prediction modes. There are two types of transform coding in HEVC, i.e. DCT -II and DST-VII. Transform block sizes vary from 8x8, 16x16 to 32x32. The Inter prediction is based on the hierarchical weighted prediction of the prediction unit (PU) level motion vector prediction. There are two loop filters within the encoder framework. One being the deblocking filter and the other being the sample adaptive offset (SAO) filter which removes the artifacts produced by transform and quantization technique. CABAC is the only entropy coding technique used in HEVC. The encoder framework of HEVC is analysed in Figure 2.

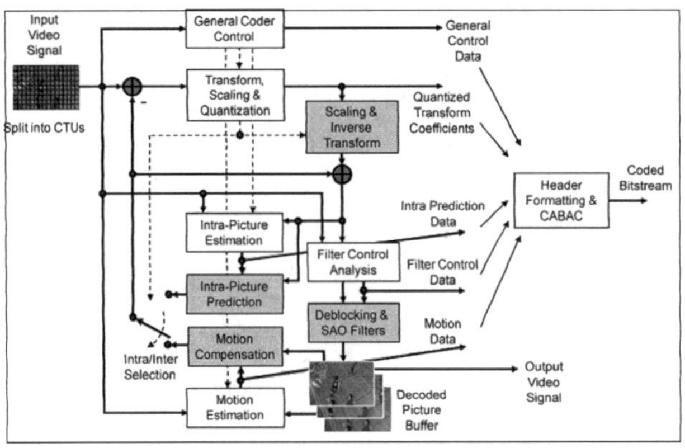

Figure 2. HEVC video coding encoder framework [1] [4]

2.3. VVC Structure

The VVC codec is the next generation video codec which also follows the hybrid motion compensated algorithm. Block structure partition is performed by Quad Tree Binary Tree (QTBT) and Ternary Tree (TT) structure which is new in VVC. The CTU 's has maximum size up to 256x256. The inter prediction is performed by cross component linear model (CCLM) and having 65 intra prediction modes as enhanced intra mode coding. Mode dependent non-separable secondary transform and adaptive multiple core transform are the two types of transform coding. Transform block sizes vary from 4x4, 8x8, 16x16, 32x32 and 64x64. Inter prediction includes the new affine motion prediction and everything else in that part is similar to HEVC. In the filter block, Adaptive loop filter is the new addiction along with deblocking filter and sample adaptive offset filter (SAO) of HEVC. Modified version of CABAC is the new technique at entropy block. The encoder framework of VVC is analysed in Figure 3.

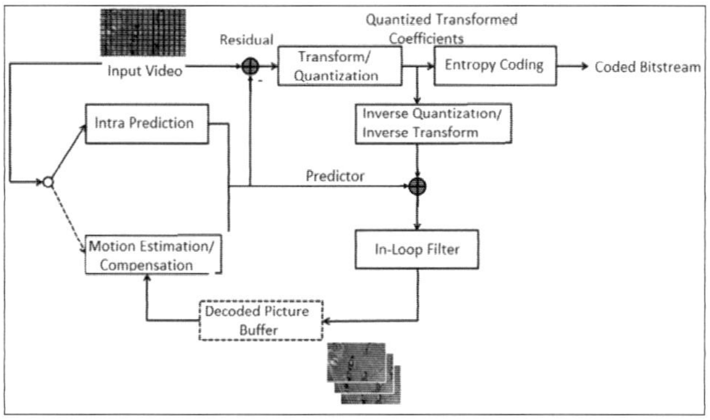

Figure 3. VVC video coding encoder framework [14]

2.4. EVC Structure

The EVC codec by the MPEG team uses quad tree (QT), binary and ternary tree (BTT) and split unit coding order (SUCO) for block structure partition with each CTU sizing up to maximum 128x128. There are two profiles in EVC, naming baseline profile and high profile to select the coding techniques. Inter prediction has Uni/Bi prediction, Affine motion prediction, adaptive motion vector resolution (AMVR), merge with motion vector difference (MMVD), history-based motion vector prediction (HMVP) and decoder side motion vector refinement. There are 33 Intra prediction modes. The transform block has DCT-II for baseline profile and Adaptive transform selection (ATS) in main profile to select between DCT VIII and DST VII. The transform size ranges from 4x4, 8x8,16x16, 32x32, 64x64 and 128x128. In-Loop filter like Deblocking filter (DBF) along with Hadamard transform domain filter (HTDF), Adaptive Loop Filter (ALF), Advanced deblocking filter is added to remove the artifacts caused by transform and quantization for smoothing the video frames. At the entropy block, there is a choice for 2-D Run Length coding, Arithmetic coding engine, context modelling and advanced coefficient coding (ADCC) based on the profile chosen for encoding. The encoder framework of EVC is analyzed in Figure 4.

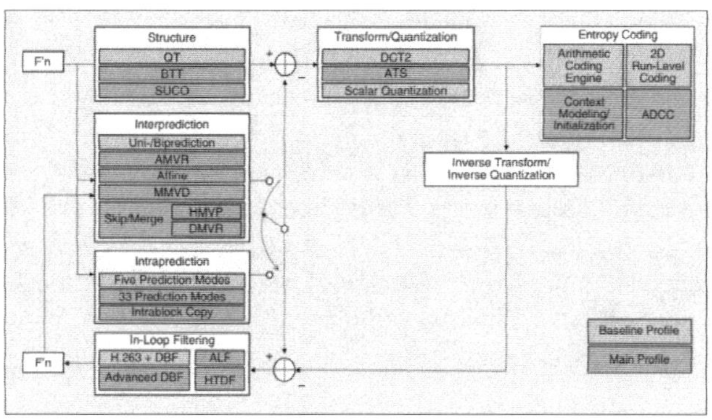

Figure 4. EVC video coding encoder framework [18]

3. Test Configurations

All the video codecs were tested for the same content with GOP = 16 (IBBBBBBBBBBBBBBI) and QP = 10, 20, 30, 40 and 50. The system used is Dell Inspiron with i7 processor/ Windows 10/ 64 bit. The test sequences used for testing is described in Table 1.

Table 1. Test sequence description [16][17]

Test sequence	Resolution	Number of frames	Original size (kB)	Source
Akiyo	176x144	300	11138	Xiph.org
SpeedBag	640x360	120	40500	Xiph.org
Shield2	640x360	120	40500	Xiph.org
MobileCalendar	720x486	360	246038	Xiph.org
Stockholm	1280x720	604	815400	Xiph.org
FourPeople	1280x720	601	811350	Xiph.org
Beauty	1920x1080	600	1822500	UVG
SunBath	3840x2160	300	3645000	UVG
Bhosphorus	3840x2160	600	7290000	UVG

3.1. H.264 [12]

Software : Joint Model (JM) – version 19.0
Encoder Commandline : lencod.exe -d encoder_main.cfg
Decoder Commandline : ldecod.exe

In the encoder commandline, -d indicated the configuration file to be added which is in the directory location 'JM\bin'. In the testing mode, encoder_main.cfg file is edited according to the parameter preferences like, 'InputFile' location on the system to be added. Input/output content parameters are edited in the source width/height and accordingly output width/height. QPI/P/BSlice are changed according to the QP value needed for testing. Other parameters to be edited in the configuration file "encoder_main.cfg" is shown.

```
#======= encoder_main.cfg File ===============

InputFile          = "C:\Desktop\akiyo.yuv"     # Input sequence
FramesToBeEncoded  = 300    # Number of frames to be coded
FrameRate          = 100.0  # Frame Rate per second (0.1-100.0)
SourceWidth        = 176   # Source frame width
SourceHeight       = 144   # Source frame height
OutputWidth        = 176   # Output frame width
OutputHeight       = 144   # Output frame height
IntraPeriod        = 16  # Period of I-pictures   (0=only first)
QPISlice           = 10  # Quant. param for I Slices (0-51)
QPPSlice           = 10  # Quant. param for P Slices (0-51)
NumberBFrames      = 15  # Number of B coded frames inserted (0=not used)
QPBSlice           = 10 # Quant. param for B slices (0-51)
```

After the encoder, the encoded binary file 'test.264' is obtained in the same folder location. After the decoder, the output file is 'test.yuv' which is the reconstructed video. Main profile is used in testing H.264 for the analysis and comparison purpose.

3.2. HEVC [13]

Software : Hybrid Model (HM) – version 16.0

Encoder Commandline : TAppEncoder.exe -c encoder_randomaccess_main.cfg -c akiyo.cfg
Decoder Commandline : TAppDecoder.exe -b str.bin -o dec.yuv -d 8

In the encoder commandline, -c indicated the configuration file to be added which is in the directory '\cfg' and '\cfg\per-sequence'. The cfg\per-sequence location file is to be created by the user under any name as the content file to be tested and includes the content parameters. Since the content to be tested in this example is akiyo, 'cfg\per-sequence' file is named as 'akiyo.cfg' as the secondary configuration file. In our testing, the main configuration file in 'cfg\' location that is 'encoder_randomaccess_main.cfg' parameters is edited. The values of 'IntraPeriod', 'GOPSize' to 16 and 'QP' parameters are edited as shown in VVC software parameters. New 'cfg/per-sequence' file to be added with parameters under the name 'akiyo.cfg' is shown.

```
#======== akiyo.cfg File I/O ===============

 InputFile            : C:\Desktop\akiyo.yuv   #Input content destination
 InputBitDepth        : 8     # Input content bit-depth
 InputChromaFormat    : 420   # Ratio of luminance to chrominance samples of the input content
 FrameRate            : 120   # Frame Rate per second of the content
 FrameSkip            : 0     # Number of frames to be skipped in input content
 SourceWidth          : 176   # Input content frame width
 SourceHeight         : 144   # Input content frame height
 FramesToBeEncoded    : 300   # Number of frames to be coded
 Level                : 4.1
```

After the encoder, the encoded binary file 'str.bin' is obtained in the same file location. After the decoder, the output file is 'dec.yuv' which is the reconstructed video. Random access main profile is used in HEVC testing for the analysis and comparison purpose.

3.3. VVC [14]

Software : VVC Test Model (VTM) – version 8.0
Encoder Commandline : EncoderApp.exe -c encoder_randomaccess_vtm.cfg -c akiyo.cfg
Decoder Commandline : DecoderApp.exe -b str.bin -o dec.yuv -d 8

VVC open source software is a test version which is still under development. The software at the encoder side uses two configuration files. The second file is the content file which is similar as HEVC as 'akiyo.cfg'. The first config file is the main file names as 'encoder_randomaccess_vtm.cfg' which specifies the parameters to be changed in order to get the output as desired as shown

```
#======== encoder_randomaccess_vtm.cfg File==============

 IntraPeriod  : 16  # Period of I-Frame ( -1 = only first)
 GOPSize      : 16  # GOP Size (number of B slice = GOPSize-1)
 QP           : 10  # Quantization parameter(0-51)
```

After the encoder, the encoded binary file 'str.bin' is obtained in the same folder location. After the decoder, the output file is 'dec.yuv' which is the reconstructed video. Random access main profile is used in test VVC for the analysis and comparison purpose.

3.4. EVC [15]

Software : EVC Test Model (ETM) – version 4.0
Encoder Commandline : evca_encoder.exe -i akiyo.yuv -q 10 -w 176 -h 144 -p 16 -f 300 -z 120 -d 8 -o str.bin –config encoder_randomaccess.cfg
Decoder Commandline : evca_decoder.exe -i str.bin -o dec.yuv

EVC software used for testing is a test model and this codec is still under development. In the Encoder commandline the following notations denotes:

'-i' - path to the input file
'-q' - quantization parameter
'-w' - source width

'-h'	- source height
'-p'	- intra period
'-f'	- number of frames
'-z'	- frame rate of the content
'-d'	- input content bit depth
'-o'	- output encoded file
'-config'	- encoder configuration file

For every test run, we need to change the '-q' value in the encoder commandline with respect to QP value for each run. The configuration file 'encoder_randomaccess.cfg' is added in the command line with no edits needed in the file. After the encoder, the encoded binary file 'str.bin' is obtained in the same file location. After the decoder, the output file is 'dec.yuv' which is the reconstructed video. Random access main profile is used in EVC testing for the analysis and comparison purpose.

4. Results and Summary

It is very essential to analyze the codec implementation and perform a comparative study looking at the performance of the codec and output quality of the encoded bitstream by both objective metric and subjective metric. Three different methods were used for the analysis of the codec implementation. They are as specified below,

4.1. Objective Quality Metric [21]:

The first method is an objective quality metric comparative study between the codecs in terms of PSNR, SSIM, Bitrate, time taken to encode, time taken to decode and compression size for each of the test sequence specified in Table 1 and keeping the QP constant to 32, GOP = 16 (IBBBBBBBBBBBBBBBI).

4.2. Rate Distortion:

The second method is the rate distortion or RD method where the rate distortion graph is obtained for a content with different QP values (10, 20, 30, 40 and 50) and same GOP =16 (IBBBBBBBBBBBBBBBI) for all four codecs.

4.3. Subjective Quality Analysis:

The third method is the subjective quality analysis where the comparisons of the codecs are observed by the visual analysis of the videos. The rate distortion method outputs for QP values (10, 20, 30, 40 and 50) are observed frame by frame and the distortion noted for codecs H.264, HEVC, VVC and EVC.

Table 2. Objective quality results of H.264 encoder and decoder software implementation

Input Video File	PSNR (dB)	*SSIM (YUV)*	MSE (YUV)	Bitrate (kbps)	Encoding Time(s)	Decoding Time (s)	Compressed o/p video size (kB)
Akiyo	36.80	0.94	15.55	106.39	323.238	0.093	39
SpeedBag	41.05	0.97	5.89	141.63	1873.631	0.451	104
Shield2	34.09	0.88	28.74	862.01	1292.539	0.668	632
MobileCalendar	34.88	0.92	24.45	2486.94	8073.856	4.539	5465
Stockholm	36.10	0.89	19.90	2027.63	38512.408	10.091	7475
Fourpeople	39.67	0.95	8.67	401.52	27572.912	8.137	1473

Table 3. Objective quality results of HEVC encoder and decoder software implementation

Input Video File	PSNR (dB)	*SSIM (YUV)*	MSE (YUV)	Bitrate (kbps)	Encoding Time(s)	Decoding Time (s)	Compressed o/p video size (kB)
Akiyo	39.250	0.98945	7.728	130.922	49.35	0.454	40
Shields2	35.947	0.98599	16.54	347.445	231.14	0.693	255
Stockholm	35.812	0.98276	17.06	4222.370	5501.18	11.149	2595
Beauty	38.6891	0.97482	8.795	2238.458	12172.03	23.354	1367
SunBath	42.921	0.99132	3.319	12965.12	37967.48	54.871	3957
Bhosphorus	39.250	0.98945	7.728	130.922	49.35	0.454	40

Table 4. Objective quality results of VVC encoder and decoder software implementation

Input Video File	PSNR (dB)	*SSIM (YUV)*	MSE (YUV)	Bitrate (kbps)	Encoding Time(s)	Decoding Time (s)	Compressed o/p video size (kB)
Akiyo	40.2677	0.9879	97.8196	83.3984	446.137	1.099	26
Shields2	36.7821	0.9890	218.2678	192.6293	1727.356	1.279	142
Stockholm	36.1117	0.9690	254.7011	2356.0132	22901.051	19.528	1448
Beauty	38.9595	0.9967	132.2072	1691.5568	101368.613	45.738	1033
SunBath	43.8083	0.9990	43.2886	9755.6192	168480.477	104.085	2978
Bhosphorus	42.0861	0.9830	64.3559	890.18	192677.252	202.441	3260

Table 5. Objective quality results of EVC encoder and decoder software implementation

Input Video File	PSNR (dB)	*SSIM (YUV)*	Bitrate (kbps)	Encoding Time(s)	Decoding Time (s)	Compressed o/p video size (kB)
Akiyo	41.43	0.995	21.262	154.576	1.518	39
Shields2	37.92	0.993	348.658	1114.427	1.581	256
Stockholm	38.99	0.983	709.307	17186.5	17.895	2615
Beauty	39.601	0.972	2079.912	40601.449	110.111	1270
SunBath	45.6415	0.999	11141.700	94910.641	98.6	3401

The objective quality metric comparison as mentioned in point one in section 4, using metrics like PSNR, SSIM, Bitrate, encoding time, decoding time and compressed output size is shown for each codec H.264, HEVC, VVC and EVC for different contents in Table 2, 3, 4 and 5. H.264 was not tested for 4K resolution videos as it does not support 4K content. These codecs were tested keeping QP=32 and GOP = 15 for random access main profile for HEVC, VVC and EVC and main profile for H.264. Keeping the quality QP constant, it was observed that the bitrate savings had a lot of variations across codecs. Bitrate for contents were observed less for VVC followed by EVC, HEVC and then H.264. Bitrate savings also concludes that encoded compression size was achieved more for VVC followed by EVC, HEVC and then H.264. This is achieved due to the various advanced algorithms in VVC and EVC compared to HEVC and H.264, like increased CTU size, advanced entropy coding, advanced motion estimation and compensation blocks. The encoding time for each content was observed to be more for VVC then followed by EVC and then HEVC as seen in Figure 5. This is due to the extensive algorithmic part CTU partitioning, motion estimation techniques, more prediction modes observed in VVC and EVC when compared to HEVC and H.264. Time consumed to decode the encoded video for different codecs were almost very close, for each content with less time difference between them.

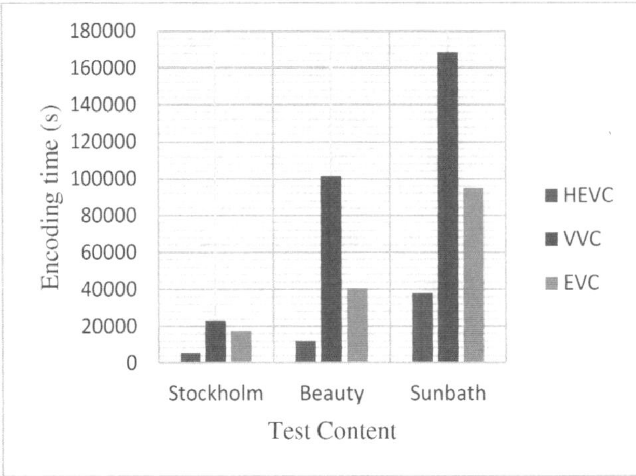

Figure 5. Encoding time for each content for HEVC, VVC and EVC

The second point in section IV performs a comparative study using rate distortion method. Here RD graph is obtained testing one content with different QP size (10, 20, 30 ,40 and 50) on all four codecs (H.264, HEVC, VVC and EVC). Other parameters of the codecs are kept constant like GOP, all filters 'ON' and main profile. For each testing scenario, PSNR and Bitrate is tabulated, and graph is obtained. Figure 6 shows the rate distortion graph of stockholm content keeping other codec parameters constant. It is observed from the graph that VVC performs better with better quality and bitrate savings, followed by EVC, HEVC and H.264 codecs.

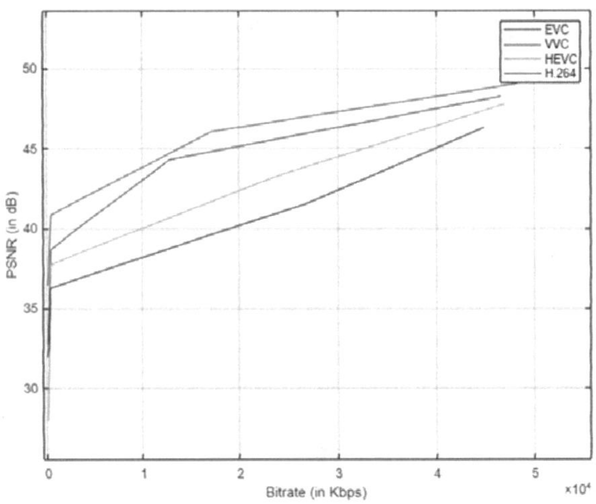

Figure 6. Rate distortion graph for Stockholm content

The third point in section IV is subjective quality analysis. The outputs of rate distortion tests are view in a media player and quality degradation is observed frame by frame. Fig. 7 shows the quality degradation in encoded output video for QP=50 and shield2 content. It was observed from the red marked area that VVC showed better quality, followed by EVC, HEVC and then H.264. This analysis was observed for other contents from Table 1 and the results for quality analysis remained same.

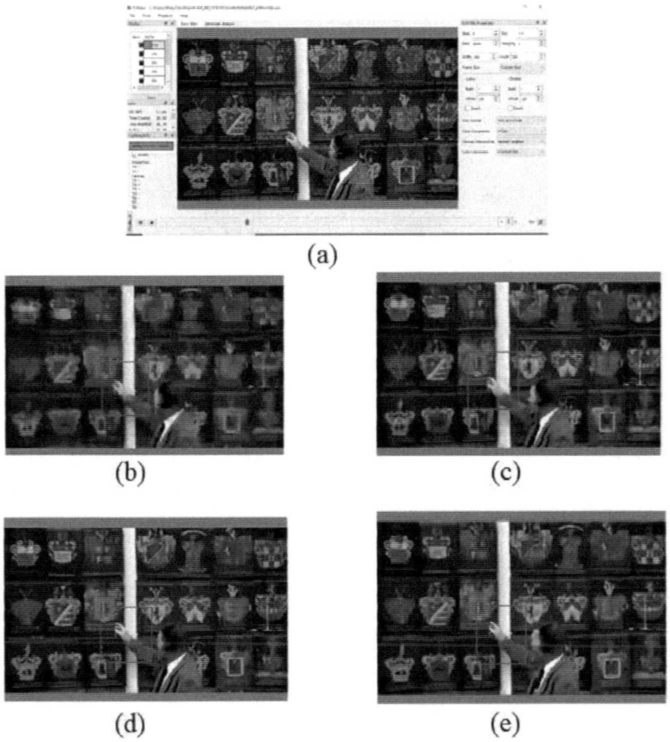

Figure 7. Subjective quality analysis of H.264, HEVC, VVC and EVC for content 'shield2' (a) Original shield2 content at frame number = 31, (b) H.264 output at QP = 50 at frame number = 31, (c) HEVC output at QP = 50 at frame number = 31, (d) VVC output at QP = 50 at frame number = 31, (e) EVC output at QP = 50 at frame number = 31

5. Conclusion

The codec analysis using the three methods of verification concluded by providing the same results in terms of codec performance with respect to four codec software implementation. The testing verified that VVC showed better bitrate savings for the encoded bit sequence, 30% – 50% better when compared to HEVC codecs keeping the video quality constant. EVC showed 10% – 30% better bitrate savings when compared to HEVC keeping the video quality constant. HEVC providing 50% better bitrate savings than H.264 for same output video quality was verified.

Complexity of the codec is another domain which opens for research to improve the codec time performance. VVC as the codec with software version 8.0 showing more time to encode the sequence when compared to EVC, HEVC, H.264. The time taken was almost double the time taken for EVC to encode a 4K resolution video. This is due to the increase in prediction modes which is almost double the modes in EVC and HEVC. This opens for avenues of research in reducing the complexity of the codec. VVC and EVC brings in a lot of applications usages to the market like supporting 4K to 8K streaming video and HDR video support. VVC and EVC also adds the development in the field of transcoding video codecs and multiplexing with other audio codecs.

References

[1] Subbarayappa S and Rao K R 2019 Overview and extensions of the high efficiency video coding (HEVC) and beyond (Versatile Video Coding) *International Journal of Emerging Technology and Advanced Engineering* **9(12)** 70-97

[2] M. Lister, *37 Staggering Video Marketing Statics for 2018*, World stream online advertising made easy, June 30, 2020. Accessed on: July 5th , 2020. [Online] Available: https://www.wordstream.com/blog/ws/2017/03/08/video-marketing-statistics#:~:text=45%25%20of%20people%20watch%20more,online%20marketers%20use%20video%20content.

[3] Sullivan G J, Ohm J R, Han W J and Wiegand T 2012 Overview of the high efficiency video coding (HEVC) standard *IEEE Transactions on circuits and systems for video technology* **22(12)** 1649-68

[4] Ohm J R, Sullivan G J, Schwarz H, Tan T K and Wiegand T 2012 Comparison of the coding efficiency of video coding standards – including high efficiency video coding (HEVC) *IEEE Transactions on circuits and systems for video technology* **22(12)** 1669 - 84
[5] Wiegand T, Sullivan G J, Bjontegaard G and Luthra A 2003 Overview of the H.264/AVC video coding standard *IEEE Transactions on circuits and systems for video technology* 13(7) 560-76
[6] Topiwala P, Krishnan M and Dai W 2018 Performance comparison of VVC, AV1 and HEVC on 8-bit and 10-bit content *Applications of Digital Image Processing XLI International Society for optics and Photonics* **10752** 107520V
[7] Ochoa D H and Rao K R 2019 *Discrete Cosine Transforms* CRC Press
[8] Kwon S K, Tamhankar A and Rao K R 2006 Overview of H.264/MPEG-4 part 10 *Journal of Visual Communication and Image Representation* **17(2)** 186-216
[9] Richardson I E 2004 *H. 264 and MPEG-4 video compression: video coding for next-generation multimedia* John Wiley and Sons
[10] ISO Website: Accessed on July 9th, 2020 [Online] Available: https://www.iso.org/home.html
[11] ITU-T Website: Accessed on July 9th, 2020 [Online] Available: https://www.itu.int/en/ITU-T/Pages/default.aspx
[12] JM Software: Accessed on July 9th, 2020 [Online] Available: https://github.com/shihuade/JM.git
[13] HM Software: Accessed on July 9th, 2020 [Online] Available: https://hevc.hhi.fraunhofer.de/svn/svn_HEVCSoftware/trunk/
[14] VTM Software: Accessed on July 9th, 2020 [Online] Available: https://vcgit.hhi.fraunhofer.de/jvet/VVCSoftware_VTM.git
[15] ETM Software: Accessed on July 9th, 2020 [Online] Available: https://gitlab.com/MPEG-5/ETM
[16] Test videos: Accessed on July 9th, 2020 [Online] Available: https://media.xiph.org/video/derf
[17] Test videos: Accessed on July 9th, 2020 [Online] Available: http://ultravideo.cs.tut.fi/
[18] Choi K et al 2019 MPEG-5: essential video coding standard *Applications of Digital Image Processing XLII International Society for optics and photonics* **11137** 1113710
[19] Topiwala P, Krishnan M and Dai W 2019 Performance comparison of VVC, AV1 and EVC *Applications of Digital Image Processing XLII International Society for optics and photonics* **11137** 1113715
[20] Subbarayappa S 2012 Implementation and analysis of directional discrete cosine transform in H.264 for baseline profile *M.S. thesis Electrical Engineering Dept. University of Texas at Arlington*
[21] Hore A and Ziou D 2010 Image quality metrics: PSNR vs. SSIM *2010 20th International conference on pattern recognition* 2366-69

Design and vlsi implementation of a decimation filter for hearing aid applications

*[1]Purushothaman V, [2]Rama devi R and [3]Manjula S

[1]Department of Electronics and Communication Engineering, Sathyabama Institute of Science and Technology, Chennai, India.
[2]Department of Electronics and Instrumentation Engineering, Sathyabama Institute of Science and Technology, Chennai, India.
[3]Department of Electronics and Communication Engineering, Rajalakshmi Institute of Technology, Chennai, India.

*Purushothamanece@gmail.com

Abstract. Roughly 10% of the total common people experience some kind of hearing misfortune, yet just little level of this measurement utilize the portable amplifier. The shame related with wearing a portable amplifier, client disappointment with listening device execution, the expense and the battery life. Using computerized signal handling the advanced portable amplifier currently offers what the simple listening device can't offer. As of now low power VLSI design is more preferable in digital designs. More and more individuals around the globe experience the ill effects of hearing misfortunes. The expanding normal age and the developing populace are the principle explanations behind this. The pulverization channel utilized for portable amplifier is planned and executed both in MATLAB and Very High Speed Integrated Circuit Hardware Description Language. The obliteration channel is structured utilizing the circulated number-crunching multiplier in Very High Speed Integrated Circuit Hardware Description Language. Each computerized channel structure is reproduced utilizing Matlab and its total engineering is caught utilizing Simulink. The subsequent engineering is equipment productive and expends less force contrasted with traditional decoding channels. Contrasted with the decoding FIR-FIR engineering, the structured decoding channel design utilizing Comb-half band FIR-FIR adds to an equipment sparing and decreases the force scattering.

Keywords: IIR Filters, FIR Filters, Digital Hearing Aids, Sigma Delta Analog to Digital Convertor

1. Introduction

In modern technology the dependency on radio frequency (RF) correspondence handsets accentuates both higher joining, to satisfy customer need for minimal effort, low-power, little structure factor individual specialized gadgets, and the capacity to adjust to various correspondence guidelines. Higher combination can be acquired by utilizing recipient structures and circuit methods that dispose of the requirement for outside parts. Using a beneficiary design that can choose channel sifting on chip at modulation band takes into consideration the customization important to adjust to numerous correspondence norms.

Many applications requires simple to-computerized change and hence become mainstream since it accomplishes expanded execution and adaptability by moving sign handling intricacy from simple to advanced hardware. Oversampling utilizes an inspecting rate that is a lot more noteworthy than the transmission capacity of the sign of intrigue. Advanced data handling systems is utilized to carry additional separating, decode and even do under sampling process. The scope of human hearing is commonly viewed as 20 Hz to 20 Kilo Hz, yet the ear is unquestionably responding to voice frequency between 1 Kilo Hz to 4 Kilo Hz. Thus it is added helpful to structure an amplifier application working inside the predefined recurrence go. Amplifiers are one of numerous cutting edge, versatile, computerized frameworks requiring power proficient structure so as to draw out battery life. Listening devices perform signal preparing capacities on sound signs. With the coming of numerous new sign preparing strategies, their necessity for higher computational capacity has squeezed power utilization.

Remote media transmission principles right now utilized all through the world have channel transfer speeds extending from 6.25 kilo Hz to 1.728 Mega Hz. A multiplexer is utilized to select one of the base band channel separating in the computerized area and carries Sigma-delta analog to digital convertor conversion with an extensive powerful range that can oblige select channels just as the ideal. It should likewise have the option to adjust to the different powerful range prerequisites and examining paces of the guidelines that are actualized. An extensive unique range sigma-delta modulator can be utilized to attain the necessities for various norms. Sigma-delta analog to digital converter is a minimal effort, minimal-transfer speed, minimal-power, increased-goal analog to digital convertor and utilized in many applications including information securing, correspondences, signal preparing and instrumentation. A portable amplifier application utilizes the increased sampling rate and comprises of a last yield goal of the analog to digital convertor. This work centers mostly around the plan and usage of the decimator.

2. Development of Digital Hearing system

Roughly 10% of the populace experiences some consultation misfortune, anyway just a little level of this class really utilizes an amplifier. There are a few components influencing market infiltration. Initially, there is the shame related with wearing a listening device. Second is client disappointment with the gadgets not living up to their desires, third is the expense related with the new advanced adaptations of listening devices.

The ongoing improvement of business listening devices with computerized signal handling abilities has permitted the improvement progressed signal preparing strategies to help the consultation weakened. The outcome for the wearer of the portable amplifier is more exact sound generation with least bending and clamor. Practically the entirety of the biggest amplifier makers have computerized listening device items available, and of the 6 million portable amplifiers sold in 2001, roughly 20% were advanced gadgets. So as to meet the little size and super low force prerequisites of listening devices, the current arrangements resort to unique reconfigurable ASIC gadgets for every amplifier structure. By upgrading to a monetarily accessible DSP approach which can be programmed according to custom needs, the listening device organizations can reduce altogether their price, subsequently a bigger segment of the populace with a reduced value, enhanced class computerized instrument.

In last few decades there has been increase in the quantity of advanced listening devices available in the market. Two well-known approaches of amplifiers that are accessible, behind the ear (BTE) and Completely in Canal (CIC). Regardless of their greater expense, they were generally welcomed by clinicians and customers.

Computerized listening devices today play out an assortment of cutting edge advanced sign handling calculations, for example, commotion decrease and input dropping. Consequently listening devices present difficulties for force and territory prerequisites (for manufacture chip) contrasted with other compact gadgets. It is essential to decide if advanced listening devices are enhanced for customers who faced hearing aid problem; it is additionally critical to concentrate on the prevalent preparing and highlights of these hearing aid gadgets. Computerized amplifiers cannot be depicted, as these devices are a different substance from simple listening devices. Computerized essentially demonstrates that the simple waveform is changed over in to a line of number for handling, help sadly; there is nothing characteristically mysterious about this cycle. There for, computerized isn't prevalent on the grounds that its advanced, but since advanced sign preparing (DSP) permits fabricates to listening devices with upgrade handling and highlights.

2.1. *Sigma-Delta analog to digital convertor*

Analog to digital converters are classified based on relying the inspecting rate. The primary variant examples the simple contribution at the nyquist recurrence fn with the end goal that $fs \geq 2B$, where fs is the testing recurrence and B is the transfer speed of the information signal. The second kind of ANALOG TO DIGITAL CONVERTORs tests the simple contribution at a lot increased operating frequencies greater than the nyquist recurrence and are brought over examining analog to digital converters, sigma-delta analog to digital converter. In sigma-delta analog to digital converter, the information data is examined at an increased testing recurrence.

$$fs = KX$$

Figure 1 depicts the overall graph and a short depiction of a sigma delta Analog to digital converters. The modulator modulates the input signal signals which are close to higher frequency band components set by the increased sampled proportion and changes over the simple information data into a heartbeat thickness balanced computerized data having both the first info data and the undesirable out-of-band data. A de-multiplexer selects one of the modulator channels having band other than base band. For a given clock frequency the modulator coordinates with the decimator simulatneously. In Figure 1, the modulator indicated is of first request with a 1-piece quantize and creates a 1-piece yield. The yield of the decimator is appeared as N-bit computerized information, where N is the yield goal of the analog to digital convertor and is subject to the increased testing proportion. The request for the decimator relies upon the request for the modulator.

2.2. *Modulator*

This block is an integral part of a sigma-delta Analog to digital converters. The order at which the modulator is operating and the level of sampling the input decides the modulator performance in the architecture. Since the modulator utilizes the guideline of oversampling the requirement for antialiasing channel is mitigated and the simple info signal can straightforwardly be tested utilizing the oversampling clock.

2.3. *Decimator*

The cycle of carefully changing over the testing pace of a sign from a certain increased data rate fs to a decreased data rate fn is called decimation. Decoding in severe sense implies decrease by 10 percent yet in signal handling capability implies a decrease in inspecting speed by any factor. Fundamentally a decimator is an advanced low In channel, which additionally plays out the activity of test rate decrease. The sigma-delta modulator does activity of commotion molding and consequently the clamor is pushed to higher frequencies so the decoding stage following the modulator can sift through this commotion over the cutoff recurrence, fn. The band restricted sign would then be able to be resampled by disposing of $K - 1$ examples out of each K tests, where K being the oversampling proportion. By averaging K tests out of the quantized sigma-delta yield, the decoding channel accomplishes a high yield goal and furthermore the recurrence of the yield information is at double the info signal data transmission which is the nyquist rate.

3. Design and implementation of structure of the proposed filter

The proposed filter structure comprises of demolition channel for incorporating with a current planned modulator along with a total sigma-delta Analog to digital converters. The utilization of multi-stage decoding channel which implies the single decoding channel is supplanted by fell channels. In this section, we are going to discuss the channel design utilized in the proposed architecture, which comprises of filter models, qualities and disadvantages. The initial phase in structuring a decoding channel is to choose which sorts of channels will be utilized and where decoding will happen. This part investigates the issues engaged with picking channel design for a listening device application. The general intensity of a few models is thought about, bringing about the three-phase engineering that is picked to execute this channel.

Decoding channels must be simple and also proficient due to the necessity of sifting which is generally carried at an increased data rate. A sigma-delta analog to digital convertor creates a piece data bits at a increased data rate of 1.28 Mega Hz. This bits are given as input to the decoding channel and are reduced by means of data rate to 20 Kilo Hz. The architecture utilize a decimation factor of 16 to down sample the bit streams from the decoding channel from 1.28 Mega Hz to 80 kilo Hz. We have picked a In band of 4 Kilo Hz in light of the fact that the ear is delicate to all voice signal inside 4 Kilo Hz band. The channel has an In band wave of 0.001, that compares to the level reaction of the channel. Since the progress band is a little level of the examining rate, the channel will have numerous taps, and the zeros will be nearer to one another making the channel be more complex to simple architecture. The traditional demolition channel is actualized by this strategy, and the suggested framework mitigates issues introduced in traditional approach by utilizing a multiple stage approach having varied data rate. The force utilization is straightforwardly relative to the quantity of taps and the working recurrence. The specification of the proposed filter channel is given in Table 1.

Table 1. Filter specifications.

Sampling frequency (Fs)	1.48Mega Hz
Decimation factor (M)	8
In band frequency	8 Kilo Hz
In band ripple	0.002
center frequency	18 Kilo Hz
Decimation filter stages(N)	5

3.1. Implementation of the Comb Filter

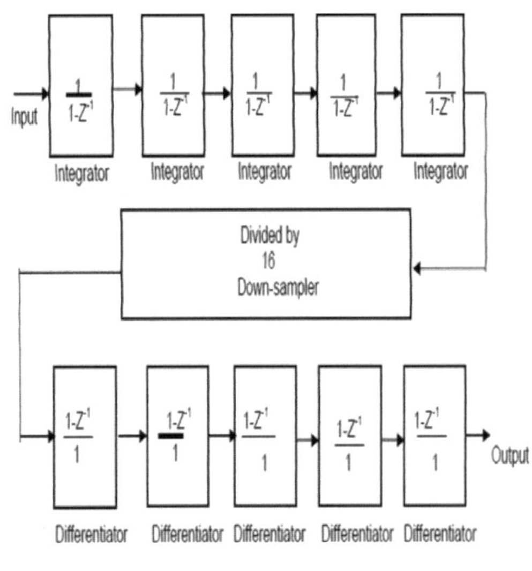

Figure 1. The multi stage filter archtechture.

IOP Conf. Series: Materials Science and Engineering 1045 (2021) 012029 doi:10.1088/1757-899X/1045/1/012029

The contribution of the decoding channel is produced from a sigma-delta modulator running at 1.28 Mega Hz. Guard channel can be intended to introduce an indent at every one of the voice channels that will avoid interference between the base-bands. In addition, decoding channel does not requires multiplier and comprises of basic activities reasonable at increased frequencies. The response of the decoding channel is a low In channel with a more keen cutoff. Because of these reasons, look over channels are utilized for the main phase of decoding.

4. RESULTS AND DISCUSSIONS.

The proposed Decimation filter architecture is synthesized using Xilinx FPGA Device.

Here 1.28 Mega Hz input signal is processed using ADC Sigma Delta modulator and obtained as Digital output .The input and output signal of ADC sigma delta modulator are shown in Figure 2 and Figure 3.The Digital input signal feeds to the FIR filter and the response of the low pass filter output is shown in Figure 4.

4.1. The ADC conversion from sigma-delta modulator

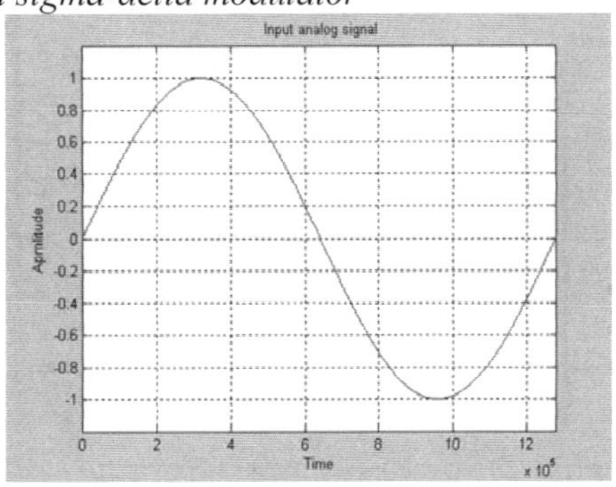

Figure 2 Input signal at 1.28Mega Hz

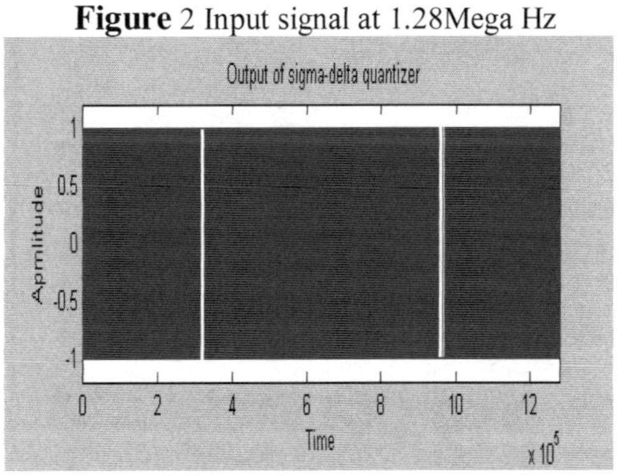

Figure 3 Output signal

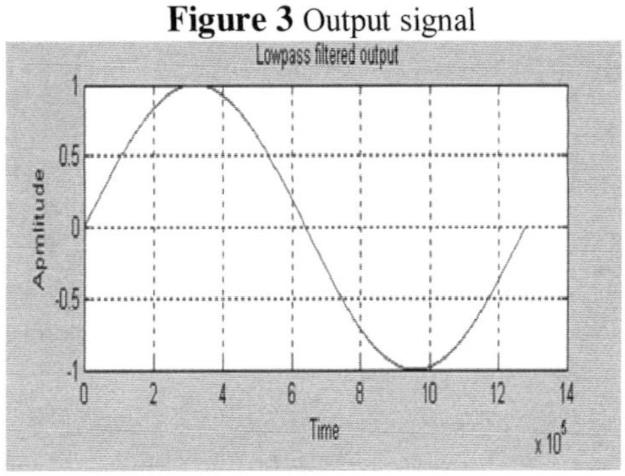

Figure 4The low pass filter output

IOP Conf. Series: Materials Science and Engineering 1045 (2021) 012029 doi:10.1088/1757-899X/1045/1/012029

4.2. The DAC conversion from sigma-delta modulator

The Digital output response of the low pass filter output is processed to the DAC sigma Delta modulator and produce the time domain output signal . The input , output signal and low pass filtered response are shown in Figure 5,6 and 7.

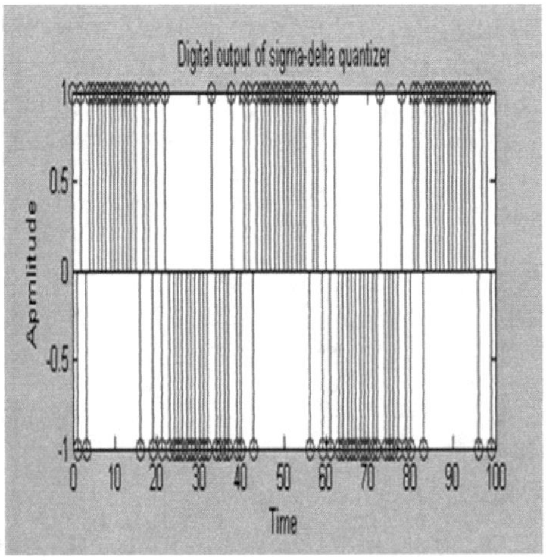

Figure 5 Input signal at 4 Kilo Hz

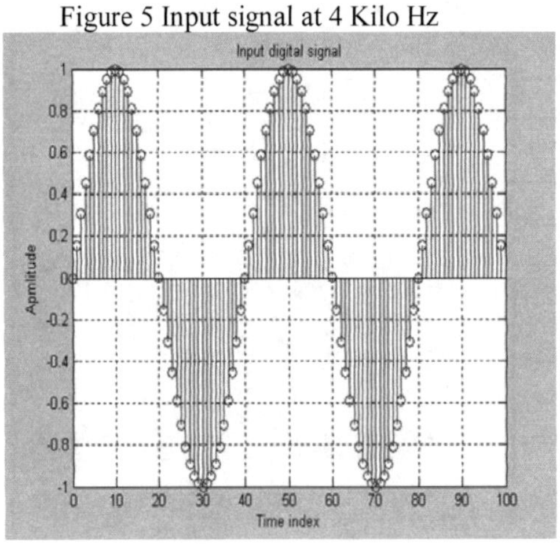

Figure 6 Output signal

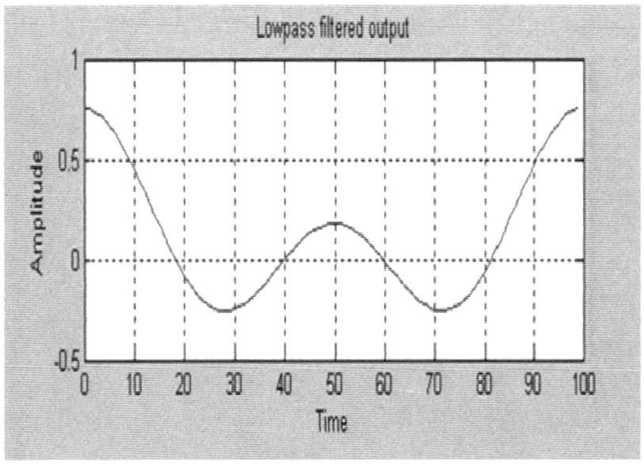

Figure 7 Low pass filtered output

5. VHDL SIMULATION RESULTS.

The Comb-half-band FIR-FIR channel plan and decoding FIR-FIR channel structure. The decoding half band FIR-FIR channel configuration utilizes less equipment contrasted with the decoding FIR-FIR channel structure. The force utilization is less utilizing decoding half band FIR-FIR channel configuration contrasted with the decoding FIR-FIR channel is shown in Table 2.

Table 2. Proposed decoding channel utilizing Comb-half-band-FIR

Cells used	Slices	Slice Flip Flops	4-LUT	logic	Shift registers
Comb-FIR- FIR	698	1574	743	658	165
Comb-half- band-FIR- FIR	635	1274	703	583	125

Table 3 depicts that the suggested decoding channel utilizing Comb-half-band-FIR-FIR. It utilizes minimal equipment and adds to equipment sparing and a force sparing contrasted with the decoding FIR-FIR design.

Table 3. Performance assessments of filter architectures

Decimation filter architectures	No of taps	Power consumption
5-stage Comb-FIR-FIR	**20**	**62mW**
5-stage Comb-half-band-FIR-FIR	**10**	**52 mW**

6. CONCLUSION

The decoding half-band FIR-FIR demolition channel is planned utilizing Matlab and checked for constant execution utilizing Simulink. The decoding channel is intended for 6-piece information input data bits. The channel comprises of 13 pieces, and the away from the pass band the signal got weak by a gain factor of - 66 dB. What's more, the appropriated number juggling multiplier is utilized for executing in Very High Speed Integrated Circuit Hardware Description Language. In particular, we look at the relative force utilization of two plans; the cell use for each structure is acquired utilizing the combination description. The suggested decoding channel design utilizes minimal equipment and adds to an equipment sparing contrasted with the decoding FIR-FIR engineering. Pulverization channels for sound applications use FIR channels on account of the straightforwardness with which direct stage can be accomplished. Be that as it may, direct stage over the whole band isn't frequently needed.

Reference

[1] Venugopal V and Abed, Khalid.H. ''Design and implementation of a decimation filter for hearing aid applications''. *Nerurkar*, S.B.8-10 April 2005 Page(s):111 –115

[2] John G. Proakis, Dimitris G.Monalikes, and D.Sharma. "Digital Signal Processing principles, *Algorithms, and Applications"*, Third Edition.

[3] P. M. Aziz, H. V. Sorenson and J. Van der Spiegel, "An overview of Sigma Delta. Converters," *IEEE Signal Processing Magazine*, vol.13, pp. 61-84, January 1996.

[4] Steven W. Smith, *The Scientist and Engineer's Guide to ''Digital Signal Processing''*, Chapter 22, Second edition, 1999.

[5] D. George, W. Sjursen, J. R. Hochschild, et al"A 1.1-V .270-μA mixed-signal hearing aid chip," *IEEE Journal of Solid-State Circuits*, vol. 37, no.12, December 2002.

[6] J. C. Candy, "Decimation for sigma-delta modulation," *IEEE Trans. Commun.*, vol. COM-34, pp. 72–76, Jan. 1986.

[7] R. E. Crochiere and L. R. Rabiner, "*Multirate Digital Signal Processing"* Prentice- Hall, 1983.

[8] L. Naviner, J.F.Naviner, "*Design and Prototyping of a ∑Δ Decimator Filter for DECT .Standard*", Proceedings of 43rd IEEE Midwest symposium on Circuits and Systems, 2000.

[9] V. Oppenheim and R. W. Schafer, '*Discrete-Time Signal Processing*. Prentice-Hall', 1989.

[10] Tormod Njolstad and Tor A. Ramstad, "*Low-power Narrow Transition-band FIR Filters*' Norwegian University of science and Technology, dept of physical electronics, 7034, trondheim, Norway.

[11] Peter Y. K. Cheung, Eric S.K.See "*A comparison OF Decimation Filter Architectures for Sigma-Delta A/D Converters*" Department of Electrical Engineering, Imperil college of science, Technology and Medicine, Londan, SW7 2BT, England.

[12] Alan N. Willson, Jr., Fellow, and H. J. Orchard, Life Fellow, "A Design Method for Half-Band FIR Filters" IEEE . vol. 45, No. 1, January 1999.

[13] Rashid Ashid Ansari "*Satisfying the Haar Condition in Half band FIR Filter*'' IEEE. vol *36*. No I . January 1988.

IOP Conf. Series: Materials Science and Engineering 1045 (2021) 012030 doi:10.1088/1757-899X/1045/1/012030

Simulation and experimental validation of hysteresis current control technique for speed control of brushless DC motor

Hetal Patel[1,3*] **and Hina Chandwani**[2]

[1]Electrical Engineering Department, C.K.Pithawala College of Engineering and Technology, Surat-395007, Gujarat, India.
[2]Electrical Engineering Department, Faculty of Technology and Engineering, The Maharaja Sayajirao University, Vadodara-390001, Gujarat, India
[3*]hetal_ramjiwala@yahoo.co.in

Abstract. Brushless DC Motors (BLDC) are finding wide applications in household appliances and electrical vehicles. Microcontroller are widely used for controlling the BLDC motors however the realization of BLDC drive requires expertise in microcontroller programming. In this paper modelling, simulation and easy hardware implementation of BLDC motor having shaft encoder using STM32F4 series ARM Cortex-M4 microcontroller is demonstrated. A simple method of controlling the BLDC motor in closed loop, using a hysteresis current controller is implemented. The hall sensors are used to find the exact rotor position for the electronic commutation and the encoder is used to measure actual motor speed. The closed loop operation consists of an inner loop which is hysteresis current controller. The gate pulses for the three phase inverter bridge are produced by comparing the actual currents with the reference currents. The outer speed loop consists of a PI controller which compares the actual speed with the reference speed to obtain smooth speed control under varied load condition. The complete hardware circuits and MATLAB/Simulink environment blocks for programming are discussed in detail. The hardware results demonstrate the easy and accurate realisation of closed loop speed control of BLDC Motor.

1. Introduction

In countries like India, with growing demand in new machine technologies, the permanent magnet motors are gaining popularity. Generally the motors used in industries are three phase squirrel cage induction motors because of rugged construction. Nowadays they are replaced by permanent magnet brushless dc motor and permanent magnet synchronous motor due to the advancement in control techniques, reduced size of motor, reduced copper losses as well as less maintenance[1-2]. These motors finds their place in many applications such as domestic, aircraft, automotive, medical, textile industry, spacecrafts etc.as they have high power density and good dynamic performance compared to DC motors and 3-phase induction motors[1-2].

The stator winding inductance prevents the phase current commutation instantly with the change in rotor position resulting in a spike in non-conducting phase current with only two conducting phases in BLDC motor. This is the major cause of producing torque ripple in BLDC motor drive along with the inverter DC link voltage. The performance and dynamic response of the motor can be improved by incorporating current control, voltage control, flux control or torque control methods. To predict and monitor the dynamic performance of the BLDC motor drive [3] proposed an advanced simulation model using hysteresis current control. [4] has implemented the closed loop control of BLDC motor using a PID controller and tested the performance of the drive before and after addition of the load.

[5] has developed a low cost control for BLDC motor drive with one current sensor and a PI controller to regulate the speed using Arduino Mega controller.

Today most of the techniques applied for controlling the speed of BLDC motor use high speed computers, fast ARM controllers, dedicated controllers and digital signal processors, which requires immense coding ability. To control the motor, complicated programming and skilled programmers are required. In some industrial applications constant speed drives are required and when load on motor is variable closed loop operation is compulsory. The mathematical modeling of BLDC motor is accomplished using MATLAB/SIMULINK. The modeled motor is used for performing speed control of BLDC motor using hysteresis current control technique and a PI controller in MATLAB/SIMULINK so that a comparison can be made with hardware implementation. Further hardware implementation is executed using a simple technique to control the speed of BLDC motor using 32 bit arm core ARM controller by interfacing it with MATLAB/SIMULINK WAIJUNG blockset. This method does not require complex coding. WAIJUNG blockset is user friendly similar to simulink blockset.

The closed loop control of BLDC motor with trapezoidal back emf with any two phases carrying current at any instant is developed in this paper sounds to be simple and user friendly. This method does not require any complex transformations like Clarke and Park which is necessary for DTC and FOC of theBLDCand PMSM. This paper discusses the detailed modelling of closed loop controlled BLDC motor drive incorporated with a PI controller and hysteresis controller using MATLAB/SIMULINK platform in section 2. The simulation results helps to validate theoretical concept. It also provides a platform to integrate the hardware implementation with the software implementation to make the drive operation simple without using any complicated programming. The detailed hardware setup is discussed in section 3. The simulation and hardware results are discussed in section 4 and 5.

2. BLDC motor modelling and Simulation

Please The BLDC motor represented by its equivalent circuit with three phase inverter circuit acting as electronic commutator for the motor is as shown in Figure1.

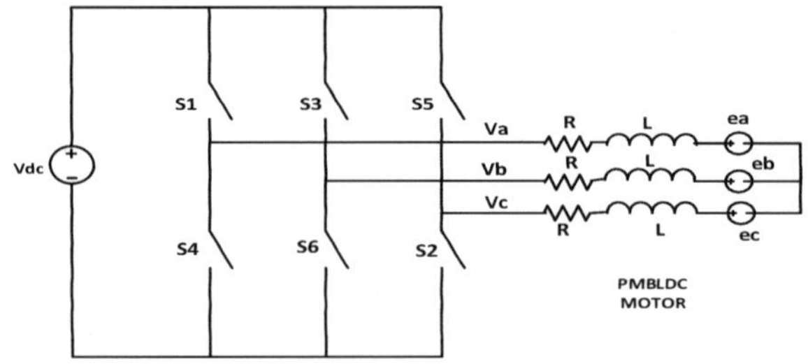

Figure 1. Equivalent circuit of BLDC motor with 3-phase Inverter as electronic commutator

Let,

V_a , V_b , V_c = Stator phase voltages
i_a , i_b , i_c = Stator phase currents
e_a , e_b , e_c = trapezoidal phase back emf
R= stator resistance per phase
L = stator inductance per phase,
M = mutual inductance

From Figure1, the three phase stator winding can be expressed in terms of motor electrical constants for a balanced 3-Φ system (ia+ib+ic=0) is given by (1) as

$$\begin{bmatrix} V_a \\ V_b \\ V_c \end{bmatrix} = R\begin{bmatrix} 1 & 0 & 0 \\ 0 & 1 & 0 \\ 0 & 0 & 1 \end{bmatrix}\begin{bmatrix} i_a \\ i_b \\ i_c \end{bmatrix} + \begin{bmatrix} L-M & 0 & 0 \\ 0 & L-M & 0 \\ 0 & 0 & L-M \end{bmatrix}\frac{d}{dt}\begin{bmatrix} i_a \\ i_b \\ i_c \end{bmatrix} + \begin{bmatrix} e_a \\ e_b \\ e_c \end{bmatrix} \tag{1}$$

From (1) three phase currents I_a, I_b and I_c are generated in Simulink. Here line to line voltages and back emf are considered for generation of stator phase currents.

The back emf of BLDC motor is trapezoidal in nature and is a functionofrotorpositionwithamagnitude E as expressed in(2)

$$\begin{bmatrix} e_a \\ e_b \\ e_c \end{bmatrix} = \omega_m \lambda_s \begin{bmatrix} f_{as}(\theta_r) \\ f_{bs}(\theta_r) \\ f_{cs}(\theta_r) \end{bmatrix} \tag{2}$$

Where,

ω_m = rotor speed in radians per sec.

λ_s = flux linkage

θ_r = the rotor position in radian

The electromagnetic torque is expressed asin (3)

$$T_e = (e_a i_a + e_b i_b + e_c i_c)\frac{1}{\omega_m}(N.m) \tag{3}$$

Equation (4) is expressed in terms of electromagnetic torque, load torque , inertia J, friction coefficient B widely known as motion Equation.

$$T_e - T_l = J\frac{d\omega_m}{dt} + B\omega_m \tag{4}$$

Electrical rotor speed is obtained by differentiating the rotor positionas given by (5)

$$\frac{d\theta_r}{dt} = \frac{P}{2}\omega_m \tag{5}$$

Based on the above Equations a simulink model describing the BLDC motor as well as the drive operation is incorporated in MATLAB/Simulink environment.

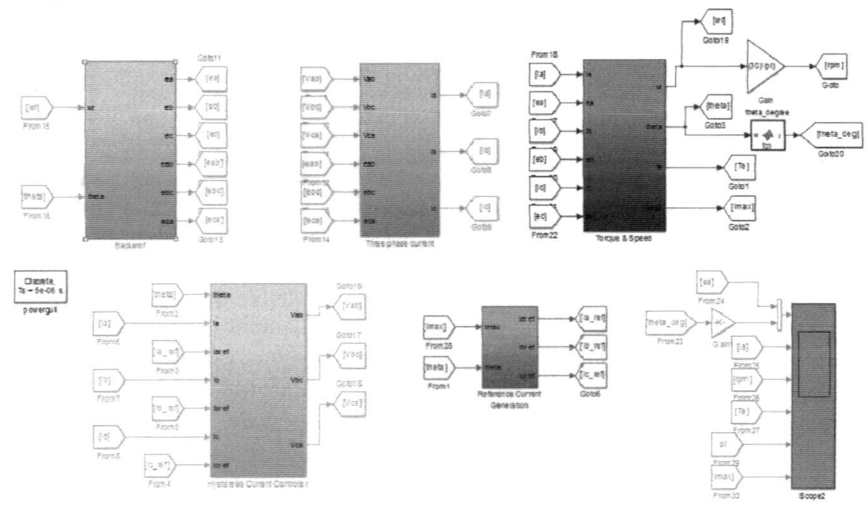

Figure 2. Simulink model of a hysteresis current controlled BLDC drive

Figure 2 shows the overall simulink model of a hysteresis current controlled BLDC motor model consisting of a back emf generation block, three phase current generation block, torque and speed control using a PI controller, reference current generation block and generation of switching logic using a hysteresis current controller block.

Equation [1] is used for BLDC motor model represented in Figure 1.The trapezoidal back emf is generated using Equation (2) where $f_{as}(\theta_r)$ is function of rotor position and is formulated using Equation (6) as shown in Figure3.

$$e_a = \begin{cases} \left(\frac{6E}{\Pi}\right)\theta r & \left(0<\theta r<\frac{\Pi}{6}\right) \\ E & \left(\frac{\Pi}{6}<\theta r<\frac{5\Pi}{6}\right) \\ -\left(\frac{6E}{\Pi}\right)\theta r+6E & \left(\frac{5\Pi}{6}<\theta r<\frac{7\Pi}{6}\right) \\ -E & \left(\frac{7\Pi}{6}<\theta r<\frac{11\Pi}{6}\right) \\ \left(\frac{6E}{\Pi}\right)\theta r-12E & \left(\frac{11\Pi}{6}<\theta r<2\Pi\right) \end{cases} \qquad (6)$$

The line voltages and back emf are used to generate the three phase currents as only two phases are conducting from Equation (1) as shown in Figure 4.

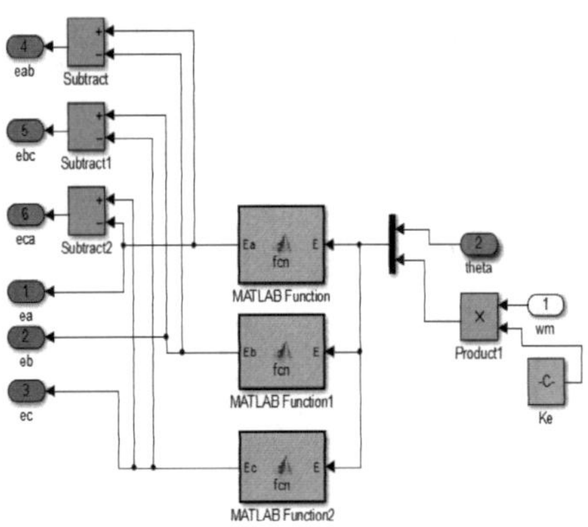

Figure 3. Back emf generation

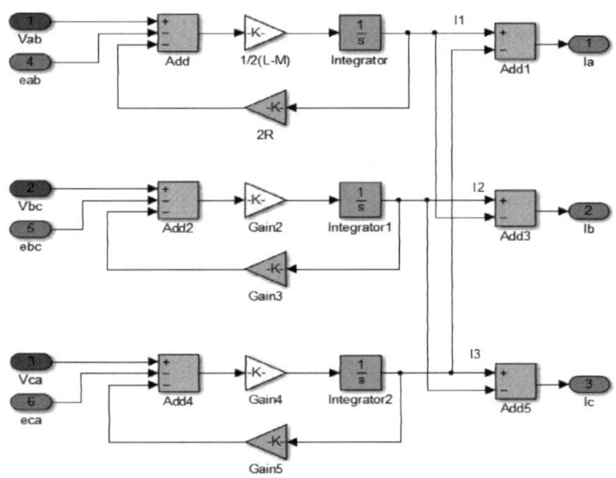

Figure 4. Three phase current generation block

The electromagnetic torque is a function of back emf, current and speed as represented in Equation (3). The electrical rotor position is obtained by integrating the motor speed. is as shown in Figure 5.A PI controller block is used to process the error produced by comparison of the set speed with the actual speed as shown in Figure5. The output of which is torque reference which when divided by torque constant Kt gives the maximum current I* which is used to generate three reference currents.Tuning of a PI controller is generally done using a trial and error method. A simple way to tune the drive is discussed using the mechanical and electrical time constants as given in Equations (6) and (7). This time constants depends on the motor parameters. Using this time constatnt a transfer function is obtained using Equation (8).

$$\tau_m = \frac{J3R}{K_eK_t} \qquad ()$$

$$\tau_e = \frac{L}{3R} \tag{7}$$

$$G_s = \frac{\frac{1}{Ke}}{\tau_m.\tau_e.S^2 + \tau_m.S + 1} \tag{8}$$

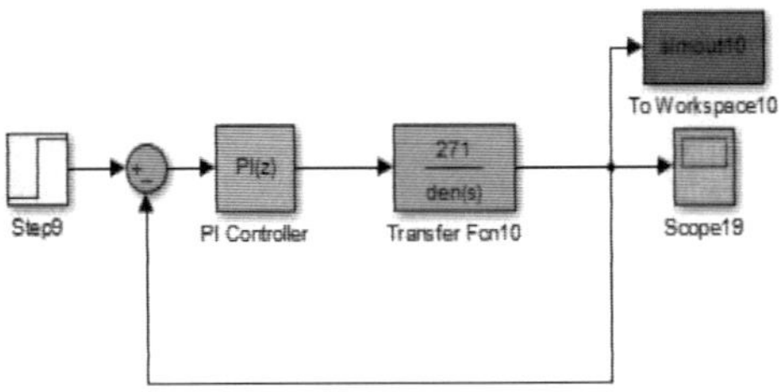

Figure 5. Transfer function based BLDC motor modelling for PI controller tuning.

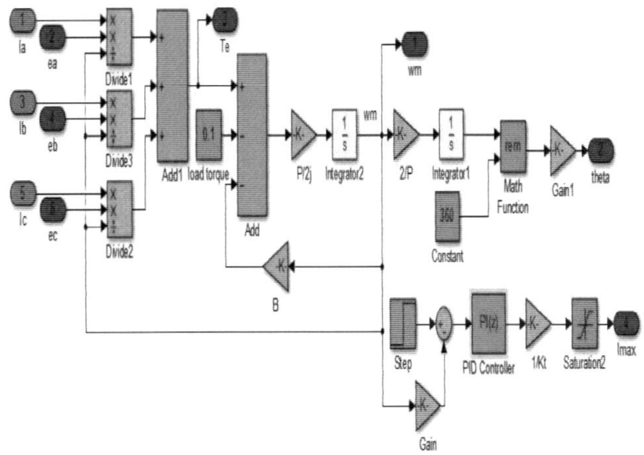

Figure 6. Torque and speed control block

This BLDC transfer function is obtained using Equation (8) is utilized as shown in Figure 5 to tune the PI controller. The tuned block is then placed in Figure 6 which process the error signal from which maximum torque is obtained. To generate the maximum current the maximum torque is divided by torque constant Kt as

Using the knowledge of the maximum current I* obtained from Figure 6, three reference currents are generated as shown in Figure 7 based on the rotor position as given in Table 1.

Table 1. Generation of three reference currents

Rotor Position	Three Reference Currents		
Θ(degree)	Iaref	Ibref	Icref
0-60	-I*	+ I*	-
60-120	-I*	-	+ I*
120-180	-	-I*	+ I*
180-240	+ I*	-I*	-
240-300	+ I*	-	-I*
300-360	-	+ I*	-I*

The reference currents are compared with the actual currents using a hysteresis current controller. A hysteresis controller is modeled with a narrow hysteresis band of ± 0.1. If the error is greater than the hysteresis band upper switch is ON and if error is less than the hysteresis band lower switch conducts to produce pulse width modulated gate pulses as shown in Figure 8 for the three-phase inverter shown in Figure 1.

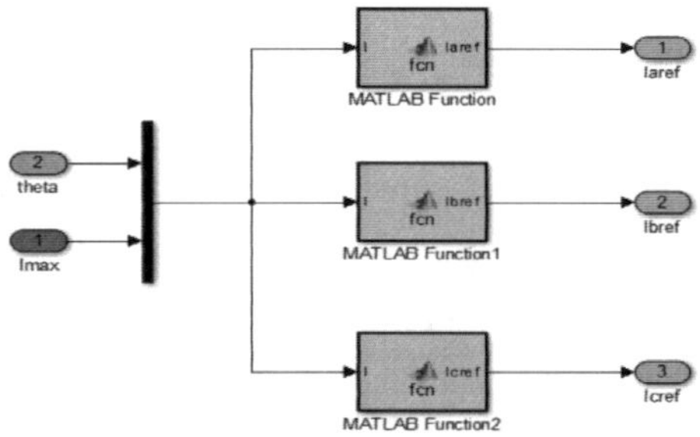

Figure 7. Generation of three reference currents

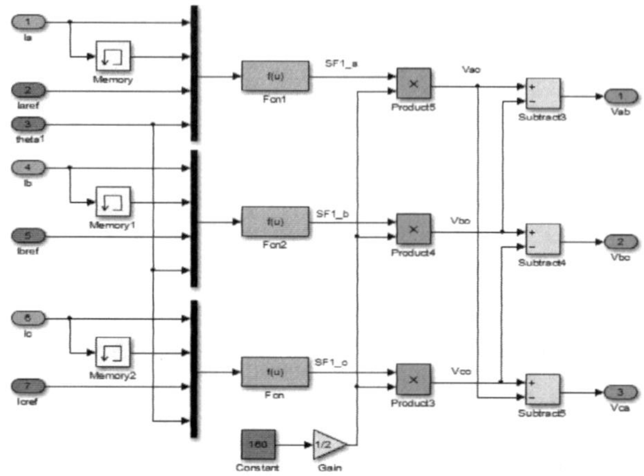

Figure 8. Modelling of a hysteresis current controller

3. System Implementation

Hardware implementation of closed loop control of BLDC motor is implemented as shown in Figure 9. The Hardware setup consists of a current sensor card, driver card, 3-phase inverter card acting as electronic commutator, BLDC motor with hall sensor/shaft encoder, STM32 discovery card.

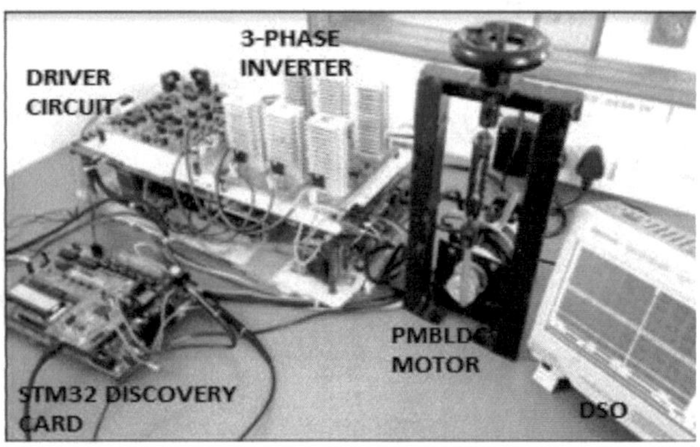

Figure 9. Hardware System with STM32 Discovery Card

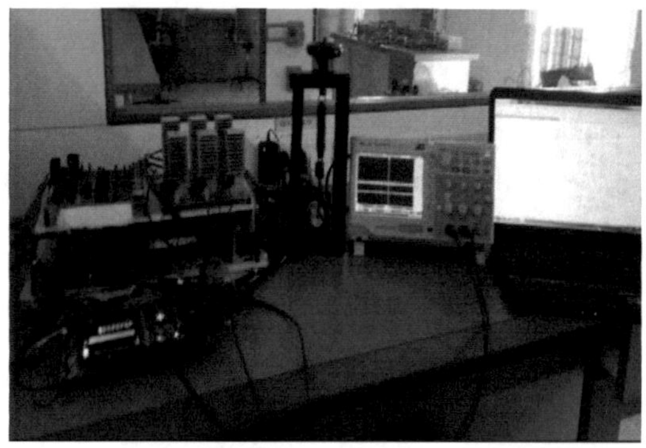

Figure 10. Interfacing Hardware setup with MATLAB/SIMULINK software

Figure 10 shows the interfacing of Hardware circuit with MATLAB/ SIMULINK WAIJUNG block set to generate gate pulses for the IGBT switches of the threephase bridge inverter circuit. BLDC motor with the parameters as given in Table 2 is used for experimentation.

For operating BLDCmotor the key factor is the acknowledgement of rotor position and energizing two phase windings. In sensored BLDC motor magnetic hall sensors are used for acknowledgment of correct rotor position. Here a shaft encoder is used for detecting the rotor position is used. In some motors hall sensors are available to provide rotor position. Generally 120^{0} hall sensing technique is used. Three different hall sensors are located at three different places in stator winding. When they are moved in bipolar magnetic field they generate TTL compatible output [6] that can be interfaced with logic circuit or controllers.

For every hall sequence there are two possible combinations of inverter switches that with one combination the motor will rotate in clockwise (forward) direction and for other it will rotate in counter clockwise(reverse) direction. The hall sequence and the respective switching combinations are discussed in brief in this paper. For changing the direction of rotation of motor the only thing required is the selection of proper switches at a particular hall sequence[7].

In BLDC motor with trapezoidal back emf, at any instant only two phases of the stator winding conducts. For closed loop operation of motor, inner current loop with low computational time than outer speed loop is used. A hysteresis controller acting as a current comparator is used to compare the actual motor current with the reference currents within a hysteresis band to produce six pulse width modulated signals for the six inverter switches. A PI controller is used to process the error produced by comparing the actual speed with the set speed[5]. The closed loop speed control of BLDC motor requires the following hardware as well as software implementation:

- BLDC motor set with Encoder/Hall sensors.
- IGBT Driver and Power Card
- STM32 Discovery Card
- ARM Cortex M4 - WJ - 32 Bit Kit
- Matlab 2014a
- MDK516a (Keil ARM IDE)
- STM32 ST-LINK Utility_v2.3.0
- WAIJUNG14_12a or higher

3.1 BLDC motor with Shaft Encoder

A permanent magnet brushless DC motor of 36 volt, with concentrated winding producing a trapezoidal back emf and quasi square wave currents with only two phases excited at a time is used for hardware implementation. A shaft encoder with resolution of 5000 PPR is mounted on the shaft. The speed of the motor is calculated from the output of the encoder using controller, Unlike the DC motor which have mechanical commutator BLDC motor requires electronic commutator. The switching of the electronic commutator depends on the rotor position. The rotor position required for producing the commutation logic for the inverter circuit is also obtained using encoder. The motor is provided with a loading arrangement which helps to study motor behavior with change in load.

Table 2. BLDC Motor Parameters

Parameters	Values
Voltage (V)	36 volt
Poles (P)	4
Rated Speed (N)	4000 rpm
Rated Torque (T)	0.32 Nm
Resistance per phase (R)	0.5 Ω
Inductance per phase (L-L)	1.65 mH
Moment of inertia (J)	$17.3*10^{-6}$ Kg-m^2
Torque Constant (Kt)	0.061 Nm/A

3.2 Operation of 3-Φ Inverter as Electronic Commutator

The output of the incremental shaft encoder is given to the QEI pins of the controller to collect the information of rotor position. There are six commutation states in one electrical cycle depending on the rotor position. To obtain correct switching combinations for the electronic commutator a simple method is discussed for motor with hall sensors. For example if the present hall sensor state is 101 and the possible switch combinations are 1&2, 2&3, 3&4, 4&5, 5&6, 6&1. Out of these six combinations only two combinations are correct. For that two combinations the motor will rotate once in forward direction and then in reverse direction. The switching pattern matches with that of 3-Φ, 120° conduction mode inverter. This can be obtained by a switch matrix at initial stage. For every commutation state or hall sequence there are two possible combinations of inverter switches that with one combination the motor will rotate in forward direction and for other it will rotate in reverse direction as shown in Table 3

Table 3. Switching combination for forward and reverse operations

Commutation States			**Switch Combinations**	
H_A	H_B	H_C	**Forward direction**	**Reverse direction**
1	0	1	S3, S4	S1, S6
1	0	0	S4, S5	S1, S2
1	1	0	S5, S6	S2, S3
0	1	0	S6, S1	S3, S4
0	1	1	S1, S2	S4, S5
0	0	1	S2, S3	S5, S6

3.3 Driver Circuit

The driver circuit is used for driving the IGBTs of Inverter card as well as to provide isolation between control circuit and power circuit. Here a Infineon made driver card is used for driving the six IGBT switches. Each switches are provided separate isolation from a coreless transformer to isolate their supply and ground terminals.

3.4 Current Sensor

Current sensors are used to measure the threephase stator current. For this current transformers are used. The measured current is fed to the ADC of STM32 discovery card. A unipolar signal is to be provided to the ADC whose operating range is from 0-3.3 volts.

3.5 STM32Discovery Card with WAIJUNG

STM32F407VGT6 microcontroller is a 32-bit ARM controller with FPU core[8]. The controller is featured with 1MB Flash memory for coding large data, 192-Kbyte RAM which helps in storing large data with LQFP100 package is used in hardware, which facilitates the closed loop control of drive. STM32 bit Arm controller has generic core which facilitates C coding. The software requirement is MATLAB 2014a which has a WAIJUNG block set. Installation of MDK516a (Keil ARM IDE) and STM32 ST-LINK Utility_v2.3.0 along with MATLAB 2014a to interface STM32Discovery card with WAIJUNG block set is required.

In this hardware implementation of BLDC drive, the hysteresis current controller and PI controller are modeled in MATLAB/SIMULINK WAIJUNG block set which generates C code automatically for STM32 ARM controller. For this the MATLAB model is interfaced with the microcontroller to dump the code generated by WAIJUNG block set.

3.6 Generation of Gate pulses for 3-phase Inverter Bridge

Generally closed loop operation of BLDC motor using 32 bit ARM Controller requires knowledge of assembly language or C coding. In this paper a simple technique is used in which no coding is required. The 32 it ARM controller is interfaced with the MATLAB/SIMULINK model. Programming of the microcontroller is done using WAIJUNG block set in MATLAB/SIMULINK library [9].

At first, target set up from the device configuration block is done as shown in Figure 11. The selection of ADC, DAC, timers etc. is done from the onchip peripherals readily available. From hardware modules the Character LCD display is selected to display set speed and actual speed. ADC is used to convert analog value in digital value. As ADC is 12 bit, hence 4095 count is available. This count can be calibrated in terms of speed, current, voltage, frequency, modulation index etc.. Maximum voltage to ADC is 3.3 volt. For an analog signal of 0 to 3.3-volt, ADC gives a 0 to 4095 count. Here three phase currents are measured by the CT. The value of current is calibrated as 5A (rms) to 3 Vpp. DAC converts digital signal to analog. Here it is used to convert angle theta in to analog value. Actual speed of motor is calculated from theta. The actual speed is compared with the set speed. A PI controller is used to process the error. The output of the PI controller is torque reference from which maximum current is obtained. This signal is used to generate three reference current based on rotor position as mentioned in Table 1.

This reference currents are compared with the actual measured currents within a narrow band of 0.05 using hysteresis controller to generate gate pulses. Advanced Timer-8 is used to generate the six gate pulses for the three-phase inverter. A dead band of 1-micro second is kept between the upper and lower switch of the same phase leg, to avoid shoot through. Execution time of the overall system as shown in Figure 11 is approximately 100 μs. WAIJUNG also facilitates the display of set speed and actual speed on the LCD display.

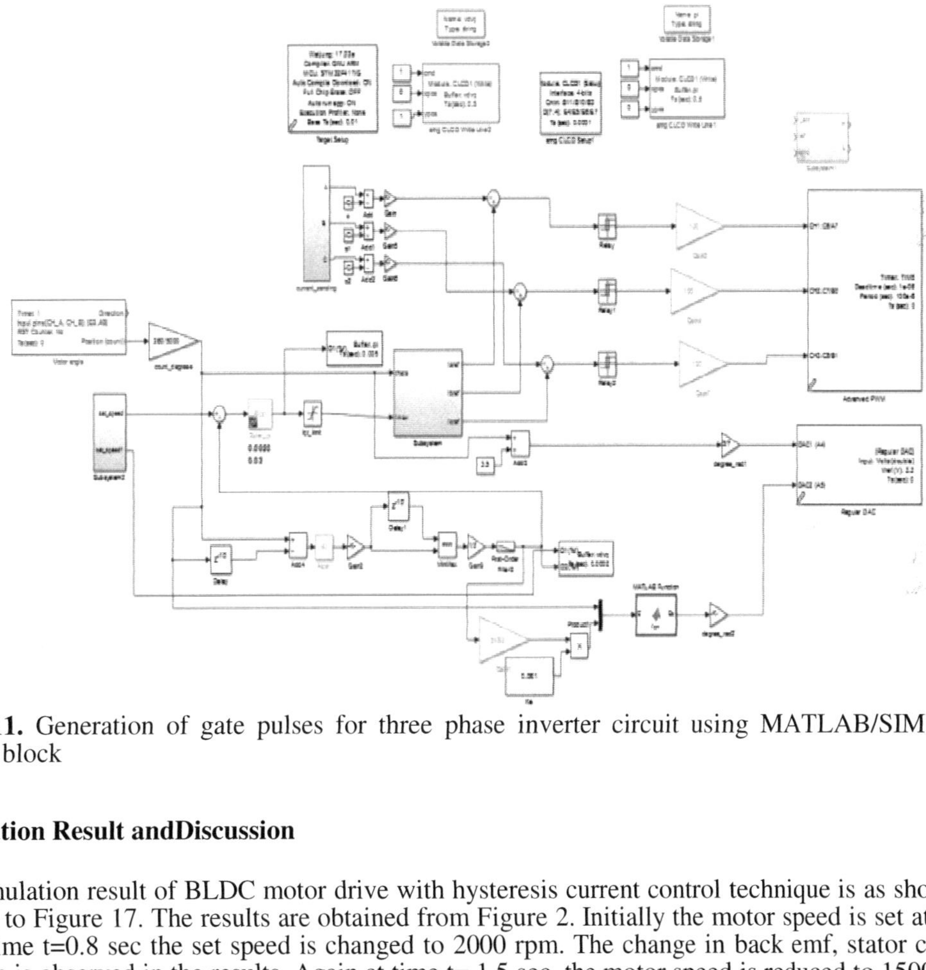

Figure 11. Generation of gate pulses for three phase inverter circuit using MATLAB/SIMULINK WAIJUNG block

4. Simulation Result andDiscussion

Simulation result of BLDC motor drive with hysteresis current control technique is as shown in Figure 12 to Figure 17. The results are obtained from Figure 2. Initially the motor speed is set at 1000 rpm. At time t=0.8 sec the set speed is changed to 2000 rpm. The change in back emf, stator current and torque is observed in the results. Again at time t= 1.5 sec, the motor speed is reduced to 1500 rpm. It is observed that the time taken by the controller to reach the set speed is very less and is not noticeable for higher speed. Figure 12 shows the variation in phase back emf with the zoomed results of phase back emf as shown in Figure 13. Figure 14 shows the variation in stator phase current with the zoomed stator phase current as shown in Figure 15. The variation in speed according to the set speed can be observed in Figure 16. The behavior of electromagnetic torque with respect to time with the applied speed is as shown in Figure 17.

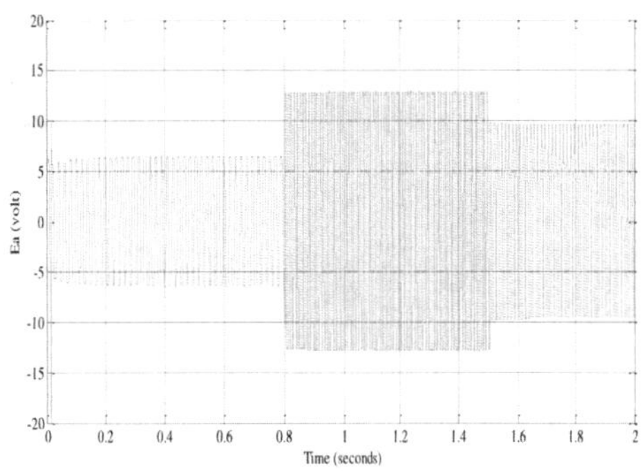

Figure 12. Variation in back emf with a speed change at 0.8 sec and 1.5 sec with a constant load of 0.1 Nm

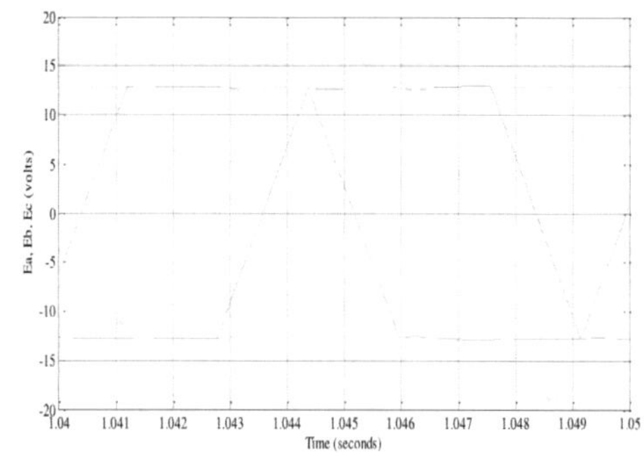

Figure 13. Zoomed view of three phase back emf at a speed of 2000 rpm

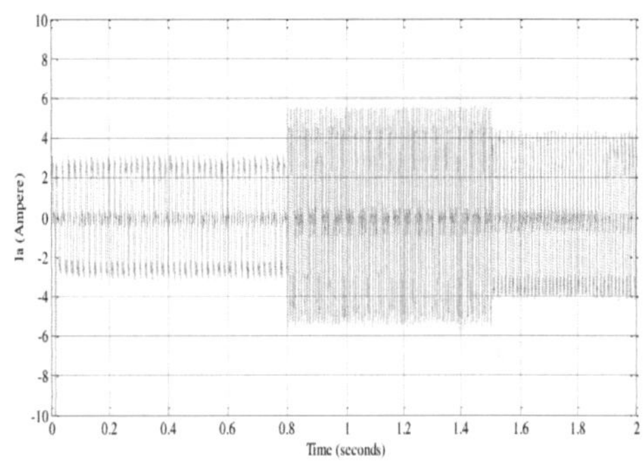

Figure 14. Variation in stator current with a speed change at 0.8 sec and 1.5 sec with a constant load of 0.1 Nm

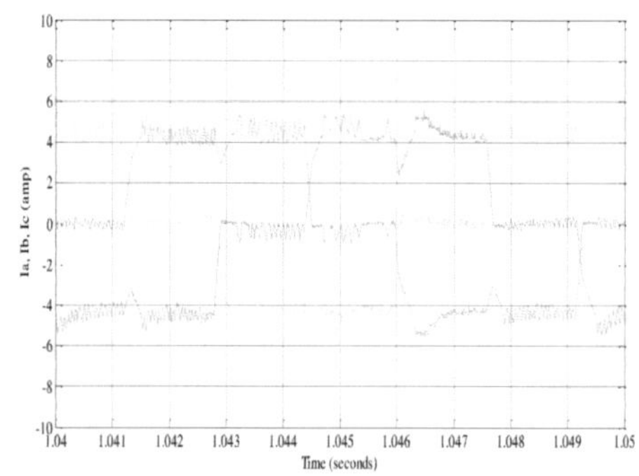

Figure 15. Zoomed view of three phase stator current at a speed of 2000 rpm

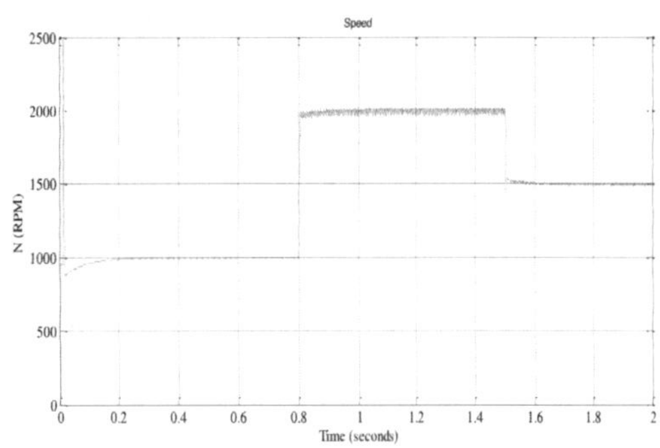

Figure 16. Applied speed change at 0.8 sec and 1.5 sec with a constant load of 0.1 Nm

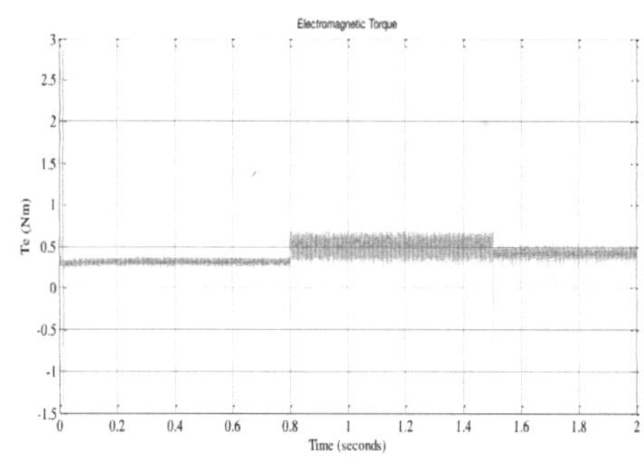

Figure 17. Variation in motor torque with a speed change at 0.8 sec and 1.5 sec with a constant load of 0.1 Nm

5. Experimental Results and Discussion

Experimental results prove the workability of the method in actual practice. It provides the actual behavior of the system and a platform to compare the simulation results with the actual results. Hardware results of closed loop control of BLDC motor are captured using DSO. The control scheme is tested on 36volt BLDC motor with motor parameters described in Table 3. Waveforms of back emf, stator current, rotor position theta as well as the variation of actual speed with the set speed can be seen in the following figures. It can be seen from Figure 18, that the motor back emf is trapezoidal in nature with quasi square wave currents. The variation of back emf and stator current with rotor position can be seen in Figure 19 and 20.

Precise tuning of the PI controller is done to allows the actual speed to vary according to the reference speed as shown in Figure 21. It can be seen that as the load vary the PI controller takes corrective actions to reduce the error and maintains the actual speed of motor equal to the set speed of 2000 rpm. A maximum loading of up to 2.5 ampere is given on the motor.

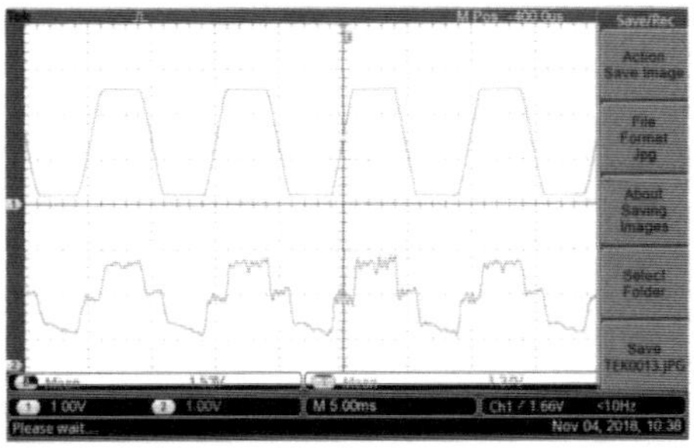

Figure 18. Back emf and Stator Current at a speed of 2000 rpm

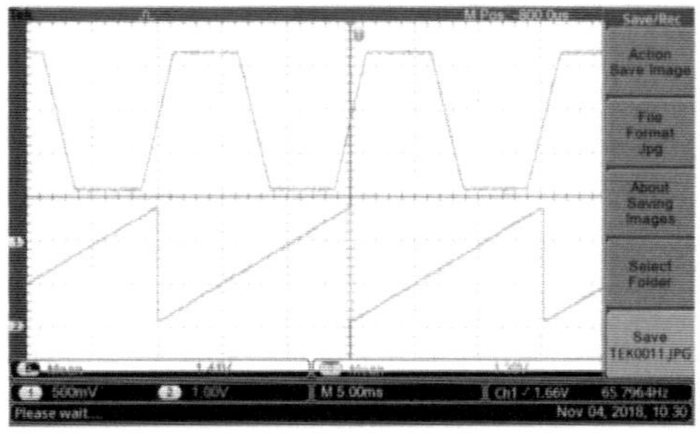

Figure 19. Back emf and Theta at a speed of 2000 rpm

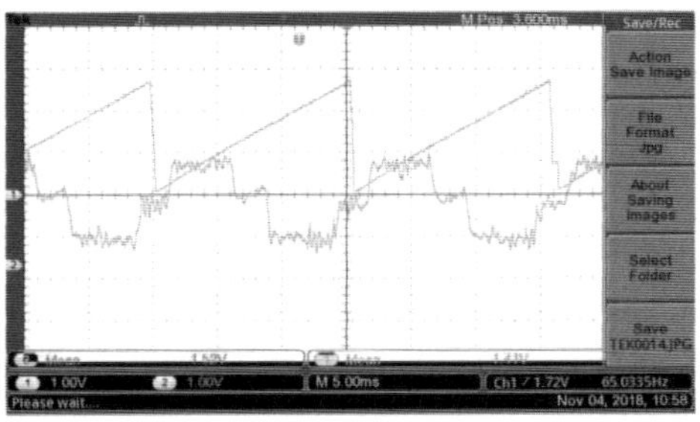

Figure 20. Theta and Stator Phase Current at a speed of 2000 rpm

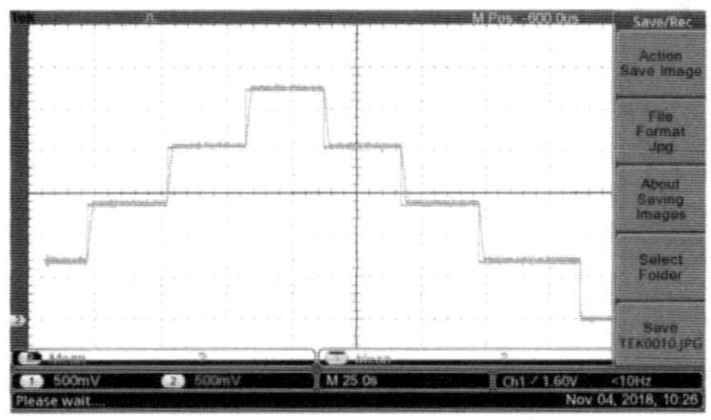

Figure 21. Variation in Actual speed with defined set speed

6. CONCLUSIONS

The popularity of the BLDC motor in the international drive market is increasing very fast especially in automobile engineering, textile industry and home appliances. This paper depicts a simple and effective speed controlling technique with MATLAB Waijung blockset which does not require 32bit coding or C programming. The BLDC motor is first modelled with actual motor parameters and then simulation of the same is carried with a hysteresis current control technique in MATLAB/SIMULINK. To validate the simulation results, hardware implementation is carried out on a 36volt motor by interfacing the hardware modules with MATLAB/SIMULINK simulation which is used to produce six gate pulses for the 3-phase inverter bridge which drives the motor using correct switching logic. The loading to the motor is given with the belt and pulley arrangement. It can be observed from hardware results that fine tuning of PI controller makes the motor to follow the set speed despite of variation in load.

References

[1] P. Pillay and R. Krishnan (1989). Modelling, simulation, and analysis of permanent-magnet motor drives, part II: the brushless DC motor drive, *IEEE Trans. on Industry Applications*, vol. 25, no. 2, pp. 274–279.

[2] T.J.E. Miller (1989). *Brushless Permanent Magnet & Reluctance Motor Drives*. Clarendon Press, Oxford, Vol.2, pp: 199.
[3] B. K. Lee and M. Ehsani (2003).*Advanced Simulation Model for Brushless DC Motor Drives*, Electric Power Components and Systems, 31:9, 841-868.
[4] Huazhang WAN (2012). *Design and Implementation of Brushless DC Motor Drive andControl System* International Workshop on Information and Electronics Engineering (IWIEE), Procedia Engineering 29 (2012) 2219 – 2224.
[5] Ghen Mohamed Dahbi, Said Doubabi, Ahmed Rachid (2020). *European Journal of Electrical Engineering*, vol. 22. No. 1, February, 2020, pp 63-69.
[6] Padmaraja Yedamale, *Brushless DC (BLDC) Motor Fundamentals*, Microchip, Application Note AN885
[7] B. K. Lee and M. Ehsani(2001). A simplified functional model for 3-phase voltage-source inverter using switching function concept," *IEEE Trans. on Industrial Electronics*, vol. 48, no. 2, pp. 309–321, April 2001.
[8] https://www.st.com
[9] http://waijung.aimagin.com

Simulation of exhaust gas energy recovery system using turbo-compounding for a motorcycle engine

Srivatsa S Joshi[1] and Aditya Deshpande[1]
[1]School of Mechanical Engineering, KLE Technological University, Hubballi, 580031, Karnataka, India

E-mail aditya@kletech.ac.in

Abstract. The project aims to recover the energy from the waste exhaust gas and utilize it for application such as battery charging of hybrid electric vehicles. The concept used for energy recovery is electric turbo-compounding which encompasses a turbine integrated with the exhaust manifold of an automobile. The exhaust gases are allowed to expand through the surface of blades of the turbine imparting the energy which can be seen by rotational action of impeller. This motion is captured by a motor generative unit which then produces electric power. The performance of the exhaust gas energy recovery system is simulated on a 390cc motorcycle engine using MATLAB Simulink. The engine was simulated using standard practical low loading and high loading drive cycles such as FTP72, FTP75 (city driving cycles) and US06 (highway driving cycle). Results from the simulation indicate a considerable amount of energy recovery from the exhaust gases of an engine under the above-mentioned driving cycles. Considering the amount of energy recovered, the next focus is to optimize the current system.

Keywords: Energy recovery, turbine, exhaust gas, motor-generative unit, MATLAB.

1. Introduction

The increasing concern on energy conservation, rising oil prices and focus on reducing emissions is leading to the development of energy-efficient technologies for automobiles at an accelerated pace. The conventional internal combustion engine normally has an efficiency between 25% and 35% under optimal conditions. This also signifies that up to 75% of energy from vehicle fuel combustion is lost as heat. This has led to the emergence of energy recovery systems which increase overall vehicle efficiency by recovering energy which otherwise would be lost as heat. For vehicle applications, designs that require minimal integration are gaining significant interest [1]. There is a wide variety of engineering concepts and designs in the market to maximize the potential of energy recovery from vehicle exhaust. These include Thermoelectric Generation (TEG), where the exhaust flow path has thermoelectric elements to generate electricity. Organic Rankine cycle systems are also being studied, and much research is being done to find a solution for waste heat recovery. In addition to these technologies, a turbo-generator system which generates electrical energy from the exhaust gas has proven as an attractive solution for energy recovery [2]. Another interesting method of energy recovery is turbo compounding which is a mechanism of attaching a turbocharger to the exhaust manifold of an automobile which contains two components, a turbine, which rotates by the action of the incoming flow of pressurized exhaust gas and a compressor connected to the turbine via a shaft, to extract the exceeding power from the turbine and assisting in compressing more airflow at the inlet thereby increasing the supply of air during high load request and improving the dynamic response of the engine [3]. There are two ways in which a turbo-compounding can produce an output, first in the form of mechanical energy by coupling a power turbine to the primary turbine, hence called mechanical turbo-compounding, second in the form of electric energy by coupling a motor-generator to the shaft of the primary turbine and hence this method is called electric turbo-compounding. In order to achieve the maximum power extraction from the exhaust gases, many researchers have worked on a low-pressure turbine placed downstream of the main turbocharger. One of the disadvantages of this architecture is that there is a back pressure to the engine that could cause a rise in pumping losses and result in higher fuel consumption.

The following report is mainly based on electric turbo-compounding for low powered gasoline engines where-in a turbine is attached to the exhaust manifold to capture the incoming flow of exhaust gases from a combustion chamber. The compressor is eliminated because, for the low powered engine, there is no need to boost the incoming air as the load on the engine will be relatively less. Based on Faraday's law of induction which states that "the total emf induced in a closed circuit is equal to the time rate of decrease of total magnetic flux linking the circuit", the shaft of the turbine is connected to a generator which converts mechanical energy to electrical energy. The pressurized exhaust gas produced as the result of combustion expands through the blades of the impeller inside a turbine causing a rotational motion. This motion captured by the generator via the turbine shaft starts producing electrical energy which can be stored in a battery or can be used for various applications.

2. Literature Survey

The aim is to simulate an exhaust gas energy recovery system which can be used to charge the auxiliary battery through the method of electric turbo-compounding which involves the integration of a motor generative unit with a turbocharged engine. One such method was proposed by Weilin Zhuge et al. [4] to optimize the electric turbo-compounding for a gasoline engine with a fixed geometry turbine (FGT). The system comprised of a turbo-generator placed parallel with the turbocharger. Throttle-1 and throttle-2 control the exhaust gas flow distribution between the turbocharger and turbo-generator. The position of FGT can be referred from figure 1. At low engine load, the throttle-1 would be partially or full closed to transfer the exhaust energy to the power turbine. At low engine speed, the throttle-2 would partly or fully close to guaranteeing enough energy flow through turbocharger. Their system was simulated under two drive cycles US06 and FTP75, producing electric power of 1.01 kW and 0.266 kW respectively.

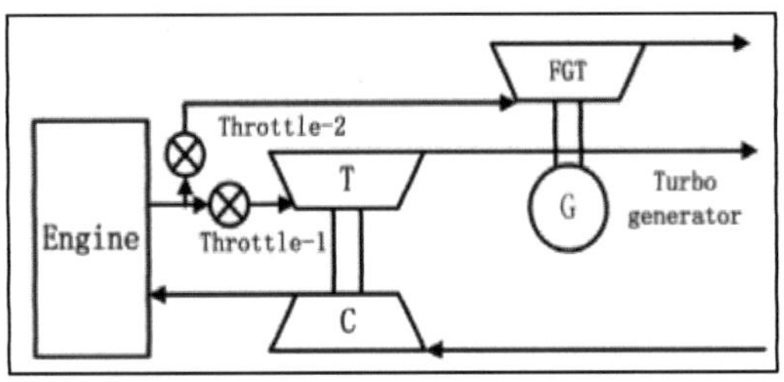

Figure 1. The positioning of the fixed geometry turbine with respect to engine exhaust [4].

M. Michon et al. [5] tested a system which comprised of a turbine, attached to the exhaust manifold of an engine. A switched reluctance motor was connected to the turbine to convert the mechanical energy to electrical energy, as shown in figure 2. To regulate the exhaust mass flow rate through the turbine, there is a wastegate. Also, the wastegate helps to protect the system in the event of a failure in the generator or the power supply to the throttle actuator. The complete turbo-generator system was tested on the cold air compressor test rig. For initial testing and verification, an air-compressor was used to run the turbine. The system was successfully tested up to 1.8kW electrical output power at 65000 rpm.

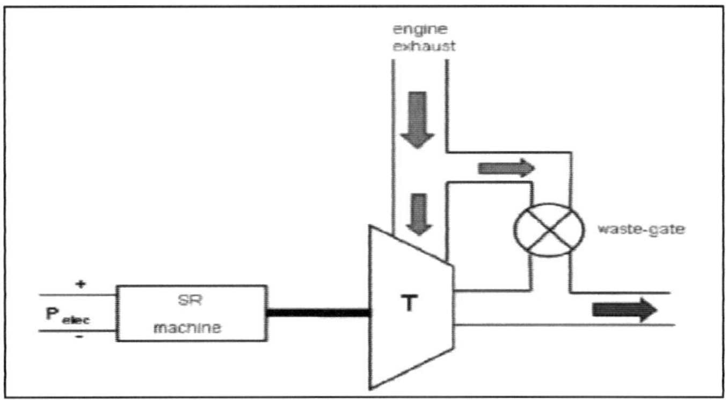

Figure 2. Switch reluctance motor coupled with turbine [5].

Chuang Yu et al. [6] proposed a thermoelectric waste heat energy recovery system for internal combustion engine automobiles, including gasoline vehicles and hybrid electric vehicles. They aimed to use the heat energy of automotive exhaust and convert it to electrical energy using a thermoelectric generator [TEG]. A power converter was used to regulate the charge to the battery using maximum power point tracking. Thermoelectricity was generated using the Seebeck effect, which states that electricity can be generated if there is a temperature difference between the two surfaces of a material. The flow of energy was as follows, an exhaust gas system, a heat exchanger, a TEG system, a power converter system and a battery pack. The maximum power obtained was 34.8W on a duty cycle at a temperature of 200 °C. They also suggested that there should be more than one TEG device connected in series to obtain more power. Figure 3 can be referred to know the positioning of a TEG device.

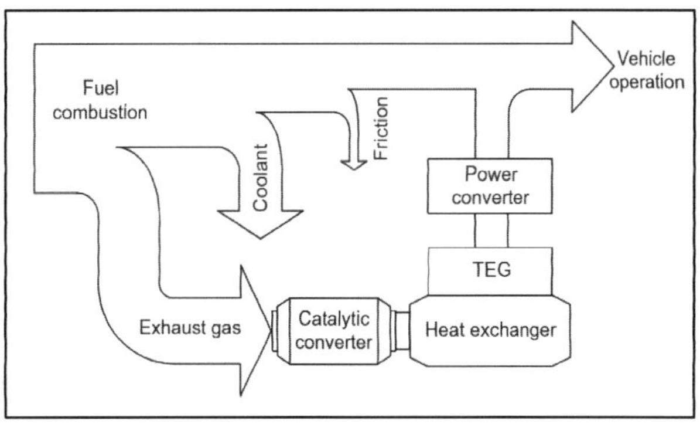

Figure 3. Schematics of a thermoelectric generator system [6].

Andre Haughton et al. [7] developed the Turbine-generator Integrated Gas Energy Recovery System (TIGERS) which consisted of an impeller, a volute and a separate back-plate, angular contact grease-filled bearings, which were in direct contact with exhaust gas which could reach up to a temperature of 900 °. A coolant system was also developed for the TIGERS because of the bearings were operating between the temperature range of 80 - 105 °C with a flow rate of up to 6 l/min and a minimal pressure drop. Figure 4 shows a bypass valve being introduced for both safety and performance control. The set-up was tested on a dynamometer, and the following observations were made, for a turbine speed of 45000 RPM, the device generated power of 1.2 kW and 0.6 kW at coolant temperatures of 80 °C and 105 °C respectively.

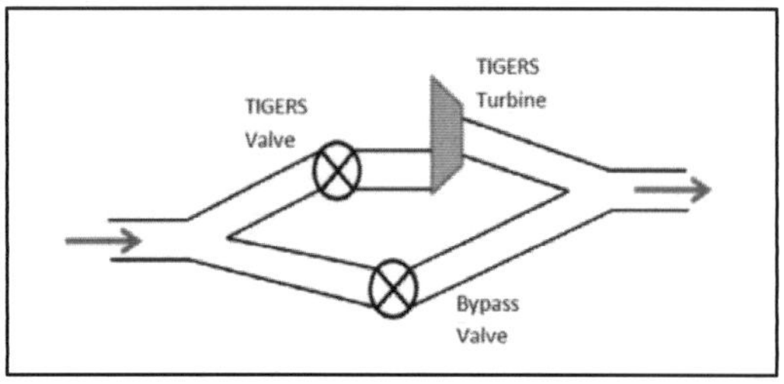

Figure 4. Arrangement of Tigers valve and bypass valve [7].

Albert Boretti [8] focused on the study of the performance of the Motor Generator Unit-Heat (MGU-H) system as used in a Formula 1 car applied to the turbocharger of a gasoline hybrid electric passenger car. The author considers the steady operation of a turbocharged engine with an MGU-H system and the performance parameters like total efficiency (the ratio of engine crankshaft power plus the turbocharger shaft power to the fuel flow power) and total brake mean effective pressure (BMEP). The MGU-H system consists of a motor-generator unit installed either on the side of the compressor or between the compressor and the turbine. Figure 5 demonstrates a series arrangement of MGU with a turbo-charger. When the turbine power exceeds the power needed by the compressor at high speeds, the MGU-H system recovers the energy by producing electric power to charge the battery. The MGU-H system increases the total efficiency and allows to control the turbocharger operating points. The use of an MGU-H during steady operation shows an improvement in total fuel conversion efficiency and an increase of 4–5% in total efficiency. This system also helps to increase the brake mean effective pressure (BMEP) by increasing the boost pressure. The system also helps reduce turbo lag considerably during accelerations. The author concludes that from turbocharging with MGU-H, both performances and fuel economy will increase and it will definitely benefit road Hybrid Electric Vehicles (HEV).

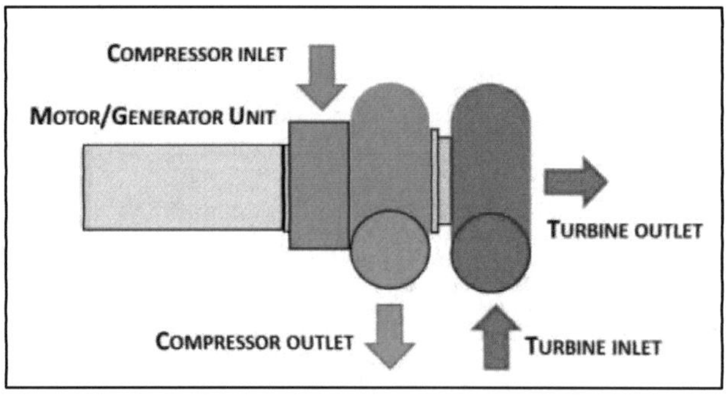

Figure 5. Coupling of the motor generative unit with turbocharger [8].

3. System Modelling

As discussed in the introduction, the modelling is done considering the physical environment through which vehicles manoeuvre. The input for the system is throttle response of the driver or engine rpm whichever is available, and the output is torque obtained for different values of input and its subsequent power generation. The overall system consists of two subsystems, first an IC engine, second a motor generative unit. The mechanical aspects of the system are modelled in the first subsystem, and the electrical aspects are modelled in the second subsystem. With reference to figure 6, a drive cycle is acting as an input parameter to the IC engine, which is a mechanical subsystem. The output of the first subsystem is fed as input to the second subsystem.

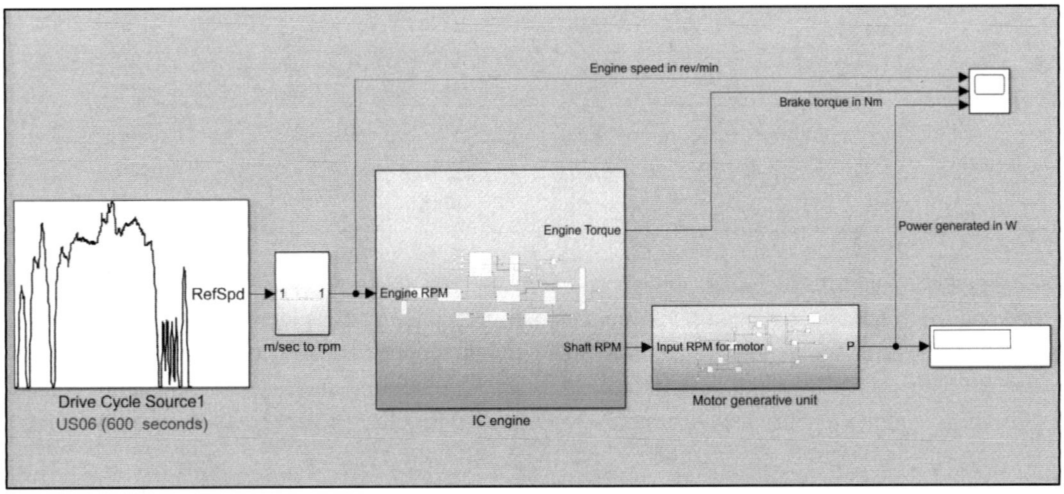

Figure 6. Overview of Electric Turbo-Compounding system.

3.1. Mechanical subsystem

This subsystem consists of automotive powertrain components. Literature survey helped to identify various blocks required, as from the real-time point of view and these blocked were connected via virtual mechanical linkages. The various blocks include a core gasoline powered engine, a throttle body, an intake manifold, an exhaust manifold, a turbocharger, an exhaust pipe, a muffler and a mechanical shaft. Figure 7 helps in understanding the system.

The SI core engine block implements a spark-ignition (SI) engine from intake to exhaust port. Different parameters govern the working of this block such as fuel injector pulse width, spark advance, intake cam-phase angle, exhaust cam-phase angle, ambient pressure, input engine speed, engine coolant temperature and air intake system. Most of the parameters were obtained when data was logged from a real engine. The input engine speed was given through city drive cycles such as FTP72, FTP75 and US06, which are the standard drive cycles. The different parameters on which the drive cycle was built is displayed in table 1.

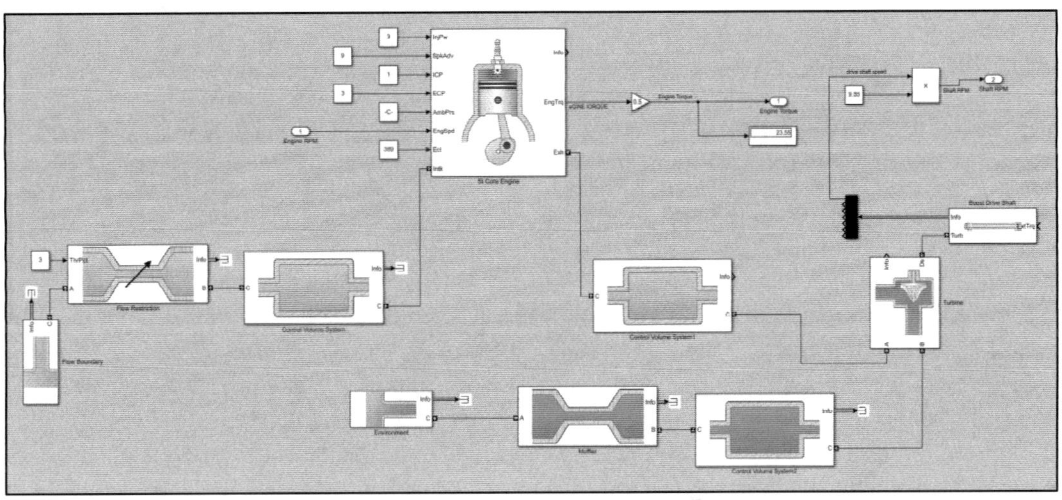

Figure 7. Modelling of Mechanical subsystem.

Table 1. Parameters of drive cycles.

	Total distance	Total duration	Average speed	Maximum speed
FTP-72	12.07 km	1369 s	31.50 kmph	91.20 kmph
FTP-75	17.77 km	1877 s	34.12 kmph	91.25 kmph
US06	12.80 km	596 s	77.90 kmph	129.2 kmph

The air intake system of the model consists of a throttle body to draw air from the environment to the intake manifold and then to engine. For initial condition, the throttle plate was 98% closed for idle engine speed. The obtained output parameter, the exhaust mass flow rate was given as input to further model of the turbine, allowing the exhaust gas to expand through the blades of an impeller to convert it into rotational motion. This kinetic energy is captured by the motor generative unit (electrical subsystem) and convert it to electric energy.

3.2. Electrical subsystem

This subsystem consists of electrical components assembled in an electric circuit to convert mechanical energy into electrical energy and is called motor generative unit. Figure 8 depicts the electrical circuit, which constitutes a motor generative unit. Electric blocks from simscape block set were used to model the motor generative unit. The different blocks involved here are an ideal angular velocity source, an ideal torque sensor, a DC motor, a load, voltage sensor, current sensor, mechanical and electrical rotational reference.

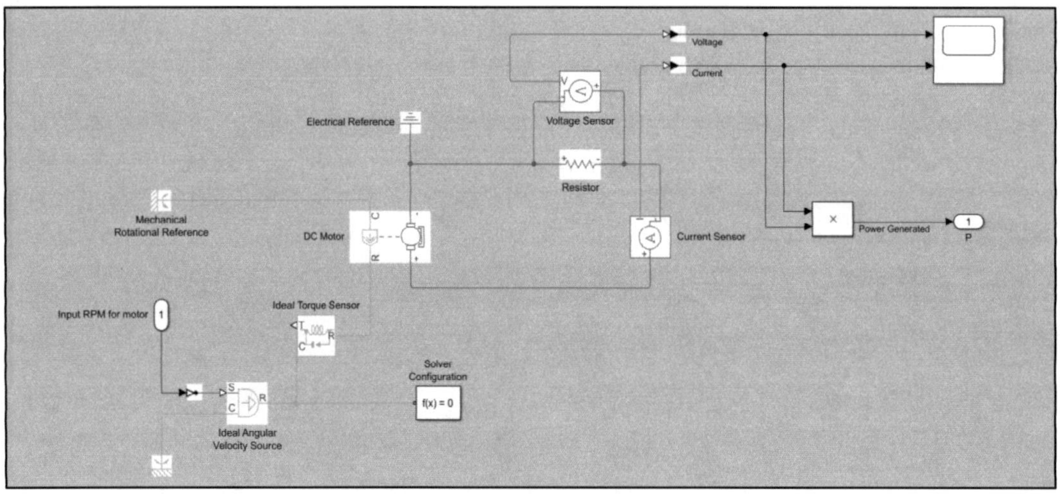

Figure 8. Modelling of the electrical subsystem.

An ideal source of angular velocity is the one that generates velocity differential at its terminals, proportional to the input physical signal. The output speed of the turbine is fed to angular velocity source to obtain its velocity differential. The ideal torque sensor is a device that converts a variable, passing through the sensor into a control signal proportional to the torque. The sensor is ideal because it does not account for inertia, friction, delays, energy consumption. The velocity differential obtained is converted to a variable through the ideal torque sensor. The variable obtained serves as an input to the DC motor shaft. The input signal starts rotating the rotor, which is eventually connected to the shaft. The carbon brush connected to commutators helps to capture this voltage. A load (in the current system a resistor of 100 ohms) connected across the terminals of the carbon brushes helps to calculate the amount of power generated by connecting a voltage sensor across the load and by measuring the current through the circuit using a current sensor. Integration of simscape block set with Simulink block is done through a PS-S converter.

4. Observation and Results

When the model was simulated against all the three drive cycles, the graphs were obtained and are analyzed. Two graphs of the vehicle's velocity and exhaust mass flow rate for city drive cycle FTP-75 is plotted below with respect to time.

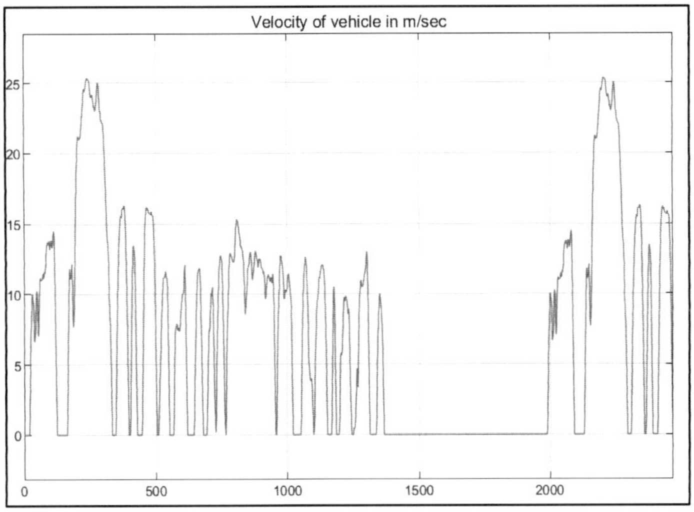

Figure 9. Graph of the velocity of the vehicle (m/s) vs time (s) for FTP-75

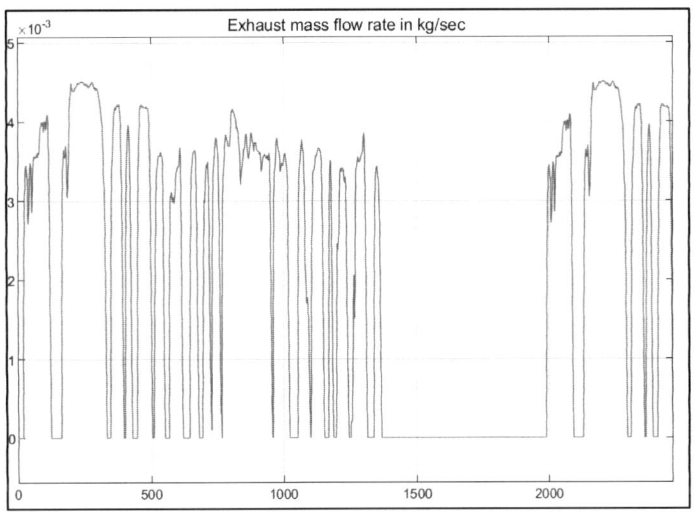

Figure 10. Graph of mass flow rate (kg/s) vs time (s) for FTP-75

From the figure 9, we can observe that the maximum speed attained by the vehicle was 26 m/s (94 kmph) at two instances, i.e. approximately at 210th second and 2210th second for which its subsequent mass flow rate was around 4.5e-3 kg/s (refer figure 10). For the rest of the duration, the speed was averaged at 10 m/s (36 kmph). The engine was kept idling between 1350th second and 1900th second to record the idle case behaviour of the mass flow rate.

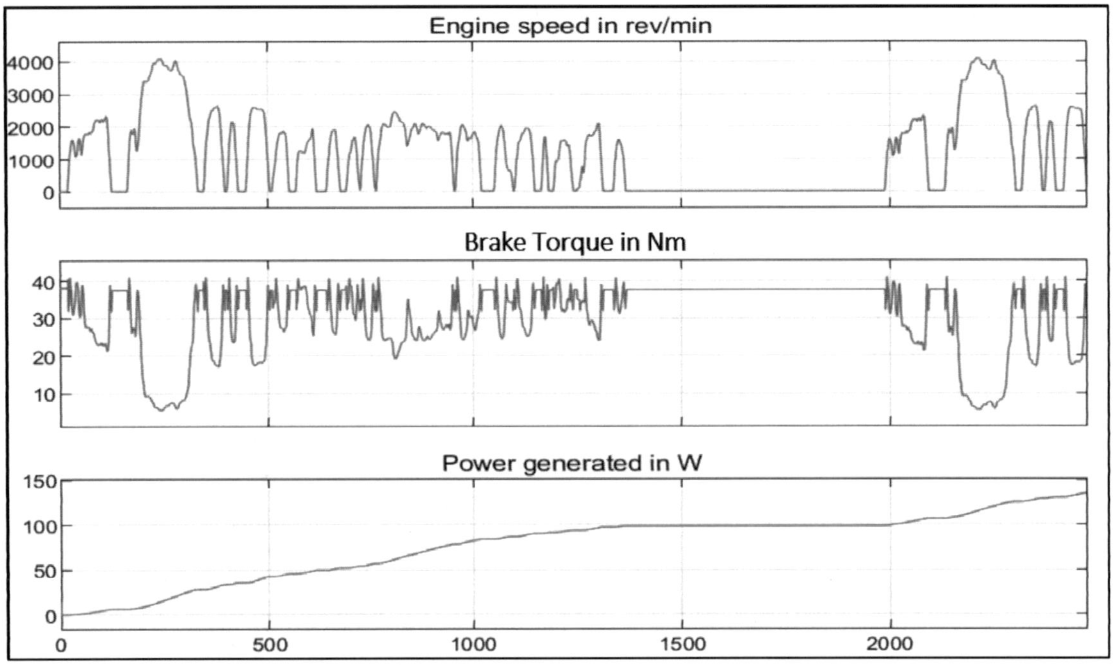

Figure 11. The cumulative variation of power for the FTP-75 drive cycle

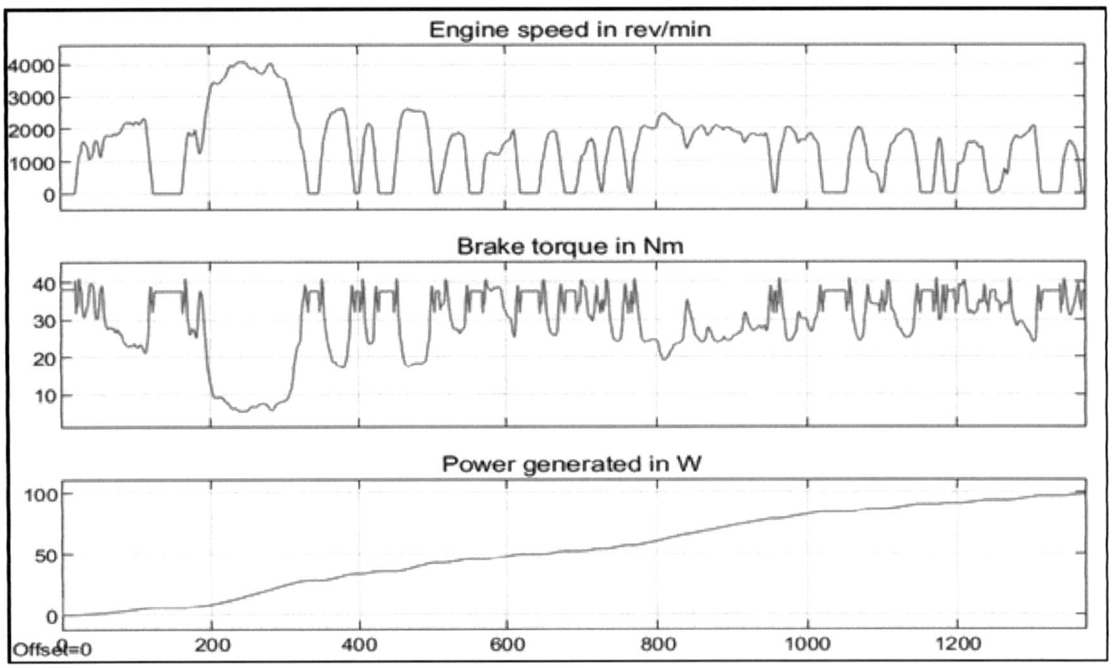

Figure 12. The cumulative variation of power for FTP-72 drive cycle.

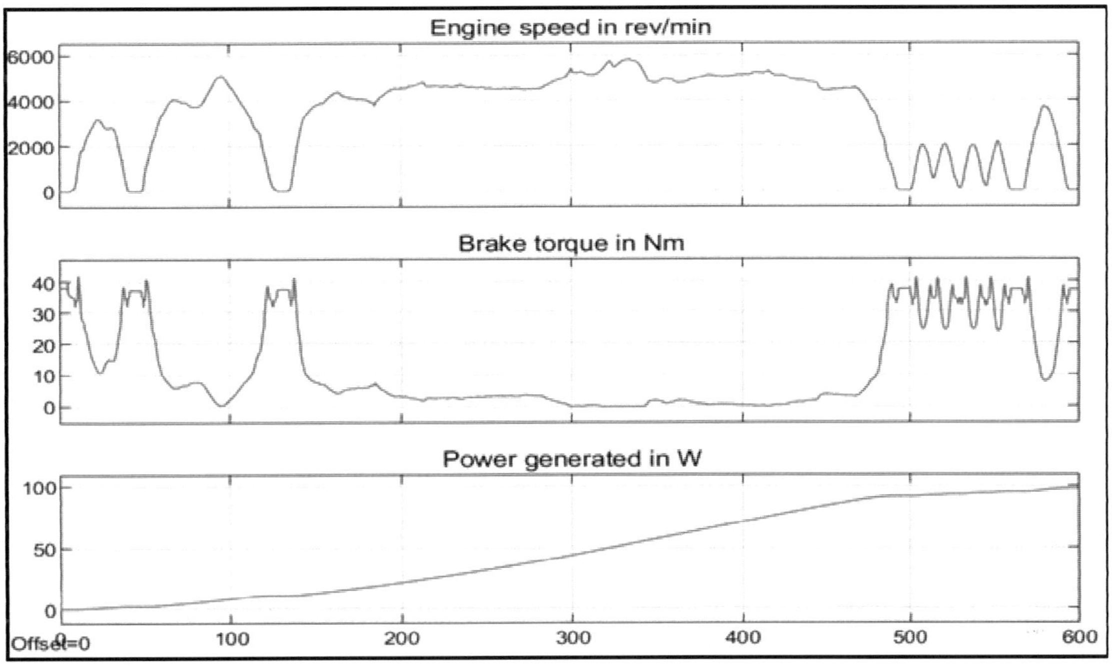

Figure 13. The cumulative variation of power for US06 drive cycle.

From figure 11, 12, 13, it can be observed that the power generated is the function of exhaust flow rate, which in turn is a function of engine speed. Also, the relation between engine speed and brake torque produced can be referred from the respective figures. The output values for different input drive cycles are tabulated in table 2.

Table 2. Comparison of various parameters for all the three drive cycles

	FTP-72	*FTP-75*	*US06*
Run time (s)	1372	2474	600
Average exhaust flow rate (kg/s)	3.5e-3	3.6e-3	4.4e-3
Average turbine shaft speed (rpm)	40,620	49,740	39,740
Exhaust power (W)	3576	3659	4901
Power generated (W)	98.65	134.7	97.47
Energy recovered (%)	2.75	3.68	1.98

The simulation was conducted in MATLAB Simulink for a vehicle of engine displacement 0.000371 m3 powered 30.4 kW at 9600 rpm with variable throttle response from different city drive cycles to generate power. The system was able to generate a maximum of 134.7 W of energy for drive cycle FTP-75 which accounts for 3.68% of total exhaust energy generated during combustion for a drive time of 42 minutes. Also, the other two drive cycles result show a positive trend in recovering energy, i.e. 2.75%

recovery for FTP-72 and 1.98% recovery for US06, respectively. The turbine shaft attained a maximum speed of 86,000 rpm. Therefore, the speed of the shaft would appear to be less than the maximum value when compared with a real-time application on the same engine due to the inertia factor.

5. Conclusion

The results from the simulation were promising to say that the electric turbo-compounding concept would greatly help in recovering the energy from the wasted exhaust gas. There was also an increase in the overall efficiency of the engine by 0.4%. The system can, not just be integrated with gasoline powered engine but also with hybrid commercial vehicles.

6. References

[1] F. Zhou, E. Dede, and S. Joshi, "Application of Rankine Cycle to Passenger Vehicle Waste Heat Recovery - A Review," *SAE Int. J. Mater. Manf.* **9(2)**; (2016). doi:10.4271/2016-01-0178.

[2] Quinn R. et al.: "*TIGERS exhaust gas energy recovery using an integrated turbo-generator system*", 2007 JUMV Conference paper.

[3] Cooper, A., Bassett, M., Hall, J., Harrington, "*HyPACE - Hybrid Petrol Advanced Combustion Engine" - Advanced Boosting System for Extended Stoichiometric Operation and Improved Dynamic Response*," SAE Technical Paper 2019-01-0325; (2019). doi:10.4271/2019-01-0325

[4] Zhuge, W., Huang, L., Wei, W., Zhang, Y. et al., "*Optimization of an Electric Turbo Compounding System for Gasoline Engine Exhaust Energy Recovery*," SAE Technical Paper 2011-01-0377, 2011.

[5] Michon, Melanie & Calverley, S.D. & Clark, R.E. & Howe, D. & Chambers, J.D.A. & Sykes, P.A. & Dickinson, P.G. & McClelland, Mike & Johnstone, G. & Quinn, R. & Morris, G. (2007). *Modelling and Testing of a Turbo-generator System for Exhaust Gas Energy Recovery.* 544 - 550. 10.1109/VPPC.2007.4544184.

[6] Chuang Yu, K.T. Chau, "Thermoelectric automotive waste heat energy recovery using maximum power point tracking." *Energy Conversion and Management*, Volume 50, Issue 6,2009.

[7] Haughton, A. and Dickinson, A., "*Development of an Exhaust Driven Turbine-Generator Integrated Gas Energy Recovery System* (TIGERS®)," SAE Technical Paper 2014-01-1873, 2014.

[8] Boretti, Albert. (2017). F1 style MGU-H applied to the turbocharger of a gasoline hybrid electric passenger car. *Nonlinear Engineering*. 6. 10.1515/nleng-2016-0069.

The influence of electromagnetic waves on conductivity tensor in the presence of laser field in quantum wells with parabolic potential for the case of electrons-acoustic phononscattering

Hoang Van Ngoc

Institute of Applied Technology, Thu Dau Mot University, No 6, Tran Van On Street, Thu Dau Mot city, Binh Duong province, Vietnam

Email:ngochv@tdmu.edu.vn

Abstract. In quantum well studies, the conductivity tensor problem is one of the fundamental problems. In this work the topic covered here is the effect of electromagnetic waves on the conductivity tensor with the presence of a laser field in a quantum well with a parabolic potential considering the case ofelectrons-acoustic phonon scattering. By using quantum kinetic equations for electrons in quantum wells with parabolic potential, the author has calculated conductivity tensor with electrons–acoustic phonon scattering in the presence of an electromagnetic wave field and a laser field. The expression tensor for conductivity shows its dependence on the frequency of the electromagnetic wave field, the frequency of the laser field, and other parameters specific to the system. From the conductivity tensor expression plotted the effect of the electromagnetic wave field on the conductivity tensor in the presence of the laser field. The quantum well is discussed and plotted here is the GaAs/GaAsAl quantum well.

1. Introduction

In recent years, research on low-dimensional semiconductors in general and quantum wells in particular has brought many benefits to science and technology, emerging new effects strongly applied in engineering [1-5]. Low-dimensional semiconductors are semiconductors in which electrons are confined in 1, 2 or all 3 directions. It is this motion restriction that the energy of electrons is interrupted according to the confinement methods [6-12]. A quantum well is a two-dimensional structure in which electrons are restricted to move in one direction. Since the electrons are confined in one direction, the wave functions, energies, and other quantities that characterize the system have many differences from the bulk semiconductor [3-5].

There are many researches on low-dimensional semiconductor in general and quantum well in particular [13-19]. Conductivity tensor is a familiar concept in materials research, conductivity tensor is also an important concept that forms other fundamental concepts. This work investigates the effect of electromagnetic waves on conductivity tensors in the presence of a laser wave field in a quantum well, but only considers the case of electrons -acoustic phonon scattering and quantum wells with parabolic potential.

Conductivity tensor expression shows the dependence of tensor on electromagnetic wave frequency, laser field frequency, laser field amplitude and characteristic parameters for low-dimensional semiconductor system. The influence of conductivity tensor on electromagnetic wave frequency, laser field frequency, laser field amplitude will be investigated and plotted for the quantum well GaAs/GaAsAl.

2. Calculating conductivity sensor for the case electrons–acoustic phonon scattering

In case of electrons–acoustic phonon scattering, the time-independent component of distribution function ofacoustic phonon:

$$N_q = \frac{k_B T}{v_s q_\perp} \tag{1}$$

and the electrons-acoustic phonon interaction constant:

$$C_q^2 = \frac{\xi^2 q_\perp}{2\rho v_s V} \tag{2}$$

here V, ρ, v_s and ξ are volume, the density, the acoustic velocity and the deformation potential constant.

The parabolic potential in a quantum well is given by:

$$\varepsilon_{n,p_\perp} = \hbar\omega_0\left(n+\frac{1}{2}\right)+\frac{\hbar^2 p_\perp^2}{2m} \tag{3}$$

with ω_0 is the frequency of the parabolic potential.

and wave function:

$$\psi(x) = \sum_{n,i}\psi_n(\vec{R}_i)W\left(\vec{r}-\vec{R}_i\right) \tag{4}$$

Electromagnetic wave field with wave vectors:

$$\begin{cases}\vec{E}(t) = \vec{E}(e^{-i\omega t}+e^{i\omega t})\\ \vec{H} = [\vec{n},\vec{E}(t)]\end{cases} \tag{5}$$

hereω is the electromagnetic wave frequency;

and a laser field:

$$\vec{F}(t) = \vec{F}\sin\left(\Omega t\right) \tag{6}$$

withF is the laser field amplitude, Ω is the laser field frequency.

The quantum kinetic equation for electron in quantum well ($\hbar = 1$) [13-14]:

$$\frac{\partial f_n(\vec{p}_\perp,t)}{\partial t}+\left(e\vec{E}(t)+\omega_H\left[\vec{p}_\perp,\vec{h}(t)\right],\frac{\partial f_n(\vec{p}_\perp,t)}{\partial \vec{p}_\perp}\right)=$$
$$= 2\pi\sum_{n,n',\vec{q}} M_{n,n'}(\vec{q})\sum_{l=-\infty}^{+\infty} J_l^2(\vec{a},\vec{q})\left[f_{n'}(\vec{p}_\perp+\vec{q},t)-f_n(\vec{p}_\perp,t)\right]\delta\left(\varepsilon_{n',\vec{p}_\perp+\vec{q}}-\varepsilon_{n,\vec{p}_\perp}-l\Omega\right) \tag{7}$$

here $f_n(\vec{p}_\perp,t)$ is distribution function of electrons;

$$f_n(\vec{p}_\perp,t) = f_0(\varepsilon_{n,\vec{p}_\perp})+f_1(\vec{p}_\perp,t);\ f_0(\varepsilon_{n,\vec{p}_\perp}) = \theta(\varepsilon_F-\varepsilon_{n,\vec{p}_\perp});\ f_1(\vec{p}_\perp,t) = -\vec{p}_\perp\vec{\chi}(t)f_0'\left(\varepsilon_{n,\vec{p}_\perp}\right); \tag{8}$$

$$f_0'\left(\varepsilon_{n,\vec{p}_\perp}\right) = \frac{\partial f_0(\varepsilon_{n,\vec{p}_\perp})}{\partial \varepsilon_{n,\vec{p}_\perp}};\ f_1(\vec{p}_\perp,t) = f_1(\vec{p}_\perp)e^{-i\omega t}+f_1^*(\vec{p}_\perp)e^{i\omega t};\ f_1(\vec{p}_\perp) = -\vec{p}_\perp\vec{\chi}f_0'\left(\varepsilon_{n,\vec{p}_\perp}\right) \tag{9}$$

$$\vec{\chi}(t) = \vec{\chi}e^{-i\omega t}+\vec{\chi}^*e^{i\omega t};\ \vec{\chi} = \frac{e}{m}\vec{E}\frac{\tau(\varepsilon)}{1-i\omega\tau(\varepsilon)}; \tag{10}$$

$$\frac{\partial f_n(\vec{p}_\perp,t)}{\partial t} = \frac{\partial}{\partial t}\left[f_0\left(\varepsilon_{n,\vec{p}_\perp}\right)+f_1(\vec{p}_\perp)e^{-i\omega t}+f_1^*(\vec{p}_\perp)e^{i\omega t}\right] = -i\omega f_1(\vec{p}_\perp)e^{-i\omega t}+i\omega f_1^*(\vec{p}_\perp)e^{i\omega t} \tag{11}$$

$\vec{h}(t) = \frac{\vec{H}(t)}{H}$ is the unit vector in the magnetic field direction; $\vec{q}$ is momentum of phonon;

$\vec{p} = \vec{p}_\perp+\vec{p}_z$ is momentum of electron;

$$\varepsilon_{n,\vec{p}_\perp} = \frac{(\vec{p}_\perp)^2}{2m} + \hbar\omega_0(n+\frac{1}{2});\ \varepsilon_{n,\vec{p}_\perp+\vec{q}} = \frac{(\vec{p}_\perp+\vec{q})^2}{2m} + \hbar\omega_0(n+\frac{1}{2});\qquad (12)$$

$J_l^2(\vec{a},\vec{q})$ is the Bessel function of real argument; $\vec{a} = \frac{e\vec{F}}{m\Omega^2}$; $\omega_H = \frac{eH}{mc}$

$$M_{n,n'}(\vec{q}) = |C_{\vec{q}}|^2 I_{n,n'}^2 N_{\vec{q}} \qquad (13)$$

$I_{n,n'} = \sum_{j=1}^{S_0}\int_0^d \exp(iq_z z)\psi_{n'}(z-jd)\psi_n(z-jd)dz$ is form factor;

Calculate with simple case: $l = 0, \pm 1$: $J_0^2 \approx 1$; $J_\pm^2 \approx \frac{(\vec{a},\vec{q})^2}{4}$;

Proceed to multiply both sides Eq. (1) by $-\frac{e}{m}\vec{p}\delta(\varepsilon-\varepsilon_{n,\vec{p}_\perp})$ and then sum over $\vec{p}_\perp$ to get expression:

$$-\frac{e}{m}\sum_{n,\vec{p}_\perp}\vec{p}_\perp\left[\frac{\partial f(\vec{p}_\perp,t)}{\partial t} + \left(e\vec{E}(t) + \omega_H\left[\vec{p}_\perp,\vec{h}(t)\right], \frac{\partial f_n(\vec{p}_\perp,t)}{\partial \vec{p}_\perp}\right)\right]\delta(\varepsilon-\varepsilon_{n,\vec{p}_\perp}) =$$
$$= -\frac{e}{m}\sum_{n,n',\vec{q}} M_{n,n'}(q)\sum_{\vec{p}_\perp}\vec{p}_\perp\left[2\pi\sum_{l=-\infty}^{+\infty} J_l^2(\vec{a},\vec{q})\left[f_{n'}(\vec{p}_\perp+\vec{q},t) - f_n(\vec{p}_\perp,t)\right]\delta(\varepsilon_{n',\vec{p}_\perp+\vec{q}} - \varepsilon_{n,\vec{p}_\perp} - l\Omega)\right]\delta(\varepsilon-\varepsilon_{n,\vec{p}_\perp}) \qquad (14)$$

Calculate the left side of equation (14):
The first term

$$-\frac{e}{m}\sum_{n,\vec{p}}\vec{p}_\perp\frac{\partial f_n(\vec{p}_\perp,t)}{\partial t}\delta(\varepsilon-\varepsilon_{n,\vec{p}_\perp}) = -i\omega\vec{R}(\varepsilon)e^{-i\omega t} + i\omega\vec{R}^*(\varepsilon)e^{i\omega t} \qquad (15)$$

here

$$\vec{R}(\varepsilon) = -\frac{e}{m}\sum_{n,\vec{p}_\perp}\vec{p}_\perp f_1(\vec{p}_\perp)\delta(\varepsilon-\varepsilon_{n,\vec{p}_\perp}) \qquad (16)$$

$$\vec{R}^*(\varepsilon) = -\frac{e}{m}\sum_{n,\vec{p}_\perp}\vec{p}_\perp f_1^*(\vec{p}_\perp)\delta(\varepsilon-\varepsilon_{n,\vec{p}_\perp}) \qquad (17)$$

The second term:

$$-\frac{e}{m}\sum_{n,\vec{p}_\perp}\vec{p}_\perp\left(e\vec{E}(t), \frac{\partial f(\vec{p}_\perp,t)}{\partial \vec{p}_\perp}\right)\delta(\varepsilon-\varepsilon_{n,\vec{p}_\perp}) = -\vec{Q}\left(e^{-i\omega t} + e^{i\omega t}\right) \qquad (18)$$

here

$$\vec{Q} = \frac{e^2}{m}\sum_{n,\vec{p}_\perp}\left(\vec{E}, \frac{\partial f(\vec{p}_\perp,t)}{\partial \vec{p}_\perp}\right)\vec{p}_\perp\delta(\varepsilon-\varepsilon_{n,\vec{p}_\perp}) \qquad (19)$$

The third term:

$$-\frac{e}{m}\sum_{n,\vec{p}_\perp}\vec{p}_\perp\left(\omega_H\left[\vec{p}_\perp,\vec{h}(t)\right], \frac{\partial f_n(\vec{p}_\perp,t)}{\partial \vec{p}_\perp}\right)\delta(\varepsilon-\varepsilon_{n,\vec{p}_\perp}) =$$
$$= -\omega_H\left[\vec{R}(\varepsilon)+\vec{R}^*(\varepsilon),\vec{h}\right] - \omega_H\left[\vec{R}(\varepsilon),\vec{h}\right]e^{-2i\omega t} - \omega_H\left[\vec{R}^*(\varepsilon),\vec{h}\right]e^{2i\omega t} \qquad (20)$$

Calculate the right side of equation (14):
Consider the case that $l = 0$:

$$-\frac{e}{m}\sum_{n,n',\vec{p}_\perp} 2\pi\sum_{\vec{q}} M_{n,n'}(q)\vec{p}_\perp J_0^2(\vec{a},\vec{q})\left[f_n(\vec{p}_\perp+\vec{q},t)-f_n(\vec{p}_\perp,t)\right]\delta\left(\varepsilon_{n',\vec{p}_\perp+\vec{q}}-\varepsilon_{n,\vec{p}}\right)\delta\left(\varepsilon-\varepsilon_{n,\vec{p}_\perp}\right)= \\ =-\frac{1}{\tau(\varepsilon)}\left(\vec{R}(\varepsilon)e^{-i\omega t}+\vec{R}^*(\varepsilon)e^{-i\omega t}\right) \tag{21}$$

Consider the case that $l=\pm 1$:

$$-\frac{e}{m}\sum_{n,n'\vec{q}} M_{n,n'}(q)\sum_{\vec{p}_\perp}\vec{p}_\perp\left[2\pi\sum_{l=\pm 1} J_{\pm 1}^2(\vec{a},\vec{q})\left[f_{n'}(\vec{p}_\perp+\vec{q},t)-f_n(\vec{p}_\perp,t)\right]\delta\left(\varepsilon_{n',\vec{p}_\perp+\vec{q}}-\varepsilon_{n,\vec{p}_\perp}\pm\Omega\right)\right]\delta\left(\varepsilon-\varepsilon_{n,\vec{p}_\perp}\right)= \\ =\vec{S}(\varepsilon)e^{-i\omega t}+\vec{S}^*(\varepsilon)e^{i\omega t} \tag{22}$$

with

$$\vec{S}(\varepsilon)=-\frac{e}{m}2\pi\sum_{n,n'\vec{q}} M_{n,n'}(q)\frac{(\vec{a},\vec{q})^2}{4}\sum_{\vec{p}_\perp} f_l(\vec{p}_\perp)\times \\ \times\left[\delta\left(\varepsilon_{n',\vec{p}_\perp+\vec{q}}-\varepsilon_{n,\vec{p}_\perp}-\Omega\right)+\delta\left(\varepsilon_{n',\vec{p}_\perp+\vec{q}}-\varepsilon_{n,\vec{p}_\perp}+\Omega\right)\right]\times \\ \times\left[(\vec{p}_\perp+\vec{q})\delta\left(\varepsilon-\varepsilon_{n,\vec{p}_\perp+\vec{q}}\right)-\vec{p}_\perp\delta\left(\varepsilon-\varepsilon_{n,\vec{p}_\perp}\right)\right] \\ =-2\pi e n_0\vec{\chi}(\varepsilon_F)\left\{\lambda\delta(\varepsilon-\Omega)-A\delta(\varepsilon-\varepsilon_F)\right\} \tag{23}$$

A and λ are calculated depending on the electron-phonon scattering mechanism. Put the expressions. (15), (18), (20), (21), (22) on (14), then unify the terms of both sides to get the system of equations:

$$\left[i\omega+\frac{1}{\tau(\varepsilon)}\right]\vec{R}^*(\varepsilon)=\vec{Q}+\vec{S}(\varepsilon) \tag{24}$$

and:

$$\vec{R}(\varepsilon)=\frac{\tau(\varepsilon)}{1-i\omega\tau(\varepsilon)}(\vec{Q}+\vec{S}(\varepsilon)) \tag{25}$$

n_0 is particle density; m is the effective mass of electron; $e=1.6*10^{-19}C$; $\tau(\varepsilon)$ is the momentum relaxation time in absence of laser radiation.

At time t = 0, the density of current [7]:

$$\vec{j}(t=0)=\int(\vec{R}(\varepsilon)+\vec{R}^*(\varepsilon))d\varepsilon=\vec{j}_l+\vec{j}_l^* \\ =\frac{4e^2n_0}{m}\frac{\tau(\varepsilon_F)}{1+\omega^2\tau^2(\varepsilon_F)}\left\{\left[\varepsilon_F-\omega_0\left(n+\frac{1}{2}\right)\right]I+\lambda\frac{\tau(\Omega)\left[1-\omega^2\tau(\Omega)\tau(\varepsilon_F)\right]}{1+\omega^2\tau^2(\Omega)}+\right. \\ \left.-A\frac{\tau(\varepsilon_F)\left[1-\omega^2\tau^2(\varepsilon_F)\right]}{1+\omega^2\tau^2(\varepsilon_F)}\right\}\vec{E} \tag{26}$$

with electrons-acoustic phonon scattering:

$$A=\frac{e^2F^2}{2m\Omega^3}\frac{\xi^2k_BT}{2\rho v_sV}I_{n,n'}^2\sqrt{2m\left(\varepsilon_F-\omega_0\left(n+\frac{1}{2}\right)\right)} \tag{27}$$

$$\lambda=\frac{e^2F^2}{2m\Omega^3}\frac{\xi^2k_BT}{2\rho v_sV}I_{n,n'}^2\left\{\sqrt{2m\left(\varepsilon_F-\omega_0\left(n+\frac{1}{2}\right)\right)}\left[\sqrt{2m\left(\Omega-\omega_0\left(n+\frac{1}{2}\right)\right)}-1\right]\right\} \tag{28}$$

here $\vec{j}(t=0)=\sigma\vec{E}(t=0)=\sigma 2\vec{E}$

so:

$$\sigma_{ik} = \frac{2e^2 n_0}{m} \frac{\tau(\varepsilon_F)}{1+\omega^2\tau^2(\varepsilon_F)} \left\{ \left[\varepsilon_F - \omega_0 \left(n + \frac{1}{2} \right) \right] \delta_{ij} + \frac{\tau(\Omega)\left[1-\omega^2\tau(\Omega)\tau(\varepsilon_F)\right]}{1+\omega^2\tau^2(\Omega)} \lambda + \right.$$
$$\left. - \frac{\tau(\varepsilon_F)\left[1-\omega^2\tau^2(\varepsilon_F)\right]}{1+\omega^2\tau^2(\varepsilon_F)} A \right\} \qquad (29)$$

with δ_{ik} is the Kronecker symbol.

Equation (29) is an explicit expression of the conductivity tensor in a quantum well with a parabolic potential and considers the case of negative phonon electron scattering. The expression shows the dependence of tensors on electromagnetic wave fields, laser fields, and other parameters that characterize low-dimensional semiconductor systems.

3. Numerical results and discussion

The parameters used in the calculations are as follows [2,3]: m = 0,0665m₀ (m₀ is the mass of free electron); ε_F = 50meV; and $\tau(\varepsilon_F) \sim 10^{-11}$ s^{-1}; $v_s = 5220\,m/s$; $n_0 = 10^{23} m^{-3}$; $\rho = 5.3\times10^3 kg/m^3$; $\xi = 2.2\times10^{-8} J$; $\omega_0 = 5\times10^{13} s^{-1}$.

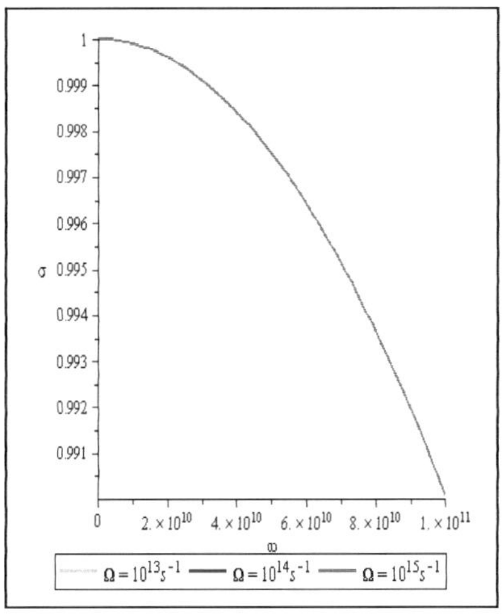

Figure 1. The dependence of σ on the frequency ω with different values of Ω while $\omega < 10^{11}$.

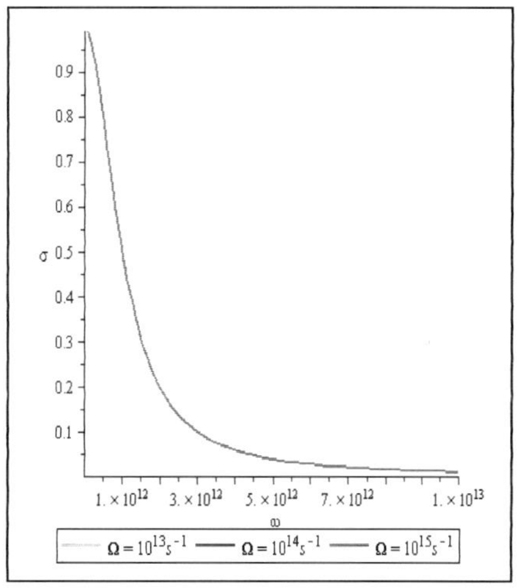

Figure 2. The dependence of σ on the frequency ω with different values of Ω when ω at frequencies from 10^{12}-10^{13}.

At frequencies of electromagnetic wave $\omega < 10^{11}$, Figure 1 shows that the dependence of the conductivity tensor on the electromagnetic wave frequency is nonlinear. As the frequency of electromagnetic waves increases, the tensor of conductivity decreases. Given the three different values of the laser field frequency for the three graphs that are almost identical, it is said that the influence of the laser field frequency on the conductivity tensor in the presence of the electromagnetic wave is very small.

Figure 2 shows the dependence of the conductivity tensor on the electromagnetic wave frequency in the range 10^{12}-10^{13} with 3 different values of the laser field frequency. From the graph it can be seen that as the electromagnetic wave frequency increases, the conductivity tensor decreases but decreases very quickly. The three plots closely coincide with the three different values of the laser field frequency, which also shows that the laser field frequency image does not affect the conductivity tensor much.

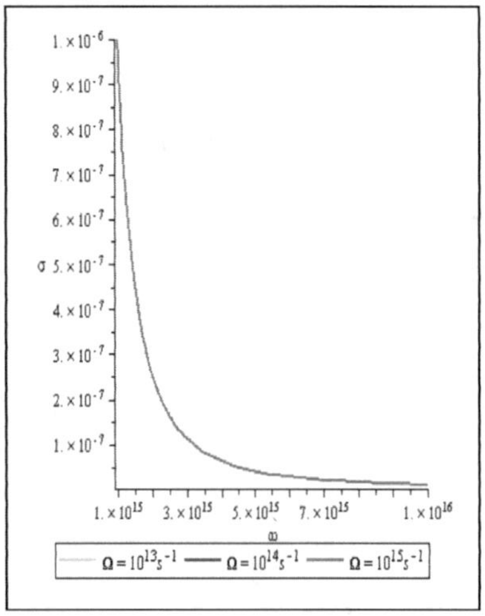

Figure 3. The dependence of σ on the frequency ω with different values of Ω while $\omega > 10^{15}$

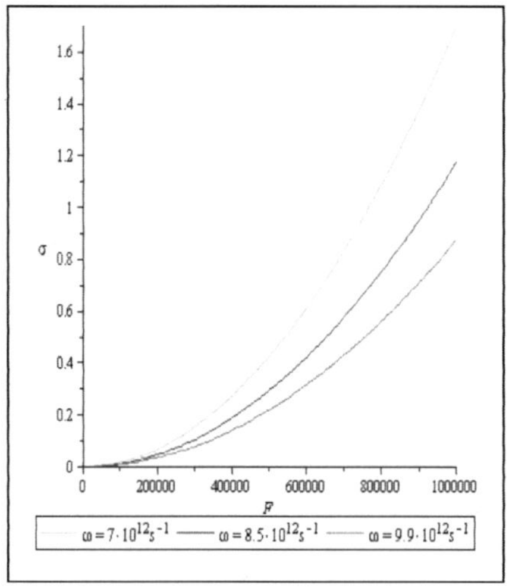

Figure 4. The dependence of σ on the amplitude F of the laser radiation.

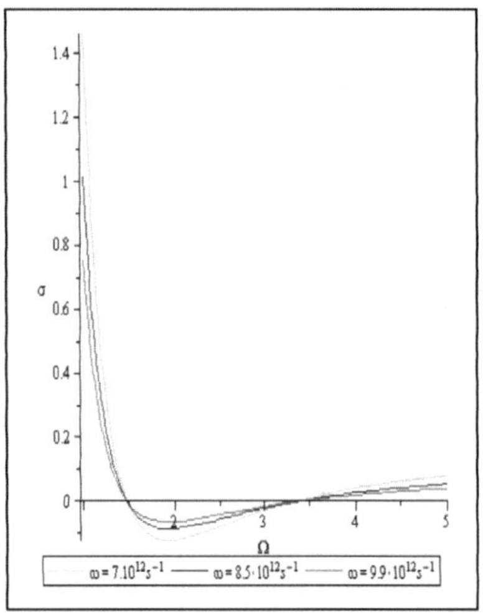

Figure 5. The dependence of σ on the frequency Ω with different values of ω at very low frequencies.

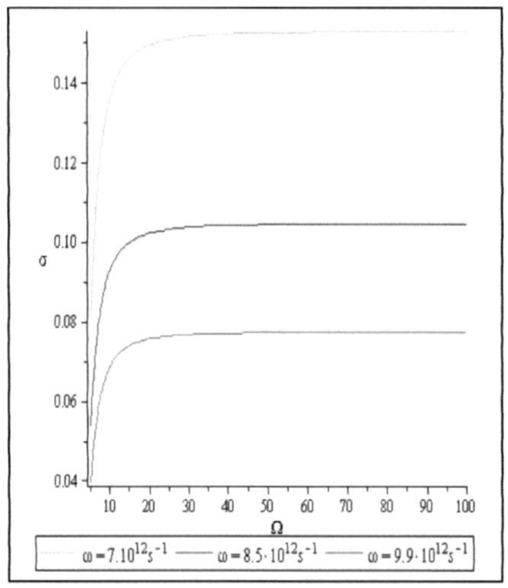

Figure 6. The dependence of σ on the frequency Ω with different values of ω when Ω at higher frequencies.

In figure 3, the dependence of the conductivity tensor on the electromagnetic wave frequency is plotted at frequencies greater than 10. The figure shows that the conductivity tensor now has a very small value close to the value equal to zero, which that is to say, the effect of the electromagnetic wave frequency on the conductivity tensor is very small.

Figure 4 shows the dependence of the conductivity tensor on the laser field amplitude, as the laser field amplitude increases, the conductivity tensor also increases. With three different values of the electromagnetic wave frequency giving us three separate graphs, it proves that the influence of the electromagnetic wave field is very large in the presence of the laser field.

Figure 5 and 6 show the effect of the laser field frequency on the conductivity tensor. At too small frequencies are not suitable for laser fields, but at high frequencies we find that the plot is straight lines parallel to the horizontal axis, which shows that the conductivity tensor has very little effect on the laser field frequency.

4. Conclusions

The paper investigated the effect of the electromagnetic wave field on the conductivity tensor in a quantum well in the presence of a laser field, and only considered the case of electron-phonon scattering. From the conductivity tensor expression shows its dependence on characteristic parameters for electromagnetic waves, laser fields, and low dimensional systems.

A graph of the dependence of the conductivity tensor on electromagnetic wave frequency, laser field frequency and laser field amplitude was also plotted. When $\omega < 10^{15}$, the graph shows that the dependence of conductivity tensor on electromagnetic wave frequency is very large, in this frequency range conductivity tensor always decreases as frequency increases. But when the frequency $\omega > 10^{15}$, the conductivity tensor decreases very quickly and has a very small value, at this time the dependence of the conductivity tensor on the electromagnetic wave frequency is very small.

The graphs also show that the dependence of the conductivity tensor on the laser field frequency is very small, the conductivity tensor is strongly dependent on the laser field amplitude, the amplitude increases, the conductivity tensor also increases very rapidly.

5. References

[1] Blencowe M. and Skik A. 1996Acoustoconductivity of quantum wires *Phys. Rev.* B 54, pp 13899-13907.
[2] A. Grinberg and Luryi 1988 Theory of the photon - drag effect in a two-dimensional electron gas, *Phys. Rev.* B 38 p 87.
[3] B. D. Hung, N. V. Nhan, L. V. Tung, and N. Q. Bau 2012 Photostimulated quantum effects in quantum wells with a parabolic potentia *Proc. Natl. Conf. Theor. Phys*, vol 37p 168
[4] S. V. Kryuchkov, E. I. Kukhar' and E. S. Sivashova 2008 Radioelectric effect in a superlattice under the action of an elliptically polarized electromagnetic wave, *Physics of the Solid State*, vol 50, No. 6 p 1150.
[5] S. Butscher, and A. Knorr 2006 Occurrence of intersubbandpolaronic repellons in a two- dimesional electron gas*Phys. Rev.* L 97, pp 197401-197405.
[6] Balandinand K.L. Wang, in Nanostructures: Physics and Technology, (Russian Academy of Sciencies, St.Petersburg, Russia) 1998 *edited by Zh. Alferov et al.*, pp.24-27.
[7] Yu. Boikov, B.M. Goltsman, V.A. Danilov 1995 Semiconductors, p 464.
[8] A. F. Khokhlov, I. A. Chuchmai, and A. V. Ershov2000Absorption Features in a- Si/ZnOx Nanostructures, *Semiconductors* 34 (3), pp 349 – 354.
[9] V. L. Bonch-Bruevich and S. G. Kalashnikov 1977 *Physics of Semiconductors* (in Russian), p 672
[10] A. I. Ansel'm 1981 *Introduction to semiconductor theory* (Moscow), p 645

[11] V. N. Lugovoi 1962 *Sov. Phys. JETP* p 1113
[12] L. I. Kats, V. P. Terjova, D. Sh. Shechter, L. Sh. Shechter 1972 *Sov. J. Informations of High education Colleges, Ser. Radiophysics* p 675
[13] Nguyen QuocAnh, Nguyen ThiTuUyen 1995 Conductivity tensor of semiconductor in variable magnetic field *VNU journal of science,* No. 01 p 19-23
[14] G. M. Shmelev, G. I. Tsurkan and É. M. Épshtein 1982 Photostumilatedradioelectrical transverse effect in semiconductors *Phys. Stat.* Sol. B, vol. 109 p 53
[15] L. I. Kats, D. Sh. Shechter 1974 *Sov. J. Informations of High education Colleges, Ser. Radiophysics*p 405
[16] AyhanOzmen, Yusuf Yakar, BekirCakir, UlfetAtav2009Computation of the oscillator strength and absorption coefficients for the intersubbandtransitions of the spherical quantum dot,*Opt.Commun*282, pp 3999-4004.
[17] Branis S. V., Li G., Bajai K. K. 1993Hydrogenic impurities in quantum wires in the presence of a magnetic field*Phys. Rev.* B 47 (3), pp 1316-1323.
[18] Boiko I. I., ShekaV. and Vasilopoulos P. 1993 Kinetics of quasionedimensional electron gas in transverse magnetic field. II Arrays of quantum wires *Phys. Rev.* B 47, pp 15809-15815.
[19] Geyler V. A., Margulis V. A. 2000 Quantization of the conductance of a three – dimensional quantum wires in the presence of a magnetic field *Phys. Rev.* B 61 (3), pp.1716-1719.

IOP Conf. Series: Materials Science and Engineering 1045 (2021) 012033 doi:10.1088/1757-899X/1045/1/012033

Effective Load Management System for Sudden Shutdown Avoidance of Stand Alone Photovoltaic Operation

Nuratiqah Mohd Isa, Amirah Nurhafizah Abu Bakar, Nurul Husna Mohd Sarif, Mohamad Shaiful Abdul Karim and Ahmad Syahiman Mohd Shah

Department of Electrical Engineering, College of Engineering, Universiti Malaysia Pahang, 26300 Gambang, Kuantan, Pahang, Malaysia asyahiman@ump.edu.my

Abstract. Stand-alone system is the system that operates independently of the electrical grid. Conventionally, the energies absorbed through photovoltaic modules are stored in batteries to reserve for night-time use. Moreover, the function of batteries in stand-alone system is to stabilize the output supplied for the load particularly during the fluctuating circumstance. Sudden shutdown of dedicated photovoltaic system during this period may occur when the storage of batteries approaches empty level which may significantly degrade the lifespan and performance of the batteries. In this study, a sequential load switching system is introduced in order to overcome the sudden shutdown of storage battery-based photovoltaic system. The developed system is mainly equipped with Arduino, MOSFETs and DC bulbs as microcontroller, high-speed switching and load, respectively. As a result, the proposed sequential load switching method is highly effective since it has successfully extended the standby time period up to approximately 27 times longer, before the system shutdowns at the cut-off voltage.

1. Introduction

Generally, photovoltaic system is categorized into two; Grid-Connected and Standalone. Standalone photovoltaic (SAPV), which also known as off-grid solar energy systems are covering approximately 80% of the photovoltaic applications utilized world widely [1]. There is high demand of standalone photovoltaic system in several countries specifically to the locations that have faced dramatical change in climate [2]. The SAPV system is designed to maintain a good performance of SAPV system components (PV, Battery and load) in order to increase the reliability and efficiency [3]. Besides, solar array is unable to store the energy by itself; a battery is required as energy storage system so that the load can be used during night-time. Typically, lead acid batteries are used due to its wide availability of sizes, low cost and well understand performance criteria [4]. Meanwhile, a charge controller functions to be capable of varying voltage and current in order to avoid the battery from being damage due to over-charging or over-discharging [5]. Due to this limitation of capacity, the system may experience sudden shutdown when the batteries drop to the cut-off voltage level [6]. Once this circumstance occurs, the PV components such as the load and battery itself are exposed to the risk of damage or malfunction.

Therefore, any intiatives in overcoming this issue are desirable since this SAPV application is greatly contributing to rapid development of secondary energy harvesting resources for human necessities, nowadays. In this paper, an effective load management system based on autonomous sequential load switching algorithm has been proposed. The main idea is to structure the switching algortihm that can feasibly extend the standby time period of the system's operation before the battery totally shutdown by itself at the cut-off voltage. Detailed methodology and finding will be discussed in the next section.

2. Methodology

2.1 Circuit Design Using Proteus

Proteus software has been used in order to simulate the overall circuit connection of hardware for this project. Figure 1 shows the overall circuit designed for this study.

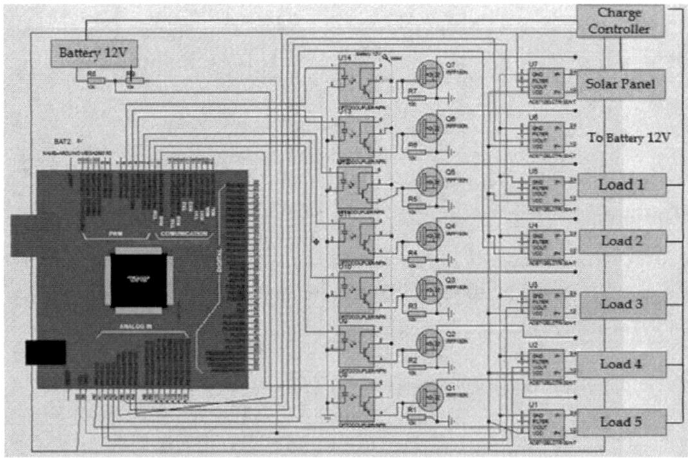

Figure 1. The overall circuit design for the project

2.2 The Block Diagram of the system

Figure 2 represents the block diagram of the developed system. Solar module (100W), battery 12V 17Ah and the load are connected to the MPPT charge controller as the charge controller acts as the regulator and protection system. Seven IXFH44N50P MOSFET switches and seven 871C optocouplers are connected to the solar module, battery and each load to undergo the switching process whether it is HIGH or LOW. MOSFET switch, and the voltage divider will be connected to the Arduino for coding process. A voltage divider is simply constructed by directly connecting the 12 V battery, to two series-connected resistors. One resistor will take 7 V and another one resistor will take 5 V. This 5 V will then be connected to the Analog input of Arduino to read the voltage. All the data that are read by the Arduino will be displayed through the serial monitor in Arduino IDE.

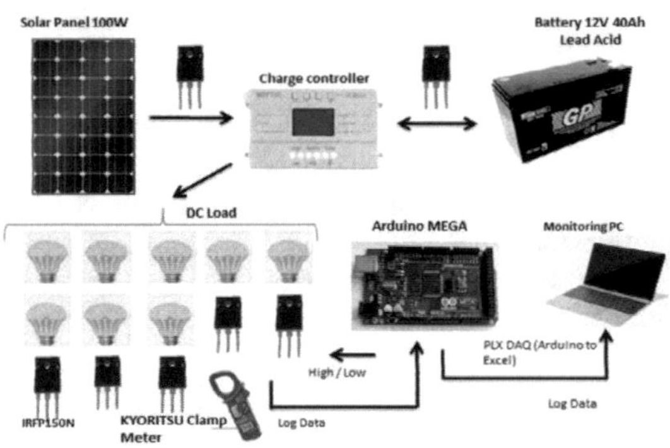

Figure 2. Block diagram of the developed system

2.3 *Load Switching Algorithm*

In this project, an algorithm of load switching will be executed in order to assist the consumer to perform a load management effectively. Figure 3 exhibits the overall flowchart of the algorithm for the system.

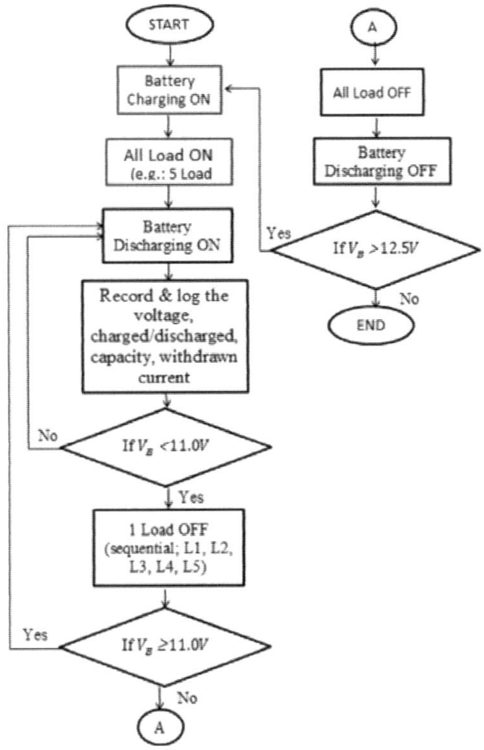

Figure 3. Flowchart of the Algorithm of Load Switching

In this study, MOSFET is trusted in protecting the switching medium and mechanism as it is successfully proven during the experiment. The ON and OFF's switching of the MOSFET includes the protection for the battery to avoid from being over-discharged. The MOSFET is set to switch off and shutdown the system when the battery voltage reaches 10.8 V. This value is the typical recommended cut-off voltage for the lead acid battery. If the lead acid battery is discharging until cut-off less than 10.5 V, it will damage the battery severely.

Firstly, the solar module absorbs the sunlight and converts it to the electrical energy which makes the battery starts to be charged. All load will be turned ON. In this system, the load that will be used are all DC lamps which are denoted as, L1, L2, L3, L4 and L5, respectively. Once all load is turned ON, automatically the battery is undergoing a discharging state. During the discharging state, all parameters such as the voltage, withdrawn current and capacity will be measured and recorded by the Arduino and then displayed on the monitoring PC. The voltage of the battery will start to decrease time by time. Once the voltage of the battery (V_B) is lesser than 11.0 V, the first load, LI will be turned OFF by the MOSFET switch as it is set to be "LOW". Consequently, as the V_B is greater than 11.0 V again, the battery will continue to discharge. After certain period of time, V_B is expected to drop below than 11.0 V again. At this point, the second load, L2 will be turned OFF and hence, the V_B will increase slightly and then, it will undergo the same discharging process again for other cases of L3, L4 and L5. Nevertheless, if the V_B remains not to be more than 11.0 V again, all load will be turned OFF. The battery will stop undergo the discharging process as there are no more load or device that is consumed by the battery. Eventually, if the V_B increases more than 12.5 V, the battery will revert automatically to the charging mode back again.

3. Results and Discussion

3.1 Conventional Battery Discharging Method

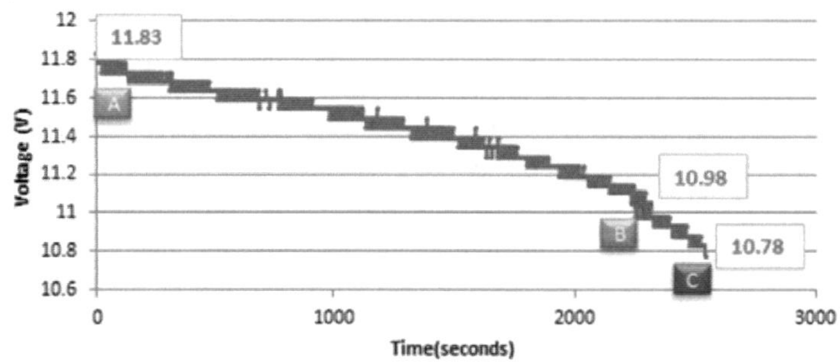

Figure 4. Battery voltage's movement during discharging using conventional method.

The condition of the system while the battery at discharging state is shown in Figure 4. In this project, the battery system will stop operate and totally shutdown when it reaches its cut off voltage that has been set to 10.80 V. Meanwhile, the standby time is set as 11.0 V. The duration taken by the battery to discharge from the standby time (11.0 V) until the cut off voltage (10.80 V) will be recorded and analyze.

The system is operating in discharging state by turning on all the loads. Figure 4 and Table 1 exhibit the result of battery voltage's movement and related data for the battery during discharging using a conventional method, respectively. The current drawn from the positive terminal of the battery was measured by using Kyoritsu clamp meter and the value displayed was 12.9 A. Eight 36W-rated DC bulbs was used here. From Figure 4 and Table 1, it can be clearly understood that the voltage is decreasing with time, starting from the initial voltage which value is 11.83 V. The overall system operated for about 42 minutes and 22 seconds. The duration taken by the battery to discharge from the standby time (11.0 V) until the cut off voltage (10.80 V) was 4 minutes and 47 seconds. This is the usual phenomenon of using load, all load will be used at one time but when it is applied on the storage battery based on standalone PV system, this method can drain the storage battery very fast and the operation of the system will shut down quickly as the user does not manage the load usage wisely. Typically, user will use and solely depend on the storage battery during night, rainy or cloudy conditions. Therefore, when the battery is empty, user will face problem or uncomfortable condition when there is no supply of electricity. If the users are doing their work or online business at home or factory, the sudden shutdown may affect their income and work. The user has to wait until the next morning when there are source of sunlight to charge the battery.

From Figure 4, state A is the initial voltage as soon as the system is integrated to the Arduino Mega. State B is when the battery reaches the standby time where the voltage level of the battery is 11.0 V and state C is when the battery reaches the decided cut off voltage of 10.80 V. Once the battery voltage reaches the cut off voltage, the systems will totally shutdown as the battery becomes empty. The duration taken between the state A and state B is approximately 37 minutes and 35 seconds. The duration taken is relatively shorter if the current drawn from the battery is higher as it drains the battery faster and pulls a greater voltage. The duration taken between state B and state C is about 4 minutes and 47 seconds. This situation occurs when the voltage and capacity of the battery have been already lesser from time to time, so the discharged current and voltage will drop quickly at the end.

Initially, the recorded voltage of the battery before connected to Arduino Mega was 12.36 V. As soon as the system integrated with the Arduino Mega, the voltage decreased abruptly from 12.36 V to 11.83 V. There was a 0.53 V of voltage drop that was being pulled suddenly and significantly by the load from the battery. Additionally, the analysis of voltage drop from the obtained curve in Figure 4 is detailed in Table 2. From Table 2, as the load draws high amount of current which is 12.9A from the battery, it affects to the decreasing of 4.29 % of voltage from the battery initial voltage. The voltage decreases to 11.63 V after 500 seconds at about 5.91 % of voltage drop. After 1000 seconds, the voltage decreases to 11.54 V with 6.63 % of voltage drop. 500 seconds, the voltage decreases to 11.39 V with 7.85 % of voltage drop. The voltage continues to decrease after 2000 seconds from the initial stage to 11.19 V with 9.47 % of voltage drop. At the end of the system, the voltage decreases to 10.78 V with 12.78% of voltage drop, eventually. Therefore, the voltage drop increases by time and it takes 12.78 % of voltage drop from the initial level of battery voltage to reach the cut off voltage of 10.80 V.

Table 1. Related data for the battery during discharging using conventional method

Voltage (V)	State	Transition	Duration	Current (A)
11.83	A	A – B	37min 35sec	12.9
10.98	B	B – C	4min 47sec	12.9
10.78	C	-	-	-
		A - C	**42min 22sec**	

Table 2. The analysis of voltage drop from Figure 4

Time (seconds)	Voltage (V)	Voltage Drop (V)	Voltage Drop (%)
0	12.36	-	-
1	11.83	0.53	4.29
500	11.63	0.73	5.91
1000	11.54	0.82	6.63
1500	11.39	0.97	7.85
2000	11.19	1.17	9.47
2540	10.78	1.58	12.78

3.2 *Proposed Method: Automatic Sequential Load Switching*

Similar to distribution of power generation, it is monitored by the usage of the load, when one line is overloaded, then the circuit breaker that acts as the protection will cut off the supply. In this project, the proposed method is implemented by switching off the load one by one by monitoring directly the voltage state of the battery or supply. Figure 5 depicts the curve of voltage's movement in the battery during discharging when the proposed automatic sequential load switching has been applied in the developed system. From here, it can be obviously suggested that the proposed method successfully manages to prolong the duration of standby time before the system shutdowns for about 2 hours, 14 minutes and 28 seconds.

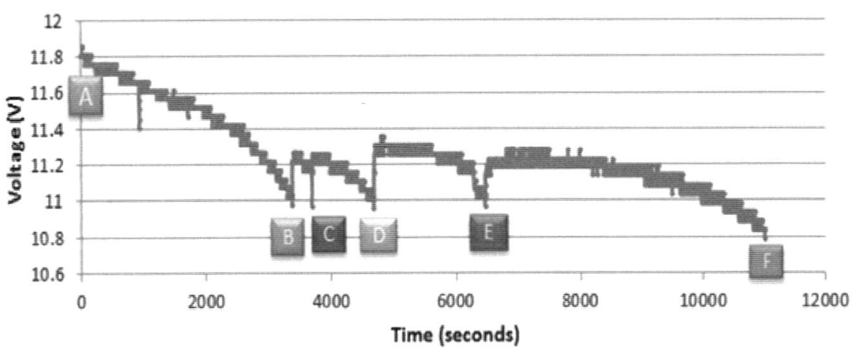

Figure 5. Battery voltage's movement during discharging using proposed method

Initially, the recorded voltage of the battery before it was connected to the Arduino Mega was 12.53 V. As soon as the system integrated with the Arduino Mega, the voltage decreased abruptly from 12.53 V to 11.81 V. There was 0.72 V of voltage drop that was being pulled suddenly and significantly by the load from the battery. Table 3 and Table 4 demonstrate the data recorded of the battery discharging using the proposed method and analysis of voltage drop from Figure 5, respectively. From Table 4, as the load draws high amount of current which is 12.9A from the battery, it affects to the decreasing of 5.75 % of voltage from the battery initial voltage. The voltage decreases to 11.71 V after 500 seconds at about 6.54 % of voltage drop. After 1000 seconds, the voltage decreases to 11.66 V with 6.94 % of voltage drop. 500 seconds after that, the voltage decreases to 11.57 V with 7.66 % of voltage drop. The voltage continues to decrease after 2000 seconds from the initial stage to 11.47 V with 8.46 % of voltage drop. Until at the end of the system, the voltage decreases to 10.79 V with 13.89 % of voltage drop, eventually. Therefore, the voltage drop increases by time and it takes 13.89 % of voltage drop from the initial voltage battery to reach the cut off voltage of 10.80 V. From Figure 5, state A is the initial voltage as soon as the system is integrated to the Arduino Mega. State B, C, D, E are when the battery reaches the standby time where the battery voltage is 11.0 V. State F is when the battery reaches the cut off voltage of 10.80 V. Once the battery voltage reaches the cut off voltage, the systems will totally shutdown as the battery becomes empty. From Table 3, it can be clearly understood that the duration taken between the state A and state B is 56 minutes and 15 seconds. Next, the duration taken between the state B and state C is 5 minutes and 18 seconds. Here, the duration taken is relatively shorter as the current drawn is still high with 8.5 A. The duration taken from the state C to D is 16 minutes and 37 seconds. This duration is a bit longer since the drawn current is lesser than before with of 5.5 A as the second load is off. The duration taken from the state D to E is 37 minutes and 4 seconds. This duration is a bit longer since the drawn current is much lesser than that of the state C to D with of 2.8 A as the third load is off. Surprisingly, the duration taken from the state E to F is 1 hour, 15 minutes and 29 seconds. This duration is relatively longer since the drawn current is lower than that of other states with the value of 1.6 A as the fifth load is turned off. Thus, the total duration for the battery to discharge starting from the state A to F is by far longer than that of conventional discharging method with 3 hours, 10 minutes and 43 seconds. From Table 3 also, the voltage increment is recorded and analysed. It is apparent that the voltage increment is approximately 0.24 V when the first load is automatically switched off. When all load is on, the discharged current and battery capacity are still higher. Thus, when the first load is disconnected from the battery, it reduces the burden from the battery and causes the voltage increases as much as 0.24 V. When the second load has been off, the voltage increment is 0.19 V. Next, when the third load is turned off, the voltage increment maintains at around 0.2 V. Since more load is disconnected from the battery, it significantly reduces the burden of the battery to carry the load. Eventually, when the fourth load is turned off, the voltage increases slightly at a rate of 0.07 V since the battery has lesser capacity already. In this experiment, a 100 W solar module becomes the source to charge up the battery. The solar module, battery and load are connected to the charge controller. The charge controller functions as a safety precaution medium to avoid over-charge and also over-discharge of the battery.

Table 3. Data recorded of the battery discharging when using the proposed method.

	State	**Duration**	**Current (A)**	**Voltage Increment (V)**
All Load ON	A – B	56min 15sec	12.9	-
4	B – C	5min 18sec	8.5	0.24
3	C – D	16min 37sec	5.5	0.19
2	D – E	37min 4sec	2.8	0.2
1	E – F	1hr 15min 29sec	1.6	0.07
	A – F	**3hr 10min 43sec**		
	B – F	**2hr 14min 28sec**		

Table 4. The analysis of voltage drop from Figure 5

Time (seconds)	Voltage (V)	Voltage Drop (V)	Voltage Drop (%)
0	12.53	-	-
1	11.81	0.72	5.75
500	11.71	0.82	6.54
1000	11.66	0.87	6.94
1500	11.57	0.96	7.66
2000	11.47	1.06	8.46
2500	11.37	1.16	9.26
3000	11.22	1.31	10.45
3500	11.22	1.31	10.45
4000	11.21	1.32	10.53
4500	11.06	1.47	11.73
5000	11.31	1.22	9.74
5500	11.31	1.22	9.74
6000	11.21	1.32	10.53
6500	11.19	1.34	10.69
7000	11.23	1.30	10.38
7500	11.23	1.30	10.38
8000	11.19	1.34	10.69
8500	11.19	1.34	10.69
9000	11.14	1.39	11.09
9500	11.09	1.44	11.49
10000	11.04	1.49	11.89
10500	10.94	1.59	12.69
10994	10.79	1.74	13.89

4. Conclusion

Based on the result, the proposed method has successfully and effectively applied in a standalone photovoltaic-based load management system by pushing the limit of the storage battery at its maximum capability. User can use any load that they want to be connected with this method. The switching mechanism by using MOSFETs are considered trustable as the safety precaution for the storage battery from over discharge is executed. By using the sequential load switching through the specific monitoring of the voltage battery's states, it is proven by the experiment that the proposed method is clearly effective since it is by far extending the standby time period before the system totally shutdowns.

Acknowledgement

The author would like to acknowledge Universiti Malaysia Pahang under Postgraduate Research Grant Scheme PGRS200317 for financially supporting the project.

References

[1] S Duryea, S Islam and W Lawrance 2001 *IEEE Industry Appl. Magazine* **7** 67-72.

[2] N Idris, A M Omar and S Shaari 2010 *Proc. 4th Int. Power Eng. Optim. Conf. Progr. Abstr. (PEOCO)* 474–479.

[3] C B-Yahya 2010 *Proc. IEEE Int. Energy Confer.* 412-416.

[4] G M Tina and A D Grasso 2014 *Energy Conversion and Manag.* **78** 862-871.

[5] B. Chowdhury and S. Rahman 1988 *Power* **3** 900–907.

[6] A N A Bakar, M R Zahariman, R Ishak and A S M Shah 2019 *J. Adv Research in Dynamical & Control Syst.* **11** 52-61.

AE signature analysis using continuous and discrete wavelet transforms to predict grinding wheel conditions

K Shivith[1] and K Rameshkumar[1]

[1] Department of Mechanical Engineering, Amrita School of Engineering, Coimbatore – 641112, Amrita Vishwa Vidyapeetham, India

Abstract. In this study, experiments were conducted in a surface grinding operation to acquire and analyze the AE signature to establish a statistical correlation between Acoustic Emission features extracted in wavelet domain with grinding wheel conditions. Grinding wear plot was established to identify the grinding wheel conditions by monitoring the wear in the abrasive grinding wheel and workpiece. Continuous and Discrete Wavelet transforms were carried out to extract the wavelet coefficients. Decision Tree based statistical models were built using discrete and continuous wavelet coefficients. The performance of J48 Decision tree and Classification and Regression Decision Tree (CART) are compared using the classification accuracy and kappa statistics measures. In discrete wavelet transforms, wavelet coefficients are extracted using four mother wavelets namely Haar, Daubechies, Symlet, and Coiflet. In Continuous Wavelet Transforms, the Morlet wavelet is used to extract the 1D wavelet coefficients using 2D scalograms.

1. Introduction

1.1. Tool Condition Monitoring (TCM) Framework

Tool condition monitoring is an indirect way of studying the tool wear and identifying the condition of the tool using the data acquired during the machining process. The tool wear may be due to many reasons like abrasion, adhesion, fatigue, dissolution, and tribo-chemical mechanisms. The tool wear or tool conditions are difficult to quantify during the process. Generally offline methods are applied to measure the tool wear and the cause for the same. In automated environments, it is essential that tool conditions need to be monitored and decisions to be taken without any manual intervention. Sensors play a vital role in TCM to capture the condition of the tool due to various failure mechanisms takes place during the machining process. Different types of sensors are available for identifying condition of the tool during the machining process. Some of the sensors that are commonly used in the tool condition monitoring process are dynamometers (for measuring force signal), vibration sensor (for measuring vibration signal), acoustic emission sensor (for measuring AE signal), temperature sensors, current sensors, optical sensors, and load cell. These sensors are used to capture the data from the machining process which will have the information about the tool wear or tool failure mechanism. The data collected by the sensor must be processed and for this, the signal processing technique is used. Features containing the useful knowledge about the tool conditions are extracted and statistical models are built using machine learning classifiers. A typical grinding wheel condition monitoring framework is shown in Table 1.

Table 1. Grinding wheel condition monitoring framework

Machining Process	Signal Acquisition / Sensors	Signal Processing Domain	Statistical Models (Classifiers)	Grinding Wheel State
Grinding	• Dynamometer • Acoustic Emission Sensor • Load cells • Motor Current sensors • Infrared sensors • Ultrasonic Sensors	• Time Domain • Frequency Domain • Time-Frequency Domain (Wavelet)	• Decision Trees • Fuzzy Logic • Neural Net • Naïve Bayes • Support Vector Machines	• Sharp – Initial condition • Intermediate steady-state condition • Worn-out condition

1.2. Acoustic Emission (AE) based Tool Condition Monitoring (TCM)

TCM is getting enormous recognition in industries nowadays due to the increased automation activities. The automation leads to increased production and reduced lead time. Moreover, in a machining processes it necessary that manual intervention to an extent need to be reduced for making decision relevant to tool replacement or repair. TCM will support the decision makers to utilise the tool effectively and efficiently without over-using and underutilizing. Thereby, the shopfloor decision maker will have a better control over the tool and effectively plan their maintenance activities and avoid catastrophic failures and quality issues. In TCM, many sensors are being used to detect the change in conditions of the tool for developing an effective TCM system. Choice of a sensor is a crucial step in the development of a TCM system. There are plenty of sensors available to detect the conditions of the tool in the machining processes such as turning, milling, drilling, and grinding etc. Some of the signatures of the cutting processes which are effectively correlated to the tool conditions are cutting force, vibration, acoustic emission, temperature, spindle current etc. The right sensor must be chosen based on the mechanism of the cutting process. The cutting tools used in the machining processes are designed for set of criterions such as metal removal rate and quality of the surface generated on the workpiece. The choice of sensors is mainly based on a) detectability of the change in condition of the tool, b) range, and span of the sensor, c) frequency range, etc.

Detailed studies have been made by researchers for selecting a suitable sensor for a typical manufacturing process. Variety of sensors have been tried in many metals cutting process namely turning, milling, drilling, and grinding. Even though there is a similarity in metal removal mechanism in these processes, but the amount of material removed in terms of chip thickness is different in these processes. The manufacturing processes can be classified into macro, micro, precision and ultra-precision and nano machining processes. In these processes the amount of chip thinness produced during the machining operation is different. While monitoring the cutting processes in these scenarios, all the sensors does not have same performance in detecting the changes in the tool conditions. Dornfeld et al. [1] observed various sensors namely encoders, vibration sensors (accelerometers), acoustic sensors and optical sensors. It is concluded in their study that piezo electric sensors (acoustic emission) have better accuracy in predicting the conditions in a precision cutting process. Better performance of AE sensors was observed compared to force, load, optical and vibration sensing sensors in a precision machining environment.

The good performance of AE sensors in precision machining process such as grinding are due to a) high sensitivity, b) identifying the defects, and failure modes at early stage c) capturing the signals in a higher frequency range (1kHz to 1MHz) d) suitability of AE based technique for real-time monitoring e) avoidance of machine disturbances taking place at lower frequencies and d) high signal-to-noise ratio. AE sensor also has a swift retort to the transitory performance of the precision machining process such as grinding. During the grinding process, AE activities such as abrasive particle wear, fracture of bonding material in the grinding wheel, thermal burns, loading and glazing etc are captured by the AE sensors in the higher frequency range without disturbing the machine vibrations due to environmental effects.

1.3. Motivation

Tool condition monitoring has now become important in every manufacturing industry. As automation has taken over the industry, tool condition monitoring has been an indispensable part of it. In such a scenario, the manufacturing industry with a variety of processes involving a variety of tools needs an efficient TCM system. The condition of the tool has a direct influence on the quality of the product being produced. It is important to maintain a good condition of the tool otherwise reliability of the product will be questioned in the market. This thought leads us to develop a good tool condition monitoring model that can predict the condition of the tool using the sensor information.

In this study, we are developing statistical models from the AE signal extracted from the grinding wheel using both the Discrete and Continuous wavelet transforms. The wavelet transforms are found to give better time and frequency resolution at the same time. Studies on building a statistical model using AE wavelet features are limited in the literature. An effort is made in this research paper to foretell the grinding-wheel condition using AE wavelet features extracted with the help of discrete and continuous wavelet transforms.

1.4. Literature Review

For developing an TCM using AE sensors, a comprehensive literature review was conducted focusing on AE signatures and their applicability in predicting the grinding wheels conditions using the machine learning models. The sensor selected for TCM of grinding process should reveal the change in the condition of the abrasive wheel. Importance of AE sensors in precision machining operations are available in [1, 2]. Liao et al. [3] developed Hidden Markov Models (HMM) using wavelet features. Clustering methods were adopted to categorize the grinding-wheel conditions. Teti et al. [4] studied machining operations and found out the advantages of using AE sensors in tool condition monitoring in micromaching processess.

A review on TCM focusing on sensors in signal acquisition, signal processing and modelling pragdiams in milling, griding, turning and drilling operations was presented bt Roth et al. [5]. Dressing operation in a grinding process is monitored using AE signals with the aid of Artificial Neural Networks by Moia et al. [6]. AE signature features are extracted, and a grinding wheel dresser was classified as 'dull' and 'sharp' in their study. In a cylindrical grinding process, Support Vector Classifiers (SVM) built with AE features performing well related to vibration features [7]. Mouli and Rameshkumar [8] carried out an AE signature analysis in surface grinding operation by acquiring time-domain AE features using CART algorithms. Grinding wheel conditins were predicted using Hidden Morkov Models (HMM) by Krishnan and Rameshkumar [9]. AE signatures were used for building the HMM models. Chen et al. [10] studied the grinding burn using wavelet packet transforms using AE signature and found that AE is predicting the grinding burn with good accuarcy. Machine learning classifiers are constructed using ensembles with wavelet features and abrasive-wheel conditions are classified [11]. Worn-out and good conditions of a abrasive wheel are studied in [12] considering various metal removal conditions using AE based wavelet approach.

Degradation assessment of a grindining wheel was studied by Benkedjouh et al. [13] using wavelet transforms. Krishnakumar et al. [14] studied AE and vibration featutes for TCM of high speed machining process. Macine learning models were built using the wavelter coefficients of AE and Vibration signatures. It was obserbed that wavelet features of AE has better resolution in predicting the tool conditions compared to vibration features of AE. Machine learning models built with fused features in time, frequncy and wavelet domain are giving better classiifcation accuarcy [15] in tool condition monitoring of high speed precision process. Off late, Alexandre et al. [16] proposeds a fuzzy method using AE features for monotoring the grinding process. Xiao et al. [17] suggested an AE based wavelet entropy-based algorithm used to select the optimal wavelet in a condition monitoring problem.

From the literature review, it is concluded that the AE sensor is best suited for detecting failure modes in precision manufacturing processes. AE emissions in grinding process are due to the change in conditions of the grinding wheel and are taking place at higher frequencies. AE sensor will be able to capture the high frequency transient signature of the process. After the signal acquisition, signal processing is carried out to identify the conditions of the abrasive wheel.

IOP Conf. Series: Materials Science and Engineering 1045 (2021) 012034 doi:10.1088/1757-899X/1045/1/012034

TCM of the grinding process is one of the emerging areas of research due to the increased automation of the grinding processes. Automation needs intelligent techniques for taking decisions without any manual intervention. Condition monitoring using sensors will lead to the effective utilization of the grinding wheel. In machining processes such as grinding, AE sensors are predicting the tool conditions with good accuracy. Machine Learning statistical classifiers were built for predicting the tool conditions using features extracted in time, frequency, and wavelet domains Time domain and frequency domain features are frequently used by the researchers to build classifiers using ML classifiers such as decision trees, SVM, ANN, etc. Very limited studies focused on ML classifies trained using wavelet coefficients for predicting grinding wheel conditions. Decision tree algorithms are simple and from the decision tree, simple If-Else rules can be generated for developing an on-line TCM system.

2. Objective and Methodology

2.1. Objectives

The objectives of this study are as follows.

- Identifying conditions of the abrasive wheel by conducting experiments
- Extracting wavelet coefficients by performing AE signature analysis in continuous and discrete wavelet domains.
- Building statistical classifiers using decision tree J48 algorithm and CART for predicting the conditions of the abrasive wheel.

2.2. Methodology

The idea in this study is to extract the AE data of the grinding wheel for the complete cycle and use that data to classify using different classifiers and identify the Sharp, Intermediate, and Dull conditions. The methodology adopted in this study is given as follows:

- Establishing an experimental set-up to identify grinding wheel conditions
- Acquire AE signature
- Selection of wavelets
- Signal Decomposition in Discrete and Continuous Wavelet Domains
- Wavelet coefficient extraction in the discrete and continuous wavelet domain
- Training machine learning classifier (Decision Tree J48 algorithm and CART) using wavelet coefficients extracted in the discrete and continuous wavelet domain
- Prediction of grinding wheel conditions

2.3. Experimental Setup

An auto-feed grinding machine was used to collect the data. The AE sensor is fixed to the workpiece. The stress waves generated due to the change in tool conditions during machining are captured. 'Micro 30 D' AE sensor with functional frequency of 100 to 350 kHz is utilized for acquiring the AE. The grinding wheel is made of Aluminum Oxide and workpiece is made up of Medium Carbon Steel. The experimental setup is as shown in Figure 1.

The set-up for conducting experiment consisting of a) Surface grinder b) Grinding wheel attachment for capturing AE signals, AE sensor, Preamplifier, DAC, AE-Win software for extracting features from the AE signature, Machine learning classifiers, and computer for data processing and data presentation. In the workpiece, the AE sensor is fixed. The output from the sensors is fed to the pre-amplifier. The amplified signal is fed to the Data Acquisition System (DAQ). From the DAQ the signal is converted into time domain signature (time Vs amplitude). AE-Win software is used to extract the AE features. Machine learning classifiers will provide the condition of the grinding wheel after training using the the AE features. The grinding wheel is subjected to ground the mild steel workpiece. Experimentations were performed with a grinding speed of 2500 rpm, a feed rate of 0.03 mm, and a depth of cut of 0.05 mm. These cutting parameters are set with the objective of maximizing the grinding wheel life.

The sensor used in the acquisition of signal consists of the piezoelectric transducer of differential type "Micro 30D" is fixed to the workpiece. AE sensor has a frequency response range of 100 – 300 kHz.

The sensor another end is attached to Pre-amplifier. The pre-amplifier another end is attached to Computer. Signals are extracted from the sensor and stored in the computer for signal processing. Feature extraction and signal processing are done in AE-Win software. For experimentation, a sampling rate of 1MSPS and a threshold of 25DB is used to collect the data. AE-Win software is used in this study for waveform processing, data storage, and display of feature date and reply of the data.

Figure 1. Auto-Feed Surface grinding machine – Experimental set-up

3. Grinding Wheel Conditions

In any TCM problem, first, it is necessary to identify the various tool conditions to be predicted by the statistical model. In an abrasive process such as grinding, abrasive wheels are used as tool to ground the workpiece to the required finish. After the few cycles of grinding operation, the abrasive grinding wheel loses its cutting property due to a) wear of the abrasive particles, b) failure of bond material and c) loading and glazing of the abrasive wheel. To continue the process, the abrasive wheel must be dressed for exposing the new abrasive grains. In a typical grinding cycle initially grinding wheel will have sharp abrasive particles. During the machining process, sharpness of the abrasive grains will gradually lose its sharpness and become dull or worn-out. In the dull or worn-out condition, the abrasive wheel must dress to regain its cutting ability. In general, grinding wheel has three states namely, a) sharp abrasive condition where the wear rate of the grinding wheel abrasive particle and work piece wear is very high b) intermediate condition where the equilibrium steady condition is established where wear rate in the grinding wheel is less and wear rate is more in the work piece and c) worn-out condition where wear in the grinding wheel is more and wear in the work piece is less. The grinding wheel life is mainly depending on the intermediate condition where the uniform wear takes place in wheel and workpiece. At the end of the steady state condition (intermediate state), or in the beginning of the third state (worn-out condition) the abrasive wheel must be sent to dressing operation to regains its cutting ability.

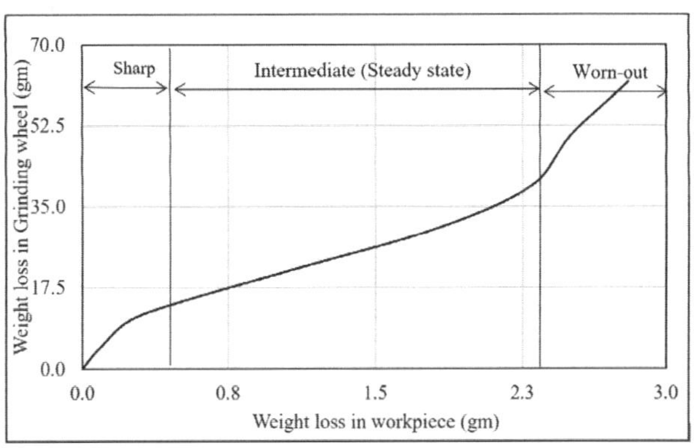

Figure 2. Grinding wear plot

Experimental approach is followed in this work to identify the grinding wheel conditions. The wear of the abrasive wheel and work piece is monitored continuously during the grinding process at various intervals of time. The weight loss in the workpiece and grinding wheel are plotted to get the grinding wheel wear plot as shown on Figure 2. From the slope of the wear plot (rate of change of wear), the grinding wheel conditions are identified. Wear data is recorded for the complete grinding cycle starting from sharp abrasive wheel condition to the worn-out condition of the tool. Three distinct zones namely sharp, intermediate, and worn-out states are identified from the wear plot. It was identified that the initial state of the grinding wheel is up to 7 passes of grinding operation. The intermediate state corresponds to pass 8 and 45. The grinding wheel lost its cutting ability after the pass 52 and identified as the worn-out condition.

4. Wavelets Coefficients Extraction in Discrete and Continuous Wavelet Domain

Wavelet based methods are used to analyse the sensor signatures in time -frequency domain. The important issue in wavelet transform is to identify the most optimum mother wavelet for the given tasks. The information content in the wavelet coefficients are not the same for all the wavelets. The time-domain signal is transformed into time-frequency domain using discrete and continuous wavelet transforms. Wavelet coefficients are extracted after decomposing the signal using the chosen wavelet. These wavelet coefficients carry information regarding the condition of the grinding wheel. Mother wavelets are characterized by properties such as compact support, orthogonality, symmetry, and vanishing moments. The mother wavelet chosen in this study for DWT is Haar, Daubechies, Symlets, and Coiflets. For CWT, the most preferred wavelet is the Morlet wavelet which is considered in this study.

4.1. Signal Decomposition in Discrete and Continuous Wavelet Domains

The DWT decomposes the signal, into a series of wavelet coefficients. multi-level wavelet decomposition is carried out to extract the discrete wavelet coefficients for all the mother wavelets considered in this study. In DWT, the AE signature is passed through a series of high pass and low pass filters for the signal decomposition. For improving the frequency resolution, multiple levels of decomposition are carried out. Decomposition is carried out using the AE signatures of sharp, intermediate, and worn-out conditions of the grinding wheel separately. In this study, it was found that two levels of decomposition produced better resolution. Decomposition is carried out for all the conditions of the grinding wheel using all the mother wavelets considered in this study. The decomposition of the dull condition signal after 2 levels of decomposition using Coiflet wavelet is shown in Figure 3. Similarly, for sharp, intermediate, and worn-out conditions, decomposition is carried out and wavelet coefficients are unearthed.

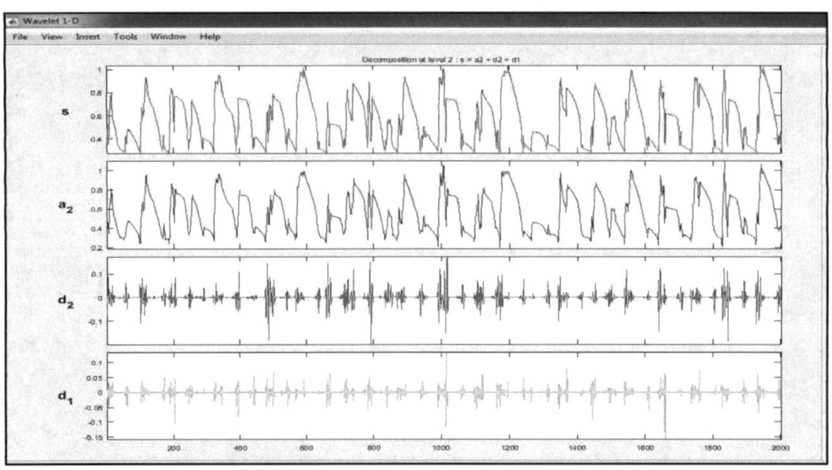

Figure 3. Decomposition of dull condition signal using Coiflet wavelet

In CWT, the values of scaling and translation factors are continuous. So, the output of the one-dimensional signal will be 2- Dimensional scalograms. From the 2D scalograms, ID features are extracted and given as an input to the J48 and CART classifiers. As an example, the Sharp grinding wheel condition scalogram obtained using the Morlet wavelet is shown in Figure 4. Similarly, scalograms of intermediate and worn-out conditions are plotted. From the scalograms, wavelet coefficients are extracted to train the J48 and CART algorithms.

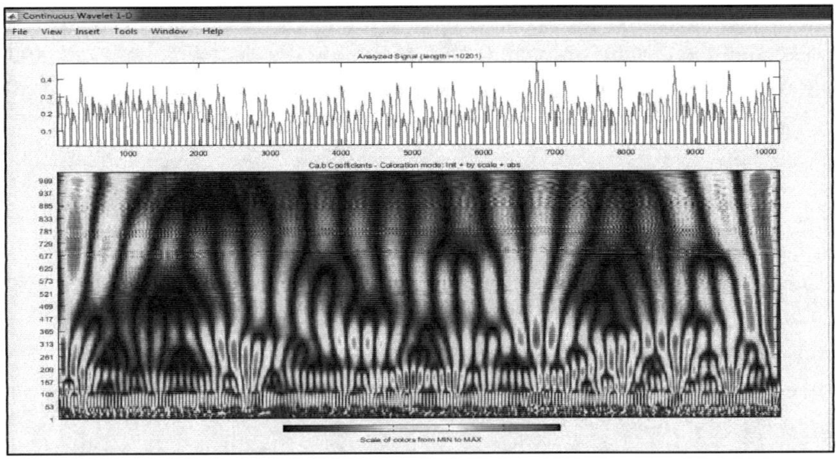

Figure 4. Scalogram of sharp condition signal using Morlet wavelet

5. Statistical Modelling and Performance of decision tree algorithms

A decision tree is a supervised learning classifier that divides data recursively using discrete or continuous data for classification or regression. The various algorithms used in the decision trees included ID3, C4.5 / J48, CART, C5.0, CHAID, QUEST, CRUISE, etc. Among these algorithms ID3, C4.5 and CART are quite popular. ID3, or Iternative-Dichotomizer (ID3), was the first decision tree algorithm proposed by Quinlan [18]. In ID3, the splitting criterion is used in information gain/entropy. ID3 handles only the categorical data and pruning is not possible. Also, ID3 is susceptible to outliers. Classification and Regression Tree (CART) algorithm handles both numerical and categorical data. Pruning can be done in CART and handles the outliers. The improvised version of ID3 is C4.5. C4.5 implementation in Java is called as J48 [19]. The advantages of J48 include a) handling continuous and discrete data, and b) the use of pruning to solve over-fitting issues. In this study, CART and J48 decision tree algorithms are used to build the statistical models using continuous and discrete wavelet coefficients. In the CART algorithm three variants viz. simple, medium, and complex decision trees are trained with wavelet coefficients.

The AE signature from the grinding wheel for the whole cycle was taken and the signal was analysed in the time-frequency domain by extracting wavelet coefficients in both DWT and CWT. In DWT, wavelet coefficients were extracted using the mother wavelets of Haar, Daubechies, Symlet, and Coiflet. In CWT, 2D scalograms were developed considering the sharp, intermediate, and worn-out conditions of the grinding wheel. From the 2D scalograms, ID features are extracted to train the classifiers. For CWT, Morlet wallets are used to generate the scalograms. J48 and CART statistical classifier models are built using the features extracted from DWT and CWT. Performances of the classifiers were compared by computing classification accuracy, and Kappa statistics. 10-fold cross-validation is performed to estimate the efficiency of the statistical model in predicting the condition of the grinding wheel.

5.1. Performance of classifiers

Classification accuracy of classifiers are computed using wavelet coefficients extracted in DWT and CWT domains are shown in Figure 5. Kappa statistics of classifiers are arrived using wavelet coefficients extracted in DWT and CWT domains are shown in Figure 6. Classification accuracy is the ratio of correctly classified instanced (data points) by the classifier to the total number of instances used for training and testing the classifier. Kappa measure will provide how good the model is performing over the performance of a model that simply guesses at random based on the frequency of each class. All the wavelets which were used with DWT and CWT are classifying the condition of the grinding-wheel with good precision. It was found that the Coiflet and Daubechies wavelets are effective in classifying the grinding wheel condition in DWT using a J48 classifier with a classification accuracy of 95.70 % and 95.30 % respectively. Compared to DWT, the CWT using Morlet wavelet giving a better accuracy of 97.06 % and making it an effective model for predicting the grinding wheel conditions. CART-complex tree trained with Morlet CWT features produces the maximum classification accuracy. This is due to the continuous values of scales and translation factors taken up by the transform while decomposing the signal. The kappa value also confirms the CART-Complex tree classifier as a good classifier.

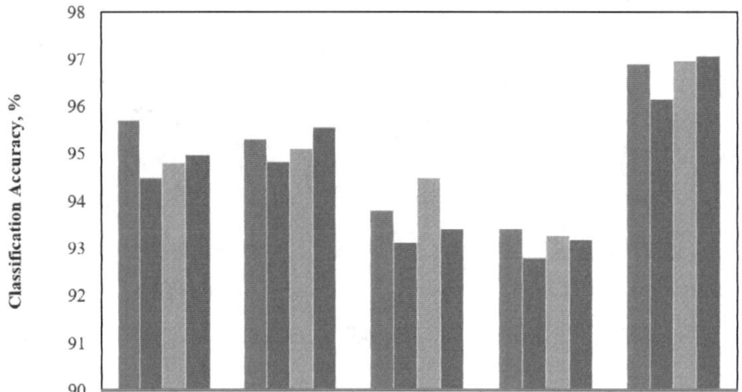

	Coiflet	Daubechies	Symlet	Haar	Morlet
J48	95.7	95.3	93.8	93.4	96.9
Simple	94.48	94.83	93.11	92.78	96.15
Medium	94.8	95.11	94.48	93.26	96.96
Complex	94.98	95.55	93.41	93.18	97.06

Figure 5. Performance of Classifiers – Classification Accuracy

Note: Coiflet, Daubechies, Symlet and Harr wavelets are used in DWT. Morlet is used CWT

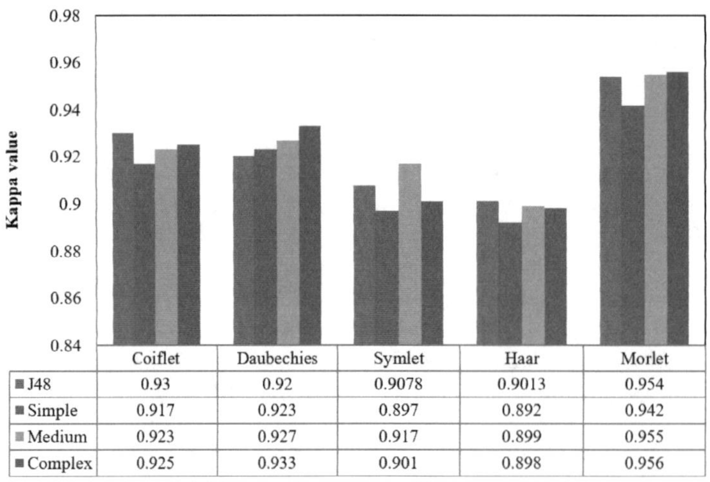

	Coiflet	Daubechies	Symlet	Haar	Morlet
J48	0.93	0.92	0.9078	0.9013	0.954
Simple	0.917	0.923	0.897	0.892	0.942
Medium	0.923	0.927	0.917	0.899	0.955
Complex	0.925	0.933	0.901	0.898	0.956

Figure 6. Performance of Classifiers – Kappa Statistics

Note: Coiflet, Daubechies, Symlet and Harr wavelets are used in DWT. Morlet is used CWT

6. Conclusions and Future Scope

In this study, it is observed that AE features extracted in the wavelet domain have good correlation with the grinding-wheel conditions. The grinding wear plot was established to identify the grinding wheel conditions. Using the established experimental set-up, AE signature is captured for the whole cycle of the grinding wheel. AE Signatures are transformed in the wavelet domain and the wavelet coefficients are extracted in the Continuous Wavelet domain and Discrete Wavelet domain. These coefficients carry potential information regarding the condition of the grinding wheel. The mother wavelet chosen in this study for DWT are Haar, Daubechies, Symlets, and Coiflets and for CWT, the Morlet wavelet is used. Using the wavelet coefficients extracted using DWT and CWT, statistical models are built for predicting the condition of the grinding wheel using the J48 and CART algorithm.

In the CWT domain, J48 and CART models trained with Morlet wavelet coefficients are producing better classification accuracy and kappa statistics compared to models trained with wavelet features extracted from the Discrete wavelet domain. The CART complex tree classifier can predict the grinding wheel conditions with a maximum classification accuracy of 97.06 % and kappa value of 0.956.

Wavelet analysis is found to be effective in predicting the grinding wheel conditions using J48 and CART decision tree algorithms. The CWT performed better compared to the DWT in classifying the grinding wheel conditions. The data and knowledge gathered in this research will be helpful for establishing an on-line monitoring system. The methodology proposed in this study may be extended to other types of grinding processes such as cylindrical grinding and centreless grinding.

7. References

[1] Dornfeld D A, Lee Y and Chang A 2003 Monitoring of ultraprecision machining processes. *The International Journal of Advanced Manufacturing Technology*. **21** 571-80

[2] Lee D E, Hwang I, Valente C M, Oliveira J F and Dornfeld D A 2006 Precision manufacturing process monitoring with acoustic emission. *In Condition Monitoring and Control for Intelligent Manufacturing*. (Springer, London) pp 33-54

[3] Liao T W, Hua G, Qu J and Blau P J 2006 Grinding wheel condition monitoring with hidden Markov model-based clustering methods. *Machining science and technology*. **10** 511-38

[4] Teti R, Jemielniak K, O'Donnell G and Dornfeld D 2010 Advanced monitoring of machining operations. *CIRP annals*. **59** 717-39

[5] Roth J T, Djurdjanovic D, Yang X, Mears L and Kurfess T 2010 Quality and inspection of machining operations: tool condition monitoring. *Journal of Manufacturing Science and Engineering*. **132** 041015-16

[6] Moia D F, Thomazella I H, Aguiar P R, Bianchi E C, Martins C H and Marchi M 2015 Tool condition monitoring of aluminum oxide grinding wheel in dressing operation using acoustic emission and neural networks. *Journal of the Brazilian Society of Mechanical Sciences and Engineering*. **37** 627-40

[7] Arun A, Rameshkumar K, Unnikrishnan D, Sumesh A 2018 Tool condition monitoring of cylindrical grinding process using acoustic emission sensor. *Materials Today: Proceedings*. **5** 11888-99

[8] Mouli D S and Rameshkumar K 2020 Acoustic Emission-Based Grinding Wheel Condition Monitoring Using Decision Tree Machine Learning Classifiers. *In Advances in Materials and Manufacturing Engineering*. (Springer, Singapore) pp 353-359

[9] Krishnan P S and Rameshkumar K 2020 Grinding wheel condition prediction with discrete hidden Markov model using acoustic emission signature. *Materials Today: Proceedings*. 2020 Jan 30.

[10] Chen X, Griffin J and Liu Q 2007 Mechanical, and thermal behaviours of grinding acoustic emission. *International journal of manufacturing technology and management*. **2** 184-99

[11] Liao T W, Tang F, Qu J and Blau P J 2008 Grinding wheel condition monitoring with boosted minimum distance classifiers. *Mechanical Systems and Signal Processing*. **22** 217-32

[12] Liao T W 2010 Feature extraction and selection from acoustic emission signals with an application in grinding wheel condition monitoring. *Engineering Applications of Artificial Intelligence*. **23** 74-84

[13] Benkedjouh T, Zerhouni N and Rechak S 2018 Tool wear condition monitoring based on continuous wavelet transform and blind source separation. *The International Journal of Advanced Manufacturing Technology*. **97** 3311-23
[14] Krishnakumar P, Rameshkumar K and Ramachandran K I 2018 Acoustic emission-based tool condition classification in a precision high-speed machining of titanium alloy: a machine learning approach. *International Journal of Computational Intelligence and Applications*. **17** 1850017
[15] Krishnakumar P, Rameshkumar K and Ramachandran K I 2018 Feature level fusion of vibration and acoustic emission signals in tool condition monitoring using machine learning classifiers. *Int. J. Progn. Health Manag*. **9** 1-15
[16] Alexandre F A, Lopes W N, Dotto F R, Ferreira F I, Aguiar P R, Bianchi E C and Lopes J C. 2018 Tool condition monitoring of aluminum oxide grinding wheel using AE and fuzzy model. *The International Journal of Advanced Manufacturing Technology*. **96** 67-79
[17] Xiao R, Hu Q and Li J 2019 Leak detection of gas pipelines using acoustic signals based on wavelet transform and Support Vector Machine. *Measurement*. **146** 479-89
[18] Quinlan J R 1986 Induction of decision trees. *Machine learning*. **1** 81-106
[19] Quinlan J R. 1996 Learning decision tree classifiers. *ACM Computing Surveys* (*CSUR*). **28** 71-2

Acknowledgments
This research is supported by ER & IPR, DRDO, ERIP/ER/0803740/M/01/1194.

Lifetime study of white organic light emitting diodes fabricated by both solution and vacuum evaporation processes

Fuh-Shyang Juang*[1], Kuo Chun Huang[1], Jeng Yue Chen[2], Lin Kuo[1], Lin Ann Hong[1]

[1]Institute of Electro-Optical and Materials Science, National Formosa University, Huwei, Yunlin, Taiwan
[2]Department of Electrical Engineering, National Formosa University, Huwei, Yunlin, Taiwan

*fsjuang@nfu.edu.tw

Abstract. In this study, a white light organic light-emitting diode is fabricated by solution processes. The focus of the research is to extend the lifetime of device. A hole transport layer doped with P-type material (F4-TCNQ) and a stepped electron transport layer TPBi (blocking holes and lowering the LUMO energy level) were employed in the device structure. And each layer thickness was optimized. At 9 V, the brightness can reach 206 cd/m^2, current efficiency 1.29 cd/A, and the device lifetime is extended from 0.8 to 8 hours. The CIE color-coordinates of white OLED samples are (0.38, 0.28).

1. Introduction

The solution process has low-cost and simple-process advantages. But the uniformity of film formation, uniformity of light emission, and the production of multi-layer structures cause the problems such as mutual solubility between solution spin-coatings. At present, there are many literatures using different structures and materials to fabricate white organic light emitting diodes (WOLEDs), and they would like to obtain good light sources and full-color displays. But they often do not work well. For example, the use of a single-layer light-emitting layer to fabricate WOLED requires a small amount of red guest to be doped into the blue-green emitting host. Due to the high efficiency of fluorescent energy transfer, the doping concentration of the red guest must be controlled at a very small level which increases the difficulty of the processes. The main purpose of this research is to fabricate WOLED by solution processes. First, a water-soluble polymer with good conductivity was employed as hole injection layer (HIL), which is not easily miscible with other solution materials. Then, a hole transport layer (HTL) and a dual-band emitting layer (EML) are spin-coated to produce a multilayer solution OLEDs. Finally, a hole blocking layer (HBL) is coated to control the current. HBL effectively avoids excessive charge accumulation at the EML/ETL interface, which may cause the device lifetime to decrease due to thermal effects [1]. Therefore, improving the efficiency of OLEDs and reducing the heat generated by the accumulated charge at the interface can effectively improve the lifetime of OLEDs.

2. Experiments

Firstly, the solution of the hole transport layer and the light-emitting layer were prepared according to the re-quired solution concentration. The hole transport layer is dissolved in Tetrahydrofuran (THF). The light-emitting materials (Host+Dopant) use chlorobenzene as the sol-vent to prepare the light-emitting layer solution. The sample bottle is placed on the hot plate, and the solution is continuously stirred at 50°C for more than 3 hours before using. The solution coating is carried out in the glove box. The procedure, HIL (PEDOT:PSS) → HTL → EML, is sequentially spin-coated. The p-type HTL material is NPB:F4-TCNQ. The appropriate spin coating speed and period were adjusted. After each layer of solution is coated, the sample was placed on a hot plate at 120°C for 20 minutes to evaporate the excess organic solvent. Then, the ETL TPBi, Alq3 and LiF, Al cathode materials were vacuum thermally evaporated. The light-emitting layer (EML) uses UBH-215 as the host material, and UBD-07 blue fluorescent dopant and DBP red fluorescent dopant as guest materials. The device structure is ITO/ spin-PEDOT: PSS (55 nm)/ spin-NPB: F4-TCNQ (4 %, 35 nm)/spin-UBH-215: 15% UBD-07: 0.25% DBP (50 nm)/ evap. TPBi (10 nm)/evap. Alq3 (30 nm)/ LiF (0.8 nm)/ Al (150 nm). The device band diagram is shown in figure 1.

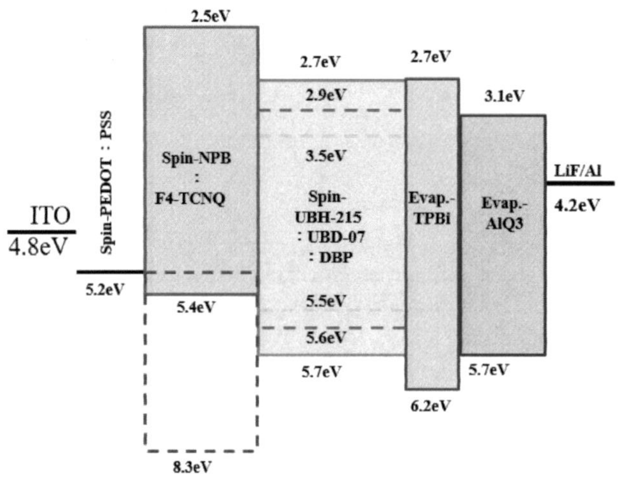

Figure 1. OLED band diagram.

3. Results and Discussion

The solution-type white light OLED structure uses UBH-215 as the host material, blue fluorescent dopant material UBD-07 and red fluorescent dopant material DBP as guest materials to make dual-band WOLEDs. Doping can greatly improve the overall lifetime and efficiency of the device. In addition, in order to avoid the occurrence of mutual dissolution between organic layers and result in the reduction of OLED efficiency [2], the ETL TPBi is deposited by thermal evaporation.

In order to improve the performance of the WOLEDs, a p-type hole transport layer (HTL) is inserted between the HIL and the EML by spin coating. Because the HOMO energy level of the hole transport material (NPB) is close to the LUMO energy level of the p-type doped material (F4-TCNQ), as shown in figure 1, the electrons at the NPB HOMO energy level will jump to the F4-TCNQ LUMO energy level, resulting in more free holes in HTL, thereby increasing the hole transport capacity [3], effectively increasing the hole transport rate in HTL. When the p-type doping concentration in HTL is 4% and the thickness is 35 nm, the device performance can be effectively improved, and the operating voltage is reduced from 7 V to 5 V at a brightness of 100 cd/m^2 [4-5]. The emitting spectra and CIE color-coordinates of WOLED with HTL doped with different dopant concentrations are shown in figure 2 (a) and (b), respectively.

In this study, TPBi was employed as both the hole blocking layer (HBL) and the ETL, and Alq3 as same as the ETL to form a stepped ETL structure, as shown in the band diagram of figure 1, which reduces the electron injection energy barrier and thus effectively promotes the electron injection into the EML to recombine with holes. It can reduce the OLED damage caused by thermal accumulation in the

EML/ETL interface. The optimal luminance and efficiency of the WOLEDs can be achieved. The device structure is ITO/spin-PEDOT: PSS (55 nm)/spin-NPB: F4-TCNQ (4 %, 35 nm)/spin-UBH-215:15%

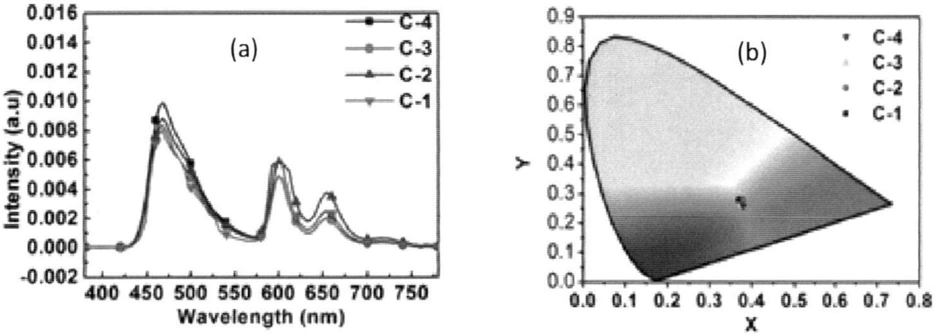

Figure 2. (a) The emitting spectra and (b) CIE color-coordinates of WOLED with HTL doped with different dopant concentration biased at 8V. C1 2% (0.37, 0.28), C2 4% (0.38, 0.28), C3 6% (0.37, 0.26), C4 8% (0.38, 0.26).

UBD-07: 0.25% DBP (50 nm)/evap.-TPBi (10 nm)/evap.-Alq3 (30 nm)/LiF (0.8 nm)/Al (150 nm). Table 1 is the parameters of each layer in WOLEDs. When the thickness of the HBL, TPBi, is fixed at 10 nm and the thickness of the Alq3 (ETL) is adjusted to 20 nm (D-1) and 30 nm (D-2), the CIE color coordinates (0.36, 0.28) and (0.36, 0.29) of the WOLEDs were obtained, respectively. According to the results shown in figure 3 and figure 4, at 10 V, the luminance of de-vice D-1 and D-2 are 225 and 260 cd/m2, respectively, and the current efficiency are 1.04 and 1.31 cd/A, respectively. As shown in figure 5, the efficiency of D-2 is better than D-1.

Table 1. Device configurations.

<table>
<tr><th></th><th>HIL</th><th>HTL</th><th>EML</th><th>HBL</th><th>ETL</th><th>EIL</th></tr>
<tr><td>NO.</td><td>PEDOT: PSS (spin)</td><td>NPB: 4 % F4-TCNQ (spin)</td><td>UBH-215: 15 % UBD-07: 0.25 % DBP (spin)</td><td>TPBi (evap.)</td><td>Alq3 (evap.)</td><td>LiF</td></tr>
<tr><td>A-1</td><td rowspan="4">55</td><td>X</td><td rowspan="4">50</td><td rowspan="2">30</td><td rowspan="2">X</td><td rowspan="4">0.8</td></tr>
<tr><td>B-2</td><td rowspan="3">35</td></tr>
<tr><td>D-1</td><td rowspan="2">10</td><td>20</td></tr>
<tr><td>D-2</td><td>30</td></tr>
</table>

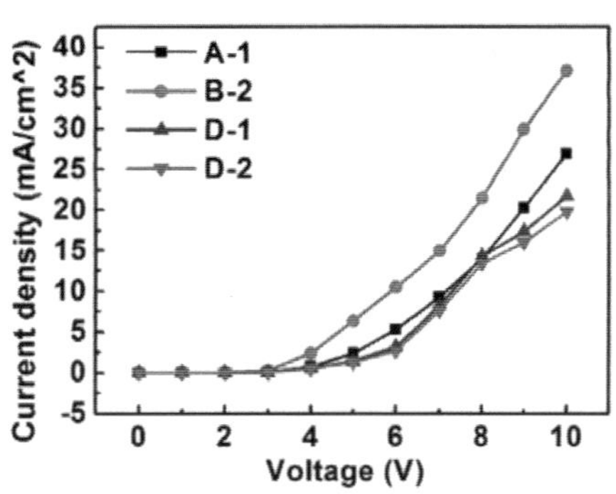

Figure 3. Current density versus voltage characteristics of OLED A-1, B-2, D-1 and D-3.

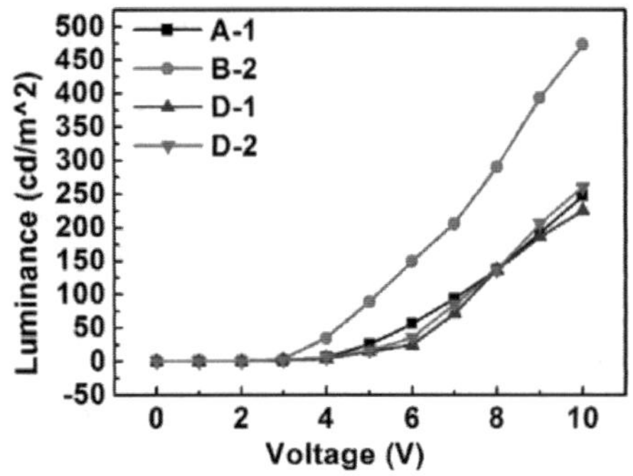

Figure 4. Luminance versus voltage characteristics of OLED A-1, B-2, D-1 and D-3.

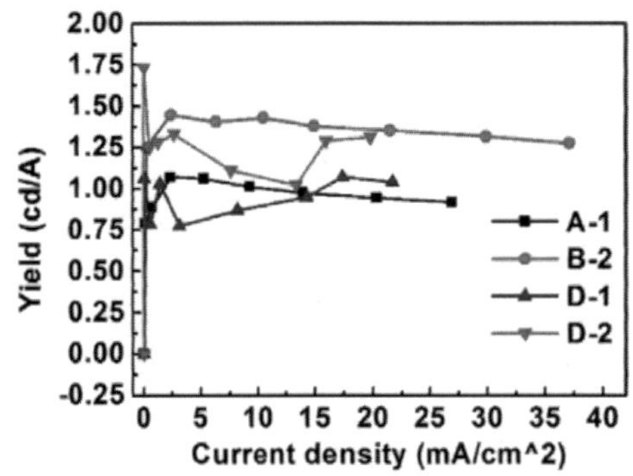

Figure 5. Luminance yield versus voltage characteristics of OLED A-1, B-2, D-1 and D-3.

It can be seen from the lifetime comparison in figure 6 that compared to A-1, B-2 WOLED have a p-type HTL inserted, which can increase the device lifetime. However, too many holes are easy to accumulate at the interface between EML and ETL (the number of electrons injected is insufficient), which causes the device to easily generate heat and wear out, thus the lifetime reduced.

Adding a stepped ETL structure on cathode side can effectively increase the injection of electrons into the EML, so that more e-h are recombined in the EML, thereby reducing the shortening of lifetime due to heat accumulation, and thus a stable WOLEDs is obtained.

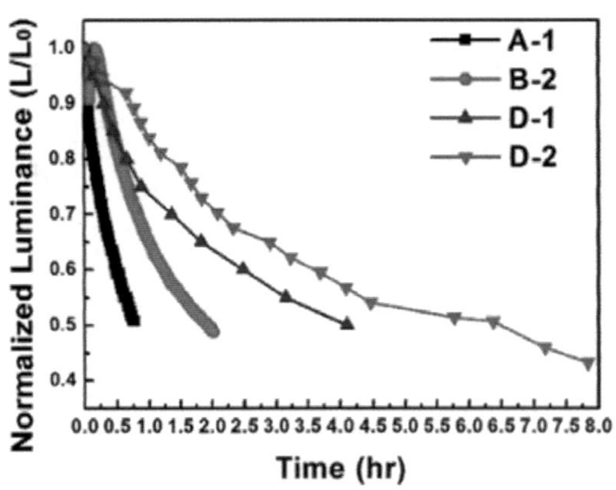

Figure 6. The measurement of WOLED lifetime.

4. Conclusions

In the WOLED structure, a HTL (NPB doped with F4-TCNQ) is employed and deposited by spin coating processes. Because the HOMO level of NPB is close to the LUMO level of F4-TCNQ, the electrons at NPB HO-MO can jump to the LUMO of F4-TCNQ. So that the HTL has more free holes, which help the holes to transport more easily inside the device, improve the conductivity of the HTL, and make the device luminance and overall lifetime is improved. The best device efficiency is obtained when the p-type doping is 4%. At a voltage of 9 V, the efficiency increases from 0.944 cd/A to 1.32 cd/A. The WOLED lifetime is increased from 0.8 hours to 2 hours.

The TPBi HBL effectively blocks the holes from flowing into the cathode directly. Finally, both Alq3 and TPBi are employed as a stepped ETL to help electrons more easily transfer from the cathode to the EML. It effectively enhances the hole-electron recombination efficiency, so that the OLED will not be damaged due to thermal effects from carrier accumulation. Although the efficiency of the WOLED is slightly reduced, but the overall lifetime has been increased from the original 2 hours to 8 hours. A more stable WOLED was obtained.

References

[1] Yang Q, Hao Y, Wang Z, Li Y, Wang H and Xu B 2012 *Synthetic Metals* **162** 398
[2] Yook K and Lee J 2011 *Organic Electronics* **12** 291
[3] Maenning B, Pfeiffer M, Nollau A, Zhou X, Leo K and Simon P 2001 *Phys. Rev.* **64** 195208
[4] Juang F, Kuo L, Lin Y, Zhang D and Tsai Y 2018 *Proc. SPIE Optics and Photonics, Organic Light Emitting Materials and Devices XXII (San Diego, California, USA)* vol 10736 (SPIE) p 107361M
[5] Juang F, Lee C, Chen J, Lin Y and Zhang D 2019 *Proc. The 2nd International Conference on Composite Materials Science and Technology (Tokyo Japan)* vol 600 (IOP Conference Series: Materials Science and Engineering) p 012018

Acknowledgements

The research is funded by the Ministry of Science and Technology (MOST) Taiwan under the project number of MOST 109-2221-E-150-014.

SCPI based integrated test and measurement environment using LabVIEW

B Aravind Balaji[1], S Sasikumar[1] and K Ramesh[2]

[1] Electronics and Communication Engineering, Hindustan Institute of Technology and Science, Chennai, Tamil Nadu, India
[2] Computer Science Engineering, Hindustan Institute of Technology and Science, Chennai, Tamil Nadu, India

Abstract. Test and Measurement processes play a vital role on the quality of hardware functionality testing. Automating these processes provides a better reduction in "time to market" and is more concerned on achieving higher test coverage area when it comes to mass production. Functional testing deals with the observation on the characteristics of the hardware to determine whether it replicates the needs for which it was actually designed. This can be achieved using set of Test and Measurement Instruments which can replicate the environmental condition in which the hardware is to be deployed. This paper provides an environment to integrate set of measurement instruments in a single platform (LabVIEW) and automate the measurement control using command language know as Standard Command for Programmable Instrumentation (SCPI). Script based control commands provide an easy configuration and remote control of instruments. And also programming interface control using Virtual Instrument Software Architecture (VISA) to control instruments using Ethernet and USB protocol.

1. Introduction

Hardware functionality test is the last passage in the assemble cycle to assess the performance of the purpose for which a hardware was structured. Functional testing duplicates the nature of the environment where the hardware is intended to work. Different properties are to be considered in resembling the environment in which they are to be deployed. The necessary generation and acquisition of those properties to create such an environment for the hardware is done using a set of different instruments that are incorporated together and automated using graphical language (LabVIEW). This research work gives a comprehensive exposure on execution of an integrated environment that gives a structure to building such test and measurement.

Conventional remote control of the instruments lack in standardization on two aspects namely programming compatibility and interchangeability. There are various instrument control methods [1] which provide a solution for this drawback. Generally control are carried out by sending a set of commands from the Test system controller(Test application) and ASCII characters are often the ones which are used for commands and the response are revert back from the instrument to the controller in ASCII form[2]. This was not standardized, each instrument contrast from one another which result in a lack of compatibility in programming same instruments of different vendors resulting in reprogramming the whole process with different commands depending upon the vendor. SCPI provides a solution for this providing a standardized script based commands followed by different instrument vendor making programming compatibility simpler [3] [4]. SCPI is simpler since it's in natural human readable form.

Similarly communication between instruments and test system controller are carried out using different communication protocol such as serial protocol(RS232), Ethernet protocol (TCP, UDP), USB protocol, GPIB [3] [4]. Interchangeability between such protocol results in reprogramming the configuration parameters during updating. VISA is an Application program Interface which provides software based link between the instrument and the controller regardless to the type of protocol in which they are connected. For instant the VISA Instruction to send the script based commands to an instrument is same regardless to the communication protocol of the instrument [3] [4] [6].
LabVIEW acts as test system controller sending such script based instructions or commands to the array of test and measurement instruments simultaneously in order to manage the test environment to the Unit Under Test (UUT) or Hardware [9]. Automated test patterns are generated to observe and evaluate the response. Test report generation is done to evaluate for further analysis.

2. Design flow of integrated environment

The figure1 shows the design flow diagram of integrated environment which is segregated of four consecutive level, Starting with the test system controller (LabVIEW) it is a graphical programming language based virtual instrumentation workbench which observes, evaluate and automates the measurements and also encapsulate the string based SCPI commands.

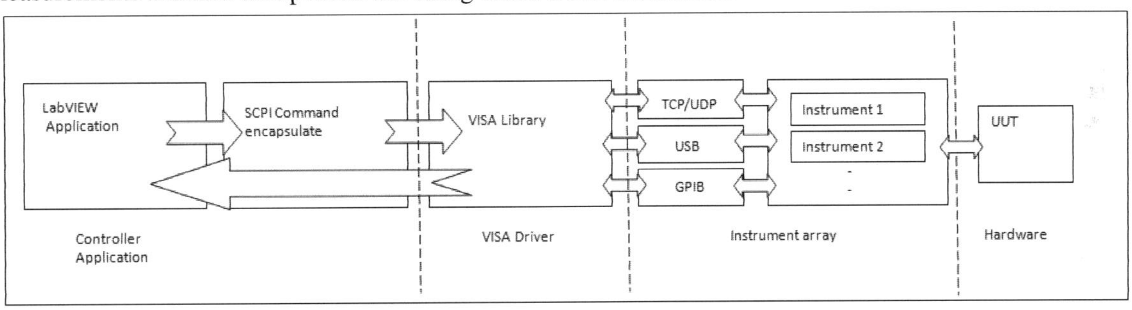

Figure 1. Design Flow of Integrated Environment.

SCPI is an ASCII-based programming language which utilizes a progressive structure knows as a tree system [10]. The related commands are gathered under a typical hub or root illustrated with an example below indicates the tree system for current and voltage configuration of a multimeter with arrow marks.

CONFigure:
→ CURRent:
→ AC:RANGe{<range>|MIN|MAX|DEF}
→ AC:RANGe?[MINimum|MAXimum|DEFault];
→ VOLTage:
→ AC:RANGe?

Where:

CONFigure - Root identifier.
CURRent - Second level identifier (under CONFigure).
VOLTage - Second level identifier (under CONFigure).
AC -Third level identifier (under CURRent and VOLTage).
Colon (:) - Isolates back to back identifiers.
Braces ({ }) - Parameter choice.
Angle brackets (< >) - specify value.
Vertical bar (|) -Splits parameters.
Semicolon (;) - Isolates identifiers within same root identifier.
Question mark (?) - Query.

VISA a software based architecture which is in form of APIs provides Write and Read function between instruments and controller application via any form of communication protocol [12]. The communication is invoked between them in the form of sessions each session connect the appropriate channel in which the resources (instruments) are connected that are indicated with unique resource name and followed by hardware array which is the set of instruments that are to be controlled and monitored. Each instrument is connected in form of different communication protocol such as Ethernet protocol (TCP /UDP), USB protocol and GPIB protocol. And finally the Hardware (UUT) is connected to the array of test instruments. The data flow can be bidirectional, for example the required properties can be queried by sending a string based SCPI command encapsulated by the test system controller (LabVIEW) to the appropriate instrument via VISA driver and the instrument response back to the application with the related data for the query via VISA driver.

3. Software design framework

Figure 2 shows the software design frame work which can be categorized with different layers out of which the VISA plays a vital role and it is the lower level/ Base layer of the framework and its acts as a back bone. Multiple VISA sessions are invoked for communication between array of Test and Measurement Instruments such as programmable AC source, DC Electronic load, multimeter, CRO etc.

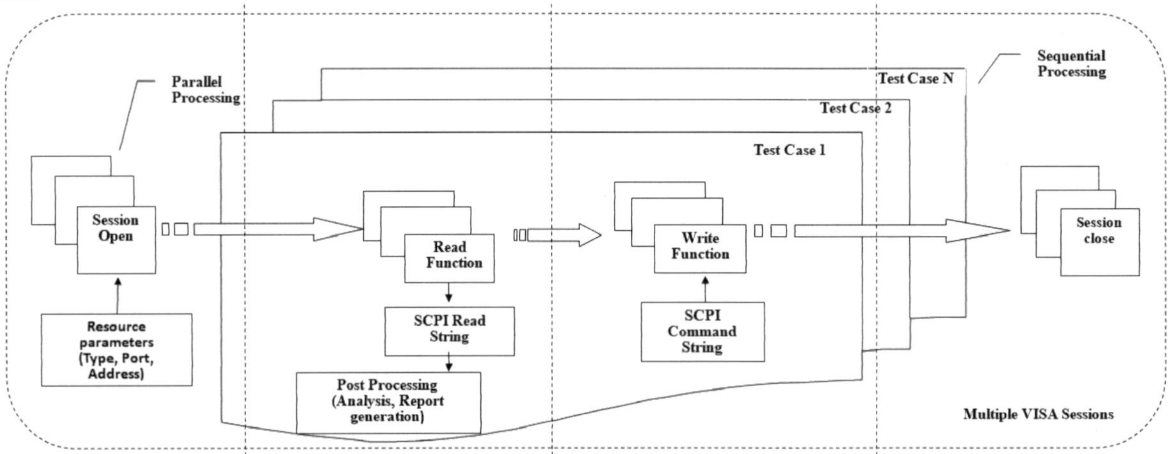

Figure 2. Design Flow of Integrated Environment.

These instruments replicate the source inputs and desired payload to the UUT and observe the response from the UUT. This frame work provides the integration of these Test and measurement instruments within a platform operating concurrently. Multiple VISA sessions are invoked by mean of parallel processing with unique channel for each instrument under the array.

VISA Library supports Write and Read functions in order to send commands and read the response back from the instrument each session starts with open session which provides a connection path between the instrument and the test system controller. This block is provided with all the required parameters such as a unique name/address, type of protocol (USB/TCP/UDP/GPIB), and Port number of the appropriate instruments to be connection. Once the connection is established, depending upon the required operation the read and write functions are used to instruct the instruments to respond. Once the response is acknowledged the session is closed, this process is repeated for all the other VISA Sessions concurrently.

The query is transferred to the instruments in form of string based ASCII codes (SCPI commands) via VISA Write functions and the response from the instruments are acquired using VISA Read function depending upon the different ways to replicate and observe the response of the UUT to evaluate it. The test cases are developed and are executed in a sequential process. There response from the instruments are monitored and compared with the desired expectation limits using windowing techniques these reading are tabulated in a report form to after each and every test cases subsequently.

4. Programming design

LabVIEW is a virtual instrumentation Automation tool used in industrial automations. Programs are based on graphical block representation and can be simply dragged, dropped and connected together.

The figure 3 below represents the design of a part of single test case algorithm illustrating the concurrent operation of sourcing 100V to the UUT using an AC source and measuring the AC voltage and current. The execution of the blocks are from left to right one after the another, parallel processing is possible by placing the block parallel to each others.

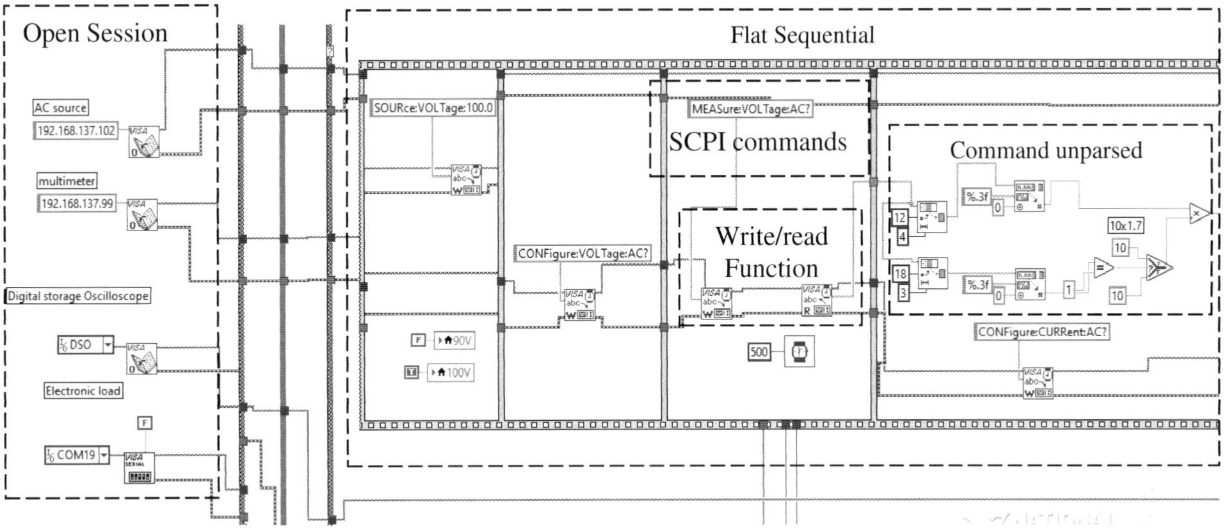

Figure 3. Test case part A.

VISA session acts as the back bone of the program in commanding the array of instruments depending upon the test cases. The session starts with VISA Open block that initiates the communication with the instruments and the resource name determines which instrument is connected. Open blocks are placed out of the loop which allows the session to be active until all the test case are executed which reduces the execution time cycle. The VISA Write and Read function are used to send and receive the strings to the instruments VISA Flush can be used to clear the I/O buffer in order to clear the previous query in the buffer which provides a better data refresh rate. The SCPI commands are parser in form string constants and are fetched to the instruments via Write function and similarly responds from the instrument are read by the Read function subsequently once the command is sent via write function and these received strings are unparsed using subset strings for further observation.

For example the figure 4 Shows the return string from AC source for a query to measure AC voltage i.e. 2.2808E+02 which mean its 2.28 x $10^2 \Rightarrow$ 228V. The return string is in form of array of strings which are to be subset or split into characters using indexing.

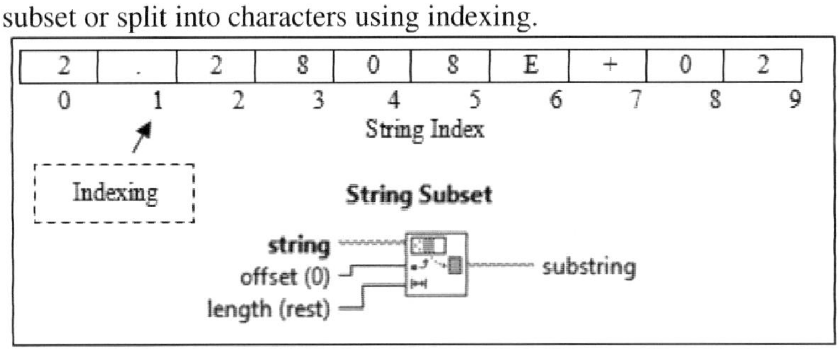

Figure 4. Subset string.

Here it has 0-9 index values, Subset block is used to split the array depending upon the offset and length of strings to be subset. Depending upon the test algorithm each process is carried out in sequential form one after another using flat Sequential within a test case. The entire program is looped under a case structure which provides a sequential operation of test cases one by one. Figure 5 illustrates remaining part of a test case algorithm.

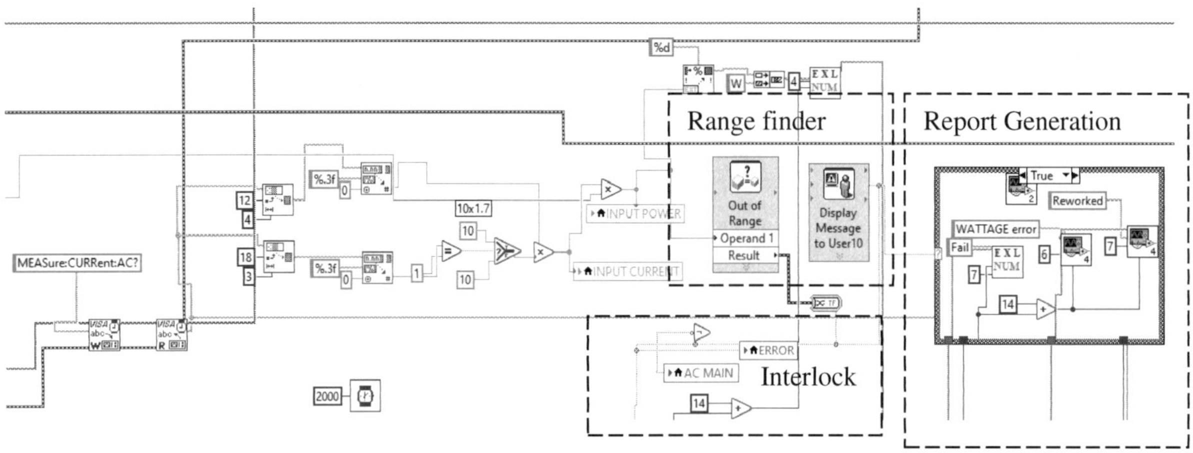

Figure 5. Test case part B.

The return string is split into sub sets and is observed if the properties are within specific range using windowing method. If the response is not within desired range the test is halted automatically. An interlock mechanism is provided to in such a way that whenever the desired range is not achieved the entire instrument array is reset to initial position and the power provided to the instruments is turned off terminating the program. This prevents the instruments from damage due to faulty UUT. The reports generation plays a vital role in the whole test process logging the response of the UUT in each and every test case throughout the test process in an excel Sheet automatically, with unique serial number for each UUT for identification of each UUTs response for further post observation query. VISA Close session is invoked at the end of the whole test process to terminate the connection.

5. Application implementation

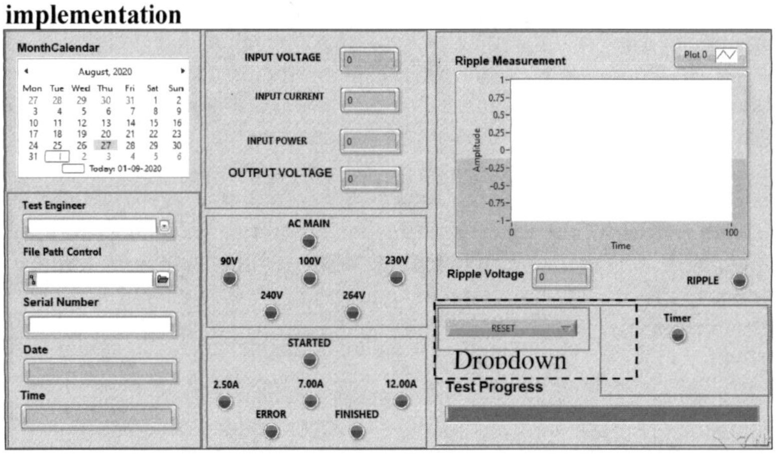

Figure 6. Graphical user interface.

With the reference of the Design Environment, The paper takes Functional testing software design of a SMPS for testing its integrity and efficiency. The design consists of a Digital Storage Oscilloscope (DSO), Multimeter, AC source and Electronic load. Cased and controlled via Ethernet and USB protocol using VISA Architecture. Test cases are developed depending upon the required characteristics to be evaluated. Test case measurement include Nominal load Current measurement, Power measurement, Ripple measurement, Calibration, Protection circuitry and MOSFET Switching Speed. Graphical User Interface (GUI) in the figure 6 is developed in such a way to provide a live visual status of the testing process using indicators and display of appropriate outputs parameters. Time stamping is provided for report generation and browser options for selecting the desired Excel sheet. Auto Rest control is provided to rest the program and instruments to initial condition. Dropdown menu is provided to choose between automated functional test and Auto reset.

6. Result
Thus the desired implementation of the functional test environment for SMPS is implemented and the test was successfully performed connecting the SMPS to the test equipment built with array of test. The figure 7 (a) shows the hardware response of the test equipment during runtime. A 24V SMPS is used as prototype for evaluating the test environment and test cases are developed to cross check the performance of the SMPS and maximum test coverage area is achieved. Fault diagnosis and interlock mechanism was cross check by testing a faulty board. Final output report was generated with appropriate reading printed in a excel sheet shown in figure 7(b) The plan strategy has perceptibly Decreased the "time to market" and shorted the development time.

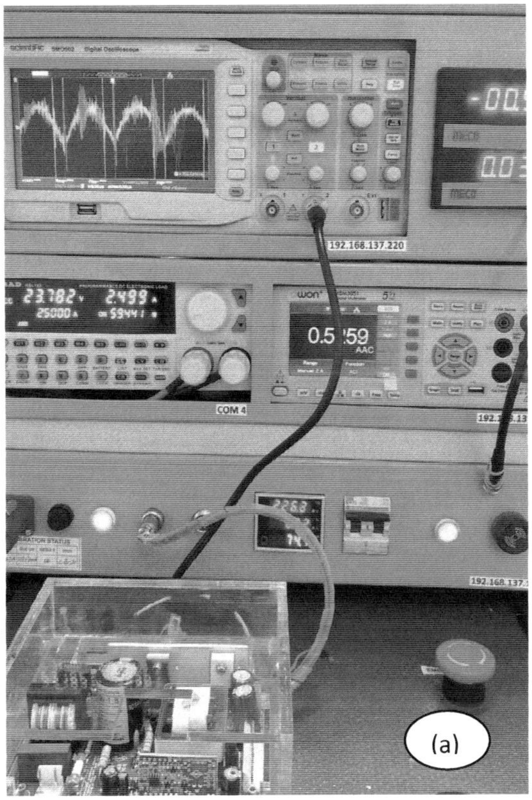

5.1 POWER UP	AC I/P = 230V	OUTPUT VOLTAGE	
(A)&(B)	DC O/P = +24V/2.5A	23.04V-24.96V	23.79V
5.3 INPUT MEASURMENTS	AC I/P = 85V	INPUT POWER	
(A)	DC O/P = +24V/3.75A	<105W	107W
	AC I/P = 140V	INPUT POWER	
(B)	DC O/P = +24V/3.75A	<105W	107W
	(NOMINAL LOAD)		
5.4 OUTPUT MEASURMENTS	AC I/P = 85V	OUTPUT VOLTAGE	
(A)	DC O/P = +24V/10mA	23.04V-24.96V	23.79V
	AC I/P = 85V	OUTPUT VOLTAGE	
(B)	DC O/P = +24V/10A	23.04V-24.96V	23.7V
	Re-Appear AC I/P = 140V	OUTPUT VOLTAGE	
(C)	DC O/P = +24V/10mA	23.04V-24.96V	23.79V
	AC I/P = 140V	RIPPLE VOLTAGE	
(D) & (E)	DC O/P = +24V/3.75A	<150mVpp	23.9V
(F)	SWITCHING PERIOD	20µSec - 24µSec	-
	AC I/P = 140V	OUTPUT VOLTAGE	
(G)	DC O/P = +24V/15A	23.04V-24.96V	23.4V

(b)

Figure 7. (a) Hardware response (b) Final report.

7. Future scope
This proposed worked can be further more enhanced by focusing more on Internet Of Things (Iot) based instrument control and fault diagnosis, prescient and prescriptive examination and data base management system based report generation for query based observation.

8. References
[1] Zhenjie zeng, Xiaofei zhu, Shiju qi, Xiaowei shen and Shangjun 2018 preliminary Exploration of Instrument control Technology *IOP conf. Ser: Mater. Sci Eng.* **394** 052064
[2] Wanzeng cai, Binwen wang and Shoulong zhang 2017 Remote Control and Data Acquisition of Multiple Oscilloscopes Using LabVIEW *IEEE trans. International conference Computer Technology Electronics and Communication* 920-24
[3] Liu guili and Kong quancun 2013 Design of Virtual Oscilloscope Based on GPIB Interface and SCPI *11th IEEE International Conference Electronic Measurement & Instruments* 294-98
[4] P Sumathi and Deol peter 2019 Instrument control through GPIB-USB Communication with LabVIEW *IEEE 28th international symposium on Industrial Electronics* 1583-88
[5] Elena daria tica , Lucian andrei perişoara and Alexandru Vasile 2016 An arduino platform for

remote control and bus testing of programmable instruments *IEEE Trans 22nd International Symposium Design and Technology in Electronic Packaging* 308- 311
[6] Wei cheng and Fenggui wang 2015 Remote automatic test system based on MATLAB using VISA over LAN *IEEE International Conference on Advanced Mechatronic Systems* 384-87
[7] Jiang chao, Xu wu-bin and Li bing 2013 Design of Instrument Control System Based on Lab VIEW *TELKOMNIKA Indonesian Journal of Electrical Engineering* **11** 6 3427-32
[8] S Reinhardt, C Butschkow, S Geissler, A Dirnaichner , F Olbrich , C E Lane ,D Schröer and A K Hüttel 2019 Lab::Measurement—A portable and extensible framework for controlling lab equipment and conducting measurements *Elsever Computer Physics Communications* **234** 216-22
[9] Chance elliott, Vipin vijayakumar and Wesley zink. 2007 National Instruments LabVIEW: A Programming Environment for Laboratory Automation and Measurement *Journal of the Association for Laboratory Automation* **12** 1 17-24
[10] SCPI Consortium, Standard commands for programmable instrumentation. (SCPI) http://www.ivifoundation.org/docs/scpi-99.pdf
[11] National Instruments Corp, USA, LabVIEW Measurements Manual [M] 2003:5-4-5-10 http://fy.chalmers.se/~f7xlh/LV_Kurs/LV_Meas_Man_2003.pdf
[12] National Instruments Virtual Instrument Software Architecture (VISA) https://www.ni.com/visa/

IOP Conf. Series: Materials Science and Engineering 1045 (2021) 012037 doi:10.1088/1757-899X/1045/1/012037

Extraction of microbial chitosan for piezoelectric application

M I ZAMLI[1], F B AHMAD[1*], and M H M AKMAL[2*]

[1]Department of Biotechnology Engineering, Kulliyyah of Engineering, International Islamic University. Malaysia (IIUM), Jalan Gombak, 53100 Kuala Lumpur, Malaysia.
[2]Department of Science in Engineering, Kulliyyah of Engineering, International Islamic University. Malaysia (IIUM), Jalan Gombak, 53100 Kuala Lumpur, Malaysia.

**email: farahahmad@iium.edu.my; maziatiakmal@iium.edu.my*

Abstract. Chitosan is a natural polysaccharide derived from chitin which is found in the fungus cell walls and crustacean shells. Chitosan has generated a great interest in piezoelectric application because of its interesting properties such as biodegradability, biocompatibility, and low toxicity. The purpose of this study is to focus on the cultivation, fabrication and characterization of chitosan thin film from fungal strain, *Aspergillus oryzae* cell walls. The fungi was cultivated in bioreactor. Fabrication of chitosan thin film via solvent casting method was optimized via one-factor-at-a-time (OFAT) with 2 parameters (drop-casting volume of solution and drying temperature). Pure chitosan dissolved in formic acid at 0.25 M concentration, dried at 60°C with 35 mL of solution volume gave the highest mechanical quality factor (3.22) and the lowest dissipation factor (0.327) for thin film fabrication. The optimized fabricate thin film was validated using fungal chitosan and shows the results of mechanical quality factor (3.68) and dissipation factor (0.248) which is comparable to conventional piezopolymer thin film. Therefore, fungal chitosan thin film obtained in this study has the potential to be used in piezoelectric application.

1. Introduction

Recently, chitosan has also been used as a natural substance for the application of piezoelectric in sensor devices [2]. Piezoelectric materials have the ability to convert the mechanical stress into electrical charge. The existence of non-centrosymmetric in chitosan's crystalline structure exert piezoelectricity [11, 13]. Crystalline structure of piezoelectric crystals consists of ions that are compacted to occupy positions in a specific repeating relationship to each other, thus exhibit piezoelectricity [17]. The conventional piezoelectric materials include piezoceramics such as lead zirconium titanate (PZT) and piezoelectric polymer such as poly (vinylidene fluoride) (PVDF), exhibit large piezoelectric effects [7]. These materials are widely used as commercialized actuators and sensors in intelligent systems and smart structures [7]. However, the high lead content in PZT has become a concern which can affect the health of humans, especially in the application of implantable biomedical devices, such as piezoelectric-based pacemakers. Piezoelectric polymers (e.g., PVDF) are more flexible, non-toxic and produce high piezoelectricity [15]. However, PVDF is also considered as a non-biodegradable polymer, the raw material is expensive and its synthesis releases toxic gases.

Bio-based functional materials have been studied for its piezoelectric behaviour [4]. Chitosan is also a biomaterial and a renewable source of natural biodegradable polymers which are also second most abundant polymers after cellulose. Commercially, it is produced from waste crustacean exoskeletons such as crabs and shrimps, as well as fungal cell wall [9].This study focused on the optimization the thin film fabrication by observing the piezoelectric coefficient, electromechanical coupling factor and tensile strength. The thin film fabricated could be further used in piezoelectric application either in energy harvesting or sensing.

2. Materials and method

2.1. Fungal cultivation and chitosan extraction

The fungi strain *Aspergillus oryzae* was grown on potato dextrose agar (PDA) in a petri dish and was incubated at 25˚C for 5 days. The pre-inoculum was cultured in the mixture of media that consist of 30 g/L glucose, 3 g/L peptone and 300 mL of distilled water [1]. The prepared media was transferred into a 2 L bioreactor with controlled parameters [14]. Extraction of chitosan from *Aspergillus oryzae* via alkaline treatment (sodium hydroxide (NaOH)) and acidic treatment (acetic acid) using method Abdel-Gawad et al. [1]. The extracted chitosan yield was calculated from the following Eq. (1):

$$Chitosan\ yield\ (\%) = \left[\frac{dry\ weight\ of extracted\ chitosan}{\text{dry weight of sample}}\right] \times 100\% \quad (1)$$

2.2. Optimization of thin film fabrication

Chitosan thin film fabrication via solvent casting method using formic acid (2.5 M) was optimized via OFAT. The parameters of OFAT were (1) drop-cast volume of chitosan solution and (2) drying temperature. The response factors analysed in this study were (1) mechanical quality factor (Q_m) and (2) dissipation factor (*tan* δ). The range for volume of chitosan solution, drying temperature and acid solvent concentration were chosen based on the previous study [12].

1 g of pure chitosan was dissolved into 99 mL formic acid solution (1% *v/v*) followed by solvent casting onto petri dish at different volumes of chitosan solution. The drying temperature and acid solvent concentration were kept constant at 60˚C and 0.25 M respectively [12]. After the volume of solution was chosen based on the best results, the drying temperature was varied (50˚C, 60˚C and 70˚C). The thin film was peeled off naturally after 3-4 hours of drying. The optimum parameters were used in the fabrication of fungal chitosan thin film.

2.3. Characterization of pure and fungal chitosan thin film

2.3.1. Chemical properties. A Fourier-transform infrared spectroscopy (FTIR) was performed in the middle infrared region (4000 cm^{-1} to 400 cm^{-1}) with a resolution of 4 cm^{-1} in the transmittance and absorbance mode for 16 scans at room temperature [8].

2.3.2. Electrical properties. Thin film sample was connected to the Keysight U1730C Series handheld LCR meter via copper wire as conductance and charged with a constant frequency (10 kHz). Mechanical quality factor (Q_m), dielectric constant (relative permittivity) and dissipation factor (*tan δ*) can be measured using a Keysight U1730C Series handheld LCR meter with the frequency at 10 kHZ.

3. Results and discussion

3.1. Optimization of solvent-casting techniques for thin film fabrication via one-factor-at-a-time method (OFAT)

The production of chitosan thin film via solvent-casting technique was optimized via OFAT with two parameters which were (1) drop-casting volume of chitosan solution and (2) temperature of oven drying; and two responses which are (1) mechanical quality factor (Q_m) and (2) dissipation factor (*tan δ*). The thickness of thin film was optimized by controlling the drop casting volume of chitosan solution.

The result (Figure 1 (a) and (b)) of chitosan dissolved in formic acid with 35 mL showed the highest mechanical quality factor (3.22) and the lowest dissipation factor (0.327). Superior piezoelectric materials, such as PZT and PVDF, are expected to possess high mechanical quality factor owing to the ability to increase dipole moments orientation which leads to the improvement of piezoelectric response within the crystal unit in thin-film [3]. The highest mechanical quality factor (Q_m) obtained was 3.22. The increasing trend in Q_m value might be related to the thickness of chitosan thin film, which was linearly increased and eventually, improved the densification and crystallinity of the chitosan thin film. The increment in crystallinity of the thin film can facilitate dipole-dipole movement. Therefore, it can lead to enhancement of piezoelectric response [15]. The reduction in the dissipation factor with increasing the chitosan thickness is likely due to the increment of localized charge carriers hence reducing the dissipation factor within the chitosan's crystalline structure [17].

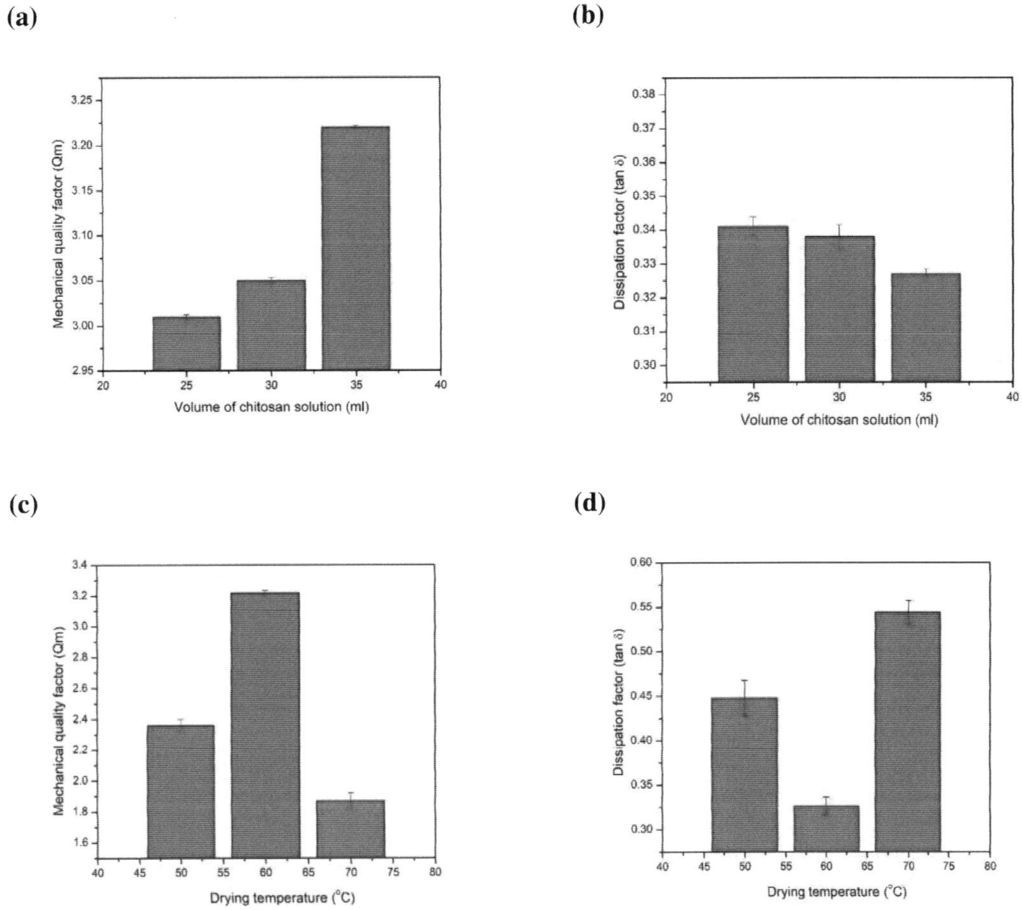

Figure 1. The results of OFAT on chitosan thin film fabrication via solvent casting method by ((**a**) and (**b**)) varying volume of drop-casting chitosan solution and on mechanical quality factor (Q_m) and dissipation factor (*tan δ*).

The drying temperature thin film was optimized by controlling the temperature of thin film drying treatment. Figure 1 ((c) and (d)) showed that the drying temperature of chitosan thin film at 60°C showed the highest mechanical quality factor (3.22) and the lowest dissipation factor (0.327). Increasing the temperature can increase the dissipation factor that may give rise to the vibrations of the lattice and some phonons that are created which interact with the charge carriers, subsequently, giving rise to electron-phonon scattering [6]. Therefore, the optimum parameters were drying temperature at 60°C with 35 mL of drop-casting chitosan solution volume.

3.2. Characterization of fungal chitosan for piezoelectric application

3.2.1. Functional group via FTIR. Based on the FTIR spectra, the distinctive functional group for amide were observed at vibration bending peaks of 1587 cm^{-1} for fungal chitosan (Figure 2) which indicated the stretching vibrations of C-O group (Amide II). The presence of bands at the peak of 3366 cm^{-1} in fungal chitosan indicate the presence of intense dimeric OH stretch. The presence of bands at peak 1587 cm^{-1} in fungal chitosan can be attributed to the C=C stretching in Amide 1 region. The bands at peak 1419 cm^{-1} in fungal chitosan is due to the presence of aromatic C=C stretch in Amide III region.

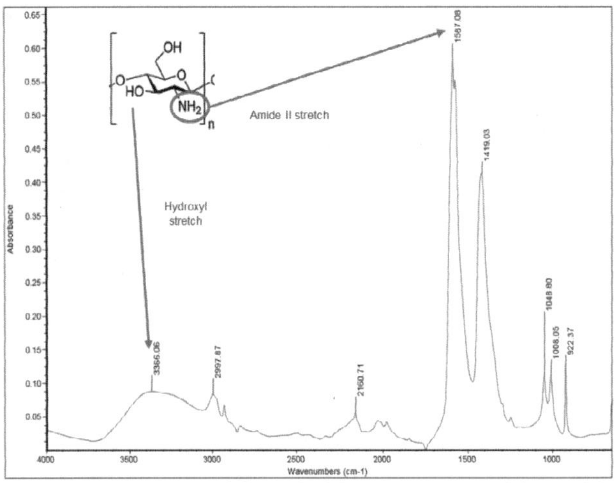

Figure 2. FTIR spectrum of fungal chitosan extracted from this study.

3.2.2. Mechanical quality factor (Qm). Q_m values obtained for pure chitosan and fungal chitosan were 3.22 and 3.68, respectively (Table 1). The fungal chitosan showed slightly higher Q_m value compared to pure chitosan. This phenomenon could mean that fungal chitosan film has slightly higher efficiency to convert mechanical force into electrical force. The difference in Q_m reading for both films can be ascribed by the surface morphology's factor that causes an abrupt decrease/increase in mechanical quality factor [5].

3.2.3. Dissipation factor (tan δ). The dissipation factor measurement was undertaken on pure chitosan and fungal chitosan thin films. The dissipation factor is a measure of loss-rate energy during the reversal of electric polarization. Table 1 showed the dissipation factor ($tan\ \delta$) of pure chitosan and fungal chitosan was 0.327 and 0.248, respectively. Higher $tan\ \delta$ value for pure chitosan likely occurred to the increment of dipole moment interaction in the pure chitosan crystalline structure. In addition, less interaction of the phonons in the lattice structure due to some factors like surface inhomogeneity will result in low electron phonon scattering, thus increasing the dissipation factor of fungal chitosan [13].

Table 1. Mechanical quality factor and dissipation factor values of fungal chitosan compared film with pure chitosan from this study and benchmarked with polyvinylidene fluoride (PVDF) film [10, 16], which has been widely used in piezoelectric application.

Piezoelectric film	Fungal chitosan	Pure chitosan	PVDF
Type of piezoelectric films	Biopolymer	Biopolymer	Synthetic polymer (conventional piezoelectric material)
Mechanical quality factor (Q_m)	3.68	3.22	~4 – 10
Dissipation factor (*tan δ*)	0.248	0.327	~0.05 – 0.3

Fungal chitosan film obtained in this study was also benchmarked with polyvinylidene fluoride (PVDF) film (Table 1), which is a conventional piezoelectric polymer [16]. The results of Q_m and *tan δ* of chitosan thin films were comparable to conventional piezopolymer thin film. Therefore, fungal chitosan thin film obtained in this study has the potential to be used in piezoelectric application.

4. Conclusion

Fungal biomass of *A. oryzae* fungi that was cultivated in bioreactor was used for extraction and deacetylation via alkaline-weak acid treatment for the production of fungal chitosan. Chitosan thin film fabrication via solvent casting method was further optimized for maximum mechanical quality factor (Q_m), in which the optimum parameters selected were drying temperature at 60°C with 35 mL of drop-casting chitosan solution volume. The characterization of chitosan thin films showed that the films were comparable with the commercial PVDF thin film. Therefore, this study revealed the potential of using fungal chitosan thin film as good biomaterial for piezoelectric application.

Acknowledgement

The authors would like to acknowledge financial assistance from Ministry of Education under Fundamental Research Grant Scheme (FRGS) (FRGS/1/2019/TK02/UIAM/02/3).

References

[1] Abdel-Gawad K M, Hifney A F, Fawzy M A and Gomaa M 2017 Technology optimization of chitosan production from Aspergillus niger biomass and its functional activities *Food Hydrocolloids* **63** 593-601

[2] Ahmad F B, Maziati Akmal M H, Amran A and Hasni M H 2020 Characterization of chitosan from extracted fungal biomass for piezoelectric application *IOP Conference Series: Materials Science and Engineering* **778** 012034

[3] Akmal M, Warikh A, Azlan U, Azam M, Anand T and Moriga T 2016 Structural evolution and dopant occupancy preference of yttrium-doped potassium sodium niobate thin films *Journal of Electroceramics* **37** 50-7

[4] Akmal M H M and Ahmad F B 2020 *Advances in Nanotechnology and Its Applications,* ed A T Jameel and A Z Yaser (Singapore: Springer Singapore) pp 63-82

[5] Amarande L, Miclea C and Tanasoiu C 2007 Evaluation of the accuracy in determining the mechanical quality factor for piezoceramic resonators *Ferroelectrics* **350** 38-47

[6] Bhuvaneshwari S, Sruthi D, Sivasubramanian V, Niranjana K and Sugunabai J 2011 Development and characterization of chitosan films *Ijera* **1** 292-9

[7] Cheng L-Q, Xu Z, Zhao C, Thong H-C, Cen Z-Y, Lu W, Lan Y and Wang K 2018 Significantly improved piezoelectric performance of PZT-PMnN ceramics prepared by spark plasma sintering *RSC advances* **8** 35594-9

[8] Hänninen A, Sarlin E, Lyyra I, Salpavaara T, Kellomäki M and Tuukkanen S 2018 Nanocellulose and chitosan based films as low cost, green piezoelectric materials *Carbohydrate polymers* **202** 418-24

[9] Kanimozhi S, Ramya D and Menaga Gandhi M 2014 Production and Optimization of Chitosan from Aspergillus niger by Solid State and Submerged Fermentation and Evaluation of its Antibacterial and Antioxidant activity *Drug Invention Today* **6**

[10] Park S-H, Lim E-C, Park K-S, Kang D-H, Han S-O and Lee D-C 1997 The dielectric properties of functional PVDF thin films by physical vapor deposition method(vol 2): IEEE) p 1144-6

[11] Pohanka M 2018 Overview of Piezoelectric Biosensors, Immunosensors and DNA Sensors and Their Applications *Materials (Basel)* **11** 448

[12] Praveen E, Murugan S and Jayakumar K 2017 Investigations on the existence of piezoelectric property of a bio-polymer–chitosan and its application in vibration sensors *RSC advances* **7** 35490-5

[13] Rajala S, Schouten M, Krijnen G and Tuukkanen S 2018 High Bending-Mode Sensitivity of Printed Piezoelectric Poly(vinylidenefluoride-co-trifluoroethylene) Sensors *ACS Omega* **3** 8067-73

[14] Rane K D and Hoover D G 1993 Production of chitosan by fungi *Food biotechnology* **7** 11-33

[15] Singh D, Choudhary A and Garg A 2018 Flexible and robust piezoelectric polymer nanocomposites based energy harvesters *ACS applied materials & interfaces* **10** 2793-800

[16] Toda M 2002 Cylindrical PVDF film transmitters and receivers for air ultrasound *IEEE transactions on ultrasonics, ferroelectrics, and frequency control* **49** 626-34

[17] Zhu X 2010 *Piezoelectric ceramic materials: processing, properties, characterization, and applications*: Nova Science Publishers)

IOP Conf. Series: Materials Science and Engineering 1045 (2021) 012038 doi:10.1088/1757-899X/1045/1/012038

Design and Implementation of FEC for CDMA System

M N Smrithi and A R Abdul Rajak

Department of Electrical, Birla Institute of Technology & Science, Pilani-Dubai Campus, Dubai International Academic City, P. O. Box No. - 345055, Dubai, UAE.

malavalli.smrithi@gmail.com and abdulrazak@dubai.bits-pilani.ac.in

Abstract. In every wireless communication the data that is transferred is at a high risk of falling under the clutches of various types of disturbances termed as noise, there are various types of noises which are thermal noise, shot noise, IF noise, burst noise, avalanche noise etc. they occur due to a number of factors it can be due to temperature, device malfunctioning and frequency interference. Due to this, the data being transferred in the wireless communication channel gets corrupted resulting in data loss. This problem is rectified by employing FEC (Forward Error Correction) method on the data being transferred. This research focuses on the implementation of various FEC codes in CDMA system to study its detailed performance analysis with the help of MATLAB Software.

1. Introduction

The major highlight of the current century is study and focus which is concerned with increasing the stability of interaction on the digital programs, is the Error Correction coding. There is no doubt that even the CDMA system needs to be protected and at the same time it should be able to detect and correct the errors when the data of information is being transferred in different types of environments, an effective and better way is included via research carried out recently [1].In the forward channel of CDMA [2] the existence of repetition of a symbol normally occurred is changed using an algorithm which is new and used for encoding. The new algorithm is named as a logic symbol repetition called "Memorizing" an encoding scheme where this new logic symbol repetition is having the capacity of making the right corrections to the bits having errors. The performance of the error correction is studied using a simulation program which was developed, it studies the bit rate of the information bits that were received.

Spread spectrum systems [3] especially CDMA systems counterpart well by applying the methods of forward error correction. This research will cover a few viewpoints of CDMA systems with coding of FEC channel. FEC/CDMA arrangements could be believed as schemes that were integrated, besides inward and decoding beyond it. In an iteration, the soft benefits of the beyond program bits are next utilized as a priori benefits for the inner CDMA decode/de-spreader [4]. This could be formulated by decoded bits (soft) obtained from the supplementary users beyond a decoder which could be compared to a scheme used for the cancellation of interference [5]. IS 95 arrangements requests display a significant gain from the repetitions and the soft output. This is believed to be spread to CDMA [19] arrangements alongside PN spreading sequences whereas the coding of FEC in 1-Dimension is given during the periods association as well as FEC coding during the supplementary dimension which is given by a logical program in the direction of CDMA program. The supplementary CDMA codes send the parity bits of this particular program. The supplementary interference [6] deterioration is hereby provoked and is outweighed by the supplementary gain of FEC coding attained by decoding iteratively [7].

2. Forward error correction.

Forward Error correction is a channel coding method which is used to control the extent of errors that take place during transmission through a communication channel. The main idea is that the sender keeps adding extra bits (redundant bits) to his/her message, this is done by using the technique of ECC (Error Correcting Code) [8].

In telecommunication and information theory, forward error correction (FEC) is termed as a system of error control for data transmission, in this case the sender will add redundant data to its message, and this is also known as an error correction code[9]. This ensures that the receiver need not depend on the sender to send any further information after the transmission, it creates the opportunity to detect the errors as well as correct them. The requirement of a back-channel is avoided in forward error correction and it proves to be greatest plus point but it requires a higher bandwidth. Due to this reason FEC is mainly used in circumstances where retransmission of data is difficult or it is difficult to incorporate any other error correcting method.

In digital signal processing, it is used as the first step and located close to the receiver. It is also observed that FEC circuit are found as in most of the analog-to-digital converting process, it is also sued for modulation and line encoding. The Bit error rate that is given by the FEC is used for fine tuning in the analog signal receiving device.

2.1 How it works.

Forward Error Correction [10] can be achieved by adding redundancy bits to the transmitted information which can be done by using a predetermined algorithm. The redundant bit used is actually an invariable complex function of many original information bits. The original information may or may not appear in the encoded output, in order to identify which codes originally belong to the output the codes are classified into two, those which are a part of unmodified input in the output are systematic, while those are not are termed as non-systematic [11].

Considering a simple example mentioned in Table 1 to understand, this could be an analog to digital converter which will sample three bits of data in every transmitted signal [12]. The transmitted bit is considered to be zero if the three samples received are majorly zero and the transmitted bit is considered to be one if the transmitted bit has majorly one in the three bit sample received. In this way it can be concluded that the easiest way for correcting the error is for the receiver to look for most frequently occurring bit in the sample of three.

Table 1. Binary Interpretation

Triplet Received	Interpreted as
000	0
001	0
010	0
011	1
100	0
101	1
110	1
111	1

2.2 Averaging noise to reduce errors.

The corruption of some symbols due to noise ruins the original data and becomes impossible for the receiver to read the data, as he/she depends on the original uncorrupted data to be sent from the sender. The effect of risk pooling ensures that the signal to noise ratio is above a particular minimum level. This enables FEC to be used as a form of error correction technique. The method also helps to reach the limit imposed by Shannon. Hence it is m stronger than other codes which are used. The method of interleaving the FEC codes data further reduces the all or nothing properties of transmitted FEC codes [21].

Like every other method the averaging noise has got its own limit, this method can work well only in narrowband data. Majority of the telecommunication system uses fixed channel code which is designed to withhold the worst expected case of error in bits, if the error rate exceeds the limit then the communication channel fails to work. However there are certain adaptive systems, which works on a fixed FEQ method until the error occurred is within the channel capacity once it exceeds then the FEQ method automatically gets shifted to ARQ method [11].

2.3 Type of FEC.

Block coding and Convolution coding [15] are the two main classifications under Forward Error Correction. Block coding works on the concept of blocks, where the size of the block is determined with a predetermined size of bits it can also be called as packet. Convolutional codes on the other hand deal with bits or symbol streams of a particular fixed length. It is to be noted that the convolution code can be converted into a block code if needed based on the situation. Decoding of Convolutional codes is mostly done using the Viterbi algorithm, apart from this there are other methods of decoding available. Similarly in the case of block coding, it comes in different versions, but the most commonly used is Reed- Solomon coding. It is used on a Compact disc even found in hard drives [16].

2.4 Averaging noise to reduce errors.

In block coding the entire message that is to be sent is divided into blocks of fixed length. Each of them will be represented as 'u' and each block will have 'k' information bits. The total number of code words possible in block coding is 2^k and there are 2*k block codes. This set of 2^k code words of length n is called (n, k) block code. The ratio is determined by the formula stated below.

$$R = k/n \tag{1}$$

Where R is called the code rate. It is defined as the total number of information bit that is entering the encoder per each symbol that is transmitted. The transmitted 'n' depends only on the k-bit. This is why the encoder does not require a memory. It is to be ensured that if the binary code can be useful only if the code rate is less than 1 or in other words the value of 'k' is lesser tan 'n'. How to choose these redundant bits to achieve reliable transmission over a noisy channel is the major problem in designing the encoder [17]. Block codes are referred to as (n, k) codes. A block of k information bits is coded to become a block of n bits. But before going any further with the details, let's look at an important concept in coding called Hamming distance. Let's say that 10 integers, 0 to 9 by a digital sequence is required. Sixteen unique sequences can be obtained from four-bit words. The first ten are assigned, one to each integer. Each integer is now identified by its unique sequences of bits as in the Table 2 below. The remaining six possible sequences are unassigned. Assigning 4-bit sequences to the 10 integers.

Table 2. Bit Sequences.

Sequence	Integer	Sequence	Integer
0000	0	0111	7
0001	1	1000	8
0010	2	1001	9
0011	3	1010	Unassigned
0100	4	1011	Unassigned
0101	5	1100	Unassigned
0110	6	1101	Unassigned

In the above example there are 16 different types of codes available, together they form the code space. Among these only 10 of these codes were used. The one that is not labeled or used is called as an unassigned code and if received by the receiver then it is discarded [22].

2.5 Convolutional coding.

A convolutional code is a kind of code which corrects the errors in the telecommunication where. The information bit is represented as 'm' and is then converted to 'n' bit-symbol. It is to note that the code rate is the information bit divided (m) by converted bit (n). The limit of code is 'k', which are the last symbol of information that is converted. Convolutional code is used in different applications including satellite and mobile communication. It is generally implemented with "Reed Solomon" which is termed to be as a hard decision code.

Before turbo code came into existence, it was used widely as it was the most closest to the Shannon Limit. Here also allows k-bits of block information series 'u' and generates a protected series (code word) 'v' of the 'n' symbol-block. Yet, each block-encoded relies upon not only the consequent k-bits information simultaneously, but also- on the 'm' prior information block. The encoder has a storage order of 'm' other sets of encoded series by 'k' input, 'n' output encoder of storage order 'm' is known as an (n, k, m) convolution code. The repetitive bit can be included by improving storage order 'm'. The usage of the memory is to accomplish efficiently transmitting over a channel that is noisy which is a significant issue in developing encoder. The integrated codes are formed from the block codes which correct the burst errors, as the outer codes by frequently using the convolutional codes as the inner codes. In the decoding algorithms, which are Viterbi decoding for the convolution codes, the errors likely arise in bursts as it is a consequence in the trellis diagram from going in the wrong path. The correction efficiency of the burst errors in the code outer is normally used to pull through from such patterns of the burst errors while the inner codes are decoded.

3. Implementation of FEC in CDMA.

Below Figure 1 is the block diagram of given CDMA [18]. Data to be transmitted is encoded by channel coding techniques. Here in this work block code and convolution code are implemented. This encoded data is further getting spread by PN Sequence, the spreaded data is transmitted to wireless channel through BPSK modulation process. This wireless channel adds AWGN to the modulated data because of AWGN, BPSK demodulator at the receiver side detects error. This data is despreaded to get back the original data by the PN sequence at the receiver. If despread data is with error, that error is getting corrected by channel decoder and hence BER is reduced.

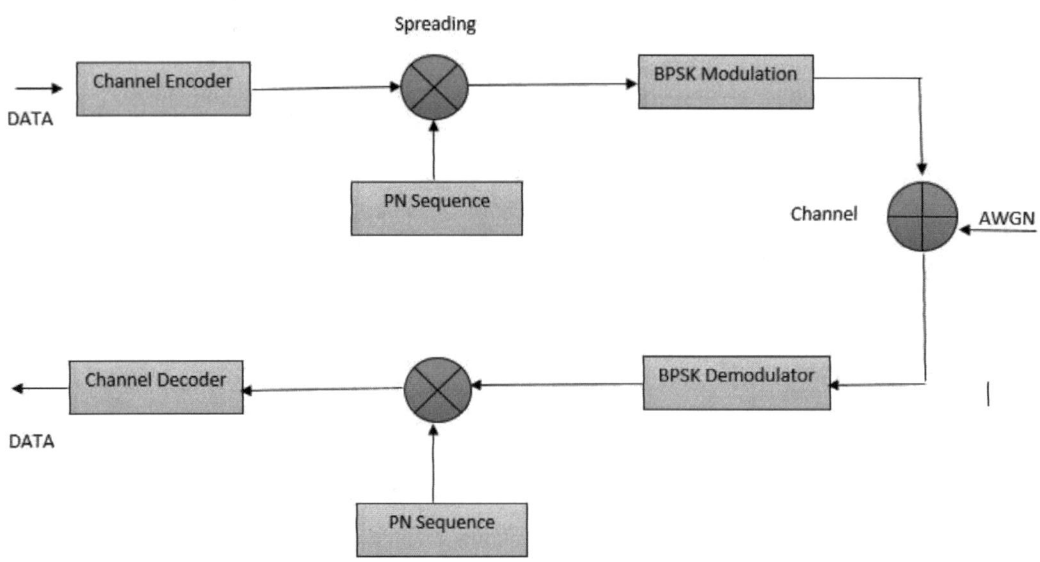

Figure 1. Block diagram of implementation of FEC in CDMA

3.1 Graphical results obtained after implementation in Matlab.

Figure 2 shows binary sequences to be transmitted. They are original binary sequences .They are not coded by FEC process and Spreading process. This data is applied to the input of FEC encoder to channel coding process.

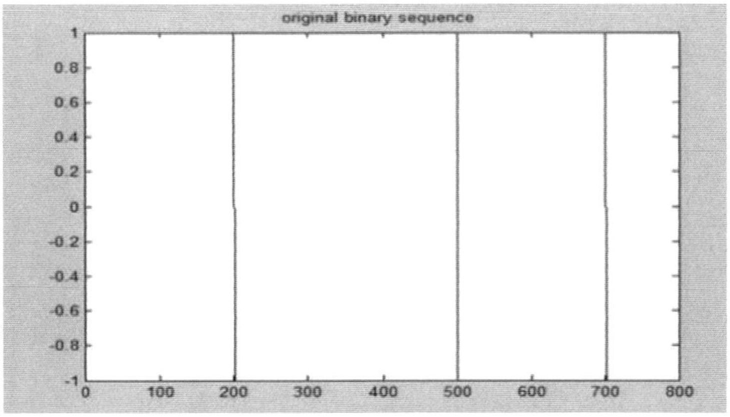

Figure 2. Original Binary Sequence

The PN sequence are available in Figure 3.These PN sequences spread binary sequence to be transmitted. The details of spreaded binary data is available as shown in Figure 3.This action leads to get coded data. After the binary sequence get spreaded, the coded data gets modulated. BPSK modulation is employed. The waveform in figure shows the details of BPSK modulation of code data and transmitted through AWGN wireless channel.

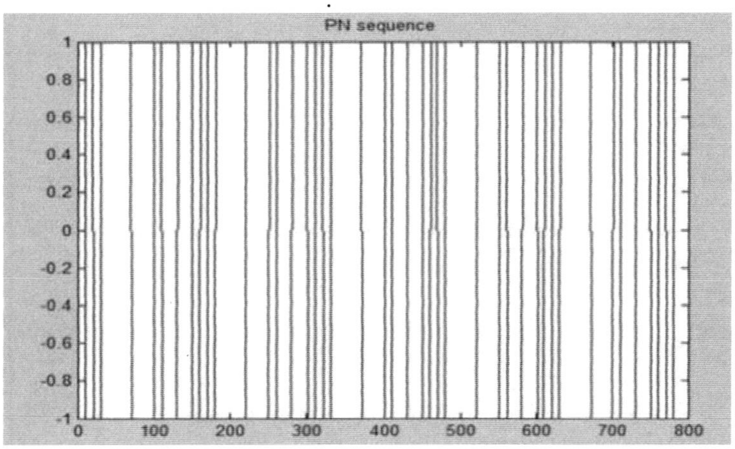

Figure 3. PN Sequence generated

Figure 4 shows the output signal of circuit employed at the receiver. Because of noise in the channel, the received signal distorted wave form.

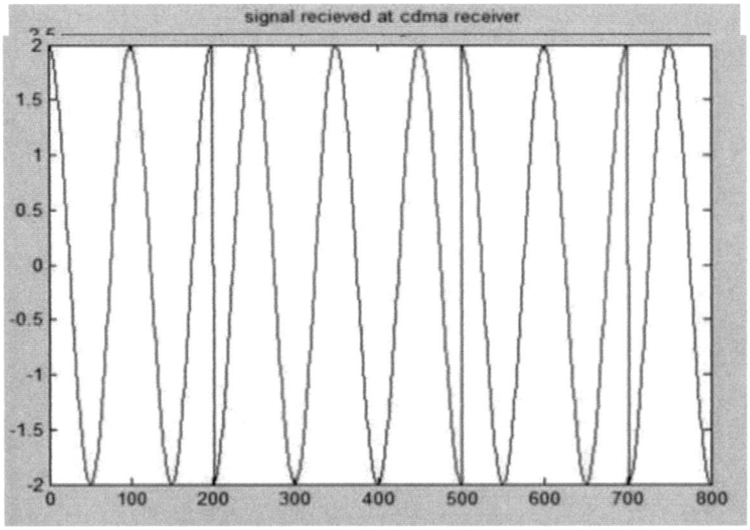

Figure 4. Desired Signal.

4. Output of FEC codes in CDMA using MATLAB.

4.1 Block Coding.

Figure 5 and Figure 6 shows bit error rate performance of CDMA system in hard decision and soft decision respectively when Block coding is employed as FEC. It is found that BER is getting decreased with the increase in SNR.

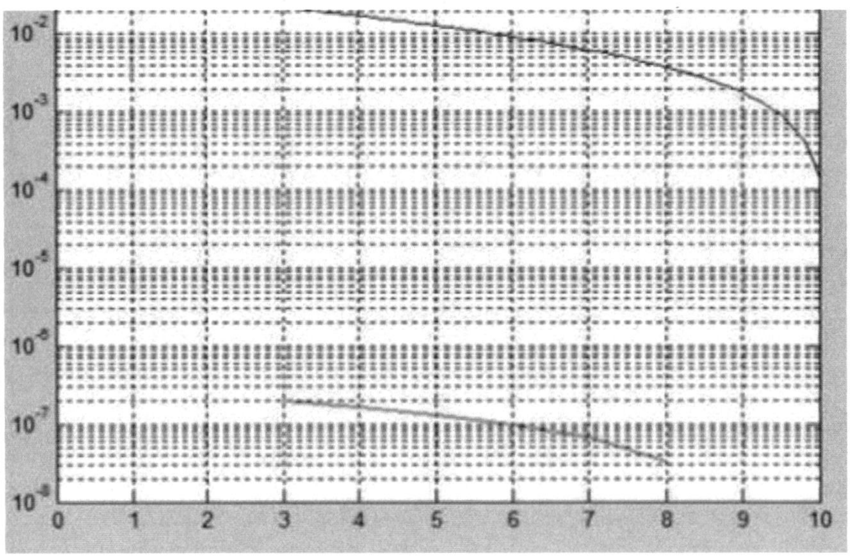

Figure 5. The plot of BER vs. SNR (Hard Decision)

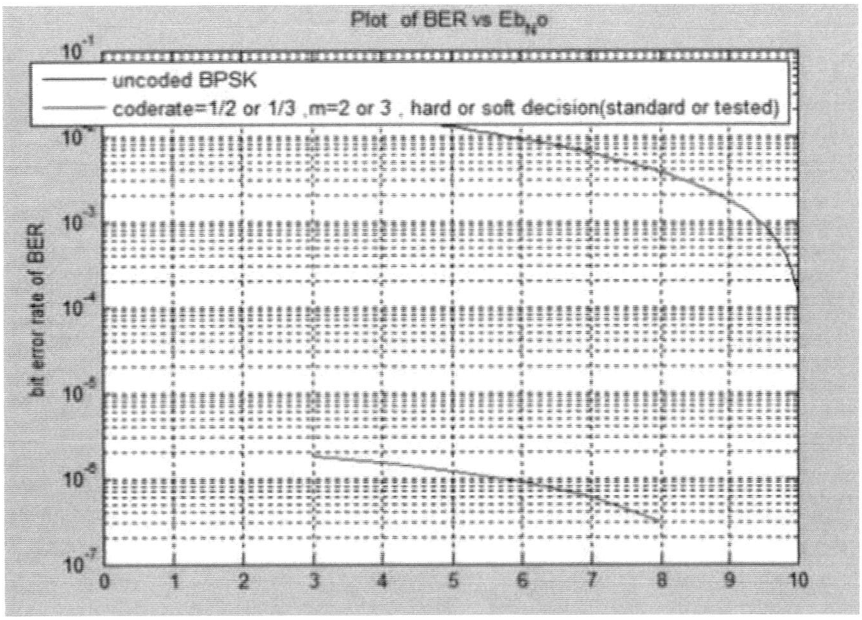

Figure 6. The plot of BER Vs. SNR (Soft decision)

4.2 Block coding results in tabular form.

From the Table 3, Table 4 and Table 5, it is found that BER is better than FEC coded CDMA than encoded CDMA. If the code rate is ½ and m=2, BER at 5 dB for coded CDMA is 9.5x10-8 using hard decision and for uncoded CDMA is $9.2x10^{-7}$ using soft decision. If the code rate is changed to 1/3, BER is $1x10^{-7}$ in the case of hard decision and $9.2x10^{-7}$ in the case of soft decision.

Table 3. The plot of BER vs. SNR for the rate ½, m=2, encoder considering hard decision Viterbi decoder.

SNR(Db)	CODED(BER)	UNCODED BPSK(BER)	DECODER USED
5	$9.5x10^{-8}$	$9.2x10^{-3}$	Hard
7	$7x10^{-8}$	$5x10^{-3}$	Hard
5	$9.2x10^{-7}$	$9.5x10^{-3}$	Soft
7	$6x10^{-7}$	$5x10^{-3}$	Soft

Table 4. The plot of BER vs. SNR for the rate 1/3, m=2, encoder considering hard and soft decision Viterbi decoder.

SNR(Db)	CODED(BER)	UNCODED BPSK(BER)	DECODER USED
5	$1x10^{-7}$	$9.2x10^{-3}$	Hard
7	$9.5x10^{-8}$	$5x10^{-3}$	Hard
5	$9.2x10^{-7}$	$9.5x10^{-3}$	Soft
7	$6x10^{-7}$	$5x10^{-3}$	Soft

Table 5. The plot of BER vs. SNR for the rate 1/3, m=6, encoder considering hard and soft decision Viterbi decoder.

SNR(Db)	CODED(BER)	UNCODED BPSK(BER)	DECODER USED
5	4x10^{-7}	9.2x10^{-3}	Hard
7	2x10^{-7}	5x10^{-3}	Hard
5	0.5x10^{-6}	9.5x10^{-3}	Soft
7	9.2x10^{-7}	5x10^{-3}	Soft

4.3 Convolutional coding.

Figure 7 and Figure 8 shows bit error rate performance of CDMA system in hard decision and soft decision respectively when Conventional coding is employed as FEC. It is found that BER is getting decreased with the increase in SNR.

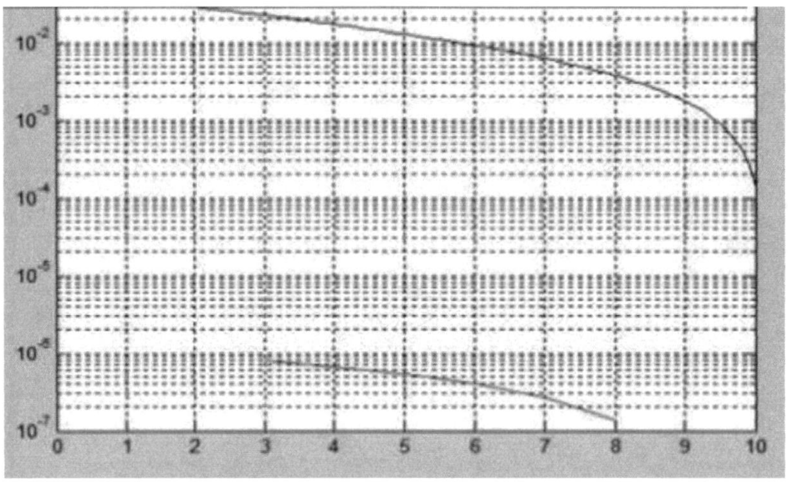

Figure 7. The plot of BER vs. SNR (Hard decision)

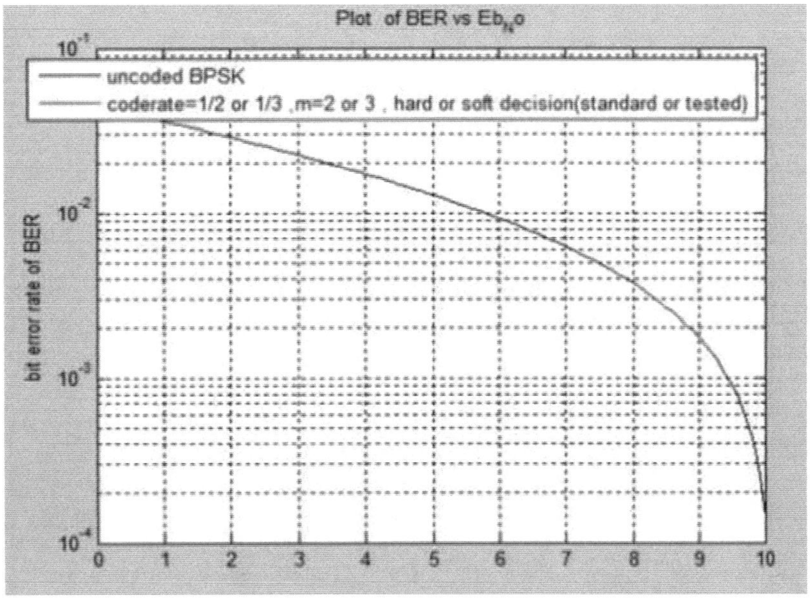

Figure 8. The plot of BER vs. SNR (Soft decision)

4.4 Convolutional coding results.

From the Table 6, Table 7 and Table 8, it is found that BER is better than Convolutional coded CDMA than uncoded CDMA. If the code rate is ½ and m=2, BER at 5 dB for coded CDMA is $2x10^{-7}$ using hard decision and for uncoded CDMA is $2x10^{-6}$ using soft decision. If the code rate is changed to 1/3, BER is $9.5x10^{-8}$ in the case of hard decision and $9.2x10^{-7}$ in the case of soft decision.

Table 6. Plot of BER vs SNR for the rate ½, m=2, encoder considering hard or soft decision Viterbi decoder.

SNR(Db)	CODED(BER)	UNCODED BPSK(BER)	DECODER USED
5	$2x10^{-7}$	$9.5x10^{-3}$	Hard
7	$1.5x10^{-7}$	$5x10^{-3}$	Hard
5	$2x10^{-6}$	$9.5x10^{-3}$	Soft
7	$9.5x10^{-8}$	$5.5x10^{-3}$	Soft

Table 7. Plot of BER vs SNR for the rate ½, m=6, encoder considering hard and soft decision Viterbi decoder.

SNR(Db)	CODED(BER)	UNCODED BPSK(BER)	DECODER USED
5	$9.5x10^{-8}$	$9.2x10^{-3}$	Hard
7	$7x10^{-8}$	$5x10^{-3}$	Hard
5	$9.2x10^{-7}$	$9.5x10^{-3}$	Soft
7	$6x10^{-7}$	$5.5x10^{-3}$	Soft

Table 8. The plot of BER vs SNR for the rate 1/3, m=6, encoder considering hard and soft decision Viterbi decoder.

SNR(Db)	CODED(BER)	UNCODED BPSK(BER)	DECODER USED
5	4x10-7	9.3x10-3	Hard
7	2x10-7	5x10-3	Hard
5	9.2x10-7	9.5x10-3	Soft
7	6x10-7	5x10-3	Soft

5. Conclusion.

The results were displayed accurately in MATLAB in regard to the implementation of CDMA. The program used a simple technology of CDMA. EC coding was used as the method of error correction. Forward Error Correction works very well with additive white Gaussian noise in short AWGN. This was achieved by s both convolutional and block encoding. In the receiver side Viterbi Decoder was used. The encoding process was demonstrated using a (2.1.3) convolutional encoder. A 3-bit input system stream was encoded as an example to show the working of this encoder.

The decoding process was depicting using both hard and soft Viterbi Decoder. The noise due to which the message got corrupted was encoded in a 3-bit input sequence and decoded and recovered using the concepts of Hamming distances and the Euclidean distances for the hard decision Viterbi decoder and the soft decision Viterbi decoder respectively. The performance factors which have affected the FEC were discussed. These included encoder memory size and the more significant factor of SNR.

To study how these performance factors, affect the working of the FEC technique, the encoder for the FEC technique was considered for different values of code rates and the number of memory registers. Two code rates were used in this study. They were 1/2 and 1/3. These code rates were used with each of the two values of memory. The decoder for the FEC technique was then tested with both hard and soft Viterbi decoder.

From Table 9, Table 10 and Table 11, it is seen that at a particular Bit error rate in the execution of Block and convolution codes, the Bit error rates of the convolutional coded data is comparatively low or equal to that of the block coded data. For example, at the code rate=1/3, m=6 and code rate =1/2, m=6 the convolutional code and the block code functions equally. But for the code rate=1/3, m=2 the convolutional code performs better when compared to block codes. So, since block code performance is low for a few code rate combinations it is recommended to use convolutional coding in the CDMA network.

Table 9. Comparison table between convolutional coded & block coded BER values at code rate=1/3, m=6.

SNR(Db)	Convolutional coded (BER)	Block coded(BER)
4	$5x10^{-7}$	$5x10^{-7}$
5	$4x10^{-7}$	$4x10^{-7}$
6	$3x10^{-7}$	$3x10^{-7}$
7	$2x10^{-7}$	$2x10^{-7}$

Table 10. Comparison table between convolutional coded & block coded BER values at code rate=1/3, m=2.

SNR(Db)	Convolutional coded (BER)	Block coded(BER)
4	$5x10^{-7}$	$9x10^{-7}$
5	$4x10^{-7}$	$8x10^{-7}$
6	$3x10^{-7}$	$6x10^{-7}$
7	$2x10^{-7}$	$4x10^{-7}$

Table 11. Comparison table between convolutional coded & block coded BER values at code rate=1/2, m=6.

SNR(Db)	Convolutional coded (BER)	Block coded(BER)
4	$5x10^{-7}$	$5x10^{-7}$
5	$4x10^{-7}$	$4x10^{-7}$
6	$3x10^{-7}$	$3x10^{-7}$
7	$2x10^{-7}$	$2x10^{-7}$

References

[1] Prof.Jeffrey.G.Andrews, *Code Division Multiple Access for Wireless Communications*, [Online]. Available: http://users.ece.utexas.edu/~jandrews/publications/cdma_talk.pdf

[2] Jing Kang, Jun She An and Bingbing Wang, July 21st 2020, *An Efficient FEC Encoder Core for VCM LEO Satellite- Ground Communications*, IEEE ACCESS.2020.3007923.

[3] Joe Khalife, Kimia Shamaei and Zaher M. Kassas, APRIL 15, 2018, *Navigation With Cellular CDMA Signals—Part I: Signal Modeling and Software-Defined Receiver Design*, IEEE VOL. **66**, NO. 8.

[4] Zhenyu Jia, Fei Yuan, Keyu Chen, En Cheng, Jianghui Li, 2018, *Research on Code Division Multiple Access Based on Chirp Multi-Carrier*, IEEE.

[5] R. Kohno, R. Meidan, and L. Milstein, Jan 1995, *Spread spectrum access methods for wireless communications*, IEEE Communication Magazine.

[6] [6 Eric E. Petrosky, Alan J. Michaels and Devin B. Ridge, August 2019, *Network Scalability Comparison of IEEE 802.15.4 and Receiver-Assigned CDMA*, IEEE Internet of Things Journal, VOL. **6**, NO. 4.

[7] Chesto and Lawrence.F, 1982, *Spread Spectrum and Frequency Management*, IEEE Military Communications Conference-Progress in Spread Spectrum Communications, Volume **2**, MILCOM.

[8] L. Barukang, G. Markarian, 2010, Packet Header FEC, IEEE.

[9] A.J.Viterbi, April, 1967, *Error Bounds for Convolutional Codes and an Asymptotically Optimum Decoding Algorithm*, IEEE Transactions on Information Theory, vol. IT-**13**,pp.260-269.

[10] Dirk Wubben, Henning Paul and Peter Rost, *Implementation and Analysis of Forward Error Correction Decoding for Cloud-RAN Systems*, Conference: 2015 IEEE International Conference on Communication Workshop (ICCW), At London, UK.

[11] Andrew J. Viterbi, Addison-Wesley, 1995, *CDMA Principles of Spread Spectrum Communication*, ISBN 0201633744.

[12] Sui Luyi, Yang Xiaohua, 2012, *Forward Error Correction*, Fourth International Conference on Computational and Information Sciences, IEEE.

[13] Theodore S. Rappaport, Prentice-Hall, *Wireless Communications Principles and Practice*, 1996, ISBN 0133755363.

[14] Pascal Frossard, March 2001, *FEC Performance in Multimedia Streaming*, IEEE Communications Letter, VOL. **5**, No. 3.
[15] Sijia Huang, Zexi Zhang, 2011, *Principles of FECs with Evaluating Different Types of FEC used in the Internet and Wireless Networks*, IEEE.
[16] Available on the internet at Blog "Universitas Udayana"
Website: http://staff.unud.ac.id/~sukadarmika/wp-content/uploads/2010/02/pert-2-types-of-codes.pdf
[17] B.Sklar, 2001, *Digital Communications Fundamentals and Applications*, 2nd edition, Prentice Hall, Upper Saddle River, New Jersey.
[18] Olanrewaju B. Wojuola, Stanley H. Mneney and Viranjay M. Srivastava, 2016, *CDMA in Signal Encryption and Information Security*, IEEE.
[19] Tiong Sieh Kiong, Tan Chang Hai, Ding J, Ismail M, 2003, *Error correction for CDMA code in mobile communication*, 4TH National Conference on Telecommunication Technology, NCTT Proceedings.
[20] Schilling, D.L Pickholtz, R.L Milstein, 1990, *Spread spectrum goes commercial*, Spectrum, IEEE Volume **27**, Issue:8.
[21] S.Haykin, Communication Systems, Fourth Edition, Wiley, 2001.
[22] A.R. Abdul Rajak, 2006, *CDMA with novel Concatenated FEC Schemes in AWGN Channel*, Asian Journal of Information Technology.

Acknowledgments
Firstly, I would like to thank Birla Institute of Technology for giving me a chance to be a part of such an exemplary institution and to perform this research which has helped me understand the topics learned by me throughout the course.

Secondly, I would like to thank the Director Prof R N Saha for his support cooperation without out which the results obtained by this paper would not be possible. I would like to thank my Professor Dr. A.R. Abdul Rajak for his time and patience in helping with this paper. Without his help, it could not have been completed.

Crystal structure and optical properties of non-vacuum solution-based processed Cu_2ZnSnS_4 (CZTS) thin-film

H S Nugroho[1], G Refantero[1], E C Prima[2], C Panatarani[4], Suyatman[1,3] , N Nugraha*[1,3], and B Yuliarto*[1,3]

[1] Department of Engineering Physics, Faculty of Industrial Engineering, Institut Teknologi Bandung, Bandung, Indonesia
[2] Department of Science Education, Faculty of Mathematics and Science Education, Universitas Pendidikan Indonesia, Bandung, Indonesia
[3] Research Center for Nanosciences and Nanotechnology (RCNN), Institut Teknologi Bandung, Bandung, Indonesia
[4] Department of Physics, Faculty of Mathematics and Natural Science, Universitas Padjadjaran, Bandung, Indonesia

brian@tf.itb.ac.id, nugraha@tf.itb.ac.id

Abstract. Cu_2ZnSnS_4 (CZTS) material has emerged as an attractive material for the absorber layer in solar cells application. CZTS has similar properties with its successful predecessor CIGS, but CZTS offers the advantage of low-cost constituents, material abundance, and non-toxicity. We fabricated CZTS thin-film using non-vacuum solution-based process and then the deposition process using the spin-coating technique in the present work. Afterward, we observed that the CZTS thin-film was successfully fabricated with kesterite structure crystal with an optical bandgap of 1.56 eV. We also confirmed that the CZTS thin-film exhibit a high light-harvesting efficiency at a low wavelength suitable for the solar cells application's absorber layer.
Keywords: CZTS thin-film, non-vacuum solution-based process, crystal structure, absorbance, bandgap energy, light-harvesting efficiency

1. Introduction

Technological and industrial advances that have developed rapidly in the last few decades have led to an increase in energy demand. However, the majority of energy sources used today come from fossil fuels that are not renewable and can negatively impact the environment [1]. Therefore, solar cells have been developed as an alternative solution to meet environmentally friendly energy needs. Among different solar cells, the third-generation solar cells have emerged as one of the most promising candidates [2]. The reason being its possibility of generating high-efficiency (31 – 41% Shockley-Queisser limit), low-cost manufacturing process, non-toxicity, and high elemental abundance [3].

The Cu_2ZnSnS_4 (CZTS) kesterite compound has been widely studied as an alternative material for absorber layer in thin-film solar cells [1, 4]. As a prospective absorber layer, CZTS offers similar properties with its excellent predecessor CIGS such as high absorbance of 10^4 – 10^5 cm^{-1}, optical bandgap near 1.5 eV, and tetragonal crystal structure [5, 6]. However, CZTS offers more advantages than CIGS, including material abundant, cheaper, and non-toxic [7]. Therefore, CZTS solar cells are categorized as third-generation emerging solar cells.

Several techniques for CZTS thin-film fabrication process have been studied including vacuum deposition [8], magnetron sputtering [9], pulsed laser deposition [10], electron beam evaporation [11], spray pyrolysis [6], electrochemical deposition [12], and sol-gel [13]. Basically, there are two categories of the fabrication process which are vacuum and non-vacuum [14]. In vacuum deposition, the layer of materials was deposited in the manner of atom-by-atom or molecule-by-molecule in a solid surface that operates at a pressure below the atmospheric pressure. Meanwhile, non-vacuum deposition usually is carried out by dissolving the material constituent in a solvent and then coated onto a substrate.

It has been widely known that CZTS material has low chemical stability [15]. Therefore, the nominal composition of the CZTS molar ratio is an important matter to be considered. Empirically, the nominal atomic composition of the CZTS material is generally arranged into a non-stoichiometric Cu-poor and Zn-Rich condition to produce high efficiency [16, 17]. In this composition, the secondary phase density of the Cu_{Zn} profound defect is reduced without reducing the p-type conductivity of the CZTS absorber [7]. This is because the presence of shallow defects V_{Cu} is still in high density. Finally, a lower density of deep defects leads to a lower recombination mechanism that improves performance.

In this work, we fabricate CZTS thin-film using a non-vacuum solution-based process. The precursor was prepared using the sol-gel method and the deposition process was carried out via spin-coating process. The CZTS precursor film was then sulfurized for the crystal growth. Afterward, the crystal structure and optical properties characterization were carried out to observe the CZTS thin-film characteristics to apply it as an absorber layer in solar cell application.

2. Experimental procedure

2.1. Substrate cleaning and synthesis of CZTS precursor

The Soda Lime Glass (SLG) with an area of 3 cm x 3 cm was used as the substrate. The substrate cleaning was done utilizing sonication procedure in acetone and IPA for 10 minutes each. The cleaned substrate was then dried fast using a dryer.

The CZTS precursor was prepared using the solution process by dissolving 1.02 gr copper acetate hydrate, 0.724 gr zinc acetate hydrate, 0.596 gr tin chloride dihydrate, and 1.568 gr thiourea into 10 mL Ethylene glycol monomethyl ether. The solution was then stirred at 50°C for 2 h using a stirring rate of 250 rpm. The Cu-poor and Zn-rich non-stoichiometric condition was obtained by arrange the Zn/Sn and Cu/(Zn + Sn) molar ratio to 1.25 and 0.86, respectively.

2.2. CZTS deposition and sulfurization

A few drops of triethanolamine and monoethanolamine were applied to the precursor solution prior to the deposition process. The CZTS deposition was then done using a repeated process of spin-coating and preheating. For 20 s, the spin-coating process was set at 3000 rpm, and then the coated precursor was heated for 5 min at 280°C. To obtain the appropriate thicknesses, this method was repeated several times. This process was repeated several times to obtain the desired thicknesses. Subsequently, the precursor film was then sulfurized using an RTP furnace at 600°C for 40 min in the N_2+S atmosphere to obtain CZTS thin-film.

2.3. Characterization

The Bruker Advanced D8 instrument with Cu radiation of 0.154439 nm was used for the crystal structure investigation. Meanwhile, the Thermoscientific evolution 201 UV-Visible Spectrophotometer was used for the optical properties investigation.

3. Results and discussions

The presence of kesterite crystal structure can be observed through the XRD pattern in figure 1. It showed the marked peaks that appear at 2θ of 28.4°, 47.26°, and 56.18°, similar to the kesterite peaks reference at JCPDS No. 26-0575. We also noticed that the absence of Cu_2S, SnS_2, and SnS secondary phases can be confirmed from the XRD pattern. Meanwhile, the ZnS secondary phase could not be confirmed because of its mutual interference peaks [18]. In this case, the formation of the ZnS secondary phase is possible [19], given that the composition taken is Cu-poor and Zn-Rich which suppresses the formation of a secondary Cu_2SnS_3 phase instead forms a ZnS phase [17].

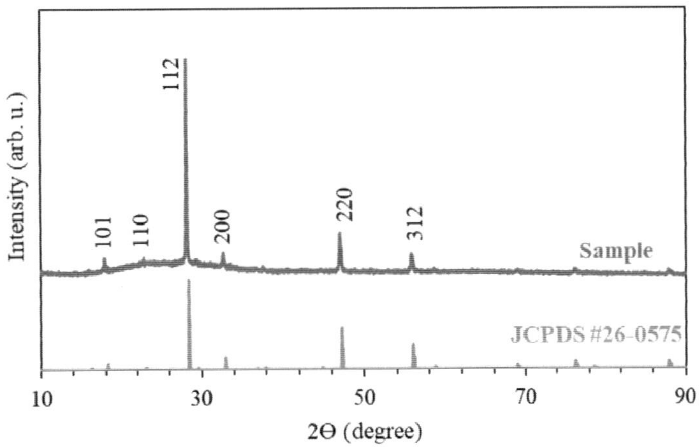

Figure 1. XRD pattern of CZTS thin-film

Using the Scherrer equation (1), the average grain size of kesterite crystals was determined [20]. K is a dimensionless constant, λ is the x-ray wavelength, β is the Full width at half-maximum (FWHM), and D is the average grain size produced in the absorber layer. In this work, the average grain size measured using Scherrer 's equation for the CZTS thin-film absorber is ~53.7 nm.

$$D = \frac{k\lambda}{\beta\cos\theta} \tag{1}$$

The absorbance level from the spectroscopy measurement of UV-Vis is shown in Figure 2. The outcome shows that, at a low wavelength range of 320-498 nm, the CZTS thin-film has excellent optical sensitivity. Meanwhile, the absorbance level tends to be constant at low values at the higher wavelength of 843 – 1080 nm. The optical bandgap of CZTS thin-film was confirmed to be 1.56 eV using the Tauc Plot method from the absorbance level, as shown in Figure 3.

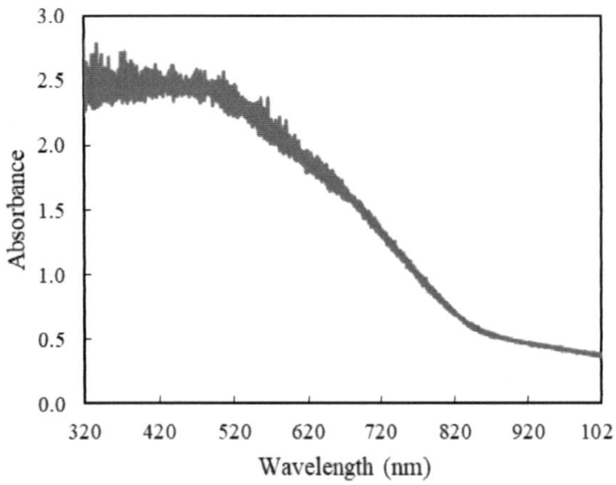

Figure 2. The absorbance level of CZTS thin-film

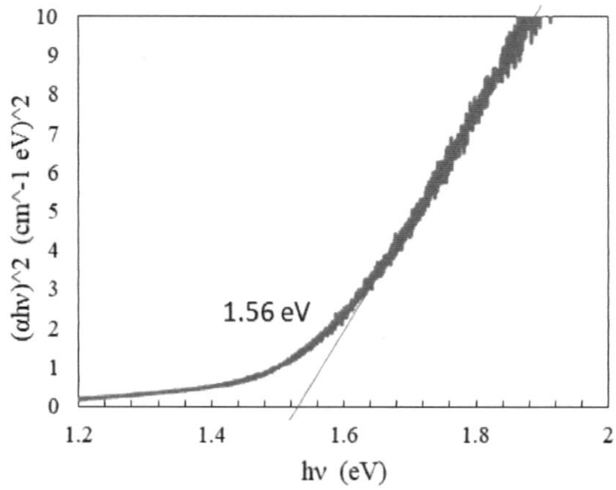

Figure 3. The optical bandgap of CZTS thin-film

The high-efficiency of light absorption at low wavelength can be confirmed using the Light-Harvesting Efficiency (LHE) calculation. LHE is the fraction of the light photon that absorbed at a certain wavelength by CZTS thin-film. The LHE was calculated using equation (2), where $\propto$ is absorbance [21].

$$\mathrm{LHE}\ (\lambda) = (1 - 10^{-\alpha}) \times 100\% \tag{2}$$

The LHE data in figure 4 confirmed the high efficiency of light absorption at a low wavelength for the fabricated CZTS thin-film in this work. It starts to decline gradually at a wavelength of 498 nm and decreases rapidly, starting at 748 nm. Overall, the CZTS thin-film absorber has a high light absorption efficiency. Therefore, it can be concluded that the CZTS tihn-film in this work has the potential to produce high Incident Photon to Current Efficiency (IPCE) in its application since IPCE is directly proportional to LHE [22].

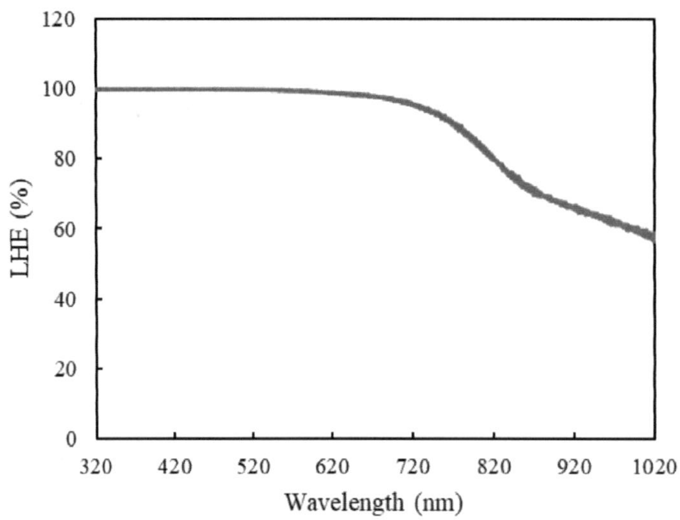

Figure 4. The LHE of CZTS thin-film

4. Conclusions

In summary, CZTS thin-film was successfully fabricated using a solution-based process. The results show that the CZTS thin-film has the kesterite crystal structure with an average grain size of ~53.7 nm. Furthermore, the optical characterization shows that the CZTS thin-film has an optical bandgap of 1.56 eV with excellent light-harvesting efficiency at a low wavelength between 320 and 498 nm. These results indicated that the CZTS thin-film fabricated in this work is suitable for the absorber layer in solar cell application.

5. Acknowledgments

This work was supported by research grant of Ministry of Research and Technology/National Research and Innovation Agency (Indonesia) (RISTEK-BRIN). The authors would like to acknowledge the support of the Advanced Functional Material (AFM) Laboratory, Engineering Physics, Bandung Institute of Technology and Research Center for Nanosciences and Nanotechnology (RCNN), Bandung Institute of Technology, for providing their facilities.

References

[1] Petroleum B 2019 'BP Statistical Review of World Energy Report' BP: London UK
[2] Pal K Singh P Bhaduri A and Thapa K B 2019 'Current challenges and future prospects for a highly efficient >20% kesterite CZTS solar cell: A review' Solar Energy Materials and Solar Cells 196 pp 138-156
[3] Khalate S A Kate R S and Deokate R J 2018 'A review on energy economics and the recent research and development in energy and the Cu2ZnSnS4 CZTS solar cells: A focus towards efficiency' Solar Energy 169 pp 616-633
[4] Tanaka K Fukui Y Moritake N and Uchiki H 2011 'Chemical composition dependence of morphological and optical properties of Cu2ZnSnS4 thin films deposited by sol–gel sulfurization and Cu2ZnSnS4 thin film solar cell efficiency' Solar Energy Materials and Solar Cells 953 pp 838-842
[5] Katagiri H Sasaguchi N Hando S Hoshino S Ohashi J and Yokota T 1997 'Preparation and evaluation of Cu2ZnSnS4 thin films by sulfurization of E□B evaporated precursors' Solar Energy Materials and Solar Cells 491 pp 407-414
[6] Li W Tan J M R Leow S W Lie S Magdassi S and Wong L H 2018 'Recent Progress in Solution-Processed Copper-Chalcogenide Thin-Film Solar Cells' Energy Technology 61 pp 46-59
[7] Huang T J Yin X Qi G and Gong H 2014 'CZTS-based materials and interfaces and their effects on the performance of thin film solar cells' physica status solidi RRL – Rapid Research Letters 0809 pp 735-762
[8] Shin B Gunawan O Zhu Y Bojarczuk N A Chey S J and Guha S 2013 'Thin film solar cell with 84% power conversion efficiency using an earth-abundant Cu2ZnSnS4 absorber' Progress in Photovoltaics: Research and Applications 211 pp 72-76
[9] Akcay N Zaretskaya E P and Ozcelik S 2019 'Development of a CZTS solar cell with CdS buffer layer deposited by RF magnetron sputtering' Journal of Alloys and Compounds 772 pp 782-792
[10] Pandiyan R Oulad Elhmaidi Z Sekkat Z Abd-lefdil M and El Khakani M A 2017 'Reconstructing the energy band electronic structure of pulsed laser deposited CZTS thin films intended for solar cell absorber applications' Applied Surface Science 396 pp 1562-1570
[11] Mkawi E M Al-Hadeethi Y Shalaan E and Bekyarova E 2018 'Substrate temperature effect during the deposition of Cu/Sn/Cu/Zn stacked precursor CZTS thin film deposited by electron-beam evaporation' Journal of Materials Science: Materials in Electronics 2923 pp 20476-20484

IOP Conf. Series: Materials Science and Engineering 1045 (2021) 012039 doi:10.1088/1757-899X/1045/1/012039

[12] Khattak Y H Baig F Toura H Harabi I Beg S and Soucase B M 2019 'Single step electrochemical deposition for the fabrication of CZTS kesterite thin films for solar cells' Applied Surface Science 497 pp 143794
[13] Hadke S H Levcenko S Lie S Hages C J Márquez J A Unold T and Wong L H 2018 'Synergistic Effects of Double Cation Substitution in Solution-Processed CZTS Solar Cells with over 10% Efficiency' Advanced Energy Materials 832 pp 1802540
[14] Ravindiran M and Praveenkumar C 2018 'Status review and the future prospects of CZTS based solar cell – A novel approach on the device structure and material modeling for CZTS based photovoltaic device' Renewable and Sustainable Energy Reviews 94 pp 317-329
[15] Siebentritt S 2013 'Why are kesterite solar cells not 20% efficient?' Thin Solid Films 535 pp 1-4
[16] Todorov T K Reuter K B and Mitzi D B 2010 'High-Efficiency Solar Cell with Earth-Abundant Liquid-Processed Absorber' Advanced Materials 2220 pp E156-E159
[17] Chen S Walsh A Gong X-G and Wei S-H 2013 'Classification of Lattice Defects in the Kesterite Cu2ZnSnS4 and Cu2ZnSnSe4 Earth-Abundant Solar Cell Absorbers' Advanced Materials 2511 pp 1522-1539
[18] Valdés M Santoro G and Vázquez M 2014 'Spray deposition of Cu2ZnSnS4 thin films' Journal of Alloys and Compounds 585 pp 776-782
[19] Xie H Sánchez Y López-Marino S Espíndola-Rodríguez M Neuschitzer M Sylla D Fairbrother A Izquierdo-Roca V Pérez-Rodríguez A and Saucedo E 2014 'Impact of SnSSe Secondary Phases in Cu2ZnSnSSe4 Solar Cells: a Chemical Route for Their Selective Removal and Absorber Surface Passivation' ACS Applied Materials & Interfaces 615 pp 12744-12751
[20] Refantero G Prima E C Setiawan A Panatarani C Cahyadi D and Yuliarto B 2020 'Etching process optimization of non-vacuum fabricated Cu2ZnSnS4 solar cell' Journal of Materials Science: Materials in Electronics 314 pp 3674-3680
[21] Cahya Prima E Yuliarto B Suyatman and Dipojono H K 2015 'Theoretical Investigation of Anthocyanidin Aglycones as Photosensitizers for Dye-Sensitized TiO2 Solar Cells' Advanced Materials Research 1112 pp 317-320
[22] Guo M Xie K Lin J Yong Z Yip C T Zhou L Wang Y and Huang H 2012 'Design and coupling of multifunctional TiO2 nanotube photonic crystal to nanocrystalline titania layer as semi-transparent photoanode for dye-sensitized solar cell' Energy & Environmental Science 512 pp 9881-9888

Investigative analysis of match-line sensing schemes in TCAM for high-speed performance

Fahim Faisal, Mirza Muntasir Nishat, Sayka Afreen Mim, Hafsa Akter and Md. Rafid Kaysar Shagor

Dept. of Electrical and Electronic Engineering, Islamic University of Technology, Dhaka, Bangladesh

{faisaleee, mirzamuntasir, saykaafreen, hafsaakter,rafidkaysar} @iut-dhaka.edu

Abstract. This paper delineates an investigation of TCAM-design techniques for both circuit level and architectural level. In the case of circuit level, low power Match Line sensing techniques and Search Line driving propositions are exhibited. At the architectural level, four processes for curtailing power absorption are introduced. In this paper, 16×16 bit TCAM is designed in 0.18μm CMOS. The hypothesized Match Line sensing scheme truncates power absorption by pruning search time and dampering voltage swing of traditional ones. With respect to the conventional CR-MLSA, the proposed MLSAs delineate the depletion of 56% and 48% for measuring energy. For simulation, 1.8V supply voltage is used.

Keywords. CMOS, Current Race Scheme, MDPA, AF, RF

1. Introduction

Ternary Content Addressable Memory, in other words, named as TCAM, is a tremendous mechanism orchestrated by the structural architecture. The complexity and the delicacy beheld the painstaking efforts of the researchers for their future progeny. Many different kinds of algorithms have been squinting wearily in the misty fogs of the industrial environment for many purposes like medical and healthcare [1-3], power electronics [4-5], and many more which can be so revitalizing that anyone cannot help rejuvenated. TCAM is based on hardware and it is a parallel look-up table in company with masking capability in bitwise level role-playing a coprocessor in any sophisticated network. This momentous sight rode us to an epiphany of thought revealing the intricacies of TCAM and how only purposeful efforts tinged with an earnest target for the well-being of the future can bring about such a wonderful creation. In this case, a flexible and adaptable processing solution is provided by a vicarious value which can be emerged as contemplation about the workflow. TCAM can explicitly be utilized in the case of network routers. Here, each address usually consists of a Network Address and Host address. Only a meaningful stride stemming from a poignant target to uplift the quality of processing will lead to a quantum leap towards success through this.

Network address proliferates in case of the size usually on subnet configuration. On the contrary, the rest of the bits are engrossed by the host address. Utilizing Ternary CAM in the routing table delineates this lookup affair quite coherently. In this case, the address can be stored denoting 'X' for the host portion. Look-up destination address sustains scrupulous routine engagement without further ado. In this regard, developed architectures are apprehending that supplemental functionality cannot but be consummated in an avant-garde way so that deep packet classification can be accelerated and forwarded in the next generation [6-8].

In this paper, the basic construction of TCAM is shown in section 2. The objective is to limit power consumption maintaining high search speed. Maximum consumption of power occurs in match-line (ML) and search-line (SL). In section 3, different match-line sensing schemes are discussed. In section 4, simulation and results and comparisons of different ML sensing schemes are portrayed. The simulations are done in HSPICE environment.

2. Ternary Content Addressable Memory

Ternary Content Addressable Memory (TCAM) is an exceptional archetype of data searching souvenir which can search its total content in a single clock cycle [9]. The significance of the term 'ternary' is that it can query and store for (i) High, (ii) Low, and (iii) Wildcard (Ternary) states.

Table 1. Tabulated Address lookup.

Entry No.	Address Binary	Output Port
1	101XX	A
2	0110X	B
3	011XX	C
4	10011	D

It is the key device to speed up packet forwarding in internet routing. IP (Internet Protocol) forwarding is the main objective of internet routers [10-11]. Speed is the main performance parameter here. TCAM is one of the fastest devices in this case. How a CAM effectuates address lookup in Table 1 is portrayed in Figure 1 [12-14].

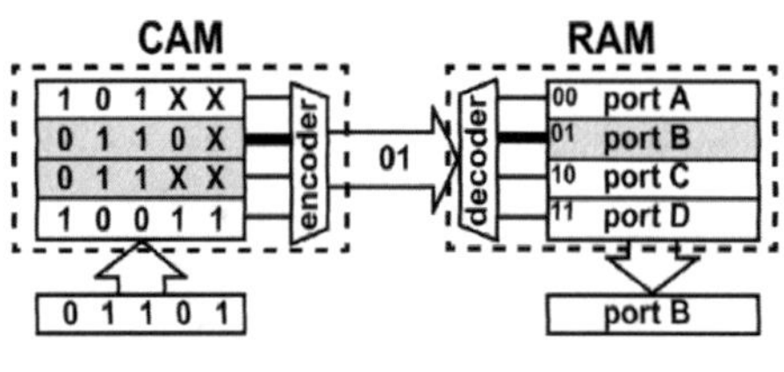

Figure 1. CAM-based implementation of Table 1.

3. Match-line (ML) Sensing Schemes

MLs and SLs are highly capacitive. Combined SL capacitance arising from a large number of columns is also very large. During the search, MLs are activated simultaneously along with SLs. The switching of these extreme capacitive lines causes vast power absorption in the time of match detection. To lessen this power loss, energy-efficient match detection in ML sensing schemes are quite handy.

3.1. Precharge-High ML Sensing Scheme:

To further materialize the development, improvements should be analyzed with more insight and an empirical-based mathematical model will help in simulating the conditions necessary to implement these improvements in a perspicacious way. Hence, substantial efforts need to be undertaken to set up a qualitative model that can weigh in the standards prevailing state-of-the-art to ensure the practical realization of making the fundamental part. At first, pre-charging is required high in the match-line, consecutively evaluating by dint of accessing NOR cells. While in case of a miss, it pulls down, in case of a match, leaves the ML high. The imposition of the ML scheme is portrayed as a schematic diagram in Figure 2 (a). SL precharge, ML precharge and ML evaluation are depicted in Figure 2 (b) [15].

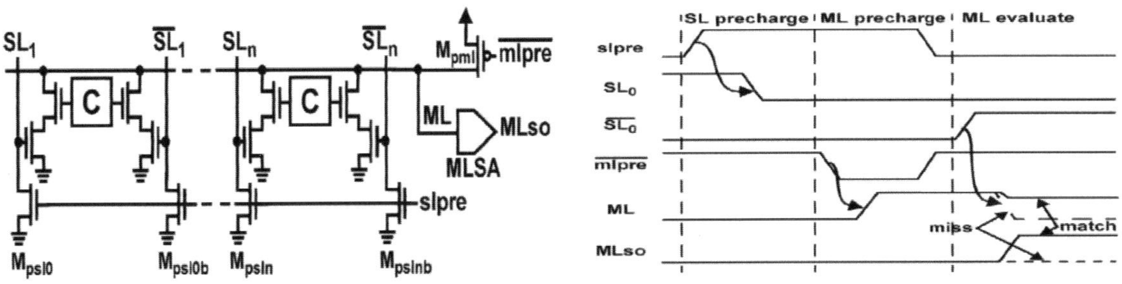

(a) (b)

Figure 2. (a) A levying Structure and (b) Respective sketch portraying corresponding alterations [6]

However, all the match lines are precharged to V_{DD}. As the number of mismatch of words is lower than match word and those mismatched ML voltages are discharged to the ground, a large amount of wastage of energy is encountered [6]. The comparison circuit suffers from charge distributing perplexity. It depends on the connection between the storage bits and the top transistors or lowest transistors in case of the pull-down path.

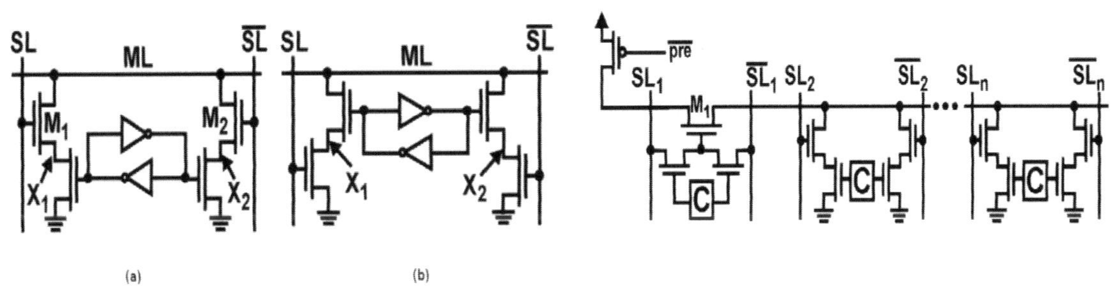

(i) (ii)

Figure 3: (i) Two possible configurations for the NOR cell [6] and (ii) Simple implementation of the selective-precharge match-line technique.

In the Figure 3 (i), gates are connected to the bottom transistors. Charge sharing occurs between ML and nodes X_1, X_2. After the match-line precharge phase during search line precharge one of the search lines is turned ON causing ML voltage V_{ML} to drop. To solve this from this charge sharing problem configuration shown in Figure 3 (ii) can be used where search bits are convoluted to those transistors which lie in top positions. As it is evident that the stored bit is constant in the course of a search operation, the charge share is annihilated.

3.2. Selective-Precharge Scheme

In the conventional scheme, the same amount of energy is required for each match-line regardless of the data pattern and match/mismatch. A selective precharge scheme can provide a good saving in power consumption. Except for the worst-case scenario selective precharge scheme is one of the most quotidian procedures utilized to conserve power on ML. In this scheme, at first, only a subgroup of TCAM cells from the cumulative word is juxtaposed and further requirement of comparison is decided accordingly. A mundane schematic of selective precharge akin to that presented in [16] is shown in Figure 3(ii).

3.3. Current Race (CR) Scheme

To overcome the problems of conventional ML sensing schemes, the Current Race (CR) scheme is proposed by [10]. CR scheme is the most versatile scheme. For the structural architecture, fundamental for the construction is used by many researchers [7], [8], [10], [12], [13], [14], [15] till date to get a modified dispatch with respect to CR.

Traditional Scheme is precharged high. On the other hand, the CR scheme is precharged low. That is the main discrepancy between them. In this way, despite charging all MLs into high, what CR scheme does is to pre-discharge them to low. Figure 4 (a) and (b) exhibit the CR scheme for word and dummy word respectively.

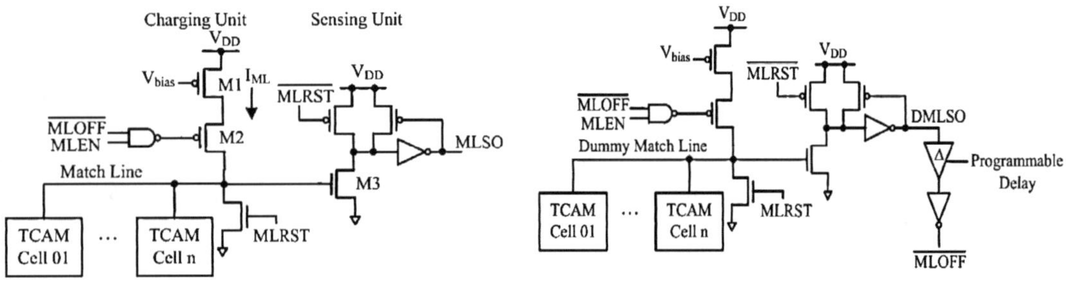

Figure 4 (a)Word in Current Race Scheme and (b) Dummy word in Current Race Scheme

4. Positive Feedback Match-line (PFML) Sensing Schemes

The state-of-the-art CR-MLSA usually charges MLs with a similar enormity of current for both matched and mismatched cases. As long as matched MLs charge to a higher voltage value, consecutively current will gradually decrease in those cases. On the contrary, mismatched ones possess a higher conductive path to ground. Whenever numerical values for mismatch will rise high, the equivalent resistance of the ML will automatically decrease. Current through ML will rise with a spiraling number of mismatches. Undoubtedly, the number of mismatched MLs are enormous. Consequently, a significant amount of energy is squandered due to the high magnitude current in these mismatched MLs. The proposed Positive Feedback Match-line Sensing Scheme solves this problem by supplying a smaller current to mismatched MLs.

4.1. Mismatch-Dependent Power Allocation (MDPA) Scheme

This tremendously excellent Scheme is hypothesized by [8]. In this scheme, to discern mismatched MLs in the time of rising charge and depleting current to these, a feedback unit adheres to the fundamental CR MLSA. The proposed one is depicted in Figure 5 (a). This feedback unit is consisting of two parts namely the level shifter and feedback circuit. All this is to enact the feedback mechanism. At first, the MLSO, voltage of ML, and voltage of VAR are pre-discharged to control the speed. MLEN begins the charge dissipation for all of the MLs. Initially, not only V_{ML} but also V_{VAR} are raised high by current I_{ML} and I_{bias} respectively.

An increment of VAR voltage will simultaneously cause a decrement of ML current. On the contrary, increased V_{ML} will cause the rising of the output of level shifter as well. Matched I_{ML} is noteworthy with respect to mismatched I_{ML}. Indistinguishable dummy word is utilized to supply ML_{OFF} signal to hegemonize the ML charging time. Although a huge amount of energy curtail is depicted in this, the level shifter along with the feedback unit always absorbs almost tantamount energy.

In case of a mismatch N1 sometimes remains weakly turn on which means there is a conducting path between V_{DD} and ground. So, for considering a vast amount of mismatches, V_{ML} is a bit of a tiny one for there are numerical paths to ground for MLs which in turn causes static power consumption. On top of that, due to the large size of the level shifter transistor, ML's capacitive loading spirals up. Again transistor count in this MLSA is higher than the conventional one. Finally, feedback action is very sensitive to V_{bias} and transistor sizes. For the above reasons, the size of the transistor is required to be tuned for modified and proper implementation.

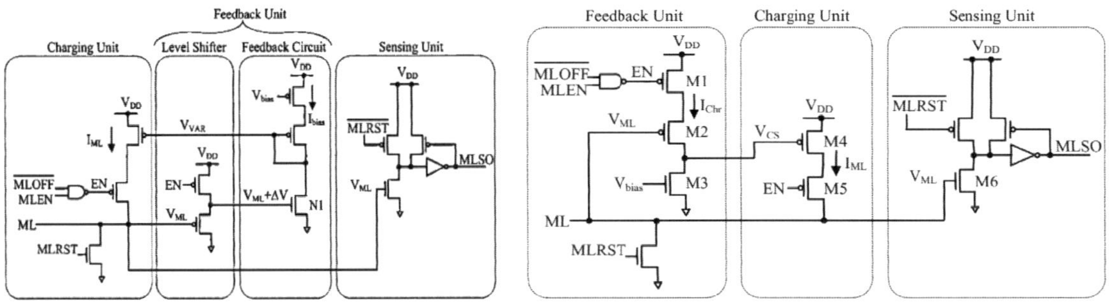

(a) (b)

Figure 5. MLSA in (a) Mismatch-Dependent Power Allocation Scheme. [8], [14] and (b) AF Scheme [7], [15].

4.2. Active Feedback (AF) Scheme

To amortize the difficulties of *MDPA* scheme, the Active Feedback (AF) ML sensing scheme is convoluted. No invariable power is consumed by this proposed one and it outperforms the MD-MLSA in numerous terms namely momentum, energy, area, stoutness, and so on [7]. Figure 5 (b) shows the proposed MLSA with active feedback.

4.3. Resistive Feedback (RF) Scheme

The proposed MLSA with resistive shielding is shown in Figure 6. With a view to decoupling, a transistor is required here which lies in a non-linear region. On the other hand, the channel outlasts the sensory point from high capacitance. Accordingly, I_{Chr} should be down-sized to hoard power notwithstanding the sacrificing of the sensing speed. The N3 channel resistance simultaneously will be reducing by dint of body effect as well as decrement of gate-to-source voltage ($V_{GS}N_3$) while the voltage of ML is spiraling high [7]. It is excruciatingly dependent upon the mismatch bit's number.

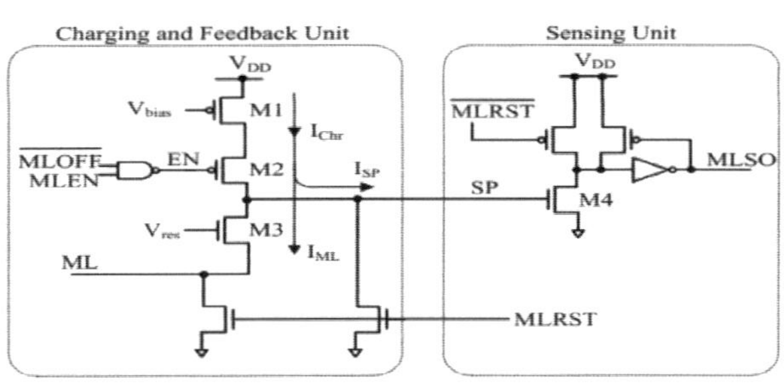

Figure 6. MLSA in RF Scheme [7], [15].

IOP Conf. Series: Materials Science and Engineering 1045 (2021) 012040 doi:10.1088/1757-899X/1045/1/012040

5. Results and Simulation

A 16×16 size array and 0.18μm technology are used. These schemes are simulated in HSPICE simulation software. In the case of State-of-the-art Current Race Scheme, Search time is calculated between Voltage of ML_{RST} and Voltage of ML_{SOD} at the half percent of the input voltage. Moreover, V_{margin} is premeditated between the connecting point of ML_1 and ML_{SO1} and the highest margin of ML_3 in the traditional scheme. These two are shown in the Current Race Scheme, Mismatch Controlled Power Allocation Scheme, Active Feedback Scheme, and Resistive Feedback Scheme in Figure 7, Figure 8, Figure 9, and Figure 10 respectively.

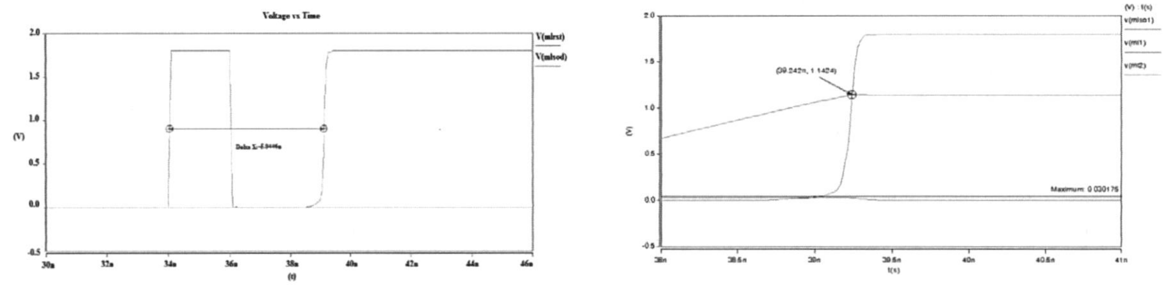

Figure 7. Current Race (CR) Scheme

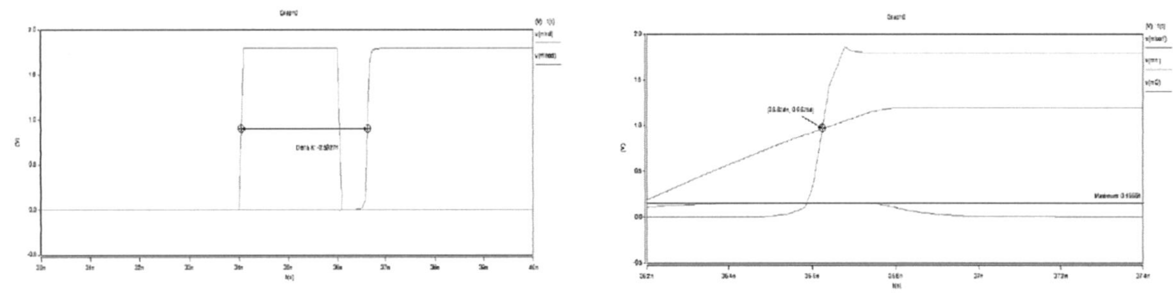

Figure 8. Mismatch-Dependent Power Allocation (MDPA) Scheme

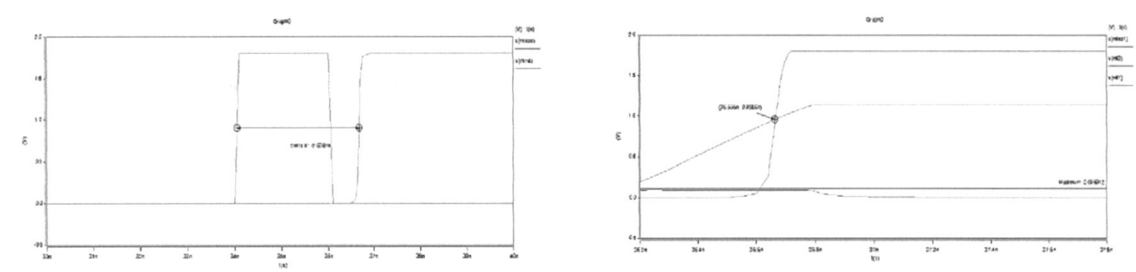

Figure 9. Active Feedback (AF) Scheme

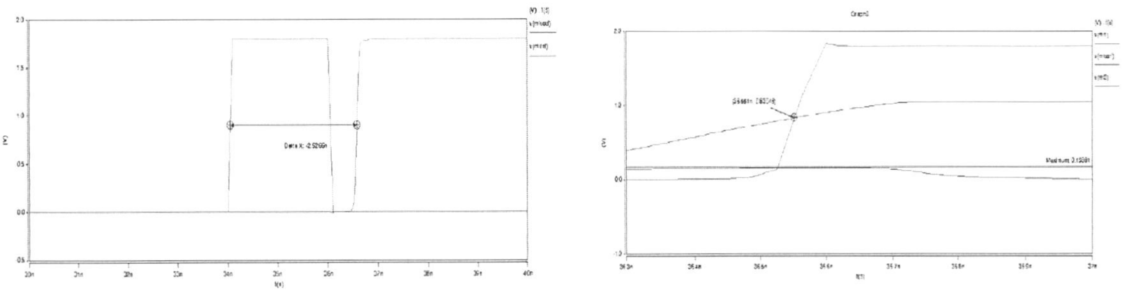

Figure 10. Resistive Feedback (RF) Scheme

Table 2. Search Time and Voltage Margin for different schemes (ns).

Name of the Schemes	Search Time (ns)	Voltage Margin (ns)
Current Race (CR) Scheme	5.0446	1.112225
Mismatch Dependent Power Allocation (MDPA) Scheme	2.5927	0.81093
Active Feedback (AF) Scheme	2.6361	0.859028
Resistive Feedback (RF) Scheme	2.5266	0.67165

From Table 2, we can see that the Resistive Feedback scheme has the lowest search time. And Current Race Scheme has the highest voltage margin. Other than CR scheme, the Active Feedback scheme has a higher voltage margin. It is very difficult to get the optimum value of both speed and noise performances.

6. Conclusion

Among these schemes, RF scheme is with the best speed performance and CR scheme is with the best noise immunity performance. But, its search time is too high (almost double than that of other positive feedback schemes). As a result, its overall efficiency is compromised. To obtain better speed, the voltage margin has to be compromised and vice versa. So, for better speed performance, RF schemes can be used and for a noisy environment, the AF scheme can be used. So, the performance of different ML schemes largely depends on the search speed also known as the search time, and noise margin also known as voltage margin. The advantage of low search time is that search speed increases which is one of the main concerns in designing TCAM. Again the advantage of a higher voltage margin is that if noise is added then it will not show a false match. The proposed positive feedback schemes improve search speed but a tiny bit of a curtailment of the voltage margin has been rendered. PFML has taken a leap-frog in its endeavor to become one of the most well-matched TCAM with a prudent plan taking into account the limitations and capacities of idiosyncratic properties. Till now, it has been remaining on a small-scale but the infrastructure does have the scope to utilize its available resources aided by further development to incorporate more criteria thereby creating a hub of knowledge and the grounds for creating multifaceted co-processors. To disseminate quality performance irrespective of difficulties, it is exigent to expend extensive efforts in the development of the existing structures. MDPA scheme improvises a significant power curtail. On the other hand, AF and RF schemes necessitate a fewer number of transistors in the schematic design. Expediating energy of the MLSAs portrays axing of 56% and 48% with respect to the traditional one.

7. References

[1] Farazi, Moshiur R., F. Faisal, Z. Zaman, and S. Farhan. “Inpainting multiple sclerosis lesions for improving registration performance with brain atlas.” In *Medical Engineering, Health Informatics and Technology (MediTec),* pp. 1-6. IEEE, 2016.

[2] Faisal F., and Nishat M. M., “An Investigation for Enhancing Registration Performance with Brain Atlas by Novel Image Inpainting Technique using Dice and Jaccard Score on Multiple Sclerosis (MS) Tissue”, *Biomedical and Pharmacology Journal* 2019,12(3)

[3] M. M. Nishat and F. Faisal, "An Investigation of Spectroscopic Characterization on Biological Tissue," 2018 4th *International Conference on Electrical Engineering and Information & Communication Technology (iCEEiCT),* IEEE, pp. 290-295

[4] M. M. Nishat, F. Faisal, A. J. Evan, M. M. Rahaman, M. S. Sifat and H. M. F. Rabbi,. "Development of Genetic Algorithm (GA) Based Optimized PID Controller for Stability Analysis of DC-DC Buck Converter" *Journal of Power and Energy Engineering* 8, no. 09 (2020): 8-19

[5] M. M. Nishat, F. Faisal, M. A. Hoque. Anik Jawad Evan, Md Moshiour Rahaman, Md Sadman Sifat and H. M. Fazle Rabbi, “Stability Analysis and Optimization of Simulated Annealing (SA) Algorithm Based PID Controller for DC-DC SEPIC Converter” *International Journal of Computer Science and Network Security IJCSNS* Vol: 19 No: 09, pp 1-8

[6] Guccione, Steven A., Delon Levi, and Daniel Downs. "A reconfigurable content addressable memory." In *International Parallel and Distributed Processing Symposium*, pp. 882-889. Springer, Berlin, Heidelberg, 2000.

[7] Srinivasan, Varad, Sanjay V. Gala, and Ketan K. Mehta. "Content addressable memory and random access memory partition circuit." U.S. Patent 5,706,224, issued January 6, 1998.

[8] Bhuvana, R., V. Prabhu, Kuppusamy PG, and M. R. Bibin. "Content Addressable Memory performance Analysis using NAND Structure FinFET." *Global Journal of Pure and Applied Mathematics* 12, no. 1 (2016): 1077-1084.

[9] I. Arsovski and A. Sheikholeslami, “Amismatch-dependent power allocation technique for match-line sensing in content-addressable memories,” *IEEE Journal of Solid-State Circuits,* vol. 38, no. 11, pp. 1958-1966, Nov 2003.

[10] Mohan, Nitin. "Low-power high-performance ternary content addressable memory circuits." (2006).

[11] Kannan, Kalapriya, and Subhasis Banerjee. "Compact TCAM: Flow entry compaction in TCAM for power aware SDN." In *International conference on distributed computing and networking*, pp. 439-444. Springer, Berlin, Heidelberg, 2013.

[12] K. Pagiamtzis and A. Sheikholeslami, “Pipelined match-linesand hierarchical search-lines for low-power content-addressable memories,” in Proceedings of the *IEEE Custom Integrated Circuits Conference (CICC),* 2003, pp. 383–386.

[13] Nakanishi, Mamoru, and Takeshi Ogura. "Real-time line extraction using a highly parallel Hough transform board." In *Proceedings of International Conference on Image Processing*, vol. 2, pp. 582-585. IEEE, 1997.

[14] F. Yu, R. H. Katz, and T. V. Lakshman, “Gigabit rate packet pattern matching using TCAM,” *IEEE International Conference on Network Protocols (ICNP),* Oct 2004, pp. 174-183.

[15] Banit Agrawal, Trimorthy Sherwood, “Modeling TCAM Power for Next Generation Network Devices,” In Proceedings of *IEEE International Symposium on Performance Analysis of Systems and Software* (ISPASS-2006).

[16] George, Silna, I. Rabeek Raja, and M. Prakash. "Exploiting a Low Power Self Timed Ternary CAM Architecture Using RWOS Mechanism." *International Journal of Engineering and Computer Science* 4, no. 04 (2015).

Evaluating RMS based continuous S-Transform with deep learning for detecting and classifying voltage sag and swell

Kamarulazhar Daud, Syazreena Sarohe, Wan Salha Saidon, Saodah Omar, Nurlida Ismail and Nazirah Mohamat Kasim

Faculty of Electrical Engineering, Universiti Teknologi MARA, Cawangan Pulau Pinang, Kampus Permatang Pauh, 13500 Permatang Pauh, Pulau Pinang

E-mail: kamarul395@uitm.edu.my, syazreena1997@gmail.com, wansalha@uitm.edu.my, saodah004@uitm.edu.my, nurlida@uitm.edu.my, nazirah261@uitm.edu.my

Abstract. Voltage sag and swell can cause serious problems like instability, short lifetime, and data errors in power quality. The objective of this paper is to present the detection and classification of voltage sag and swell. S-Transform is used as a base to detect the triggering point of disturbances using Root Mean Square (RMS) method. This paper also presents the type of sags and swells by applying the features into Extreme Learning Machine (ELM) neural network approach in MATLAB. In addition, ELM method is compared with Support Vector Machine (SVM) and Decision Tree method to observe the best classification between these three methods. The accuracy of the classifications was displayed in percentages. It was verified that the detection using RMS and classification using ELM are possible because the results are clearly showing the advantages of the RMS in detecting and ELM for classifying the power quality problems.

KEYWORDS: Root Mean Square (RMS), power quality (PQ), deep learning, s-transform, PQ disturbances.

1. Introduction

Electric power is the crucial type of energy as it requires continuous flow and cannot be conveniently stored. Power quality disturbances such as transient, harmonic, interruption voltage sag, and swell have negative impacts on a power supply based on the research studies [1][2]. To assure the quality of power is not an easy task. Therefore, it is necessary to know the root of disturbances based on detecting and classifying the power disturbance to maintain the quality of the power system.

Voltage sag and swell is an essential subject in Power Quality analysis and reason for the deterioration of power quality [3]. In [4][5], described that voltage sag happens when decreases in RMS occur within 0.5 cycles to 1 min voltage at power frequency. In fact, voltage sag is a short duration on the reduction of rms voltage. This is because voltage sag is measured from a reduction of rms voltage from below 0.1p.u. to above 0.9 p.u. of nominal voltage.

Meanwhile, voltage swell happens reversely from voltage sag. At the duration of 1.5 cycles to 1 min, increase in RMS voltage in between 1.1 p.u. to 1.8 p.u. nominal voltage. Voltage swell normally associate with the system fault condition. Energizing a large capacitor bank contributes to the presence of voltage swells based on the studies done by [6].

1.1 S-Transform

In previous research papers carried out by the [3][7][8], the Wavelet Transform (WT) was used as a signal processing in detecting the voltage sag and swell. WT can identify the locations containing the frequency domain. The problem arises as WT is complicated to compute, and the process is not accurate because the information of frequency disappeared at a certain time. Delay processing can also happen in WT. Therefore, WT is not suitable for extraction features.

To solve this problem, S-Transform was presented to overcome the imperfection of WT. S-Transform is an extension of WT and Short Time Fourier Transform (STFT). ST also is known as a Stockwell's transform, and ST has a better technique in Power Signal Processing. [3] Describes ST is based on moving and scalable localizing Gaussion windows. It has the benefit of transforming signal from time domain to two-dimensional frequencies following a similar Fourier frequency domain.

Based on previous research by [3], feature extraction of disturbance signals was extracted based on ST and produce S-matrix.

There are various methods of achieving the final equation of S-Transform. Based on the paper [9], the equations related to ST are as follow:

The Continuous Wavelet Transform (CWT), W(τ,d) of a function h(t) is defined as [9]:

$$W(\tau,d)=\int_{-\infty}^{\infty}h(t)w(t-\tau,d)dt \tag{1}$$

Equation (1) determine the expansion factor of d, which inverse from frequency f. The expansion determines the width of the wavelet.

ST is obtained by multiplying CWT with a phase factor as defined by [9]:

$$(\tau,f)=e^{i2\pi ft}W(\tau,d) \tag{2}$$

Where the mother wavelet is defined as

$$W(\tau,d)=\frac{f}{\sqrt{2\pi}}e^{-(\frac{t^2f^2}{2})}e^{-j2\pi ft} \tag{3}$$

Thus, the final form of the continuous ST is obtained as

$$S(\tau,d)=\int_{-\infty}^{\infty}h(t)\frac{f}{\sqrt{2\pi}}e^{-\frac{(\tau-t)^2f^2}{2}}e^{-j2\pi f\tau}dt \tag{4}$$

And the width of the Gaussian window is

$$\sigma(f) = T = 1/f \tag{5}$$

S-Transform of a discrete series of data creates a complex matrix (ST-matrix) in which the row corresponds to the frequency and the column corresponds to the time [7][10].

1.2 RMS Method

A previous study by [3][4][11] describes the power quality disturbance has a wide range of frequency, magnitude, and duration. There are various ways to detect power disturbances. One of the methods used was Fast Fourier Transform (FFT). The performance using FFT for detection was carried out using MATLAB Simulink and generation code. Based on the results noticed, the FFT could detect the magnitude frequency components, but the duration information cannot be provided by FFT. The problem in this section is that to detect sag and swell the time information is important to distinguish for the data used in classification. FFT is not suitable for detecting sag and swell, but it is more convenient to detect the Harmonic disturbance. Therefore, RMS method is presented in this research paper to overcome the problem and conduct the detection of voltage sag and swell.

Other than that, RMS method was used in this research paper because RMS is the best technique to detect short and long-duration variation disturbances. The detection method is based on the calculation of rms value stated in the article [6][12]. This method is one of the most broadly applied in detecting a disturbance in the power system because of its simplicity, speed in calculation, and less memory requirement. The RMS value is defined as:

$$V_{RMS} = \left(\frac{1}{N}\sum_{i=1}^{N} vi^2\right)^{1/2} \tag{6}$$

Where N is the sample per cycle, v_i is the nth sample of the recorded voltage waveform.

1.3 Deep Learning Classifier

Research conducted by [13], the classification of power quality disturbance using deep learning approach for image file classification. The researchers used image files of the three-phase PQ instead of using the sampled data PQ event. Even though the result classification was useful, but the processing time is prolonged.

This paper proposed Extreme Learning Machine (ELM) in classifies the data power disturbance sag and swell. ELM is also known as a single Feed-Forward Neural Network. According to [14], ELM's structure consists of 3 essential layers: the input layer, single hidden layer, and output layer. In fact, [14] mentions the weights that connect hidden nodes to outputs can be trained very fast because in ELM method, input weights are generated randomly. In contrast, analytical methods are used to calculate the output weights, causing the learning process to become extremely fast. Other advantages are ELM has excellent results in generalization compared to feed-forward networks with the traditional back-propagation learning algorithm. Research carried out by [15] stated the link that connects the hidden layer to the output could be customizable. Further analysis was carried out to study different types of deep learning to identify which method produces the best accuracy in classification signals.

2. Methodology

The methodology was proposed to ensure the reliability and validity of the result obtained.

2.1 Flowchart of proposal Method

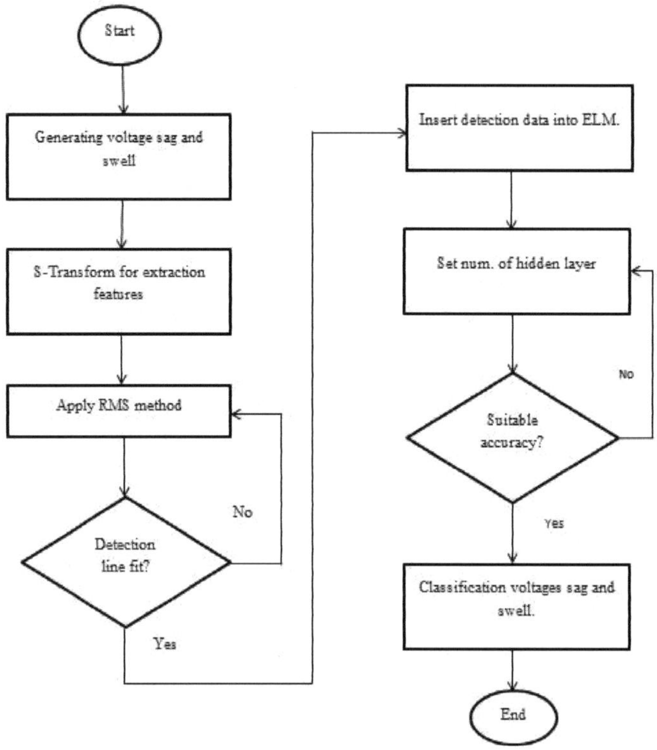

Figure 1. Flowchart of proposed method.

Figure 1 shows the methods to analyze the voltage sag and swell. A method analysis is fully carried out via programming by using MATLAB m-files. The system diagram consists of two stages, which are detection and classification.

2.2 Detection using RMS method based on S-Transform

First, sag (7) and swell (8) signals algorithm is written in MATLAB m-files based on their respective equation shown below:

y = (1-alpha*((Heaviside (t-t1)-Heaviside (t-t2))))*(sin ((2*pi*(50)*t) +w)); (7)

Where:

- alpha = decrease level of rms voltage ranged from 0.1 p.u. to 0.9 p.u.,
- t = 0:0.0001:0.4,
- t1 = sag initiation,
- t2 = sag recovery or clearance,
- Heaviside = unit step function in MATLAB
- w = phase angle jump 50°, 100°, 170°

y = (1+alpha*((Heaviside (t-t1)-Heaviside (t-t2))))*(sin((2*pi*(50)*t)+w)); (8)

Where:

- alpha = increase level of rms voltage ranged from 0.1 p.u. to 0.8 p.u.,
- t = 0:0.0001:0.4,
- t1 = swell initiation,
- t2 = swell recovery or clearance,
- Heaviside = unit step function in MATLAB
- w = phase angle jump 60°, 126°, 160°

Then, RMS method was applied to the m-files and the proper value of RMS was set before ST can process the input data. After that, the ST was applied to the m-files as a base for signal processing. ST extracts the features from the disturbance signals and produces S-matrix that contains data of standard deviation, variances, and mean. These data were collected and arranged in Excel toward the classification process later. The main reason to extract the features is to simplify the complexity of the data. The data of standard deviation contain real and imaginary. Therefore, by removing the standard deviation, the only real part will be highlighted and used as an input for the classification process.

After all the coding for detection was completed, the data was run. The value of phase angle in equation (7) and (8) applied is not the same for each sample in order to produce different sag and swell event for the comparison in the future. If the detection line does not fit the signal, the equation of RMS must be corrected to get a good result detection sag and swell signal. Along with that, the magnitude and duration of sag swell can be obtained from the waveform.

2.3 Classification using Extreme Learning Machine

Extreme learning machine is the learning algorithm used for a single-layer feed-forward network architecture useful for a single hidden layer determined by [15]. At the detection section, it had mentioned that the generated data was arranged inside Excel, and it consists of 300 inputs, which were mean, variance, and standard deviation with 1 output or target. The output or target was a binary function where output **0** is a normal signal, while **1** is a disturbance signal. These generated data were important to distinguish so that the ELM will know how to read the data. The data arrangement was saved as txt file and ready to be uploaded into the ELM. After the data was completely arranged, the data was import and loaded into the Neural Network. The hidden layer neuron number was set to 20 and the data ready to be trained.

The results classify presented in the form of a confusion matrix. Figure 2 is an example of a confusion matrix. The confusion matrix consists of three different colors. For the green box reporting, the correct classification, while red specifies misclassification and grey, determines overall accuracy. Simulation results and the comparison with the other types of neural network like SVM and Decision Tree were presented in the next section.

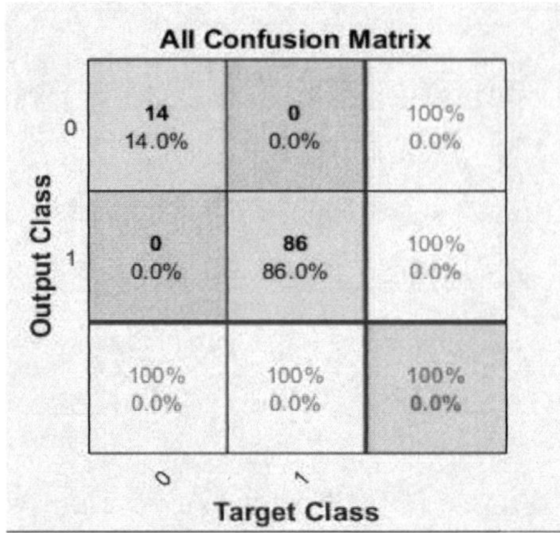

Figure 2. Example of the confusion matrix.

3. Result and discussions

3.1 Detection Voltage Sag and Swell using RMS based on S-Transform.

A blue line in figure 3 is specified as a detection line or mean line. The red line shows the input data.

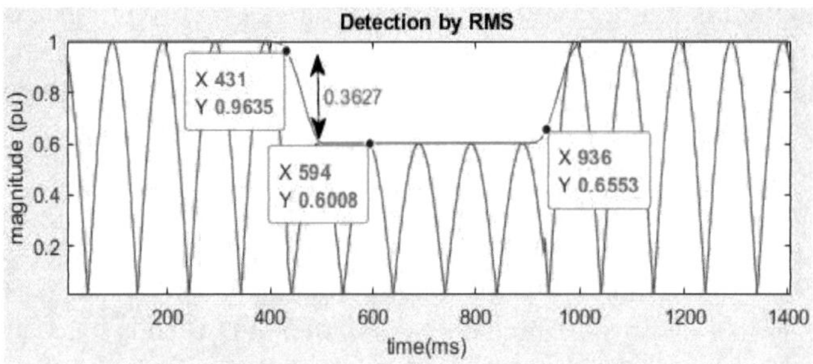

Figure 3. Detection of sag from sample 1.

Table 1. Comparison of the starting moment detection of voltage.

Sample	Phase angle, w	Sag initiation /ms	Sag recovery /ms	Duration /ms	Sag peak amplitude /pu	RMS initiation detection magnitude /pu	Distances /pu
1	50°	431	936	505	0.6008	0.9635	0.3627
2	100°	432	932	500	0.5999	0.9664	0.3665
3	170°	432	931	499	0.6004	0.8685	0.2681

Based on table 1, each sample used the same alpha value, which was 0.4 but different applied phase angle. Sample 1 had the longest duration detection refer to figure 3 and followed by sample 2 and sample 3. For the RMS line detection magnitude, sample 2 with 100° begins to detect at magnitude 0.9664 p.u., which had a long 0.3665 p.u. distance to reach the 0.5999 p.u. sag peak amplitude. Furthermore, refer to figure 3 and table 1 for the sample 1 with 50° phase angle had 0.3627 p.u. gap line detection to reach the 0.6008 p.u. of sag peak amplitude. Other than that, sample 3 with a 170°

phase angle gets to detect sag disturbance at 0.8685 p.u. magnitude, which was shorter distance 0.2681 p.u. to reach 0.6004 p.u. sag peak amplitude.

The conclusion can be made from the data collected above; the improved phase angle significantly improves the detection line. The RMS detection line was stable and faster for sample 3 compared to the other two samples. This is because sample 3 with 170° phase angle had the shorter duration detection and lower distance magnitude to reach the sag peak amplitude. For all samples, the detection line form is a straight line.

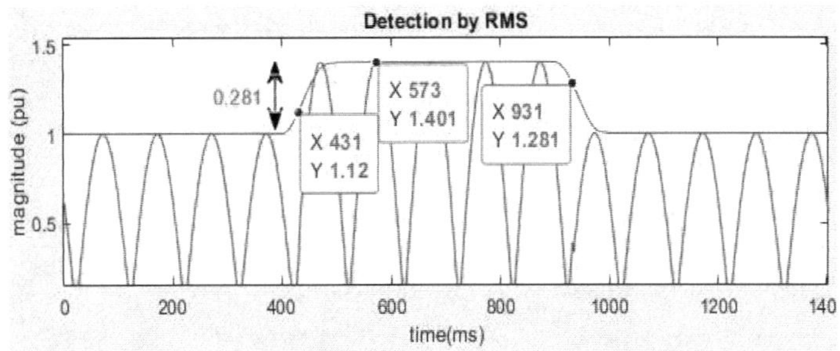

Figure 4. Detection of swell from sample 1.

Table 2. Comparison of starting moment detection of voltage swell.

Sample	Phase angle, w	Swell initiation ms	Swell recovery /ms	Duration /pu	Swell peak amplitude /pu	RMS initiation detection magnitude /pu	Distance /pu
1	60°	431	931	500	1.401	1.120	0.281
2	126°	433	927	494	1.401	1.138	0.263
3	160°	431	933	502	1.402	1.038	0.364

According to the data collected in table 2, all samples had the same value of alpha and different phase angles applied. The shape of the detection line is a straight line for those three samples. Based on table 2 and figure 4, the RMS line detection for sample 1 happens at 431ms with magnitude 1.120 p.u. The distances for the detection line to reach 1.401 p.u. swell peak amplitude is 0.281 p.u. Move to sample 2 with 126°, the distance obtained for RMS line detection to reach 1.401 p.u. peak amplitude swell is only 0.263 p.u. The duration for detection only takes 494ms, which is faster than sample 1 and sample 3 duration. For the last sample 3 with 160°, swell detection occurred at 431ms with 1.038 p.u. magnitude, and it took 0.364 p.u. distance magnitude for the detection to reach the swell peak amplitude of 1.402 p.u. From the data obtained, the improve phase angle gives better results detection. It can be verified that the sample 2 with 0.4 alpha and 126° phase angle had the best detection because the distances for line detection to reach swell peak amplitude is shorter.

3.2 Classification using deep learning

Table 3. Overall percentage accuracy on the classification of sag and swell.

Type of disturbance	Data size	method					
		ELM		SVM		Decision Tree	
		correct	Misclassified	Correct	Misclassified	Correct	Misclassified
sag	1400	96.7%	3.3%	94.9%	5.1%	93.4%	6.6%
swell	1400	97.1%	2.6%	95.7%	4.3%	95.8%	4.2%
Average percentage		96.9%	3.0%	95.3%	4.7%	94.6%	5.4%

Table 3 and Table 4 shows the comparison of ELM, SVM, and Decision tree method in classifying the disturbances sag and swell. From table 3, ELM has the highest average percentage correct classification followed by SVM method and Decision Tree method. The correct classification for ELM method in sag is 96.7%, with the misclassified of 3.3%. Meanwhile, swell get 97.1% correct and only 2.6% is misclassified. For SVM method, the correct classification for sag and swell is 94.9% and 95.7%. Moreover, the correct classifications for the Decision Tree method in sag and swell can be seen that there were not so many different percentages obtained from SVM. Decision Tree had the highest average percentage of misclassification with 5.4% compared to SVM and ELM.

Table 4. Comparison of trained time on sag and swell.

Method / Disturbance	Training Time/ sec		
	ELM	**SVM**	**Decision Tree**
sag	0.050	20.313	17.172
swell	0.160	4.959	25.503

From table 3, the average percentage accuracy for ELM, SVM, and Decision Tree was discussed, and ELM obtained the highest correct classification. However, in table 4, the given training time for the ELM was shorter than SVM and Decision Tree. Sag data that trained under ELM were only taking 0.050 sec, while SVM and Decision Tree take longer than this, which is 20.313sec and 17.172sec. The same goes for the swell training time. This gives an advantage for the ELM method in terms of computational speed.

Therefore, by comparing these three methods, it was found out that in terms of classification performances, ELM was more accurate as well in terms of training time. ELM was the fastest method to train the data. Thus, ELM had the best overall classification technique for sag and swell signal.

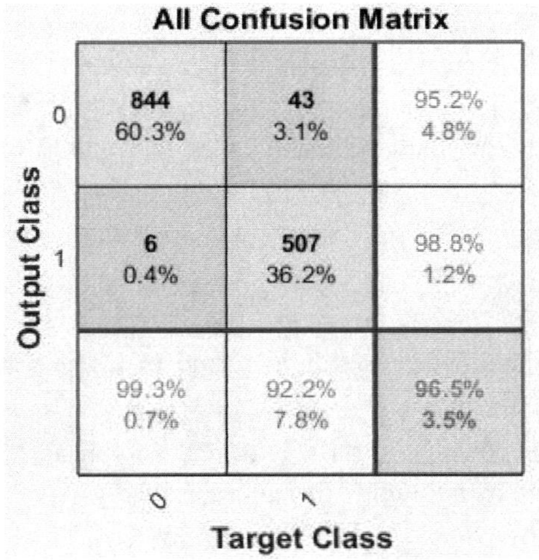

Figure 5. Sag classification by ELM method from sample 3.

Figure 5 explained:

- Class 1 stands for sag data, while class 0 stands for the normal signal.
- 507 out of 550 sag data, which was correctly classified. The rest is misclassified under output class 0, which carried 3.1%.
- Only 6 data out of 850 normal data was misclassified under output class 1, which carried 0.4%.
- The total correct classification for sag is 96.5%.

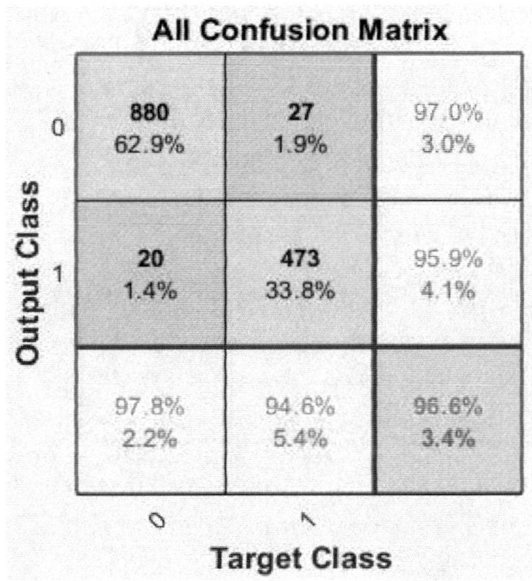

Figure 6. Swell classification by ELM method for sample 2.

Figure 6 explained:

- Class 1 stands for swell data, while class 0 stands for the normal signal.
- 473 out of 500 sag data, which was correctly classified. The rest is misclassified under output class 0 which carried 1.9%
- Only 20 data out of 900 normal data, which was misclassified under output class 1 which carried only 1.4%.
- The total correct classification for swell is 96.6 %.

4. Conclusion and recommendation

In this paper, the extraction features were analyzed using mean, variances, and standard deviation from voltage sag and swell signals by using S-Transform. This paper also included analysis detection of voltage sag and swell using RMS method by comparing the data sample to determine which sample has the best performance. As a result, the initiation detection magnitude with lower gap magnitude to reach peak sag or swell is the best.

Besides that, it is observed that ELM correctly classifies the sag and swell class with high accuracy. The ELM is compared with the other classifying methods, which are SVM and Decision Tree. It is found out that the ELM gives the best result in terms of classification and computational time. Therefore, all objectives were achieved.

Also, to improve the analysis performance, the scope of disturbances is not just limited to the sag and swell signal only. It can be done by applying other types of disruption as well.

Acknowledgment

The author wish to thank Universiti Teknologi MARA, Cawangan Pulau Pinang and Faculty of Electrical Engineering Universiti Teknologi MARA, Cawangan Pulau Pinang for their endless technical assistance and funding.

References

[1] M. I. Muhamad, N. Mariun, and M. A. A. Radzi 2007 The effects of power quality to the industries *2007 5th Student Conf. Res Dev.* (SCORED)
[2] S. Bhattacharyya and S. Cobbe 2011 Consequences of Poor Power Quality – An Overview (*Power Qual)*
[3] M. J. Dehghani 2009 comparison of S-transform and wavelet transform in power quality analysis (*World Acad. Sci. Eng. Technol* vol 38) pp 399-402
[4] N. R. Nashad, M. J. Islam, R. M. Rahat, M. T. A. Begum, and M. R. Alam 2017 A simplistic mathematical approach for detection and classification of power quality event (*ECCE 2017-Int. Conf. Electr. Comput. Commun. Eng.*) pp 698-703
[5] S. Kamble and C. Thorat 2014 Voltage Sag Characterization in a Distribution Systems: A Case Study (*J. Power Energy Eng* vol 02 no 04) pp 546-553
[6] R. Ingale and L. Tawade 2013 Detection and Comparison of Power Quality Disturbances using Different Techniques (*Int. J. Comput. Appl.* vol 75 no 18) pp 48-53
[7] F. Zhao and R. Yang 2007 Power-quality disturbance recognition using S-transform (*IEEE Trans Power Deliv* vol 22 no 2) pp 944-950
[8] N. A.B. Ramlee 2017 Detection of Voltage Sag and Voltage Swell in Power Quality Using Wavelet Transforms 2017 (*World Acad. Sci. Eng. Technol. Int. J. Electr. Comput. Energ. Electron. Commun. Eng 11.7* vol 11 no 7) pp 865-870
[9] Y. Wang 2010 The Tutorial: S Transform (*Grad. Inst. Commun. Eng. Natl. Taiwan Univ. Taipei* 2010) pp 1-23
[10] J. Ma, J. Zhang, L. Xiao, K. Chen, and J. Wu, Classification of Power Quality Disturbances via Deep Learning IETE Tech. Rev. (Institution Electron. Telecommun. Eng. India vol 34 no 4) pp 408-415
[11] V. C and S. D. V. S. S. Siva 2010 Techniques used for Detection of Power Quality Events – a Comparative Study (*IEEE* no 3) pp 307-312
[12] B. Taheri and F. Razavi 2018 Power swing detection using RMS current measurements (*J. Electr. Eng.Technol.* Vol 13 no 5) pp 1831-1840
[13] E. Balouji and O. Salor 2017 Calssification of power quality events using deep learning on event images 2017 3rd Int. Conf. Pattern Anal. Image Anal. IPRIA 2017 no August 2018) pp 216-221
[14] N. D. Hoang and D. Tien Bui 2017 Slope Stability Evaluation Using Radial Basis Function

IOP Conf. Series: Materials Science and Engineering 1045 (2021) 012041 doi:10.1088/1757-899X/1045/1/012041

Neural Network, Least Squares Support Vector Machines, and Extreme Learning Machine 1st ed (*Elsevier Inc)*
[15] F. Ucar, J. Cordova, O. F. Alcin, B. Dandil, F. Ata, and R. Arghandeh 2019 Bundle extreme learning machine for power quality analysis in transmission networks (Energies vol 12 no 8)

IOP Conf. Series: Materials Science and Engineering 1045 (2021) 012042 doi:10.1088/1757-899X/1045/1/012042

A non-uniform excitation Antenna array with defected ground structures for Dual-band 5G mm-wave Applications

Shashi Kumar D[*] and S Suganthi[1]
RF and Microwave Research Laboratory, School of Engineering and Technology, CHRIST University, Bangalore, India.
[*]Corresponding author E-mail : shashiphd2015@gmail.com
[1]E-mail : drssuga@gmail.com

Abstract. In this paper, the authors present a non-uniform excitation antenna array with defected ground structures for 28GHz and 40GHz mm-wave applications. The proposed antenna array is simulated using HFSS Ver. 18.2 to analyze the radiation and reflection characteristics. Authors used RT Duroid (ϵ_r=2.2) with a thickness of 0.782 mm to design the antenna array. To improve the gain and bandwidth, Defected Ground structures are used in the proposed design. Use of Dolph Chebyshev distribution of excitation amplitudes miniaturized the antenna array. The proposed design exhibited a fractional BW of 23.33% and 5.56% at 28GHz and 40 GHz respectively. A peak gain of 8.88 dB and 7.7 dB at 28 GHz and 40 GHz respectively is observed in the simulation results.

1. Introduction

The rapid advances in wireless communication technology are enabling the replacement of LTE and its derivatives with the emerging robust 5G technology. 5G is popularly known for its characteristics like low latency, higher bandwidth, higher capacity and support to mobility. The emerging 5G technology attracted many researchers as the mm waves play a significant role in many industrial and scientific applications. Few among them include radio astronomy, remote sensing, radar systems.

The large bandwidth is one of the critical requirements for 5G technology to decrease latency and support higher data rates. The explosive growth in connected devices such as laptop, smartphones, wearable devices and home appliances etc. created excessive demand for large bandwidth to support higher data rates. 4G technology is unable to address some of the major problems of present wireless communication systems, that include poor coverage, poor quality over crowded channels and bad interconnectivity [1]. The physical dimensions of the mm-wave antenna are tiny. The major challenge for the present researchers in mm-wave antenna design is to maintain precision in fabrication and to get an efficient antenna. Therefore, it is vital to adopt new techniques for designing antennas for 5G mm-wave communication. A variety of approaches are reported in the literature for improving gain and bandwidth. To achieve the notched frequency band of 30-35 GHz, Haraz et al. [2] created an L shaped slot in the feed line.

Arizaca et al. [3] reported a parallel fed array with uniform excitation to demonstrate the improved gain. In [4], Rahayu et al., presented a non-concentric triangles DGS antenna to show bandwidth improvement. Khattak et al. [5] achieved dual-band behaviour using the elliptical slot in a circular patch antenna array. In [6], Nanae et al. designed a 2 × 2 rectangular patch antenna array with U slot to achieve wideband behaviour. To obtain the wideband array response, Sharawi et al. [7] utilized the concept of Connected Antenna Arrays (CAA).

In this paper, the authors developed a 1 x 4 non-uniform (Dolph- Chebyshev) amplitude excitation array resonating at 28 GHz and 40 GHz. The designed antenna array is miniaturized and exhibits good resonance and radiation characteristics at desired mm-wave frequencies.

2. Design of Antenna Array

A non-uniform excitation corporate fed antenna array is designed with excitation coefficients a_1=1, a_2 = 1.6678, a_3=1.6678 and a_4=1. Since the non-uniform excitation corporate feed network has multiple microstrip segments with different impedances, impedance matching is crucial to achieving better radiation and resonance characteristics. While designing a non-uniform excitation array, the major problem one can encounter is creating an efficient feed network with microstrip segments that could be realizable using available fabrication techniques.

As we know, the width of the microstrip segment to be created will become un-realistic if the impedance of the microstrip is more than 200 Ω. To avoid this complexity, after every stage power division, authors used Quarter Wave Transformers (QWT's) appropriately to scale down the feed point impedances for achieving better impedance matching. The feed network developed to create a 1x4 Dolph-Chebyshev excitation array with feed point impedances 50Ω, 100Ω, 66.69Ω and 33.99Ω is shown in figure 1.

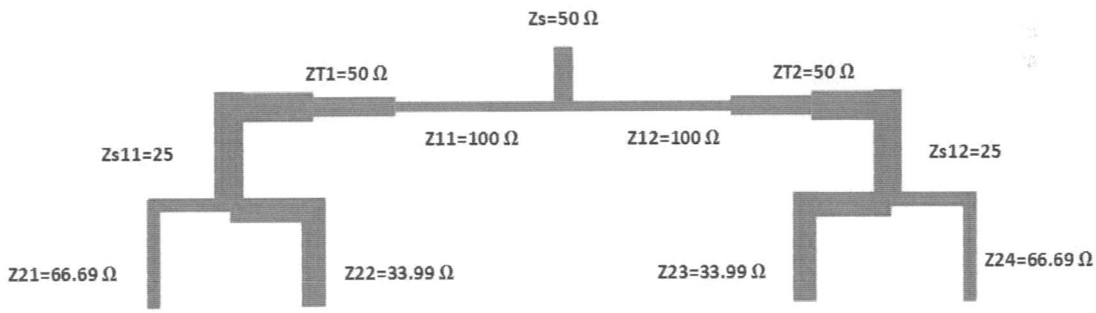

Figure 1. Adopted feed with Non-uniform excitation (Dolph-Chebyshev distribution)

2D surface current distributions were studied at the desired 28 GHz and 40 GHz resonant frequency. Based on the analysis made out of surface current distribution plots, defect on the ground plane is introduced by creating six circular rings, each 0.5 mm width. The geometry and specifications of initial antenna array and the defected ground antenna array are depicted in figure 2 and figure 3 respectively.

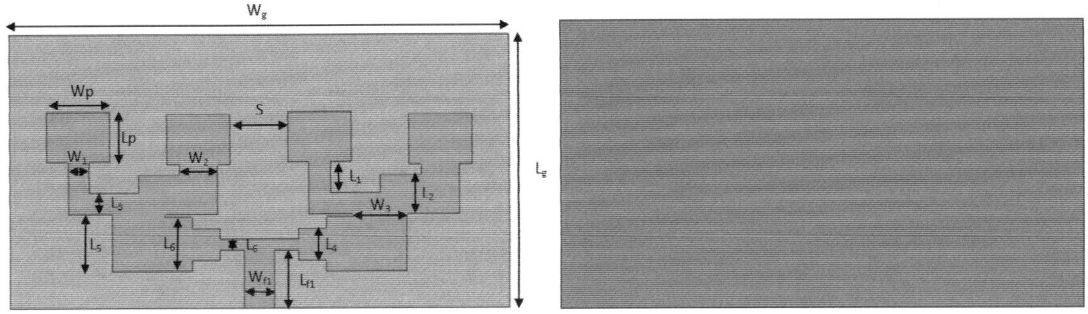

(a) (b)

Figure 2. Simulated antenna array without DGS (a) Top view (b) Bottom view

IOP Conf. Series: Materials Science and Engineering 1045 (2021) 012042 doi:10.1088/1757-899X/1045/1/012042

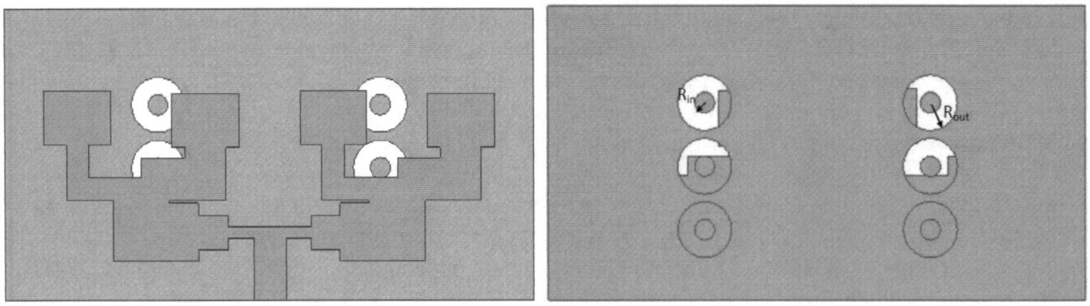

(c) (d)

Figure 3. Simulated antenna array with DGS (a) Top view (b) Bottom view

The physical dimensions of the patch element, DGS and feed network are presented in table 1.

Table 1. Parameters of the designed antenna array

Parameter	Size (mm)	Parameter	Size (mm)
Lg	16.74	Wg	31.56
Lp	3.1	Wp	4
W1	1.33	W2	2.41
W3	3.37	Wf1	1.93
S	3.61	Rout	1.581
L1	1.85	L2	2.42
L3	1.33	L4	1.93
L5	3.5	L6	3.37
L7	0.62	Lf1	3.61
Rin	0.581		

3. Results and Discussion

Simulations results are plotted using HFSS Ver. 18.2. The S_{11} and 2D radiation patterns are analysed to understand the resonance and radiation behaviour of the designed antenna arrays. From the results presented in figure 4, one can make out that both designs reveal dual-band nature. The antenna array without DGS is providing a -10 dB bandwidth of 3.78 GHz and 2.21 GHz bandwidth at 28 GHz and 40 GHz band respectively. After introducing the defects in the ground plane the current distribution is changed at 28 GHz, as a result of this, the bandwidth is increased from 3.78 GHz to 6.13 GHz, maintaining same bandwidth at 40 GHz with a resonant frequency shift closer to desired 40 GHz.

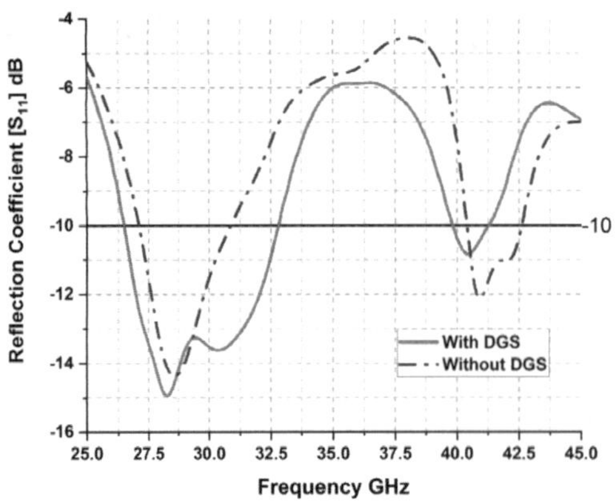

Figure 4. Comparison of Reflection coefficient

IOP Conf. Series: Materials Science and Engineering 1045 (2021) 012042 doi:10.1088/1757-899X/1045/1/012042

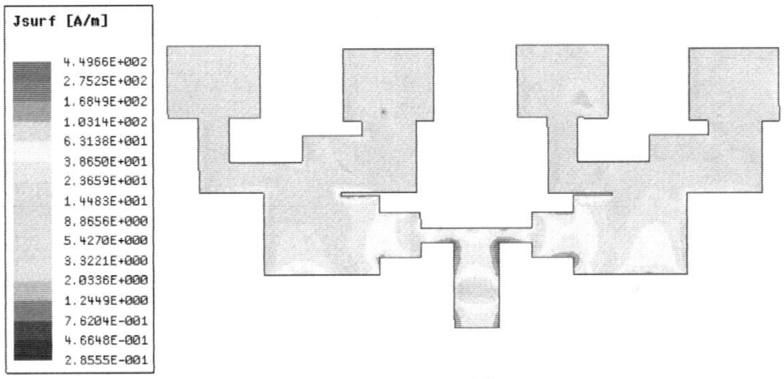

(a)

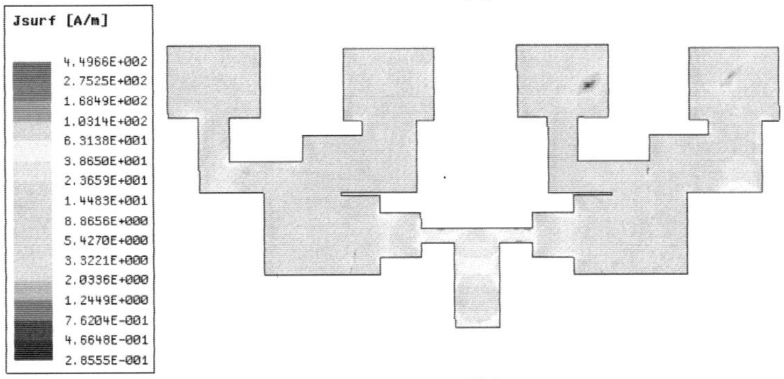

(b)

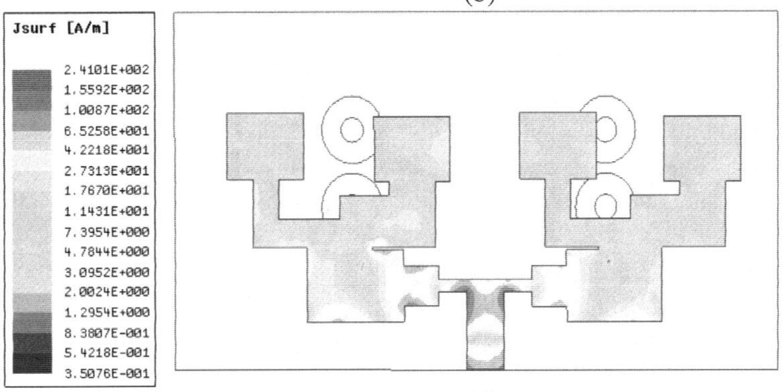

(c)

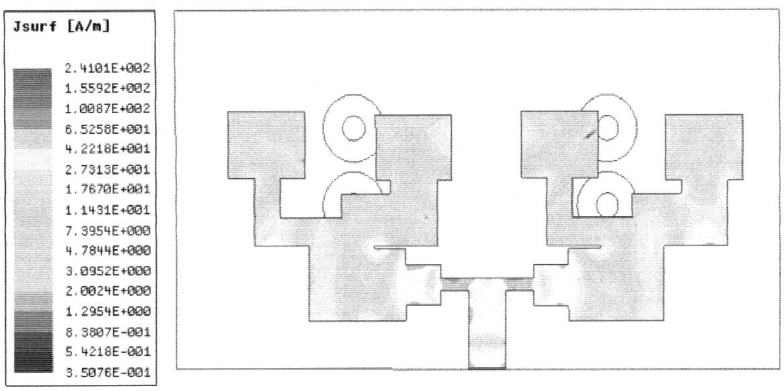

(d)

Figure 5. Current density plots (a) at 28 GHz without DGS (b) at 40 GHz without DGS (c) at 28 GHz with DGS (d) at 40 GHz with DGS

Figure 5 depicts the surface current densities at 28 GHz and 40 GHz mm-wave frequencies. From the figure, one can observe that the concentration of current near the T junction of 100 Ω and 50 Ω microstrip line segments of feed network without DGS is more at 28 GHz compared to that at 40 GHz. The defect created in the antenna array improved the current distribution at the junction and enhanced the - 10 dB bandwidth at 28 GHz from 3.78 GHz to 6.13 GHz.

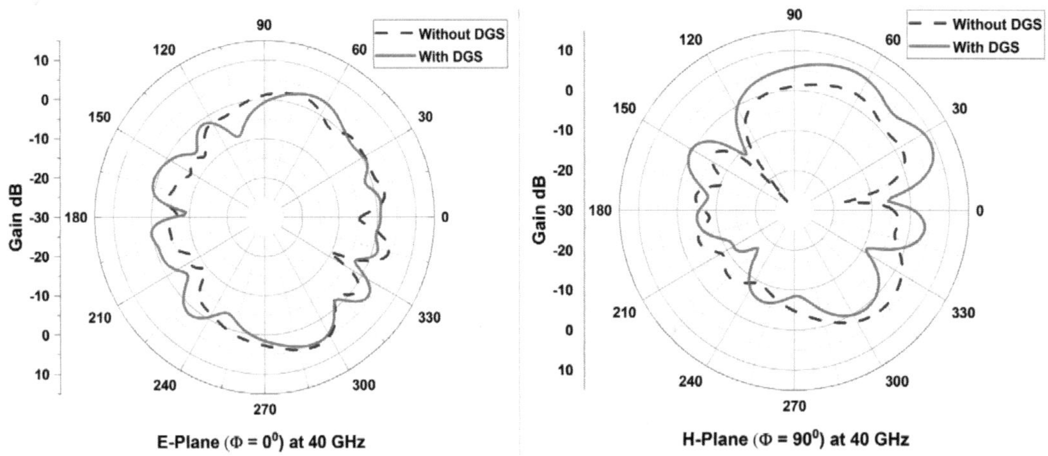

(a) (b)

Figure 6. Two dimensional radiation patterns at 40 GHz (a) E Plane ($\phi = 0^0$) (b) H Plane ($\phi = 90^0$)

The 2D radiation patterns of antenna array without DGS and with DGS are compared at 40 GHz, and the same is plotted in Figure 6. For $\phi = 0^0$ and $0 \leq \theta \leq 360^0$ (E plane pattern), the plot shows almost omnidirectional and for $\phi = 90^0$ and $0 \leq \theta \leq 360^0$ (H Plane pattern) shows the significant power concentration between 25^0 and 120^0. From the traces of E plane and H plane radiation patterns, one infer that antenna array with DGS is having the highest peak gain of 8.88 dB.

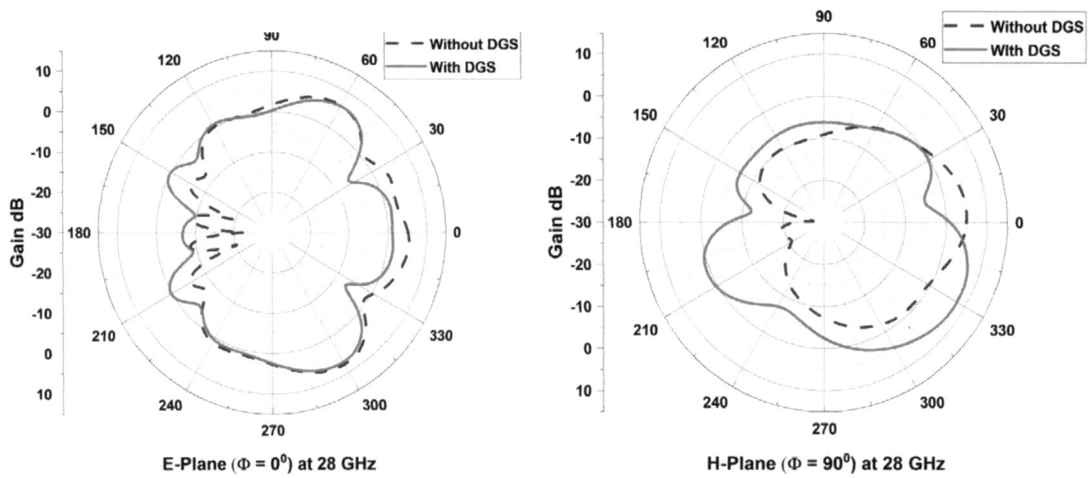

(a) (b)

Figure 7. Two dimensional radiation patterns at 28 GHz (a) E Plane ($\phi = 0^0$) (b) H Plane ($\phi = 90^0$)

Figure 7 shows the 2D radiation patterns at 28 GHz. There is not much change seen in the E plane radiation pattern at 28 GHz, compared to the radiation pattern plotted at 40 GHz. But one can observe a significant change in the H plane radiation pattern of 28 GHz, compared to 40 GHz radiation pattern.

This is due to change in the current density at 40 GHz caused by the creation of defects on the ground plane of the antenna array. The observed peak gain at 28 GHz is 7.77dB. The simulation results of antenna array without DGS and antenna array with DGS are contrasted and presented in table 2.

Table 2. Comparison between the antenna arrays

Parameter	With DGS		Without DGS	
	40 GHz	28 GHz	40 GHz	28 GHz
S_{11} (dB)	-11.41	-13.14	-7.58	-13.46
Gain (dB)	8.88	7.7	7.89	7.77
BW (GHz)	2.21	6.13	2.21	3.78
Overall Size (mm)	31.56 x 16.74			

Comparing the tabulated results, one can observe that the antenna array with DGS shows larger -10 dB bandwidth at 28 GHz, without changing the bandwidth at 40 GHz. At the same time, the DGS used in the array helped to improve the gain by almost 1 dB at 40 GHz maintaining the same overall size.

4. Conclusions

A dual-band low profile antenna array is designed to operate at 28 GHz and 40 GHz. DGS method is adopted to improve the gain at 40 GHz and bandwidth at 28 GHz. Table 3 Presents the comparison table of antenna arrays developed by the authors with closely matched antenna arrays available in the literature. Observing Table 3, one can infer that the proposed antenna arrays in this paper are smaller in size compared to identical antenna arrays exist in the literature. Among the antenna arrays reported, the proposed antenna array with DGS is exhibiting larger bandwidth. Since the proposed antenna arrays are having sound radiation and resonance characteristics at 5G mm-wave frequency bands, these antennas can be used for applications at 28 GHz and 40 GHz mm-wave wireless bands.

Table 3. The comparison of proposed work with similar works reported in the literature

Reference cited	f_r (GHz)	ε_r	S_{11}(dB)	Gain (dB)	BW (GHz)	Overall Size (mm)
With DGS	28	2.2	-13.14	7.7	6.13	31.56 x 16.74 x 0.787
	40		-11.41	8.88	2.21	
Without DGS	28	2.2	-13.46	7.77	3.78	
	40		-7.58	7.89	2.21	
[3]	28	2.2	-16.37	17	0.308	30.65 x 39.3 x 0.127
[4]	28	2.2	-31.61	7.18	4.127	35 x 25 x 1.57
	38		-31.06	10.1	3.152	
[8]	25	28	-45	8.6	0.66	25.9 x 27.6 x 7.078
	28		-25	10.2	0.76	
	38		-43	6.5	1.02	
[9]	28	4.5		8.85	5.47	36 x 74 x 1.6

IOP Conf. Series: Materials Science and Engineering 1045 (2021) 012042 doi:10.1088/1757-899X/1045/1/012042

References

[1] C. X. Wang, F. Haider, X. Gao, X. H. You, Y. Yang, D. Yuan, H. Aggoune, H. Haas, S. Fletcher and E. Hepsaydir, 2014 Cellular architecture and key technologies for 5G wireless communication networks, IEEE Comm. Magazine. **52** 2 122-130.

[2] O. M. Haraz, M. M. M. Ali, S. Alshebeili and A. Sebak, 2015 Design of a 28/38 GHz dual-band printed slot antenna for the future 5G mobile communication Networks, *IEEE Int. Sym. on Antennas and Propagation & USNC/URSI Nat. Radio Sci. Meet.* 1532-1533.

[3] D. N. Arizaca-Cusicuna, J. Luis Arizaca-Cusicuna and M. Clemente-Arenas,2018 High Gain 4x4 Rectangular Patch Antenna Array at 28GHz for Future 5G Applications, *IEEE XXV Int. Conf. on Electron. Electric. Eng. and Computing* 1-4.

[4] Y. Rahayu and M. I. Hidayat, 2018 Design of 28/38 GHz Dual-Band Triangular-Shaped Slot Microstrip Antenna Array for 5G Applications, *2nd Int. Conf. on Telematics and Future Gen. Networks* 93-97.

[5] M. I. Khattak, A. Sohail, U. Khan, Z. Barki, and G. Witjaksono, 2019 Elliptical Slot Circular Patch Antenna Array with Dual Band Behaviour for Future 5G Mobile Communication Networks, *Prog. In Electromag. Research. C*, **89** 133-147.

[6] Y. Nanae . and C. Seo. “A 28-GHz Wideband 2×2 U-Slot Patch Array Antenna, *J. of electromag. eng. and sci.* **17** 133-137.

[7] M. S. Sharawi and M. Ikram, 2018 Slot-based connected antenna arrays for 5G mobile terminals, *2018 Int. Workshop on Ant. Tech. (iWAT)* 1-3.

[8] S. Dellaoui, A. Kaabal, M. E. Halaoui, A. Asselman, 2018 Patch array antenna with high gain using EBG superstrate for future 5G cellular networks, *Procedia Manufacturing* (Elsevier B.V), **22**, 463–467.

[9] Janam Maharjan, Dong-You Choi, 2020 Four-Element Microstrip Patch Array Antenna with corporate-Series Feed Network for 5G Communication, *Int. J. of Antennas and Propagation*, **2020** 1-12.

Hand gesture recognition on python and opencv

Ahmad Puad Ismail[1], Farah Athirah Abd Aziz[1], Nazirah Mohamat Kasim[1] and Kamarulazhar Daud[1]

[1] Faculty of Electrical Engineering, Universiti Teknologi MARA (UiTM), Cawangan Permatang Pauh, Pulau Pinang, Malaysia

Abstract. Hand gesture recognition is one of the system that can detect the gesture of hand in a real time video. The gesture of hand is classify within a certain area of interest. In this study, designing of the hand gesture recognition is one of the complicated job that involves two major problem. Firstly is the detection of hand. Another problem is to create the sign that is suitable to be used for one hand in a time. This project concentrates on how a system could detect, recognize and interpret the hand gesture recognition through computer vision with the challenging factors which variability in pose, orientation, location and scale. To perform well for developing this project, different types of gestures such as numbers and sign languages need to be created in this system. The image taken from the realtime video is analysed via Haar-cascaded Classifier to detect the gesture of hand before the image processing is done or in the other word to detect the appearance of hand in a frame. In this project, the detection of hand will be done using the theories of Region of Interest (ROI) via Python programming. The explanation of the results will be focused on the simulation part since the different for the hardware implementation is the source code to read the real-time input video. The developing of hand gesture recognition using Python and OpenCV can be implemented by applying the theories of hand segmentation and the hand detection system which use the Haar-cascade classifier.

Keywords-hand gesture, human computer interaction (HCI), contour, convex hull, convexity defects, gesture recognition, python, openCV.

1. Introduction

In a day-to-day life, hand gesture recognition is one of the system that can detect the gesture of hand in a real time video. The gesture of hand is classify within a certain area of interest. Designing a system for hand gesture recognition is one of the goal of achieving the objectives of this project. The task of recognizing hand gestures is one of the main and important issues in computer vision. With the latest advances in information and media technology, human computer interaction (HCI) systems that involve hand processing tasks such as hand detection and hand gesture recognition.

The first step in any hand processing system is to detect and locate the hand in the real-time video from the webcam. The detection of hand is challenging because of variation in pose, orientation, location and scale. Also, different intensity of light in the room adds to the variability. In the process of detection of hand, according to Mohamed [1], hand gesture recognition generally involves multiple levels such as image acquisition, pre-processing, feature extraction and gesture recognition. Image acquisition involve capturing image in the video frame by frame using a webcam. The captured images go through the image pre-processing process which involves color filtering, smoothing and thresholding. Feature extraction is a method that extracting features of the hand image such as hand contours while gesture recognition is a method to recognize hand gesture by extracting the features.

In this study, designing of the hand gesture recognition is one of the complicated job that involves two major problem. Firstly is the detection of hand. User hand is detected by using webcam in real-time video. The problem would be the unstable brightness, noise, poor resolution and contrast. The detected hand in the video are recognized to identify the gestures. At this stage, the process involves are the segmentation and edge detection. According to Karishma [2], with various information of image like color, hand posture and shape based (shape of hand) in a real time would affect the recognition of gestures. Another problem is to create the sign that is suitable to be used for one hand in a time. The extraction of hand need to be followed to determine each number and sign used. According to Amiraj [3], the extraction of hand involves the contour and convexity defects. Convexity defect gives an issue on how to calculate the depth of the defects. Some of the defects have far greater depth than others so to identify the depth would include some equations.
The list of objectives that will need to be achieve for this project: (1) to establish a complete system for detecting, recognizing and interpreting hand gesture recognition through computer vision using Python and OpenCV, and (2) to create the numbers and sign languages of hand gesture shown in the system that will meets the name of the project.

This study comprises on how to implement a complete system that can detect, recognizing and interpreting the hand by using Python and OpenCV in any intensity of light, pose or orientation of hand. In order to accomplish this, a real-time gesture based system is developed. In the proposal stage, designing a system that could detect the hand through the contour of the hand. The contour of the hand refers to the curve for a two variables for the function along which that function has a contact value that is not changed. Besides, to detect the appearance of hand in a frame of the real-time video, I will be using the Haar-cascade Classifier to track the appearance of hand before the image-processing is done. The result of the appearance of hand that the system could detect will be validated for analysis. For hardware implementation section, a wired web camera is used thus the hand gesture recognition can be done and implemented.

For this project, the prototype itself has its own limitation in terms of hardware implementation. First of all, the video capturing device used for this project can only configure up to 640 by 480 pixel resolutions. This might be the disadvantage of getting the hardware to properly work for detecting the hand. To solve this, the web camera must be in a fixed position up to 30 inches (75 cm) to help capture the video more efficiently. Another limitation to this project is the variation in plane and pose. Variations can be classified as rotation, translation and scaling of camera pose [4], [5]. The direction of the web camera is important to obtain a good video without any shadowing effects and the intensity of light in a room is enough. According to Ray [6], by selecting the correct light intensity and direction, the shadows can be reduced to a minimum.

2. Methodology

Hand gesture recognition project is done using Python programming language and OpenCV [7], [8] as library. Python programming language produces simple and easy system code to understand. Also, Python package used here is Numpy [9], [10]. The image that is captured using web camera will be processed in a region called as Region of Interest (ROI) where act as a region of wanted area while ignoring the outside region, called background [11].

2.1. Hand Segmentation Methods

According to Nobuyuki Otsu [12], segmentation is done to convert gray scale image into binary image so that only two object in image which one is hand and other is background. Otsu notes again, this algorithm is used for segmentation purpose and gray scale images are converted into binary image consisting hand or background. A very good segmentation is needed to select an adequate threshold of gray level for extract hand from background for example there is no part of hand should have background and background also should not have any part of hand. In general, the selection of an appropriate segmentation algorithm depends largely on the type of images and the application areas. The Otsu segmentation algorithm was tested and found to give good segmentation results for the hand gesture. Otsu algorithm is nonparametric and unsupervised method of automatic threshold selection. The basic theory of segmentation of hand is by using the equation as follow:

$$p_i = \frac{n_i}{MN}, \qquad p_i \geq 0, \quad \sum_{i=1}^{L} p_i = 1 \tag{1}$$

where the pixels of a given picture be represented in L gray levels$[1, 2, 3, \ldots \ldots \ldots \ldots \ldots, L]$ in a digital image of size $M \times N$, and number of pixels at level i is denoted by ni and the total number of pixels in the image is $MN = n1 + n2 + n3 + n4 \ldots \ldots \ldots \ldots \ldots nL$. The normalized histogram has components of p. The formula above was adapted from Rafael [13]. To apply the formula (1), the image of each frame that has been captured from real-time video are considered as pixels and being compared to each other of the frames. As each frames contains equal value of colour in a pixel, the pixel is then considered as eliminated or removed. These applied to other pixels within each frame, and each frames are compared with respect to time. As any of the pixels between two frames do not match with each other, the pixel at the first frame retain its contain. Otsu's thresholding method involves iterating through all the possible threshold values and the measurement of spread for the pixel level for each side of the threshold is calculate for example pixels that either fall in foreground or background. The aim is to find the threshold value where the sum of foreground and background spreads is at its minimum. Figure 1 shows how the hand image is obtained from the real-time video in a gray scale filter and then the image is converted for Otsu's algorithm that showing the thresholded image.

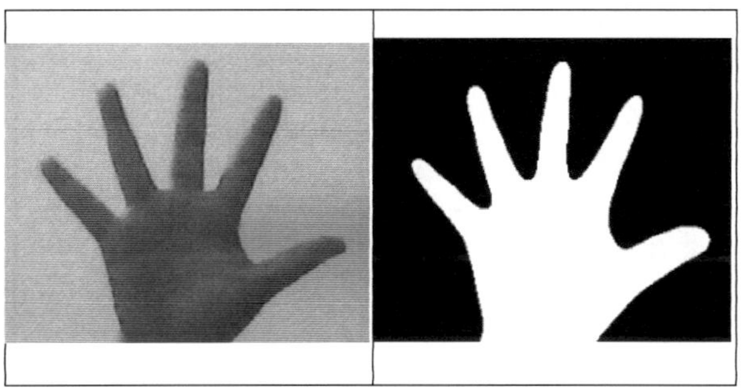

Figure 1. Comparison between gray scale image and thresholded image

The limitation of the Otsu's thresholding method is that as I previously mentioned, this algorithm only consider a bimodal image where bimodal image is an image whose histogram has two peaks. For that image, Otsu's does take a value in the middle of those peaks as threshold value. Then, it automatically calculates a threshold value from image histogram for bimodal image. This only work when a frame capturing a frame from video and converting the RGB format into gray scale filter.

2.2. Track Gesture Of Hand Using Haar-Cascade Classifier

In order for the program to continuously detect the hand at all times, especially when the hand is in static or not moving around, the program needs to learn the hand image. This is where the machine learning comes in. In this project, the machine learning technique was performed to track the gesture of hand before the image processing is done. As suggested by Sander [14], the suitable machine learning technique in term of image processing technique that is much available to use is called as cascade classifier. According to Viola [15], the cascaded classifier are actually the program features for the machine to know the exact object that will need to classify in each frames of images. One example that will be use in this project is Haar-cascaded type. Viola [15] again suggested that Haar-cascaded classifier was the simple classifier for machine learner to detect any object.

IOP Conf. Series: Materials Science and Engineering 1045 (2021) 012043 doi:10.1088/1757-899X/1045/1/012043

2.3. Region of Interest (ROI)

Region of Interest, or commonly abbreviated as ROI, is a region that will be considered as the region of wanted area while ignoring the outside region, called background. As proposed by Sander [14], to detect the availability of hand recognition, the detected region must be located the same as the region of interest. In other words, as the classifier tracks the wanted object, in this case the bounding rectangle around the hand contour, the area of hand must overlap with the region of interest. By using the conditional method of overlapping region, the system can identify whether the region selected, in this case the rectangle around the hand contour, is overlap with other active region, in this case the hand itself.

By developing this theoretical part of ROI to the simulation and hardware implementation, the process of detecting the appearance of hand will be much simplified compared to the other method like in [16]. Figure 2 shows the criteria for ROI to overlap that can be considered as overlapping ROIs. These criteria are needed to make sure that the program can identify the detection of hand gesture.

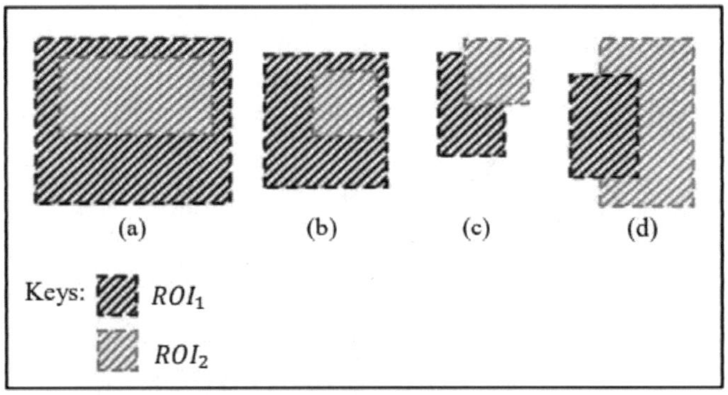

Figure 2. Overlapping Region of Interest (ROI) [16]

For this project the Region of Interest (ROI) is used to see the gesture of hand because the implementation for the ROI is suitable to detect the overlapping two regions on it. Besides, the system will detect only two outputs for each hand gesture recognition. This can be explained in Table 1 as follow:

Table 1. Application of Region of Interest (ROI).

Region of Interest (ROI)	Region implemented for
ROI_1	An active region, in this case the hand itself.
ROI_2	The rectangle sub window on screen. This region is static or not movable outside the frame of the realtime video and it highlights the area that involves.

2.4. Implementation on Simulation and Hardware prototype

The detection of hand gesture was calculated through the space consumption within the area between the convex hull and contour of hand. Convex hull is apply to find the edge of user hand by forming descriptor in which the smallest convex set contains the edges. In this project, the OpenCV library files that is suitable for any of image processing techniques are used.

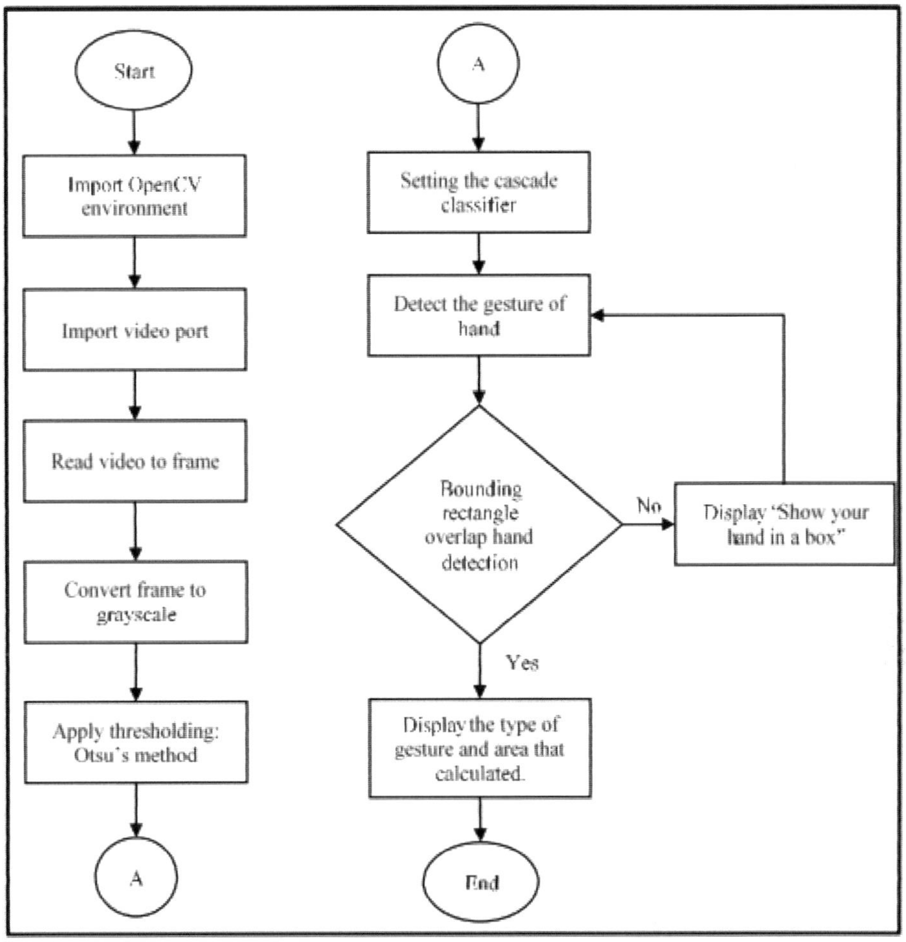

Figure 3. Flowchart of Python file for hand gesture recognition

By referring to the flowchart in Figure 3, generating a Python file for both simulating and hardware prototype can be implemented. The main program action to detect the appearance and the gesture of hand is by using the theory of Region of Interest (ROI). For programs to track the rectangle around the hand contour, the program required to identify whether the selected area of interest, referring to the case of the project, the rectangle will overlap with the detection of hand. Uncertainty it is overlap, it is considered as showing the type of gestures such as number or sign and the area within the area of region of interest. Uncertainty there is not overlap (outside the region of interest), the program can be considered as "Show your hand in a box". To plan the region of interest for this project, a group of coordinate will be recognized to present the area of bounding rectangle around the hand contour. For this project, there is an area of bounding rectangle, therefore there is a region of interest to be implemented with each type of gestures. After generate the Python main file, now the simulation begins where the detection of hand will be tested thus recognizing the type of gestures shown based on the selected area of interest. To run the Python file, the OpenCV environment must be link and run through the command window.

Then, the same procedures applied for the hardware prototype. For this prototype, USB video web camera will be (a), (b), (c), and (d) Keys: *ROI*1 *ROI*2 used as Nayana [17]. The source code in the Python core need to be modified in the video capture section to change the camera from the laptop web camera to the USB Video web camera. The procedures of detecting and analysing the gesture of hand using the USB video web camera remains the same during the simulation part. Notice that OpenCV environment is used once again to detect the hand via Haar-cascade classifier.

3. Results And Discussion

The explanation of the results will be focused on the simulation part since the different for the hardware implementation is the source code of the real-time input video. These results are analysed throughout the project scope. At first, the result of Haar-cascade classifier will be explain in detail to show how the hand is detected before image processing. Then, the simulation part begins where the area of interest that has been calculated. The Python file is actually the command file to run the specified program onto the Python command script. Table 1 shows the details of the generated Haar-cascade classifier.

Table 2. Explanations of Haar-cascade classifier file

Points	Explanations
(1)	This section covers the type of classifier to be used in this project. For this case, it will use the OpenCV and Haar-cascade classifier methods.
(2)	This is the part where the window size plays a role. The window size means the size of the specified hand. By default, it is fixed to 24 by 24 pixels.
(3)	The variation of the threshold and window detection are set. Generally, this will be based on the type of camera that will be used to detect hands and recognize gestures.
(4)	Each of the posture, about 10 times of repetition to analysing the cascade classifier.

In this project, the total data collected around 10 times of repetition with different scales for each posture to determine that the bounding rectangle could detect the hand gesture or not. Figure 4 shows how convex hull is apply to find the edge of user hand.

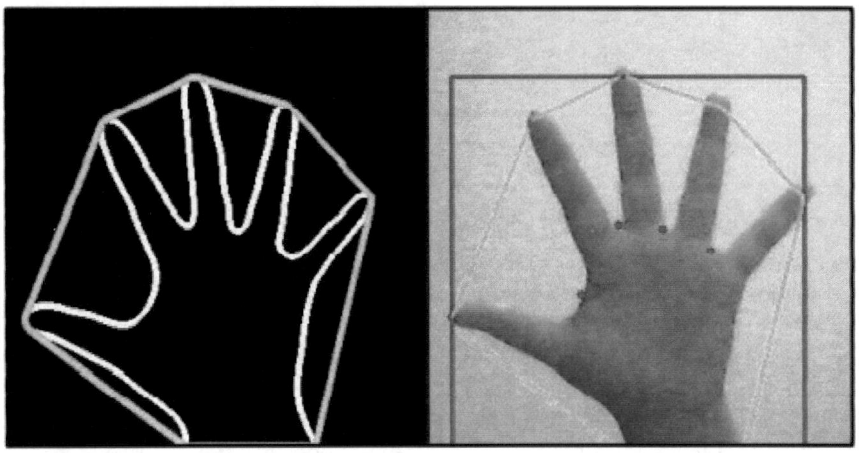

Figure 4. Left figure shows the background (in black), the hand-shape region (in black), the hand contour (in white) and the convex hull (in green). On the right, convexity defects (in red).

Initially, convex hull compares with hand contours to detect convexity defects where defects are the point where both are different. Firstly, vertices of the convex hull are calculated where they change direction. Then, the hand contour segment between consecutive pairs of consecutive vertices of convex hull and finds that pixels in each segment maximizes the distance from the convex hull. This maximum distance is called the defect depth where the dots are called convexity defects.

The proposed hand gesture recognition requires a webcam to capture the user's gesture, preferably with clear background. The test is done under normal lighting, the system can recognize and track the user's hand movement. The segmentation process is able to detect user's movement and count the contours. Each signal differentiates using different ROI. Figure 5, Figure 6 and Figure 7 shows the output when execution of Python file.

IOP Conf. Series: Materials Science and Engineering 1045 (2021) 012043 doi:10.1088/1757-899X/1045/1/012043

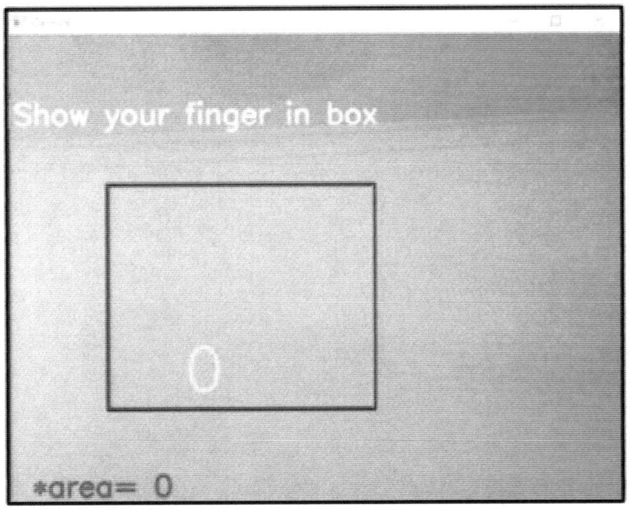

Figure 5. Image captured by web camera

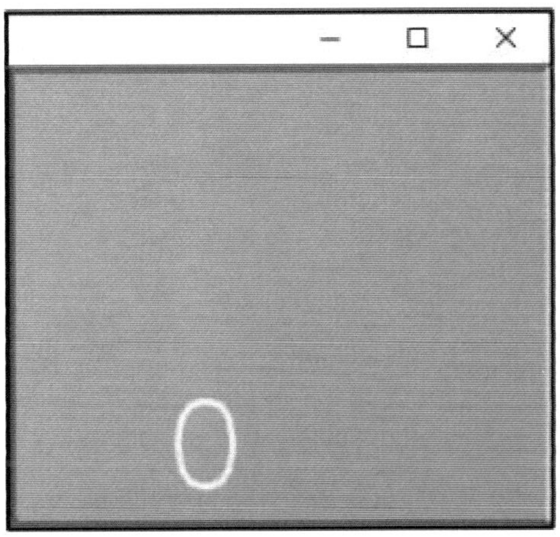

Figure 6. Bounding box without any gesture shown

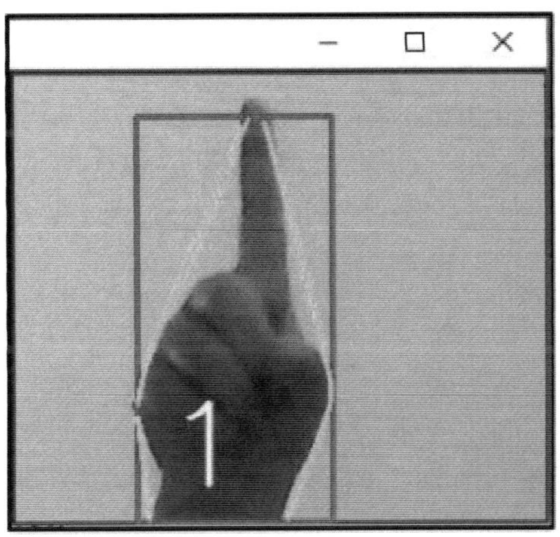

Figure 7. Recognition of hand showing number 1

All of the numbers and signs were tested until number 5. The experimental results are tabulated in Table 3.

Table 3. Analysis table for Region of Interest

Events	Area in Region of Interest, ROI		
	Area of Hull, H	Area of Contour, C	Area = H - C
0	62879	62879	0
1	13569	10876	2693
Good	13045	10821	2224
2	20646	13744	6902
Gun	20216	13804	6412
3	24362	17113	7249
OK	25468	16922	8546
4	25781	15395	10386
Rawr	20395	14148	6246
5	28383	16822	11561
Stop	33220	19047	14173

To evaluate the area of recognition proposed, an area by Region of Interest is taken and calculated in pixels value. Based on observation, most of area of number gestures have larger area compared to gestures recognition of sign.

Table 4. Analysis table for gesture recognition

Events	Performance of Haar-cascade Classifier		
	Hits	Missed	False
1	10	0	0
Good	10	0	0
2	10	0	0
Gun	10	0	0
3	10	0	0
OK	10	0	0
4	10	0	0
Rawr	10	0	0
5	10	0	0
Stop	10	0	0

Whilst in Table 4, the total data collected around 100 samples with different scales for each posture. Each of the posture with different scale were captured with 10 times to analyse the gesture. Haar-cascade Classifier determine whether the box in the frame could detect the hand gesture or not. Based on all of the different sign, the box in the frame were recognizing all of the sign.

However, there are some limitations of the program, i.e. the lighting environment changes are still influential in the segmentation process, in particular the process of background removal. Also, the user's hand and the webcam must be in a fixed position so the webcam will capture the image with the exact area to identify the gesture.

4. Conclusion

As a conclusion based on the result of the project, it can be seen that developing the hand gesture recognition using Python and OpenCV can be implemented by applying the theories of hand segmentation and the hand detection system which use the Haar-cascade classifier. To summarize it, this system has accomplished several objective in this project: (1) manage to establish a complete system for detecting, recognizing and interpreting hand gesture recognition through computer vision using Python and OpenCV, and (2) able to create the numbers and sign languages of hand gesture shown in the system that will meets the name of the project.

For the future recommendation, this system will include the execution of additional gestures that will allow any users with different skin colour and size of palm to perform more functions easily. The current system only uses the right hand with specific area in ROI to perform gestures. Therefore, the desired enhancement of the technique may possibly using both of user hands to perform different signs with computer operations. Additionally, background subtraction algorithms can be used for more effective performance. Implementing the Graphical User Interface (GUI) will be done in the future where the users know how to translate the gesture from its meaning to the sign or number and vice versa.

5. Acknowledgement

The authors wished to thank Universiti Teknologi MARA (UiTM), Cawangan Pulau Pinang, Malaysia for the instrumentations and experimental facilities provided as well as Faculty of Electrical Engineering UiTM Cawangan Pulau Pinang for all the support given during this research.

6. References

[1] S. M. M. Roomi, R. J. Priya, and H. Jayalakshmi 2010 *Hand Gesture Recognition for Human-Computer Interaction* (J. Comput. Science vol. 6) no. 9 pp. 1002–1007.
[2] S. N. Karishma and V. Lathasree 2014 *Fusion of Skin Color Detection and Background Subtraction for Hand Gesture Segmentation* (International Journal of Engineering Research and Technology) vol. 3 no 1 pp 13–18.
[3] A. Dhawan and V. Honrao 2013 *Implementation of Hand Detection based Techniques for Human Computer Interaction* (International Journal of Computer Applications) vol. 72 no. 17 pp 6–13
[4] C. Von Hardenberg and F. Bérard 2001 *Bare-hand human-computer interaction* (Proceedings of the 2001 workshop on Perceptive user interfaces) pp 1–8.
[5] K. Nickel and R. Stiefelhagen 2007 *Visual recognition of pointing gestures for human–robot interaction* (Image Vis. Comput.) vol. 25 no. 12 pp 1875–1884
[6] R. Lockton 2002 *Hand gesture recognition using computer vision* (4th Year Proj. Rep.) pp 1–69
[7] Itseez 2017 *Open Source Computer Vision Library itseez2015opencv* (Online) Available: https://github.com/itseez/opencv (Accessed: 13-Oct-2017)
[8] Itseez 2017 *The OpenCV Reference Manual itseez2014theopencv 2.4.9.0* (Online) Available: http://opencv.org (Accessed: 13-Oct-2017)
[9] S. van der Walt, S. C. Colbert, and G. Varoquaux 2011 *The NumPy array: a structure for efficient numerical computation* (Comput. Sci. Eng.) vol. 13 no. 2 pp 22–30
[10] Travis E. Oliphant et al 2017 *NumPy Developers numpy 1.13.0rc2.* (Online) Available: https://pypi.python.org/pypi/numpy (Accessed: 13-Oct-2017).
[11] S. Gokturk C. Tomasi B. Girod C. Beaulieu 2001 *Region of Interest Based Medical Image Compression with Application to Colon Ct Images* (International Conference of the IEEE

Engineering in Medicine and Biology Society) vol 23 pp 575-578
[12] N. Otsu 1979 *A Threshold Selection Method from Gray-Level Histograms* (IEEE Trans. Syst. Man. Cybern.) vol. 9 no. 1 pp 62–66
[13] R. C. Gonzalez 2016 *Digital image processing* (Prentice hall)
[14] S. Soo 2014 *Object detection using Haar-cascade Classifier* (Inst. Comput. Sci. Univ. Tartu) vol. 2 no. 3 pp 1–12
[15] P. Viola and M. Jones 2001 *Rapid object detection using a boosted cascade of simple features* (IEEE Comput. Soc. Conf. Comput. Vis. Pattern Recognition. CVPR 2001) vol. 1 pp 1511-1518
[16] M. Nahar and M. S. Ali 2014 *An Improved Approach for Digital Image Edge Detection* (Int. J. Recent Dev. Eng. Technology) vol. 2 no. 3
[17] P. B. Nayana and S. Kubakaddi 2014 *Implentation of Hand Gesture Recognition Technique for HCI Using Open CV* (International Journal of Recent Dev) vol. 2 no. 5 pp 17–21

Fault Diagnosis And Troubleshooting On 2.1kW Grid-Connected Photovoltaic Power System In UiTM Pulau Pinang

Muhammad Faris Fahmi Mohd Fadzil[1] , Ahmad Asri Abd Samat[2], A. Idzwan Tajudin[3],S. Omar[4], Kamarulazhar Daud[5]

[1,2,3,4,5]Faculty of Electrical Engineering
Universiti Teknologi MARA, Cawangan Pulau Pinang
Kampus Permatang Pauh, 13500 Permatang Pauh

farisfahmi117@gmail.com[1], ahmadasri759@uitm.edu.my[2], aimi.idzwan@uitm.edu.my[3], saodah004@uitm.edu.my[4], kamarul395@uitm.edu.my[5]

Abstract. Grid-connected photovoltaic power systems always have a connection to a utility grid via a suitable inverter because a PV array only produces a DC power. This paper presents fault diagnosis and troubleshooting on 2.1 kW grid-connected photovoltaic power system in UiTM Pulau Pinang as a case study network. Grid-connected PV systems has a solar panels that produced a power needed in the day time, however the electricity supplied ongoing by distribution network operator during day night or whenever the solar panels produced low electricity because changes of weather (cloudy or rainy day). The case study network was selected because of a fault frequently occur and make the system unstable. The functionality of the case study network has been tested and troubleshooting has been carried out to identify the cause of the problems. Troubleshooting process had been done on the DC and AC side of the system by creating a troubleshooting table. The DC side of the system free from a fault, as proved by all the equipment in a good condition and PV array able to produce desired output voltage. The fault was occurred in the AC side of the system. The inverter has failed to produce desired output.

Keywords. Grid-connected PV system, photovoltaic (PV), PV faults, troubleshooting

1. Introduction

A photovoltaic (PV) system is a power system designed to supply usable solar power by means of solar energy. This system used solar panel to absorb and convert sunlight to electricity. The solar PV industry begin to popular and rapidly develop whenever the world facing an oil crises in 1970s. Low cost in solar PV generation is one of the main reason rapid development of PV industry[1]. The use of solar PV system has grown significantly in Malaysia after Renewable Energy Act gazette in 2011[2]. The use of the renewable energy to generate electricity such as solar system will reduce greenhouse gas emissions to the environment.

Nowadays, grid-connected PV system mostly used rather than off-grid or stand-alone system. A grid-connected PV power system is a renewable electricity generating power system using solar energy that is connected to the utility grid. A grid-connected PV system consists of solar panels, one or several inverters, a power conditioning unit and a grid connection equipment's. Stand-alone PV system normally used for small residential and commercial rooftop. Unlike stand-alone PV power systems, a grid-connected system rarely includes battery bank, because they are very expensive. The grid-connected PV system will supply the excess power that is not being used by the connected load to the utility grid.

The grid-connected PV power system in UiTM Pulau Pinang has a power of 2.1kW that consist 12 PV modules in a series connection. The PV modules applied for the case study has a maximum power of 175 W with the model name NT-R5E3E. A voltage and current are 35.4 V and 5 A respectively at maximum power of the module. These 12 series connected PV modules connected to Growatt 300MTL inverter for conversion of DC to AC voltage before delivered to the grid system. Growatt 3000MTL inverter has an input voltage range from 100V to 500V maximum. If the inverter received a voltage between the given range, it would convert to 240 V AC source and deliver to the grid system.

Upon analysing the characteristic of the fault on the case study network during monitoring process, an indicator light at the inverter turn red showing that the system is in a fault condition and it display 'No AC Connection'. Therefore, troubleshooting process was conducted to search for a location of the fault that make the system malfunctioning. The troubleshooting table was created to troubleshoot the grid-connected PV system which consist of DC and AC side.

The objectives of this project are to test the functionality of the 2.1kW grid-connected photovoltaic power system in UiTM Pulau Pinang and identify the cause of malfunctional of the system. Lastly, to identify the solution needed to fix the problem in the system.

In PV solar system, like any others electrical system, the lifespan of the system is shortened if the system is not well maintained[3]. Fault detection and diagnosis methods are indispensable for the system reliability, operation at high efficiency, and safety of the PV plant[4]. Therefore, it is very important to study the types of fault that might occur in PV system and a troubleshooting methods to make sure the system operate efficiently.

Faults in the PV system can be identified in two sides of the system which is DC and AC side and , the interface between these two parts are DC/AC inverter that connected to the grid[5]. The faults that may occur in DC side of the grid-connected PV system are fault in PV array, earth or grounding fault, bridging fault, open circuit fault, mismatch fault and fault in cables. Meanwhile, the faults in the AC side of the system occurs on the inverter and unbalanced voltage or grid outage for AC part. The AC output power will become low and DC output power remains the same when there is a fault in the inverter[5]. The classification, simulation, and discussion of all possible faults in both the AC and DC side of a grid-connected PV system are represented in[5].

Fault analysis in solar PV arrays is a fundamental task to increase system reliability, efficiency, and safety[6]. Faults in the PV system may cause damage PV modules or cables, as well as lead to DC arching hazard and even fire risks without proper protection. Two type of faults inside PV arrays usually cause overcurrent back feeding into the faulted modules are line to line and ground faults[6]. A PV array ground fault is an electrical pathway between one or more array conductors and earth ground[7]. Ground or earth fault occurs when the circuit develops an unintentional path to ground. In order to protect the array during ground fault event, a ground fault protective device (GFPD) is used to detect ground faults current[7]. Line to line fault also known as bridging or short circuit fault in the PV system exists when there is accidental connection occur between two different potential points in the string of PV array or in cables.

Several research works have been proposed to detect and diagnose PV system faults. For instance, authors in [8] suggest a diagnostic method based on I-V characteristic analysis and Artificial Neural Networks (ANN). This method allows the detection and localization of faults occurring in PV cells, PV modules, PV strings, and bypass diodes. Fault diagnosis method for PV system based on operating voltage-window has been proposed in[9]. In this method, it is only based on the measurement of operating voltage of PV strings and ambient temperature. The Grid Connected PV System Troubleshooting Tree is proposed by authors in[3]. This troubleshooting tree gives better periodic maintenance and preventive maintenance. In addition, minor faults are repaired at the time of periodic maintenance and better solar array output is obtained.

This paper presented a design of case study, 2.1kW grid-connected PV power system using MATLAB Simulink and troubleshooting process of 2.1kW grid-connected PV power system in UiTM Pulau Pinang. The system is designed same as the grid-connected PV power system in UiTM Pulau Pinang to analyze the system performance and comparing the result after troubleshooting/repairing process.

2.0 Methodology

Figure 1 shows the process flow of research methodology in order to indicate overall methodology of the project. It On the early stage, literature study has been taken into main consideration of the research on the grid-connected PV power system and to study the type of faults that might occur in the PV system. MATLAB Simulink is used to model the system design of 2.1kW grid-connected PV power system and analyses the system performance. After the simulation process completed, a test on the PV system will be conducted on the case study network by measuring output voltage of main components which is PV array, surge protection device (SPD), and inverter. Troubleshooting process has been carried out on the grid-connected PV system to find the location of the faults and a solution to fix the problem. whenever, the location of the faults has been identified, repairing process will be conducted to make the system fully operated. the final step taken, by analysing the system performance and comparing the new system updated with previous system.

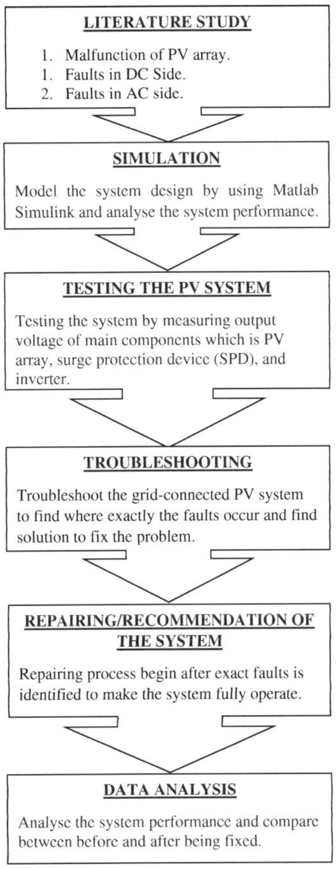

Figure 1: Process Flow of Research Methodology

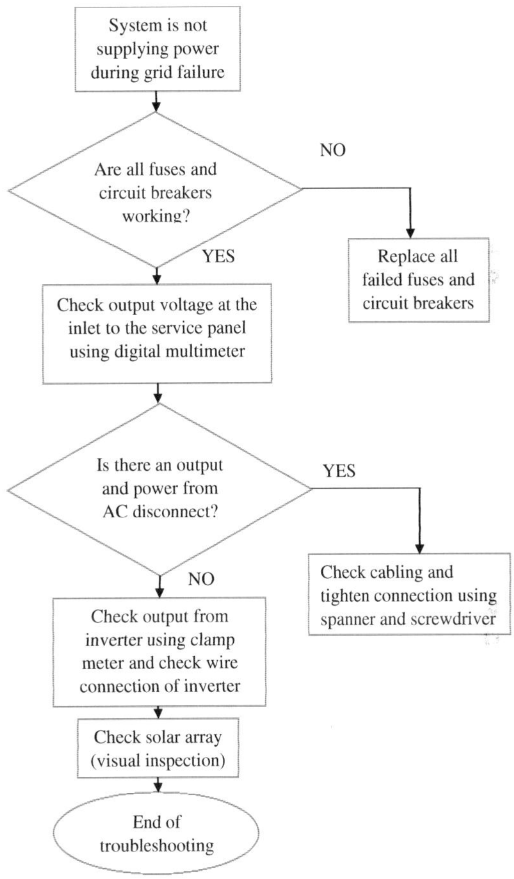

Figure 2: Grid Connected PV System Troubleshooting Flowchart.

2.1 Grid Connected PV System Troubleshooting Flowchart

Figure 2 shows the Grid Connected PV System Troubleshooting Flowchart which is the process to troubleshoot the grid-connected PV system.

2.2 Troubleshooting

Troubleshooting process is a process or procedure to identify the exact location of the faults that might occur in the grid-connected PV system.

IOP Conf. Series: Materials Science and Engineering 1045 (2021) 012044 doi:10.1088/1757-899X/1045/1/012044

2.2.1 Faults in DC Side

The faults that might occur in DC side of the system are name as solar array fault, earth or grounding fault, short circuit fault, open circuit fault and mismatch fault.

a) Solar Array Fault

Perform visual inspection of the PV array to check if there is any kind of blockage such as shading or dirt on the module that prevent sunlight to be absorbed by the PV array. Usually, visual inspection is the first step in deciding whether further tests should conducted on PV module[10]. After that, check for any loose wires on each connections of the system. If there is any loose wire, tighten the connections using spanner and screwdriver. The output voltage reading for all PV string needed to be checked using digital multimeter and identify if any of the strings give low output reading. When the faulty string is identified, the flawed module can be identified by covering each module and observe the changes in the voltage reading.

b) Earth or Grounding Fault

Earth or grounding fault can be identified using two methods: by measuring the voltage and insulation resistance.

i. Identify the fault types by Measuring the Voltage

Before measuring the voltage to troubleshoot grounding fault, the inverter must be disconnected from the supply. Next, use digital multimeter to measure voltage between positive terminal to ground potential, negative terminal to ground potential and positive to negative terminal. There is a ground fault occur in the system if the value of voltage reading between positive terminal to ground and negative terminal to ground is approximately equal to the voltage reading between positive to negative terminal. The location of the ground fault can be determined via the ration of two measured voltage. If the ground fault cannot be identified by measuring the voltage, measuring the insulation resistance should be necessary.

ii. Identify the fault types by Measuring the Insulation Resistance (IR)

To measure the insulation resistance, the theoretical value of the insulation resistance needs to be calculated for each string or the exact value can be obtained from datasheet of the PV module. Disconnect the inverter from any supply before setting up the short circuit device at the PV strings and then connect the IR tester device by setting the test voltage. The test voltage must be not exceeding the maximum PV modules voltage. Record the value of the insulation resistance and compared with value given from datasheet. Repeat the same step for others string. If there is a different of insulation resistance reading of the string compared to the datasheet value, it means that ground fault occurs in that string.

c) Short Circuit and Open Circuit Fault

This type of fault occurs when a disconnection problem appear in the PV string or more[11]. To detect this type of fault, perform visual inspection at every cable to check the insulation of the cable for mechanical damage, water ingress or corrosion. Then, check the connection between PV modules, connectors at junction box or breaks to identify if there are breaks in cable or the insulation of the cables. Lastly, check for any loose wires at every connection of the system. If there is any loose wire, tighten the connection using spanner and screwdriver.

d) Mismatch Fault

When the electrical parameters of one or group of cells are changed from other, the mismatches in PV modules will occur[5]. Therefore, perform visual inspection of the PV array to check if there is any kind of blockage such as shading or dirt on the module that prevent sunlight to be absorbed by the PV array.

After that, check the output voltage reading for all PV string using digital multimeter and identify if any of the strings give low output reading. Then, check the solder bond between the cell and contacted ribbons if there is any soldering damage. Lastly, check if there is discoloration, delamination and transparent layer crack of the PV modules.

2.2.2 Faults in AC Side

Two types of fault might occur in AC side of the inverter; inverter and grid fault which is unbalanced voltage or grid outage.

a) Inverter Fault

Troubleshooting process for inverter fault need to be done using digital clamp meter. Measure the DC input voltage and AC output voltage using this device. The AC output voltage must be at 240V and 50Hz as this system connected to utility grid. If the inverter cannot produce 240Vac, check for blown fuse, tripped breaker and broken wires of the inverter. Replace all blown fuse, circuit breakers and broken wires of the inverter.

3.0 Results and Discussions

The simulation results consist of two part which is the simulation results of PV array and simulation results of 2.1 kW grid-connected PV power system.

3.1 Simulation

The 2.1 kW PV array system was implemented in MATLAB Simulink as shown in Figure 3. The type of PV array that being used in this simulation was Sharp NT-175U1. The PV array modules can producemaximum and minimum power, 175W and157.5W respectively. This PV modules was chosen for the simulation because its nominal ratings were same with Sharp NT-R5E3E that used in UiTM Pulau Pinang grid-connected PV system. The nominal ratings of these two modules was tabulated in Table 1. In this simulation, the PV array was set to 12 modules in series connection which was same as the case study network. The irradiances and temperature were set to 1000 W/m² and 25°C respectively which was the Standard Test Condition (STC) for PV array. The output voltage of the PV array in this simulation is 501.71V while the exact value is 465V.

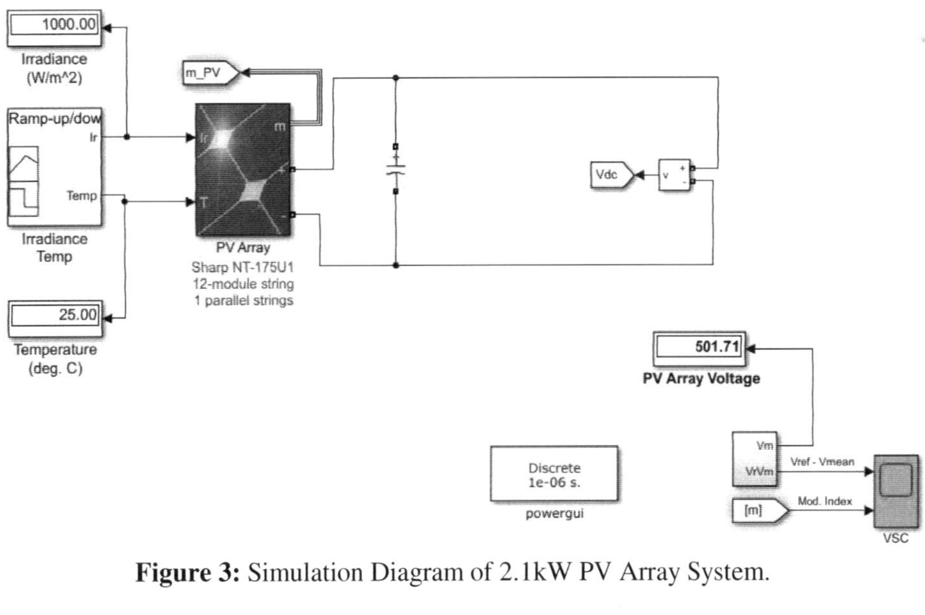

Figure 3: Simulation Diagram of 2.1kW PV Array System.

Table 1: Nominal Ratings of NT-R5E3E and NT-175U1.

	NT-R5E3E (UiTM)	NT-175U1 (Simulation)
Power	175W	175W
Open circuit voltage (Voc)	44.4V	44.4V
Short Circuit Current (Isc)	5.4A	5.4A
Voltage at Point of Maximum Power (Vmp)	35.4V	35.4V
Current at Point of Maximum Power (Imp)	5A	5A
Weight	17kg	16kg

3.2 2.1kW Grid-Connected Photovoltaic Power System

The 2.1kW grid-connected PV power system was designed using MATLAB Simulink as shown in Figure 4. The PV array was connected to inverter to convert to desired AC voltage and frequency to be delivered to the grid. The inverter then delivered 2.1kW of power to the grid.

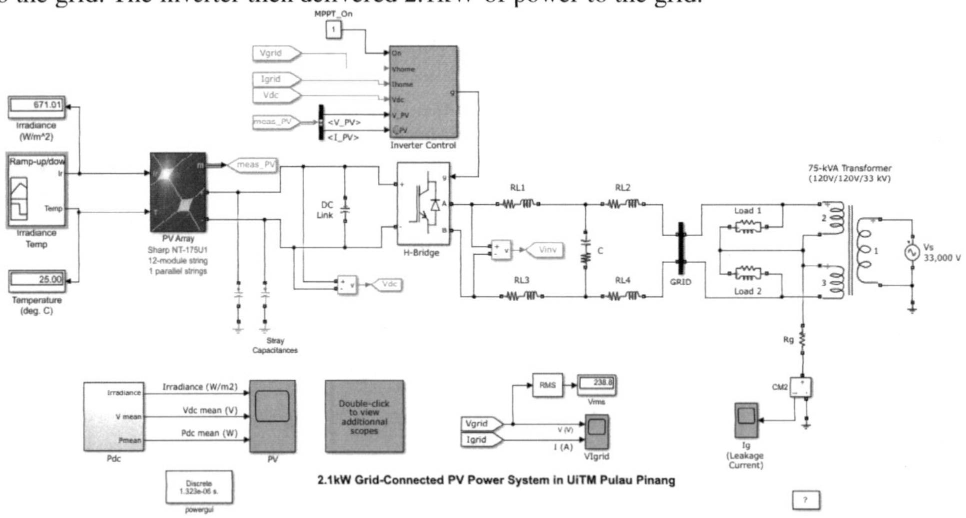

Figure 4: Case Study, 2.1kW Grid-Connected PV Power System modelled in MATLAB Simulink

The irradiance was measured using Hand Pyranometer 4890.20 to get the exact irradiance that received by PV array. The irradiance reading was taken at 10:00, 12:00, 14:00 and 17:00 to analyse the output voltage of PV array throughout the day. The irradiance reading was taken on 11 November 2019. The irradiance reading and PV array output voltage was tabulated in Table 2. It shows that the PV array output voltage increased when irradiance reading increased. Then, the irradiance data was transferred to the simulation to get the output voltage of the PV array and the inverter. Figure 5(b) shows the output of PV Array that varied from 425V to 280V. The output voltage of PV array in the simulation was a bit lower than the exact value that was measured using digital multimeter as shown in Table 2. This output voltage then enter the inverter as input and converted to AC voltage. The Growatt 3000MTL inverter that being used in UiTM grid-connected PV system have input range from 100V to 500V. This means that the inverter can convert the PV array output voltage to AC voltage to be delivered to the grid for the whole day. Figure 5(c) shows that the output voltage of the inverter that is 240V. This output then delivered to the grid.

Table 2: Irradiance and PV Array Output Reading

Time	Irradiance (W/m²)	PV Array Output Voltage (V)
10:00	692	448
12:00	985	467
14:00	766	459
17:00	423	305.4

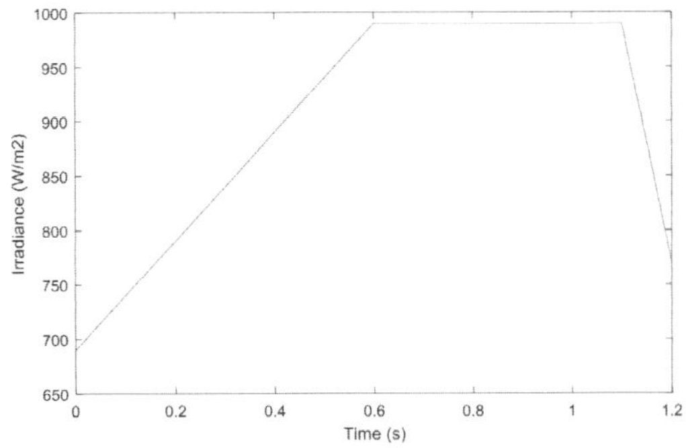

(a)

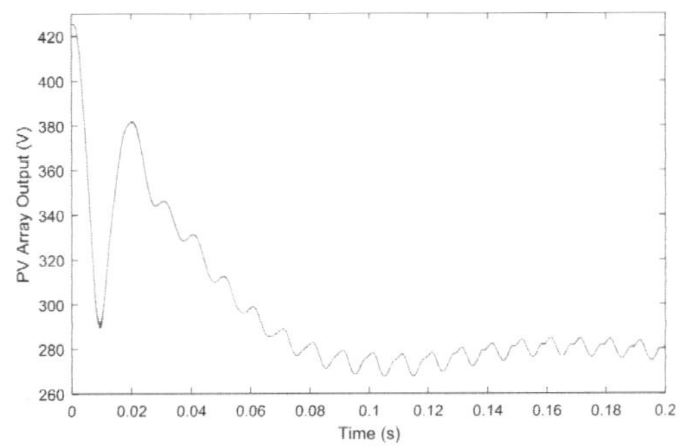

(b)

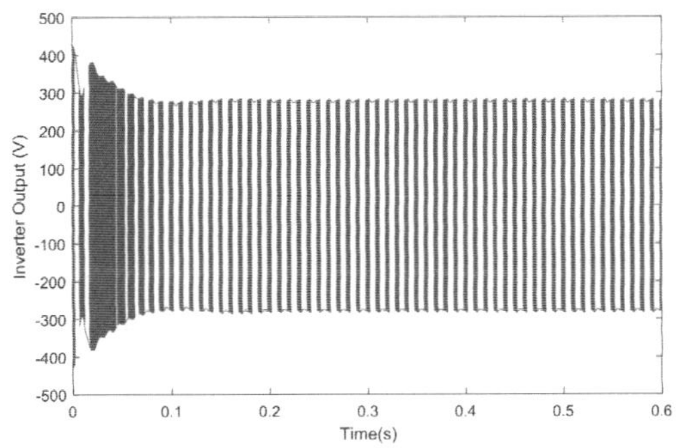

(c)

Figure 5: (a) Irradiance, (b) PV Array Output Voltage, (c) Inverter Output Voltage.

3.3 Troubleshooting

In order to troubleshoot the 2.1kW grid-connected PV system in UiTM Pulau Pinang, troubleshooting table is created to list down all the equipment that need to be troubleshoot and to record the results of the troubleshooting process. Table 3(a) and 3(b) shows the troubleshooting table for DC and AC side of the system.

Table 3: (a) DC Side Troubleshooting table, (b) AC Side Troubleshooting Table.

(a)

Name	Condition/Reading
Cleaned PV modules (dirt, bird dropping, fungus)	In good condition.
PV array structure	In good condition.
Trees and shading	No shading.
PV array cabling (mechanical or electrical damages)	No mechanical or electrical damages detected.
Cables connection (tightness and corrosion)	In good condition.
PV array output voltage	465V

(b)

Name	Condition/Reading
Cleaned inverter	External and internal of inverter clean.
No insects inside inverter	No
Cable connection inside inverter	In good condition
All fuses and circuit breaker operate correctly	Fuse blown
Cabling from Inverter to Grid	In good condition
DC input voltage of inverter	465V
AC output voltage of inverter	0Vac

The equipment used to perform this troubleshooting process was digital multimeter, digital clamp meter, set of screwdriver, spanner and plier. After performing troubleshooting process, the DC side of the system was operating in good condition. There was no fault occur on the DC side of the system. The PV array was clean from dirt, bird dropping, or fungus as shown in Figure 6(a). Its structure also in good condition. There was no cracking or bending detected when performing the visual inspection of the PV array. Figure 6(b) shows the output voltage of the PV array is 465V, so it means the PV array was in good condition and can produce desired output voltage. All the PV array cabling were free from mechanical or electrical damages and the cables connection was in good condition as shown in Figure 6(c). It can be concluded that no faults occur in the DC side of the PV system.

(a)

(b)

(c)

Figure 6: (a) PV Array, (b) Output Reading of PV Array, (c) Cables and Connection

In the AC side, the inverter was not producing output. When troubleshooting the inverter, the inverter was clean and all cable connection inside the inverter was in a good condition. There was no disconnected cable or loose cable connection but the fuse inside the inverter has blown. The fuse had been replaced with higher rating fuse that was 20 A/500 V. Its original fuse rating was 20 A/250 V. After replacing the fuse and connected the inverter to the grid, there was spark inside the inverter and the AC breaker tripped. After inspections, all cabling from inverter to the grid was in good condition. There were no mechanical or electrical damages that leads to short circuit fault and the grid voltage also in good condition that was 240Vac. This can be concluded that the inverter has failed and need to contact Growatt Sdn. Bhd. customer service for further repairing process.

4.Conclusions

This paper presented all procedures of troubleshooting process for DC side and AC side of the grid-connected photovoltaic power system. The 2.1kW grid-connected photovoltaic power system in UiTM Pulau Pinang is simulated and analysed using MATLAB Simulink. In order to get accurate result of the simulation, the irradiance reading is taken throughout the day and the reading was used in simulations to analyse the output voltage of PV array and output voltage of the inverter. For the troubleshooting process, it can be concluded, there were no fault occur in the DC side of the system. All equipment in the DC side was in good condition and the PV array can give desired output voltage. The fault was occurred in the AC side of the system which is the inverter has failed. Growatt Sdn. Bhd. Customer service need to be contacted for further repairing process to make the grid-connected photovoltaic power system in UiTM Pulau Pinang works again. In order to make the 2.1kW grid-connected photovoltaic power system in UiTM Pulau Pinang works again, only the inverter need to be troubleshoot. After the system works again, scheduled maintenance should be done in order to prevent the system from malfunctioning and the system can operate for longer time. The maintenance should be done at least once in a month.

5. Acknowledgement

The authors gratefully acknowledge to Universiti Teknologi MARA, Cawangan Pulau Pinang, specifically Faculty of Electrical Engineering for the facility that has been provided to complete this project and their support along the journey to accomplish this research.

6. References

[1] A. Chouder and S. Silvestre, "Automatic supervision and fault detection of PV systems based on power losses analysis," *Energy Convers. Manag.*, vol. 51, no. 10, pp. 1929–1937, 2010.

[2] T. M. N. T. Mansur, N. H. Baharudin, and R. Ali, "Sizing and cost analysis of self-consumed solar PV DC system compared with AC system for residential house," *Indones. J. Electr. Eng. Comput. Sci.*, vol. 10, no. 1, pp. 10–18, 2018.

[3] A. Saxena and S. K. Mishra, "Review of System Maintenance and Troubleshooting of Solar PV Grid Connected System," *Int. J. Emerg. Sci. Eng.*, vol. 1, no. 4, pp. 67–68, 2013.

[4] A. Mellit, G. M. Tina, and S. A. Kalogirou, "Fault detection and diagnosis methods for photovoltaic systems: A review," *Renew. Sustain. Energy Rev.*, vol. 91, no. February, pp. 1–17,

2018.

[5] R. Mahalakshmi, M. Karuppasamypandiyan, A. Bhuvanesh, and R. J. Ganesh, "Classification and Detection of Faults in Grid Connected Photovoltaic System," *Int. J. Sci. Eng. Res.*, vol. 7, no. 4, pp. 149–154, 2016.

[6] Y. Zhao, J. F. De Palma, J. Mosesian, R. Lyons, and B. Lehman, "Line-line fault analysis and protection challenges in solar photovoltaic arrays," *IEEE Trans. Ind. Electron.*, vol. 60, no. 9, pp. 3784–3795, 2013.

[7] J. Flicker and J. Johnson, "Photovoltaic ground fault detection recommendations for array safety and operation," *Sol. Energy*, vol. 140, no. December, pp. 34–50, 2016.

[8] Y. Chouay and M. Ouassaid, "An intelligent method for fault diagnosis in photovoltaic systems," *Proc. 2017 Int. Conf. Electr. Inf. Technol. ICEIT 2017*, vol. 2018-Janua, pp. 1–5, 2018.

[9] N. Gokmen, E. Karatepe, S. Silvestre, B. Celik, and P. Ortega, "An efficient fault diagnosis method for PV systems based on operating voltage-window," *Energy Convers. Manag.*, vol. 73, pp. 350–360, 2013.

[10] S. R. Madeti and S. N. Singh, "A comprehensive study on different types of faults and detection techniques for solar photovoltaic system," *Sol. Energy*, vol. 158, no. June, pp. 161–185, 2017.

[11] K. Abdulmawjood, S. S. Refaat, and W. G. Morsi, "Detection and prediction of faults in photovoltaic arrays: A review," *Proc. - 2018 IEEE 12th Int. Conf. Compat. Power Electron. Power Eng. CPE-POWERENG 2018*, pp. 1–8, 2018.

IOP Conf. Series: Materials Science and Engineering 1045 (2021) 012045 doi:10.1088/1757-899X/1045/1/012045

Optimum Location of DG for Loss Reduction with Ant Colony Algorithm

Saodah Omar, Muhammad Nasrullah Abdul Manan, Muhammad Naqib Mat Siam, Ahmad Asri Abd Samat and Kamarul Azhar Daud

Faculty of Electrical Engineering, Universiti Teknologi MARA, Cawangan Pulau Pinang, Kampus Permatang Pauh 13500 Permatang Pauh, Penang, Malaysia.

E-mail : saodah004@uitm.edu.my, ayoyanas@gmail.com, naqib6301@gmail.com, ahmadasri759@uitm.edu.my, kamarul395@uitm.edu.my

Abstract : The ability of the Distributed Generation (DG) to solve problems such as power system deregulation and power demand problems appropriate to its purpose, which is to inject electricity in a distributed manner at a point close to the load, causes the distributed generation to become the latest trend in electricity generation technology. Proper position of distributed generation is necessary in order to achieved maximum benefit from DG, which could be due to an incorrect allocation of DG sources to the power network would not only result in increased power losses, but could also jeopardize the operation of the system. This paper introduces an ACO-algorithm for optimal location of DGs using a real network in one of a rural area of Malaysia. The method is used to determine the effectiveness of DG by comparing the losses of power and the improvement of the voltage profile. As for the confirmation to the ACO method, another method known as brute force method is use to compare the data gain as validation purposes.

Keyword— DG, ACO, Power losses, Voltage profile, optimisation

1. INTRODUCTION

Supply, transmission and use of electrical power is the process in which the electrical system is identified as a network system with an electrical component [1]. The increased interest in DG is due to the modernization of the power grid and the tight restriction on the building of new bulk substations and transmission lines. That may be attributed to the growing of society, which has contributed to a high level of electrical energy use. The increase in demand for energy allows the distributed generation to emerge as a new option for the supply of electricity to the consumer. DG is a concept that refers to electricity generation near the point of consumption [2], [3]. In this advanced technology era, demand for electricity continues to rise with an increase in industrialization and population, especially in rural areas. The reason for this is that the substation power supply is a radial network in rural areas, therefore the demand for electricity continues to increase as the population increases. The configuration of the radial network that is intended to be fed from one end and the loads are connected to the other end. The arrangement of the radial network will cause the voltage to decrease as it leaves the substation [4].

Constraints on the new construction or expansion of existing large conventional power plants have exacerbated environmental issues and the drawbacks of modern power stations. These problems paved the way for a new technology known as distributed generation to address the supply and demand gap problem [5]. Typically, distributed generation is a small unit that produces electricity between 1KW and 50MW. The architecture of this energy-size system is such that it can be installed near to the consumer or the electrical distribution network. Suitable with its functionality, the presence of power generation distribution is designed specifically to promote system efficiency such as power quality, stability, power flow, voltage profile, reliability and protection. Without question, the implementation of DG could offer major benefits in the field of power distribution, but, like other technologies, DG also has its own advantages and disadvantages. Installation of DGs must take place at a particular and precise location in order to allow maximum use of DGs [6]. This is the problem for the engineer to solve, as the improper installation of DG could create an undesired result that could jeopardize the entire system. Insufficient allocation of DGs could lead to increased losses, harmonics and voltage flickering of the distribution network [7].

Several algorithms and methods have been developed based on the previous research papers to address the allocation of DG to the Distribution Network (DN). The variations between these algorithms relate to the formulation of the problem, the intended technique and their assumptions. These methods mentioned include Genetic Algorithm (GA), Grey Wolf Optimization (GWO), Ant Colony Optimization (ACO), the NEPLAN Software, and Imperialist Competitive Algorithm (ICA). GA have been used to reduce network losses and regulate the voltage [8][9]–[11], GWO is used to improve the voltage and reduce active power loss [5], the NEPLAN Software is used to determine the location of DG and optimum location of switch pair considering the power loss at each bus [12][13][14] , ACO is utilised for cost and power loss minimization [6] and ICA is used for enhancement of voltage profile and power loss reduction [5][15].

In this paper, ACO is implemented in one of Malaysia's rural network systems and the scientific contribution of this paper is to prove the algorithm can be implemented in the network, that power loss is minimized, that the voltage profile can be improved and DG's optimal position can be achieved

The paper's organization is as follows. Section II described the methodologies of the research and flow of the research to achieve the objective. Section III discussed about the experimental result of the optimal location of DG using ACO. Lastly, section VI concludes on the proposed methods and future suggestions of this study.

2. MATERIAL & METHODOLOGY

General Aspect. The ACO algorithm is a class method that is used to solve complex combination optimization problems. The ACO algorithm method performs the scheduling of three tasks. The first step consists of the initialization of the pheromone route. In the iteration (second) step, each ant constructs a complete solution to the problem according to the probabilistic state transition law. The state change law depends primarily on the state of the pheromone. The third stage changes the quantity of pheromone where the pheromone update laws are implemented in two stages. The first stage is the evaporation process where a fraction of pheromone evaporates and then there is a reinforcement process, increasing the amount of pheromone on the path with high quality solutions. The method is iterated until the criterion of stopping is reached.

Designation of the ACO (Matemathically):

1. The Initialisation of Pheromones [5]
$T_{ij}(0)$ refers to the initial pheromone in each pathway. During this process, the $T_{ij}(0)$ pheromone concentration between cities i and j is 0.

2. Kth is refer to the ant probability (K= 1,2,3….) choosing the next path through this equation[4]:

$$P(k) = [\tau_{ij}(t)]^{\alpha *}[\eta_{ij}]^{\beta} / \Sigma_j [\tau_{ij}(t)]^{\alpha *}[\eta_{ij}]^{\beta} \quad (1)$$
if j ε τ(k) =0 otherwise
ij

Where τ (t) is the pheromones left between i and j

η_{ij} Is the expectation the path that choose by ant through i and j
Where η ij = $1/d_{ij}$

3. Pheromone update [5]:
Total pheromone of each is updated by this equation

$$\tau_{ij}^{k+1}(t) = (1-\rho) \times \tau_{ij}^{k}(t)+\Delta\tau_{ij} \qquad (2)$$

Where ρ is positive number between 0&1

2.1. Formulation of the problem
Real power losses without DG[4] calculated by using (3). Simple distribution network shown in Figure 1. The network consists of two substations without DG.

$$P_{Loss}=\Sigma_{i=1}^{n} (I_{Li}^{2})\, R_i \qquad (3)$$

Where I_{Li} = line current
N = total branch number in system
R = resistance in i
Id= current load demand
Where I_{Li}=I_d

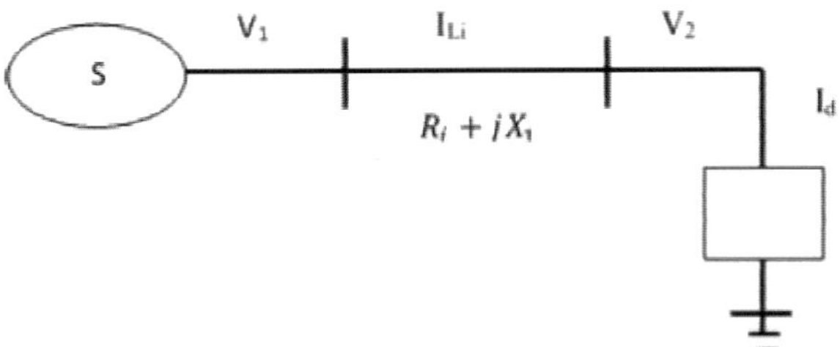

Figure 1. Example of distribution network without DG

Line power losses indicated by the (3) could be reduced by reducing the branch current of the distribution network by adding the DG to the distribution network as shown in Figure 2. DG located at substation number 2. From the basic (3), following Kirchhoff Current Law, power loss for the network calculated by using (4). By assuming DG located as shown in Figure 2.

$$P_{loss}= \Sigma_{i=1}^{n} (I_{Li2} - I_g^{2})R_i \qquad (4)$$

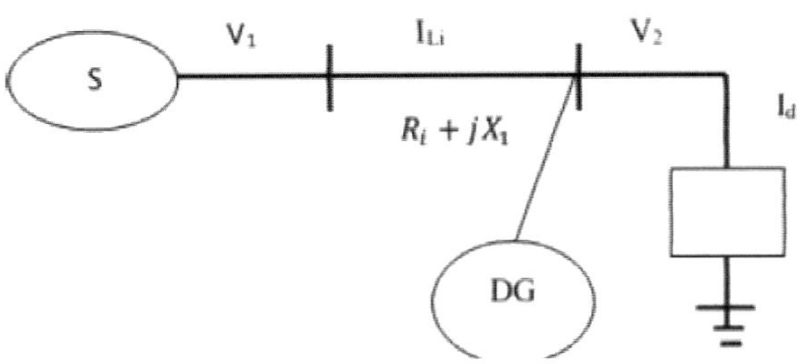

Figure 2. Example of distribution network with DG

The algorithm is designed to reduce the total power loss of the network while meeting certain device variable constraints. The inequality constraint are the real and reactive power limits and voltage limits

$$V_{min} < V_j < V_{max}$$

IOP Conf. Series: Materials Science and Engineering 1045 (2021) 012045 doi:10.1088/1757-899X/1045/1/012045

Where j is the bus number of distribution network

$$P_{min} < P_{dg} < P_{max}$$

$$Q_{min} < Q_{dg} < Q_{max}$$

The DG output in KVA real and reactive power should satisfy the limit.

Improving Voltage:

$$V_1 = E - IZ_1 \quad (5)$$

$$V_2 = E - (IZ_1 - I_2Z_2) \quad (6)$$

$$V_i = E - \Sigma_{i=1}^{n} I_iZ_i \quad (7)$$

As the load are connected away from substation voltages are decrease at the load.

2.3. ACO application to allocate DG

The equation (8) represents a stochastic model of the movement of ants, which assigns the probability of each ant to choose a direction.

$$P_1 = (m_1+k)^h / [(m_1+k)^h + (m_2+k)^h] \quad (8)$$

Where: m_1 & m_2 are the routes 1 and 2 that the ant has recently used.
k & h are constant define according to problem characteristic

Transition Rule. The action of each ant performing its travels is mainly determined by a transition rule which is restricted by two factors which are visibility and the effect of pheromone. The rule is shown through this equation:

$$P = (\tau^{\alpha} \eta^{\beta}) / [\Sigma \quad (\tau^{\alpha} \eta^{\beta})] = \tau^{\alpha} \eta^{\beta} \quad (9)$$

Where Pi is the probability of a DG connected to bus i of the system.

α&β are the parameter algorithm represent influence pheromone & visibility typically values α=0.9 & β=0.1 . Eventually, a roulette algorithm is implemented such that each of the possible solutions is likely to be chosen.

Pheromone. There are two categories of matrices τ that the pheromone implement in the algorithm. This matrices apply in order to determine the location (bus) of the DG and DG size. The first matrix referring to DG position that has a size of number of bus (NB) row, where NB is the total number of device busses and m columns, where m is the number of DGs to be located. The second pheromone matrix makes it possible to find optimal DG sizes. Every column in the array represents a device bus, and each row corresponds to a potential size value to locate.

DG position

DG size

$$\begin{bmatrix} X & X & X \\ X & X & X \\ X & X & X \end{bmatrix}$$

Where X is the power losses

Visibility. Visibility $\eta\ [i]$ serves as a mechanism that guides ants in the creation of good solutions. As for the pheromone matrix, the visibility is applied to each bus I in the row vector η[i] with the column c. Every column I is aligned with a specific generation value;

1. The location of DG. In particular, the proposed DG location algorithm is used to measure the lines connected to each bus and to determine a certain bias (favouritism) the loss values in those lines are calculated. In this case, the node with the highest number is the one with the best chance of being selected.
2. DG Scale. In the case of scale, no particular approach was used to support any kind of solution, and it was only assumed to assign equal chance to all solutions via a normal function.

Pheromone Update. Two operations are taking place during pheromone update;

1. The first pheromone update occurs in every iteration. At this point, both pheromone matrices are completely changed with the concentration on ants produced during this iteration.

$$\tau_{ij}^{\alpha} \leftarrow (1-p).\tau_{ij}^{\alpha} + \sum_{k=1}^{m} \Delta\tau_{ij}{}^{k} \quad (10)$$

Where; ρ is the evaporation rate
τijk is the amount of pheromones left by the ant k in (i,j) edge
and

$$\Delta\tau_{ij}{}^{k} = \begin{cases} \frac{losses_{w.o.GD}}{L_K}, if(i,j) is\ used \\ 0 \quad in\ other\ case \end{cases} \quad (11)$$

where: $losses^{w \cdot o \cdot GD}$ is a constant in system correlating to system power loss without DG.
L_k is the cumulative length of the direction the ant follows fitness function

2. The update will take place at the end of each iteration. In order to facilitate a quest in the particular area of the problem, the change will only be carried out with the best overall solution. To do so, the component of size and location are increased by 5% for each iteration while retaining the best overall solution.

The main steps of the proposed ACO algorithm are shown below.

- Step 1 (search space):
  The design is a framework that is ideal for seeking a solution to the problem. In this case, the search space is the 18 bus of the DN 's real rural network. Ant must read the DN data prior to the implementation of DG. The solution to the problem will be generated after the ants have completed their decision-making process for the sub-route forming a path.

- Step 2(ACO initialization):
  Beginning with the ACO algorithm, the constant value τ0>0 are initialize from the pheromone values of the edges in the search space . Initialisation causes the ants to choose their paths randomly and thus the solution space is more efficient.

- Step 3(ant dispatch):
  In this process, the ant is dispatched and the solution is built on the basis of the pheromone level on the edges. In this step, each ant will start a tour of the home colony and choose one of the states in the next stage to travel according to the following equation (7). After each ant end the tour ends, a new DG placement solution is created, which must be evaluated using the fitness function.

Where the $\tau_{ij}(t)$ is the total pheromone deposited at edge ij with iteration t and Δi is the set of available edges which ants can select in state i

- Step 4 (fitness function):
  In this stage, the fitness tour created by the ants is evaluated by their fitness function. The fitness function of the problem is defined as the power load flow plus the penalty factor for the infeasible solution. In order to speed up the convergence properties of the algorithm and, at the same time, to use the details that might still be useful to the rejected tours. This penalty tour increased from zero to a very high value across iterations.

- Step 5 (pheromone update):
  The goal of the pheromone value update is to increase the pheromone value of the solution components found in the fitness function solution. Furthermore, from a practical point of view, pheromone evaporation is required in order to avoid the convergence of an algorithm that is too fast towards a sub-optimal area. It introduces a useful method of forgetting, preferring the discovery of new areas in the search space. The equation for update pheromone are as equation (10) & (11).

- Step 6 (convergence determination):
  Step 3 to 5 will proceed until the number of the iteration counter exceeds the predefined maximum number which is calculated experimentally. The best tour to choose from among all iterations means an optimal DG solution.

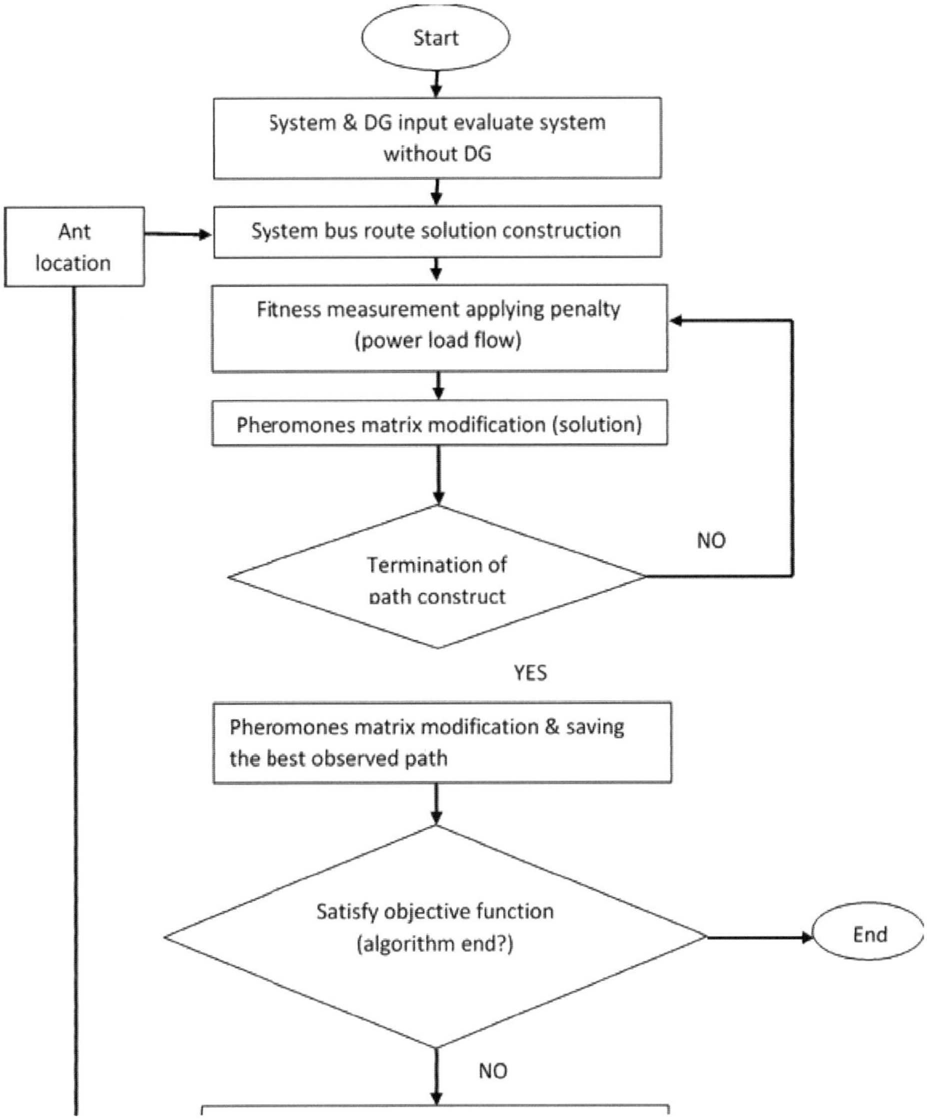

Figure 3. Flowchart of this the overall research

The purpose of algorithm is show in the flow chart of the overall research work in Figure 3

3. MATHEMATICALLY CALCULATION OF DG

Equation (12) and (13) shown a calculation of P if the network having 4 loads as shown in Figure 4.

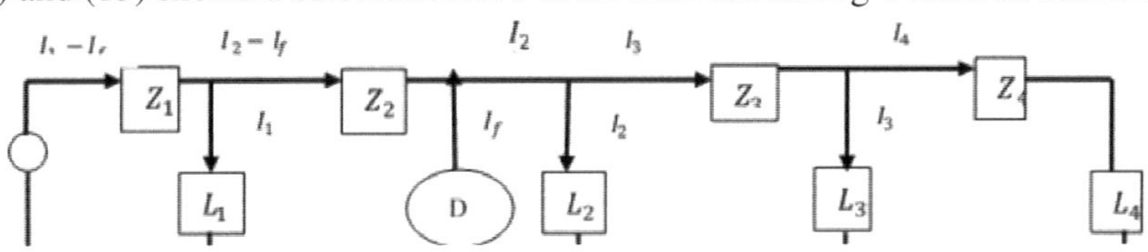

Figure 4. Example of network with 4 load, DG located at substation no 2

$P_T = \sum(I_3 - I_4)^2 R_1$

Let 4 impedance & 4 load, DG at bus 2

Therefore; $P_T = \sum_{i=1}^{4}(I_i - I_f)^2 R_i - I_f^2 R_3 - I_f^2 R_4$ (12)

3.1. Brute Force Method

In order to ensure that the ACO result is accurate, the data gain from the ACO method is compared to another collection of data gain by using the brute force method. The brute force method could be considered as simplest meta-heuristic method. It is also known by a different name, such as exhaustive search, generation and testing or direct search. It is an exhaustive evaluation of the target function with all possible input values. This type of approach is useful in order to solve the discrete type of problems, such as finding an optimum allocation of DGs in the network. This method is considered to be the most effective options available because of its ability to deliver the most reliable result. That is because brute force is a method that will consider any possible value in order to obtain a certain outcome. Although it is not an efficient solution approach, due to if the number of potential candidates increases, the number of possible solutions will increase rapidly. Therefore, the method of brute force is only applicable to a small number of candidates but the results obtain are still accurate due to all possible tests.

Apart from its simplicity and precision, the main reason for proposing brute force in this project is to confirm that the performance generated is identical to the main method used by ACO as an optimizing tool to locate DG. Using this method, it will make it possible for researchers to have a complete audit of the results. Although brute force is a non-artificial intelligent method, this confidence stems from the fact that brute force is a method that uses all potential candidates to be tested to produce ideally matched to a single or unique solution outcome.
According to the brute force rule, the total number of possible conditions could be determined as follows.

$$PC = BusNo^{GenNo} \quad (14)$$

Where *PC*; possible condition
GenNo: number of generator for optimization
BusNo: the number of distribution buses in network

4. RESULTS & DISCUSSIONS

Output results of the proposed algorithm are obtained from the program MATLAB. The distribution network of one of the rural cities in Malaysia shown in Figure 2 was used to test the efficiency of the algorithm. The system uses a regulated voltage of 11 kV, applied power of 100 MVA without voltage regulators and a bank capacitor installed. The power factor is 0.85 according to the standard used by Tenaga Nasional Berhad (TNB). Total net load is 11.64 MW, spread across three categories of customer that are industrial, commercial and residential consumers. The optimum position of DG sources in the distribution system is evaluated using the proposed ACO-based algorithm. The final solution is also compared to the initial system condition without DG. In order to confirm that the data gain from the ACO method is correct, the data gain from using the brute force method is use to compare as the confirmation that the specific bus is suitable to locate the DG. Through the ACO method the outcome gain shows that there is a significant improvement show in power loss outcome after DG is implemented via ACO method. The data resulting from power loss after DG has been installed by ACO method is shown in Table 1. In this case the algorithm was performed a total of 5 ant generation with 20 iteration in order to gain the optimal location of the DG. The confirmation of the data gain from ACO to show that the suitable place to locate the DG is verified with the data gain from the Brute Force which is shown in Table 3.The product of the graph line showing the correct DG allocation position through the power loss improvement via before and after the DG is install is shown in Figure 3 & Figure 4. In the figure it can be seen that the installing the DG at the suitable location could improve the total power loss. This can be seen at bus 8 where its shows that it has significant improvement of power loss on both real power and reactive power. This result is further verified when the data of graph line of power loss from Brute Force method shown in figure 5&6 also produce the same result where it show that bus 8 is the most suitable place to locate the DG.

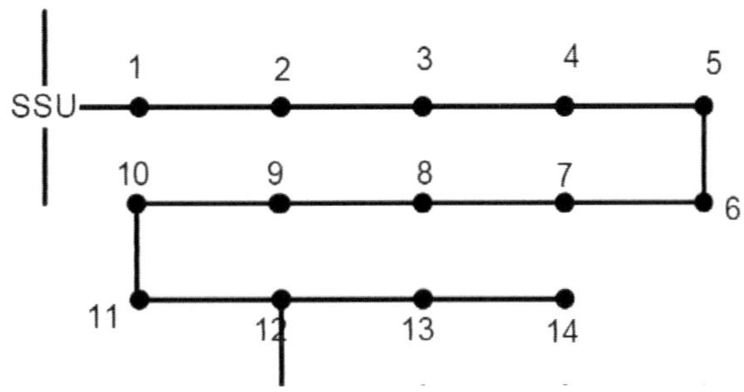

Figure 5. Part of rural area distribution network system

Table 1 shows real power loss at each bus before and after DG is installed via ACO. The suitable Dg position selected by ACO at bus 8

Table 1 Power loss at each bus when apply 710KW DG at location no 8

Bus	Real Power Loss Before DG	Reactive Power Loss Before DG	Real Power Loss After Install DG	Reactive Power Loss After Install DG
1	12.3851	17.0510	4.3830	6.0343
2	4.4666	6.3150	1.4592	2.0630
3	7.8007	10.7386	2.3608	3.2499
4	6.1665	8.4890	1.7691	2.4354
5	3.4363	4.7355	0.9965	1.3733
6	2.5262	3.4805	0.8462	1.1659
7	1.9293	2.6559	0.9190	1.2651
8	1.2352	1.7004	1.0621	1.4621
9	1.3902	1.9138	1.3902	1.9138
10	0.1931	0.2658	0.1931	0.2658
11	0.6966	0.9590	0.6966	0.9590
12	0.0773	0.1064	0.0773	0.1064
13	0.0193	0.0266	0.0193	0.0266
14	0.0204	0.0280	0.0204	0.0280
15	0.0309	0.0426	0.0309	0.0426
16	0.0137	0.0189	0.0137	0.0189
17	0.0034	0.0047	0.0034	0.0047
TOTAL	**42.3910**	**58.5316**	**16.2410**	**22.4148**

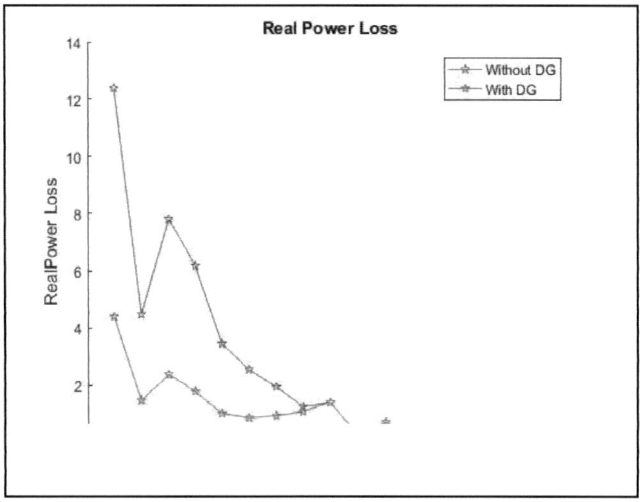

Figure 6. Real power loss in case of the real rural network before and after using ACO.

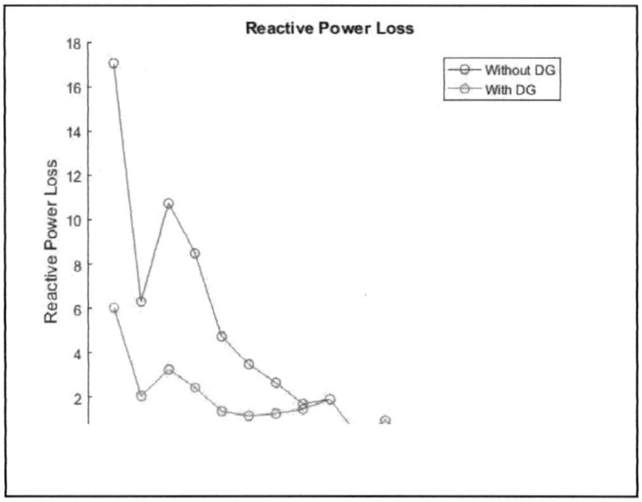

Figure 7. Reactive power loss in case of the real rural network before and after using ACO.

Figure 6 and 7 shows the power loss graph line before and after DG is implemented via ACO method

Table 2 show the confirmation of the result gain from ACO and Brute Force method. Extend your table explanation here

Table 2. Optimum location of DG using ACO and Brute Force methods

Method	Position of DG at Bus suggested from the method	Size DG (kW)	Total Power Loss (kW)
Without DG (Normal condition)	NA	NA	42.3910
Brute Force	8	710	16.2410
ACO	8	710	16.2410

Figure 8 and 9 shows the line graph after using DG is locate in the network. Based on the data gain it shows that the most suitable place to locate DG is at bus 8 this is because there is a significant improve in real and reactive power losses at bus 8 compared to when the DG is located at other buses.

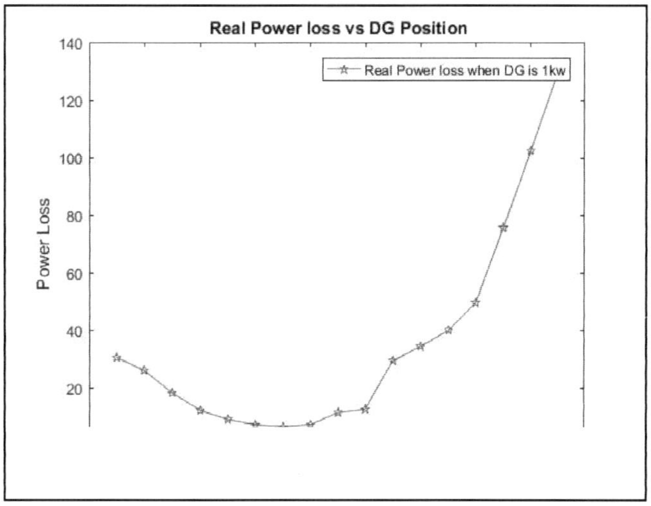

Figure 8. Real power loss in case of the real rural network using Brute Force method

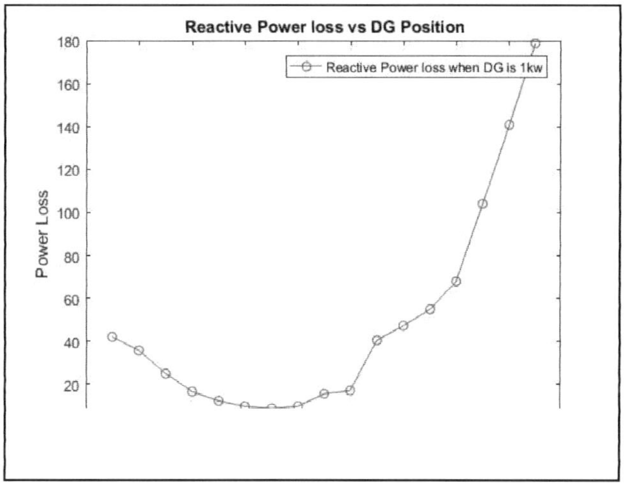

Figure 9. Reactive power loss in case of the real rural network using Brute Force method

5. CONCLUSION

This paper introduced a method for finding the optimum position of the distributed generation. In this paper, an algorithm is proposed to be tested in one of the rural real network in Malaysia where the specialization of ACO was used as an alternative solution method to solve the optimizing problem. In order to observe the efficiency of the DG positioning, a collection of data on power loss and voltage stability was taken, reported and compared with the data set before the DG was integrated in the network. From the data gain shows that implementing DG into the network could improve the DN in term of power loss and voltage stability which shows that the DG is effective to reduce a power in a case study network. Furthermore, the objective of this project to gain optimal allocation for DG using ACO is achieved. The result from the ACO method is confirm through another algorithm via Brute Force which produce the same result as the ACO method does.

ACKNOWLEDGEMENT
The authors gratefully acknowledge to Universiti Teknologi MARA, Cawangan Pulau Pinang and Faculty of Electrical Engineering for the facility that provided to complete this project and the supported.

REFERENCES

[1] R. P. Nath and V. Nishanth Balaji, "*Artificial Intelligence in Power Systems*." Accessed: Aug. 14, 2020. [Online]. Available: www.iosrjournals.org.

[2] L. I. Dulău, M. Abrudean, and D. Bică, "*Optimal Location of a Distributed Generator for Power Losses Improvement*," Procedia Technol., vol. 22, pp. 734–739, Jan. 2016, doi: 10.1016/j.protcy.2016.01.032.

[3] R. Payasi, A. Singh, and D. Singh, "Review of distributed generation planning: objectives, constraints, and algorithms," *Int. J. Eng. Sci. Technol.*, vol. 3, no. 3, Jul. 2011, doi: 10.4314/ijest.v3i3.68430.

[4] S. V. Reddy and S. Swathi, "Optimal Allocation and Sizing of multiple Distributed Generation in distribution network by Ant Colony Search Algorithm," vol. 3, no. 1, pp. 59–63, 2014.

[5] P. Kaur, S. Kaur, and R. Khanna, "*Optimal placement and sizing of DG comparison of different techniques of DG placement*," Feb. 2017, doi: 10.1109/ICPEICES.2016.7853653.

[6] "(PDF) Optimal Number Size and Location of Distributed Generation Units in Radial Distribution Systems Using Grey Wolf Optimizer." https://www.researchgate.net/publication/313841863_Optimal_Number_Size_and_Location_of_Distributed_Generation_Units_in_Radial_Distribution_Systems_Using_Grey_Wolf_Optimizer (accessed Aug. 14, 2020).

[7] H. E. A. Talaat and E. Al-Ammar, "*Optimal allocation and sizing of Distributed Generation in distribution networks using Genetic Algorithms*," 11th Int. Conf. Electr. Power Qual. Util., Accessed: Aug. 14, 2020. [Online]. Available: https://www.academia.edu/22379559/Optimal_allocation_and_sizing_of_Distributed_Generation_in_distribution_networks_using_Genetic_Algorithms.

[8] M. Heidari, A. Nekoubin, R. Heidari, and M. Jafari, "*Optimum locating and Sizing of Distributed Generation Based on Artificial Ant Colony Algorithm*," 2013. Accessed: Aug. 14, 2020. [Online]. Available: www.isca.me.

[9] H. Dezaki, H. Abyaneh, A. Agheli, and K. Mazlumi, "Optimized Switch Allocation to Improve the Restoration Energy in Distribution Systems," *J. Electr. Eng.*, vol. 63, no. 1, pp. 47–52, Jan. 2012, doi: 10.2478/v10187-012-0007-9.

[10] N. Jiravanstit and S. Chaitusaney, "*Protective device placement for reliability improvement by considering anti-islanding protection of PV generation system*," in Electrical Engineering/Electronics, Computer, Telecommunications and Information Technology (ECTI-CON), 2012 9th International Conference on, 2012, pp. 1–4, doi: 10.1109/ECTICon.2012.6254255.

[11] M. A. Velasquez, A. Cadena, and C. Tautiva, "*Optimal planning of recloser-based protection systems applying the economic theory of the firm and evolutionary algorithms*," in IEEE PES ISGT Europe 2013, Oct. 2013, pp. 1–5, doi: 10.1109/ISGTEurope.2013.6695350.

[12] S. Omar, S. Robson, A. Haddad, H. Griffiths, and N. Harid, "*HV distribution network optimum supply restoration algorithm*," in Proceedings - 2016 51st International Universities Power Engineering Conference, UPEC 2016, Jul. 2016, vol. 2017-January, pp. 1–6, doi: 10.1109/UPEC.2016.8114010.

[13] S. Omar et al., "*HV distribution network supply restoration*," in 2015 50th International Universities Power Engineering Conference (UPEC), Sep. 2015, pp. 1–6, doi: 10.1109/UPEC.2015.7339971.

[14] J. M. and W. K. Edward Coster, *Distributed Generation.* InTech, 2010.

[15] J. W. Feltes et al., "Some Considerations in the Development of Restoration Plans for Electric Utilities Serving Large Metropolitan Areas," *IEEE Trans. Power Syst.*, vol. 21, no. 2, pp. 909–915, May 2006, doi: 10.1109/TPWRS.2006.873128.

Control Structures Implementation to Allow High Penetration of a VSC within an Isolated Power System

D Curto, S Favuzza, R Musca, M Navarro Navia, and G Zizzo

Department of Engineering, Università degli Studi di Palermo, Palermo, Italy.

Abstract : Recently, the increasing occurrence of power instability caused by alternative energy sources has attracted the possibility of implementing Energy Storage System (ESS), capable of supplying the energy necessary to improve the energy quality within the grid. Therefore, the attention of researchers has been drawn on the behaviour that has the connection of the ESS, based on internal controllers such as Virtual Synchronous Machine (VSM) and a Classical Control Cascade (CCC), for different ESS power ranges. In this context, this paper deals with the implementation of an ESS represented by a Voltage Source Converter (VSC) employing control techniques of VSM and CCC simulated in PScad environment in three cases: in the first case, the power system in the steady-state is seen during the description of the controls; in the second case, the power system in the presence of a fault; in the third case, the power system with a load at the PCC. Finally, a fusion CCC-VSM using a grid impedance estimator is implemented with different current reference controls considering Short Circuit Ratio (SCR) values that represent the weak and robust grid

1. INTRODUCTION

From the last decade, Renewable Energy Sources (RES) are increasing the rated power capacity concerning the Conventional Power Plants (CPP), generating a deficit in the energy quality and grid stability [1]. Due to the intermittent and aleatory RES production, the electrical grid can be exposed to sudden changes of active and reactive power flowing [2][3][4]. To limit this issue, the adoption of Energy Storage System (ESS) was proposed as active power support for the grid. In [5], the possibility to incorporate the ESS through a Voltage Source Converter (VSC) was investigated, considering flexible AC transmission systems devices such as a Static Synchronous Compensator (Statcom) [6]. Some authors reported valuable studies on Statcom connected to energy storage units [7],[8]. In the following, the integration of a Statcom and an ESS is expressed as E-Statcom. As examples, the E-Statcom topology is investigated to support energy production from wind turbines [9] and photovoltaic panels [10]. Traditionally, VSC is used for applications that involve sinusoidal voltage with a variable output. In [11], a VSC generates a set of alternating voltages that can be controlled in amplitude and phase as it happens for the synchronous machines but, unlike them, E-Statcom has no inertia, since it has a virtually instantaneous response and it does not alter the system impedance significantly.

The control topology plays an essential role in the robustness of the converter formation. A Classical Control Cascade (CCC) is an internal loop that feedbacks among themselves. In [13], a Virtual Synchronous Machine (VSM) is proposed and fused to a classical CCC. Control algorithms for grid-connected converters in a simulated grid environment are validated considering a load connected to the Point of Common Coupling (PCC). Besides, the grid information is implemented by the algorithms; a grid impedance used to give stability and quality for the energy produced at the VSC output. To make the control model real, delays are introduced through delay compensation techniques to increase the quality power and reduce the output current oscillations. The relation between the grid and the VSC is expressed by the Short Circuit Ratio (SCR), that is the ratio between the grid short

circuit power and the converter rated power. The SCR, the grid characteristics and the voltage value are part of the facility's control requirements [14].
In this paper, the proposed controls are tested on an ideal grid representing an isolated power system of a small island. The simulations are carried out in the presence of a fault and a load at PCC for different values of SCR, to study both cases of weak and strong grids. The control is examined in the case of voltage regulation and power quality, as a function of the active power regulation, is also assessed by considering the effect of the ESS; simulations are performed in the PScad software. The remaining part of the paper is organised as follows: Section 2 gives the working principle of the power system model; Section 3 reports the control design of CCC, VSM and the fusion CCC-VSM; Section 4 provides the results obtained from the models implemented in PScad and, finally, Section 5 contains the conclusion of this paper.

2. MODEL OF THE POWER SYSTEM

In [15], an E-Statcom regulates the PCC voltage within the specified limit with reactive power support. A simple model is represented in Figure 1 with two VSC: the first is E-S.1, based on VSM, using controls that provide virtual inertia and damping; the second is E-S.2, based on CCC, using feedback current control through coordinate transformation. The controllers have as the main goal to generate an AC-voltage $Vvsc$ through internal control loops. These are connected at the PCC through an RL-filter, composed of a resistance Rf and an inductance lf, connected with a grid impedance Zg, modelled by a resistance Rg and inductance lg (see Figure 1).

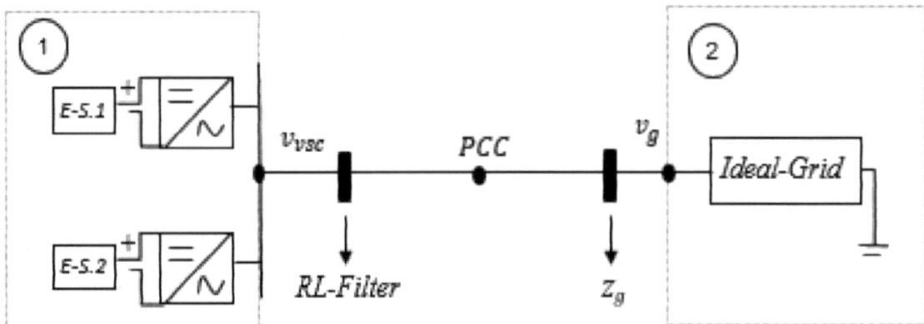

Figure 1. Model of the power system.

The CPP is modelled as an ideal grid with a total rated power of 24.5 MVA, a rated phase to phase voltage of 10 kV, rated frequency of 50 Hz and an E-Statcom, having the characteristics of a VSC. On the other hand, the power delivered by the transmission line is studied for the following reference active power of E-Statcom: 0.2, 0.5, 0.7, 1 p.u. based on E-S.1/E-S.2 internal controls. The relation between the ideal-grid and the VSC is expressed by the Short Circuit Ratio $SCR{=}Ssc/Svsc$ where, $Svsc$ is the grid short circuit power and Ssc is the converter rated active power. Therefore, varying the SCR value from 1 to 10 it is possible to analyse the behaviour of Zg.

Table 1. Grid-Z for varying SCR

SCR	lg [H]	Rg [Ω]
1	0.0001	0.041
3	0.0003	0.122
5	0.0005	0.204
7	0.0008	0.286
10	0.00108	0.408

3. CONTROL DESIGN

This section provides a brief overview of the controls.
3.a VSM-Structure: This control allows a compatible grid integration of E-Statcom even in a weak, isolated grid since it mimics a Synchronous Generator (SG) [16] [17]. The goal is to create the VSM-

angle phase θvsm from Equation (1) allowing the choice of the oscillation damping and the inertia support that will be used by the converter [18] [19].

$$2H_{vsm}\frac{dw_{vsm}}{dt} = P_{ref} - P_{vsc} - Dw_{vsm} \tag{1}$$

In Equation (1), $Pref$ is the input active power, $Pvsc$ is the VSC output active power, $Hvsm$ is the virtual mechanical time constant (measured in seconds), D is the virtual mechanical damping, $wvsm$ is the VSM angular frequency, δr is the power angle of VSM and wo is the rated angular frequency.

Simulations are performed for VSC active power references of 20% and 100% (dashed line). The following parameters are considered: constant inertia levels H 0.5 s (red line), 3 s (light blue), 7 s (medium blue), 10 s (raised blue), damping levels (D= 30, 50, 100).

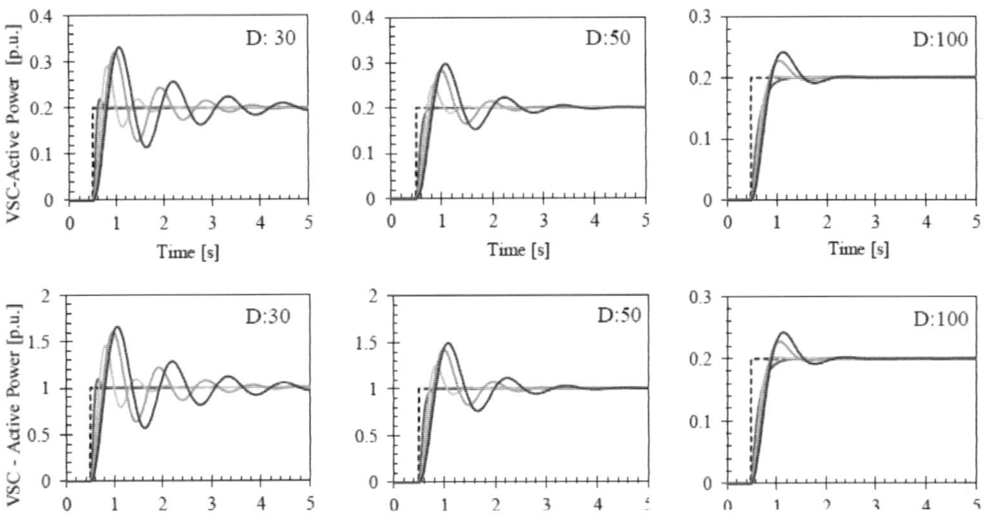

Figure 2. VSM response for different H and D values at the VSC power step equal to 20% and 100%.

Figure 2 shows that waveform improves with high damping values. Note that the damping is opposed to the inertial response. Therefore, the mean value of inertia is a good compromise.

Below, in Table 2 the analysis of overshoot $Se\%$ and settling time TαƐ is shown. Indeed, Table 2 verifies the similarity of the dynamic response of VSM with an average D between 50 to 100. It is possible to obtain a low power fluctuation. Therefore, the phase of the oscillatory component is estimated 70, thus providing active damping.

Table 2. Different response characteristics from VSM.

Reference		VSC Active Power		
		20%	50%	100%
$S_e\%$	30	60.50%	60.50%	60.50%
	50	42%	42%	42%
	100	13.90%	13.90%	13.90%
TαƐ	30	2,9	1,9	1,1
	50	2,9	1,9	1,1
	100	2,9	1,9	1,1

According to the preliminary results from simulations, the average value of inertia 7s is adopted. In the following, the paper is mainly focused on optimal strategy from the mix of active power $Pvsc0$ from Equation (2) in steady-state and Equation (1).

$$P_{vsc}^{0} = \frac{v_{vsc} v_g \sin \delta_{v0}}{\mathrm{x}} \tag{2}$$

In Equation (2), $vvsc$ and vg are respectively the magnitude voltage of the converter and of the grid in p.u., Xf is the converter reactance in p.u. The input and output ratio of the power imbalance transfer function is named $VDvsm$. To find this, it is necessary to study the response of the controller after a small variation of the reference power $\Delta Pref$ around the steady-state condition, based on Equations (1) and (2) can be expressed as:

$$2\mathrm{H}_{vsm} \frac{\mathrm{d}\Delta \mathrm{w}_{\mathrm{vsm}}}{\mathrm{dt}} = \Delta P_{ref} - \frac{\mathrm{d}P_{vsc}}{\mathrm{d}\delta_{\mathrm{v}}} \Delta\delta_{\mathrm{v}} - \mathrm{D}\Delta \mathrm{w}_{\mathrm{vsm}}$$

$$\frac{\mathrm{d}\Delta\delta_{\mathrm{v}}}{\;} = \mathrm{w}\ \Delta\mathrm{w} \quad - \mathrm{w} \tag{3}$$

$$\frac{\mathrm{d}P_{vsc}}{\mathrm{d}\delta} = \frac{v_{vsc} v_g \cos \delta_{\mathrm{v}}}{\mathrm{x}} = \mathrm{K}_{\mathrm{P-vsm}} \tag{4}$$

after some mathematical passages:

$$VD_{vsm} = \frac{\Delta P_{vsc}^{1}}{\Delta P_{ref}} = \frac{\frac{\mathrm{w_o K_{P-vsm}}}{2H_{vsm}}}{s^2 + \frac{D}{2H} s + \frac{\mathrm{w_o K_{P-vsm}}}{2H}} = \frac{w_n^2}{(s^2 + 2ℶ w_n s + w_n^2)} \tag{5}$$

where the poles from Equation (5) are: $-ℶwn \pm wn\sqrt{ℶ2-1}$, where ℶ is the damping ratio and wn is the natural frequency. Next, the VSM improvement is simulated as a function of a disturbance, considering that a step active power of 0.7 p.u. at the instant t = 5 s is activated. Virtual inertia of 7 s, it would be a good compromise among the three possibilities of inertia previously studied. The small-perturbation is a double-phase short circuit applied at the instant t = 10 s with values of short-circuit resistance Rsc=0.5 Ω.

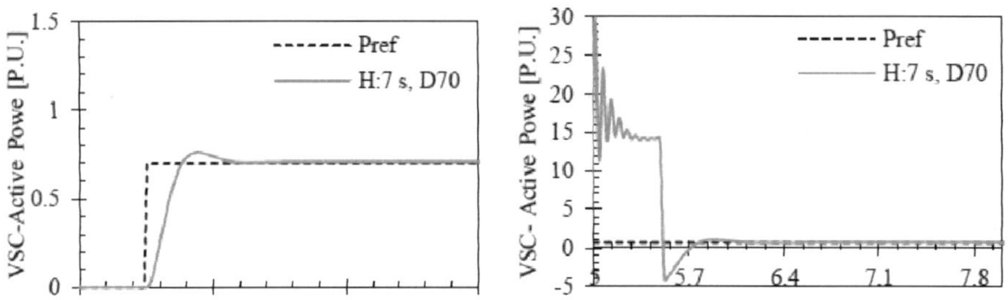

Figure 3. Damping Control Loop.

E-S.2 improvement from Figure 3 the E-Statcom's active power trend is constant in steady-state, then when is present the step-fault, it increases the E-Statcom's active power to 30 times. So far, E-S.2 is not capable of guarantee overcurrent protection to the VSC.

3.b CCC-Structure: Cascaded loops comprise the integration of an inner loop involving the secondary loop and secondary outer loops involving the primary loop [20-22]. These are a Phase-Locked Loop

IOP Conf. Series: Materials Science and Engineering 1045 (2021) 012046 doi:10.1088/1757-899X/1045/1/012046

(PLL) to track the grid voltage angle [23-25]; a Current Vector Control (CVC) used to protect the converter from overcurrent [26] [27]; a Reference Control (RC) used to obtain the converter input of the CVC and coordinate change control abc/dq and dq/abc [28] [29]. Figure 4 shows the electrical circuit of the power system.

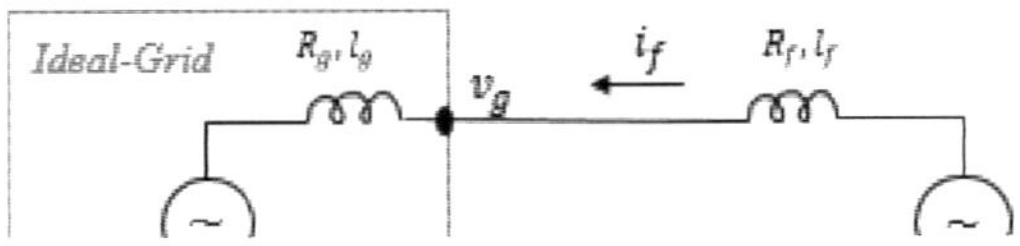

Figure 4. The classical cascade control scheme.

Where $vvsc$ is AC-voltage source represents the VSC and v_g is grid voltage. The dynamic Equation that describes Figure 4 can be given in the stationary reference frame by

(6)

where subscript dq represent the multi-system variability due to two inputs and two outputs. v_g represents the grid voltage. $if(dq)$ represents the current output of the VSC in dq coordinates. R_f and L_f are the filter resistance and inductance, respectively. w_0 represents the natural frequency. Equation 6 in terms of the Laplace domain as

(7)

where ec represents the inputs. Equation (7) is resolved through the following transfer matrix.

(8)

Assuming that:

$$Y(s) = G(s)U(s)$$

(9)

Figure 5 represents the block diagram, where $G(s)$ is a transfer function. Respectively, the structure has input reference denominated as ec and output denominated as yt. It is doesn't see but ec is formed from references currents in dq and yt is given in terms of voltage in dq. Respectively, $G(s)$ is an internal transfer function parallel to the system controller $G(s)$.

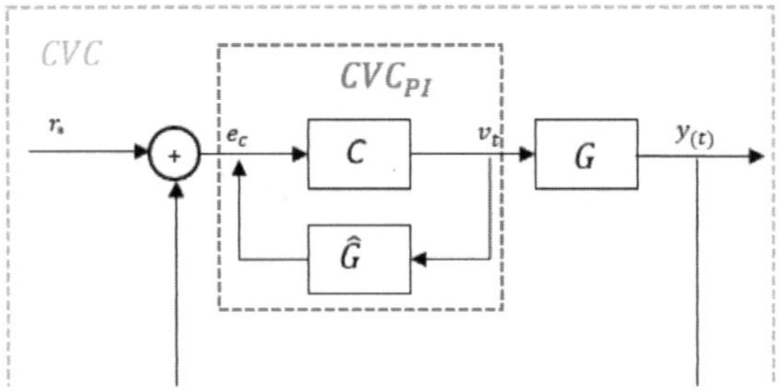

Figure 5. The current classical control without decoupled.

In other words,

$$CVC_{PI}(s) = \frac{C(s)}{1 + C(s)\hat{G}(s)} \qquad (10)$$

Applying matrix properties

$$CVC_{PI}(s)\big(1 + C(s)\hat{G}(s)\big)(I + C(s)\hat{G}(s))^{-1} = (I + C(s)\hat{G}(s))^{-1}C \qquad (11)$$

where I is the identity matrix of $C(s)$; $G(s)$ and $G(s)$ are internal transfer functions. Figure 6 is necessary to maintain the entire closed-loop stable, which means that all transfer functions must be permanent. On the other hand, to avoid the errors in steady-state should be used as an integral action; therefore, it is necessary that $G(s)$=$G(s)$. In [30] a calculation method of $C(s)$ through an H2 optimisation procedure is proposed to achieve the condition of stability according $Gc(s)$=$G(s)C(s)$. In this case $C(s)$= $G(s)-1$, giving $G(s)$=I. This result should be readjusted adding a detune for the optimal controller with a low-pass filter represented as $L(s)$= ($\alpha_{cvc}s$+/α_{cvc}), where αcvc represents the CVC bandwidth. The $L(s)$ is used in all diagonal elements of the matrix. Then, considering matrix properties maybe select $L(s)$=($\alpha_{cvc}s$+/α_{cvc})I . $L(s)$ represents replaced in Equation (9).

$$CVC_{PI}(s) = \left(I + \frac{\alpha_{cvc}}{s + \alpha}\right)^{-1} G(s)\frac{\alpha_{cvc}}{s + \alpha} = \frac{\alpha_{cvc}}{s}G(s) \qquad (12)$$

$$\frac{\alpha_{cvc}}{s}\begin{bmatrix} sL_f + R_f & 0 \\ 0 & sL_f + R_f \end{bmatrix} = \begin{bmatrix} \alpha_{cvc}L_f + \frac{\alpha_{cvc}R_f}{s} & 0 \\ & \alpha_{cvc}R_f \end{bmatrix}$$

Feedback controllers are based on PI controllers [31].

$$v_{vsc}^{dq}(\mathrm{t}) = v_{pcc(dq)}(\mathrm{t}) \pm w_o l_f i_{pcc}^{dq}(\mathrm{t}) + CVC_{PI}\big(i_{dq}^{*}(\mathrm{t}) - i_{pcc}^{dq}(\mathrm{t})\big)$$

$$v_{vsc}^{dq}(\mathrm{t}) = v_{dq-pcc}(\mathrm{t}) \pm w_o l_f i_{pcc}^{dq}(\mathrm{t}) + \propto_{cvc} l_f(i_f^{dq*}(\mathrm{t}) - i_{pcc}^{dq}(\mathrm{t})) + \propto_{cvc} R_f \int^{t} \big(i_f^{dq*}(\mathrm{t}) - i_{pcc}^{dq}(t)\big)dt \qquad (13)$$

In [28], [29], two block diagrams represent the dq input of the CVC: the first one is an active power close loop with the RC tuning in Equation (14), and the latter is a reactive power close loop with the RC tuning in Equation (15).

$$P_{PC} = \frac{P_{vsc}}{P_*} = \frac{P_{PI}}{P_{PI} + {}^{1}/_{v_{pcc}}} \quad (14)$$

$$Q_{PC} = \frac{v_{pcc}}{v_*} = \frac{Q_{PI}}{Q_{PI} + {}^{1}/_{(-X_f)}} \quad (15)$$

where PPI is a first-order low-pass filter with a cut off frequency $ki{-}p$; therefore, the bandwidth will be represented as $\propto P{=}ki{-}p$. To improve the system dynamically, it is necessary a $\propto P$ value like $10\alpha cvc/$. On the other hand, QPi is a first-order low-pass filter with a cut off frequency $ki{-}Q{=}\propto QXf/$. It is essential to know that Xf is evaluated on the greatness of the grid. Therefore, it is necessary to choose a value within the range of oscillations required by the controls. The fast-responding feedforward power controller assumes that the PLL angle always aligns with the $vpcc$.

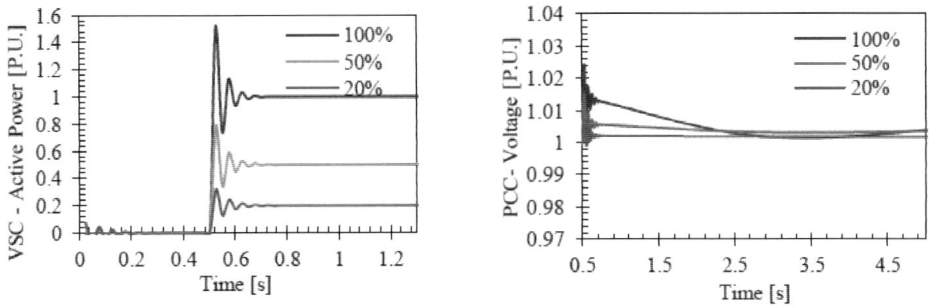

Figure 6. CVC response for different active power step.

3c) CCC-VSM: [32] proposes a fusion between VSM and CCC as shows in Figure 7. The block diagram has seven internal controllers, which are: a swing equation denominated as SC, used to create θvsm; a damping controller denominated as DC, used to improve the response of active power of the VSC as function of the pole transfer function; a current control denominated as CVC, used to protect the converter from overcurrent; a reference control denominated as RC, used to obtain the converter input of the current control and coordinate change controls abc / dq and dq / abc. On the other hand, according to the diagram $Ppcc$, $ipcc$, $Vpcc$ are firstly measured. The input of controls abc/dq and dq/abc are consequently evaluated. Note that these controllers use θvsm for generate idq and Vdq, which are the input of CVC. Moreover, $Ppcc$ is used as input of SC. Therefore, in the diagram, two types of variables are observed. First, variables denominated as complementary, are: Xf, lf and $w0$. Secondly, variables denominated as reference are: $Pref$, $Href$.

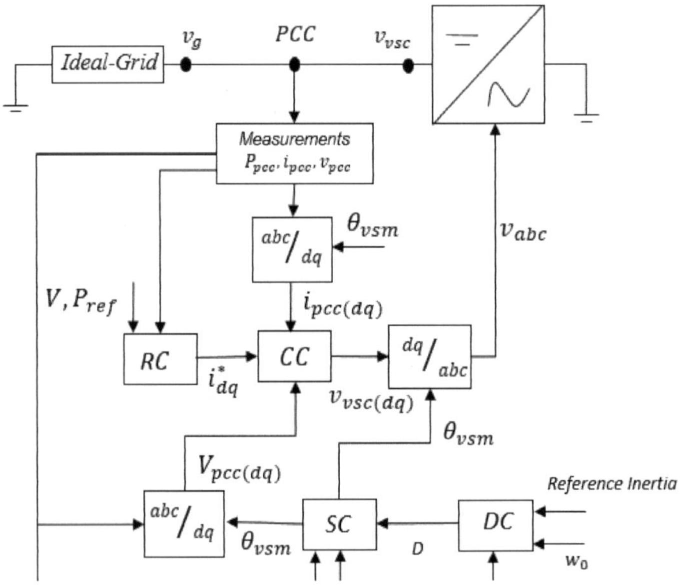

Figure 7. VSM-CCC fusion.

4. RESULTS AND DISCUSSIONS

4.a No load at the PCC: A perturbation is applied to the grid, represented by a double-phase short circuit at the instant t = 15s, with a short-circuit resistance Rsh=10 Ω. The following assumptions are made:

- A step reference active power of 0.7 p.u. is active at the instant t = 5 s. In this context, it would be a good compromise among the three possibilities examined in Section 3.
- $Vpcc$ and frequency: a step reference of 1 p.u. is active at the instant t = 5 s.
- E-S.2 is analysed with the inclusion of a compensation angle $\theta comp$ introduced by PWM using synchronous sampling [32]. The $\theta comp$ is given by $w0Td$, where the time delay is Td=1.5Ts and Ts represents a sampling time.
- E-S.1 is analysed with VSM-CCC with $Hvsm$=7s ,ב=0.707, D=70. The initial active power reference is zero and at the instant t = 0.5 s.
- the following delays are introduced: one sample delay between the calculation and the VSC-Voltage sending values; half sample between the calculation and VSC active power transmitting values. Regarding the delay values, they depend on the sampling rate of the converters: the modern converters for E-Statcom work with a sampling rate of about 1 or 5 kHz.
- Both controllers are analysed in the case of a robust grid (SCR = 10) and weak grid (SCR = 5).
- In both controllers, to improve power oscillation, a lead-lag is added to the measure of $Ppcc$.

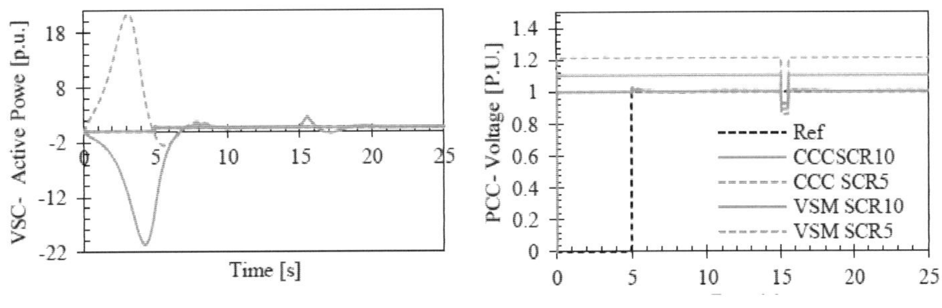

Figure 8. Controllers response.

Figure 8 shows that:

- E-S.2 control has the same VSC's active power trend for different SCR values.
- E-S.1 control has a variable VSC's active power trend that depends on grid condition SCR.
- E-S.1 control response speed is slower than E-S.2.
- E-S.2 is the best choice for an ideal-grid.
- The power responses at the time of failure are E-S.2. an overshoot of 0.2 p.u., while E.S-1 an overshoot of 200 p.u.
- PCC-voltage is 12% over the allowed range for the E-S.1 with SCR = 5. For the rest of the simulations, it remains without a problem during the small-disturbance.
- In the case of SCR, E-S.1 is not capable of supporting a weak grid. Therefore, PCC voltage is more significant than 1.15 p.u.

4.b Load at the PCC: in this case, the scheme in Figure 1 is assumed, considering a shunt load connected at the PCC through a grid impedance ZTH composed by a resistance RTH and an inductance lTH. Then, Equation (11) in continuous time is

$$v_{vsc}^{dq}(t) = v_{dq-pcc}(t) \pm w l_{TH} i_{f}^{dq}(t) + C_{PI-TH}\left(i_{dq}^{*}(t) - i_{load}^{dq}(t)\right) \quad (16)$$

where $idq*$ is the reference current calculated with Equations (14) and (15); $iloaddq$ is the measures of current in dq coordinates; $CPI{-}th$ is a transfer function of a proportional-integral PI circuit where, $kp{-}th{=}\propto cvc(th)lTH$ is the proportional gain and $ki{-}th{=}\propto cvc(th)RTH$ is the integrator gain, indicating $\propto cvc(th)$ as the closed-loop bandwidth of the CVC. Based on these criteria, the simulation is carried out in the presence of a perturbation located between the R-l filter and Zth. The setting parameters are: a step active power of 0.7 p.u., activated at the instant t = 5 s, $Hvsm$ is set to 7 s, D to 70 p.u., and $\propto cvc{-}th$ to 1215 rad/s, with a perturbation of a double-phase short circuit applied at the instant t = 15 s, with Rsh=0.5 Ω,010 Ω and SCR=10.

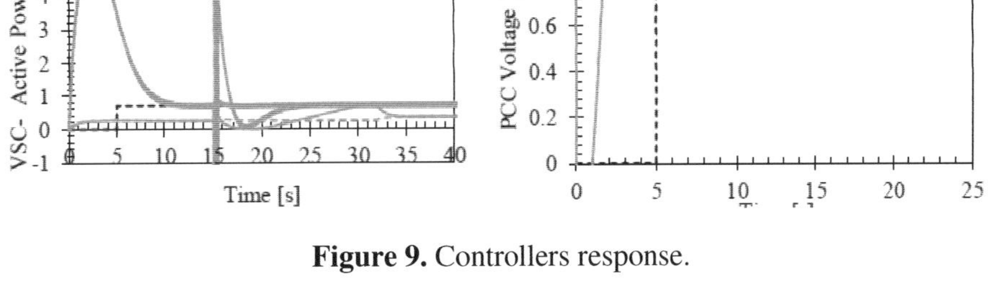

Figure 9. Controllers response.

Figure 9 shows the VSC's output active power with a Grid-TH, demonstrating worse performance for both controllers. It would be from at least two perspectives:

- first, the increase in the impedance seen by the VSC opposes the passage of currents;
- second, an effect on the power imbalance given at the time of the small-perturbation is that the d-axis inductance variation is small. However, cross-saturation may affect the q-axis inductance; as a consequence, the voltage grid had a variation of about 30%. It causes the d-axis current a variation of 3% in the q-axis and of about 5% in the d-axis. Because of this, efforts should be made to improve the dynamic robustness of CVC and RC.

4.c Controller performance utilising a Grid Impedance Estimator GIE: The grid impedance estimator $ZGIE$ consists of an indirect modification of the PCC that adds greater dynamic stiffness near the synchronous frequency. On the other hand, it is shown in Subsection 3b that the VSC's output power deteriorates due to coupling in the dq coordinates. In search of a solution, Equation (13) is rewritten in terms of an equivalent impedance $Zt(Rt,lt)$ where lt is the sum of the VSC filter inductance lf and the estimated grid inductance lg and Rt is the sum of the VSC filter resistance Rf and the estimated grid resistance Rg [13]. For the improvement of control with VSM-CCC is proposed to change RC for RC-2 as a function of Equation (6).

$$V = v \ldots + (GIE - CC) \qquad (17)$$

where $vvsc$ is VSC voltage, $vpcc(dq)$ is the measured VSC output voltage in dq-frame, $GIEPI$ the transfer function based on Equation (17). A simulation to test the internal control utilising GIE with an RC2 is proposed in Figure 10. The fault is located between Rth and lth in this case. The setting parameters are the same as in Figure 9.

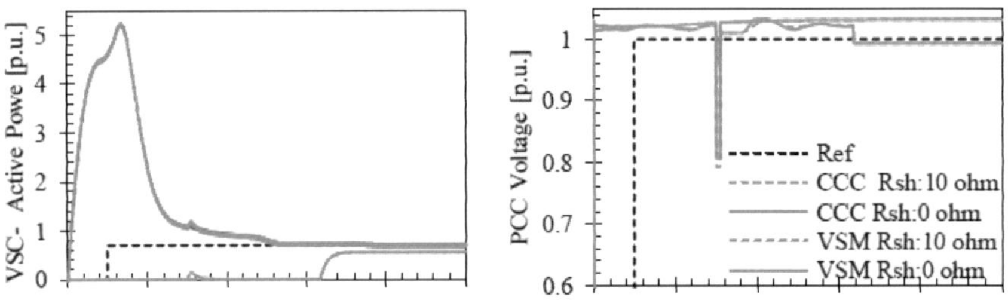

Figure 10. Controllers response with GIE.

The trends in Figure 10 reveal the following aspects:

- The settlement time Taε of the PCC-voltage trend was evaluated as the time taken by the response to definitively enter a band between ±3% of the steady-state value after the step voltage; in this case, it is set ε% = 0.3 of the steady-state value that corresponds to 0.3 kV. These values are not placed in tables because after the fault the voltage is within this range.
- The controller response in the previous Sections shows that the mix between E-S1 and E-S2 with the strategy based on virtual impedance is an exciting proposal because it allows implementing virtual inertia and damping in the isolated micro-grid leading to advantages for grid stability.
- The simulation results show that the virtual impedance has a visible improvement. System stability and current sharing ability are strongly enhanced.
- The controls require almost 15 s to stabilise in the event of extreme failure resistance.
- E-S.2 allows the VSC to produce only 78% of the reference at 26 s means that it is slower than E-S.1.

5. CONCLUSION

The dynamic response of the two controls is good at steady state. In particular, CCC is faster than VSM. Below, Table 3 reports a comparison of the two internal controls within the VSC.

Table 3. Comparison of VSM/CCC.

Description	CCC	VSM
Possibility of high-frequency instabilities	x	x
Increased RoCof	x	
Emulates inertia		x
Provides synchronising power and voltage reference		x
No capability of regulating voltage	x	
Control succeeds considering delays, even in case of low SCR		x
Possibility of modelling in RMS system studies		x
Contributes to RoCof		x

In case of a short-circuit, when an overcurrent is observable within the grid, it is necessary to implement a new control strategy. Therefore, various CRs, current limiters and some low-pass filters are studied to obtain the best internal collegiate controls. Considering the above results, the second RC option is chosen for the VSM-CCC model with the Load on the PCC, because it allows implementing virtual inertia and damping in the isolated micro-grid leading to advantages for grid stability.

The results of this study will be used for facing the issue of enhancing the penetration of RES-based generators in small islands, a very hot topic, as demonstrated by the recent literature [33]-[36].

REFERENCES

[1] J. Dakovic, M. Krpan, P. Ilak, T. Baskarad, I. Kuzle "Impact of wind capacity share, allocation of inertia and grid configuration on transient RoCoF: The case of the Croatian power system", *International Journal of Electrical Power and Energy Systems*, Volume 121, October 2020.

[2] E. Bargiacchi, S. Frigo, G.Spazzafumo "Energetic and energetic analysis of an innovative plant for the production of electricity and substitute natural gas", *Energy Procedia*, Vol. 148, 2018, Pages 312-319.

[3] J. A. Sa'ed, S. Favuzza, M. G. Ippolito, F. Massaro, "*An Investigation of Protection Devices Coordination Effects on Distributed Generators Capacity in Radial Distribution Systems*". Proceedings of 4th IEEE International Conference on Clean Electrical Power - ICCEP 2013, Alghero (Italy), June 11-13, 2013, pp. 686-692, Print ISBN: 978-1-4673-4430-2, CD ISBN 978-1-4673-4429-6, DOI 10.1109/ICCEP.2013.6586928.

[4] J. A. Sa'ed, S. Favuzza, M. G. Ippolito, F. Massaro "*Investigating the Effect of Distributed generators on Traditional Protection in Radial Distribution Systems*". Proceedings of IEEE Powertech 2013, Grenoble (France), June 16-20, 2013, pp. 1-6, Print ISBN: 978-1-4673-5667-1, DOI 10.1109/PTC.2013.6652100

[5] Huang Yuhui, Liu Dong, Liao Huaiqing "Site and Size Selection Strategies of Energy Storage System Based on Power Supply and Storage Capability Index", International Conference on Renewable Power Generation (RPG 2015), 2015.

[6] L. S. Xavier, W. C. S. Amorin, A. F. Cupertino, V. F. Mendes, W. C.do Boaventura and H. A. Pereira "Power converters for battery energy storage systems connected to medium voltage systems: a comprehensive review", BMC Energy, 2019.

[7] K. Sundararaju, A. Nirmal Kumar "Cascaded Control of a Multilevel STATCOM for

Reactive Power Compensation", *International Journal of Innovative Research & Development*, vol 3, July 2014.

[8] A. Bharadwaj and S. Maiti "*Modular Multilevel E-STATCOM using Supercapacitor Based Energy Storage System*", International Conference on Power, Energy and Environment: Towards Smart Technology (ICEPE), 2018.

[9] A. Bharadwaj and S. Maiti "*Analysis and control of modular multilevel converter-based E-STATCOM to integrate large wind farms with the grid*", IET The Institution of Engineering and Technology, 2019.

[10] S. M. Muyeen, R. Takahashi, T. Murata, J. Tamura, M. H. Ali "*Stabilisation of Wind Farms Connected with Multi-Machine Power System by Using STATCOM/BESS*", Electrical Machines and Systems, 8 October 2007.

[11] S. Kharjule "*Voltage Source Converter*", International Conference on Energy Systems and Applications (ICESA), 2015.

[12] A. Bharadwaj and S. Maiti "*Modular Multilevel Converter based Hybrid Energy Storage System*", IEEE PES Asia-Pacific Power and Energy Engineering Conference (APPEEC), 2017.

[13] J. A. Sa'Ed, D. Curto, S. Favuzza, R. Musca, M. Navarro Navia, Gaetano Zizzo "*A Simulation Analysis of VSM Control for RES plants in a Small Mediterranean Island*", 2020 IEEE International Conference on Environment and Electrical Engineering and 2020 IEEE Industrial and Commercial Power Systems Europe (EEEIC / I&CPS Europe), Madrid (Spain), 9-12 June 2020.

[14] R. P. Rodriguez, A. V. Timbus, R. Teodorescu, M. Liserre, F. Blaabjerg, "Flexible Active Power Control of Distributed Power Generation Systems During Grid Faults", in *IEEE Transactions on Industrial Electronics*, vol. 54, no. 5, pp. 2583-2592, Oct. 2007.

[15] S. Feng, X. Wu, P. Jiang, L. Xie and J. Lei, "Mitigation of Power System Forced Oscillations: An E-STATCOM Approach", IEEE ACCESS, Vol. 6, 2018.

[16] H. P. Beck, and R. Hesse, "Virtual Synchronous Machine" 2007 9th International Conference on Electrical Power Quality and Utilisation, pp. 1-6, 2007.

[17] O. Mo, S. d Arco and J. A. Soul "Evaluation of Virtual Synchronous Machines with Dynamic or Quasi-Stationary Machine Models" IEEE transitions on Industrial Electronic, Vol. 64, No. 7, 2016.

[18] M. Torres, L. A. C. Lopez "A Virtual Synchronous Machine to Support Dynamic Frequency Control in a Mini-Grid That Operates in Frequency Droop Mode" Energy and Power Engineering, May 2013.

[19] S. d Arco and J. A. Soul "Equivalence of Virtual Synchronous Machines and Frequency-Droops for Converter-Based MicroGrids" IEEE transitions on Smart Grid, January 2014.

[20] Yi Zhou, Zhixin Miao, Yin Li, Lingling Fan "Design Robust Cascade Control Structure for Voltage Source Converters", North American Power Symposium (NAPS), 2017.

[21] Seborg, D.E., Edgar, T.F., Mellichamp, D.A.: "Process dynamics and control" (Jhon Wiley & Sons, 2010, 3, edn).

[22] Ali Akbar Jamshidifar, Dragan Jocvic "*3-Level Cascaded Voltage Source Converters Controller with Dispatcher Droop Feedback for Direct Current Transmission Grids*", IET Generation, Transmission & Distribution, 10 October 2014.

[23] Siyu Gao, M Barnes "*Phase-Locked Loops for Grid-Tied Inverters: Comparison and Testing*"

[24] Jim Ogren "*PLL design for Inverter Grid Connection, Simulations for ideal and non-ideal grid conditions*", UPPSALA Universitet, 2011.

[25] Rutvik Desai, Smit Patel, Priyanka Patel "*PLL Based Method for Control of Grid-Connected Inverter for Unbalanced Grid Frequency*", Chhotubhai Gopalbhai Patel Institute of Technology, 2017.

[26] J. Wei, "*Review of Current control strategies in Modular Multilevel Converter*", Master Thesis at Norwegian University of Science and Technology, Department of Electric Power Engineering.

[27] L. Harnefors, L. Zhang and M. Bongiorno "Frequency-domain passivity-based current

controller design", in IET Power Electronics, vol. 1, no. 4, pp. 455-465, December 2008.

[28] F. Guo, T. Zheng, Z. Wang "*Comparative Study of Direct Power Control with Vector Control for Rotor Side Converter of DFIG*", 9th IET International Conference on Advances in Power System Control, Operation and Management (APSCOM 2012).

[29] Elena-Daniela, Iona Fagarassan "*Voltage - Reactive Power Control in Renewables Power Plants*", IEEE, 2016.

[30] H. Kwakernaak "*H2-optimization — Theory and applications to robust control design*", Annual Reviews in Control, Vol. 26, No.1, 2002, Pages 45-56.

[31] S. d'Arco, J. A. Suul, O. B.Fosso "Automatic tuning of Cascaded Controllers off Power Converters Using Eigenvalue Parametric Sensitives", *IEEE transactions on Industry Applications*, Vol, 51, N.2, 2015.

[32] Anil Bharadwaj and Suman Maiti "*Modular Multilevel Converter based Hybrid Energy Storage System*", IEEE PES Asia-Pacific Power and Energy Engineering Conference (APPEEC), 2017.

[33] D. Curto, S. Favuzza, V. Franzitta, R. Musca, M. A. Navarro Navia, G. Zizzo "Evaluation of the optimal renewable electricity mix for Lampedusa island: The adoption of a technical and economical methodology", *Journal of Cleaner Production*, Vol. 263, 2020, article 121404.

[34] M. Bongiorno, S. Favuzza, M. G. Ippolito, R. Musca, G. Zizzo. "*Inertial response of isolated power networks with wind power plants*", 2019 IEEE Milan PowerTech, 23-27 June 2019, Milan (Italy), DOI: 10.1109/PTC.2019.8810574.

[35] S. Favuzza, M. G. Ippolito, R. Musca, M. Navarro Navia, E. Riva Sanseverino, G. Zizzo, M. Bongiorno. "*An Analysis of the Intertial Response of Small Isolated Power Systems in Presence of Generation from Renewable Energy Sources*", 2018 IEEE 4th International Forum on Research and Technology for Society and Industry (RTSI), pp. 1-6, 10-13 September 2018, Palermo (Italy), DOI: 10.1109/RTSI.2018.8548401.

[36] S. Favuzza, M. G. Ippolito, R. Musca, M. Navarro Navia, E. Riva Sanseverino, G. Zizzo, M. Bongiorno. "*System Stability of a Small Island's Network with Different Levels of Wind Power Penetration*", 2018 IEEE 4th International Forum on Research and Technology for Society and Industry (RTSI), pp. 1-6, 10-13 September 2018, Palermo (Italy), DOI: 10.1109/RTSI.2018.8548355.

Smart Soft-Switching Switchover Circuit for Dual-Power PV-grid System Energy

[1]I R Ibrahim, [2]M A F M Nasir, [3]M N M Hussain and [4]A M Omar.

[123]UniversitiTeknologi MARA, Cawangan Pulau Pinang, Malaysia
[4]Faculty of Electrical Engineering, Universiti Teknologi MARA, Malaysia

E-mail : [1]intan121@uitm.edu.my, [2]aliffaisal61161@gmail.com, [3]najib830@uitm.edu.my, [4]Maliki_omar@uitm.edu.my

Abstract : Abstract—The dual-power PV-grid system is introduced to reduce the installation cost of a PV energy system. However, in the implementation of the dual-power PV-grid system, there is an interruption occurs during the transition of power process. To avoid the harmful effect during the transition, a small delay is inserted between the transition. The delay affects sensitive electronic equipment or industrial processes relying on constant electrical power. An intelligent control and switching scheme are required to solve the transition and synchronization problems. The objective of this project is to develop the soft-switching switchover circuit for dual-power PV-grid energy system. A new integration of hardware components and software developments is proposed to monitor PV current, to decide the power supply source, and to transfer the supply power at zero voltage and current. The proposed system involves three circuits, supervisory controller, switchover circuit and soft-switching circuit. Simulation and hardware implementation of the system proves by utilizing smart soft-switching switchover circuit, problem encountered during the transition of power supply could be eliminated

Keywords—dual-power PV-grid system, soft-switching, supervisory controller, switchover circuit, PV energy system

1. INTRODUCTION

It is estimated energy demand in Malaysia is expanding in average of 4.8% annually to pace with development progress. There is an urgent need to diversify our energy resources to ensure sustainability and security in energy supply. Sustainable green technology application is seen as most adopted solution by many countries in the world to address energy and environmental issues. Malaysia has set a target to achieve advanced economy by 2020 with unconditionally recognized the importance of environmental sustainability as part of a comprehensive socio-economic development plan. Malaysia is determined to reduce the greenhouse gasses (GHGs) emission intensity of its Gross Domestic Product (GDP) by up to 40%. [1]

Solar photovoltaic (PV) energy has been the focus in RE development in Malaysia. Large allocation had been dedicated to the energy system in the 10th and 11th Malaysian Plan. However, the PV energy system alone is not technically viable for continuous power supply for varying load because of its unpredictable nature which dependent on the solar radiation and only exists during the daytime[1-4]. It is necessary to have multi-input supplies of hybrid PV system to provide 24 hours continuous power supply[5, 6] .

The main constraint for widespread use of PV electricity in Malaysia is high installation cost, low efficiency, and reliability of the system. In the dual-power PV-grid system, the load will be supplied by two energy system: PV-battery system and grid system. During the day or on-demand tariff rates, the system will be supplied by PV-battery system and during the night or an off demand lower tariff, the system will be connected to the normal grid system. The project was introduced to manipulate the lower tariff rate at off-peak period in the dual tariff system[7-11]. The system is aimed to reduce the battery autonomy. By reducing the battery autonomy, the number of PV panels and batteries will be reduced, thus reducing the capital installation cost of the PV energy system. The system also allows capability of future expansion and interconnection[12, 13].

The main problem encountered in the dual-power PV-grid system is the transition process due to fast-changing solar irradiation and weather conditions. To respond to these conditions, the dual-power PV-grid system should be able to switch the power supply based on the rapid changing weather condition. The switching activities instigate power losses, switch stress, electromagnetic Interference (EMI) and heat to the system. To avoid voltage and current spikes during switchover process, a small delay should be introduced in between the transition of the supply. The delay, even though very small, produces power interruption to the sensitive electronic equipment such as computers, air-conditionings or industrial processes relying on constant electrical power. In some cases, the equipment and process should be reset and reconfigured. It is most likely to cause a significant loss of data as well as profit. In power conditioning system, the ideal output of an inverter should be pure sinusoidal symmetrical voltage of desired magnitude and frequency. However, under variation of solar irradiance, temperature and load, the practical output is sometimes distorted and contains harmonics.[14-16] For smooth transition of supply, the output signal from the inverter should be in phase with the grid signal. In the worst-case condition, it could become disastrous if the output signal from the inverter is out of phase from the grid[11, 17].

This paper discusses the proposed smart soft-switching switchover circuit. The circuit is implemented in the dual-power PV-grid system to solve the problem during the transition of power supply. It consists of a supervisory controller and a soft-switching switchover circuit. A supervisory controller monitors the PV current and load current to decide which source of power supply will be connected to the load; PV-battery or grid system. It triggers the control signal to switchover circuit to change the connection of the power supply. Soft-switching circuit will be activated to allow transition of power supply occurs at zero-voltage timing.

2. DUAL-POWER PV-GRID ENERGY SYSTEM

The dual-power PV-grid system is combination PV-battery system and grid system which alternately supplying a load. The transition of power supply form PV-battery system to grid system and vice versa is controlled by the switchover circuit as depicted in Figure 1.

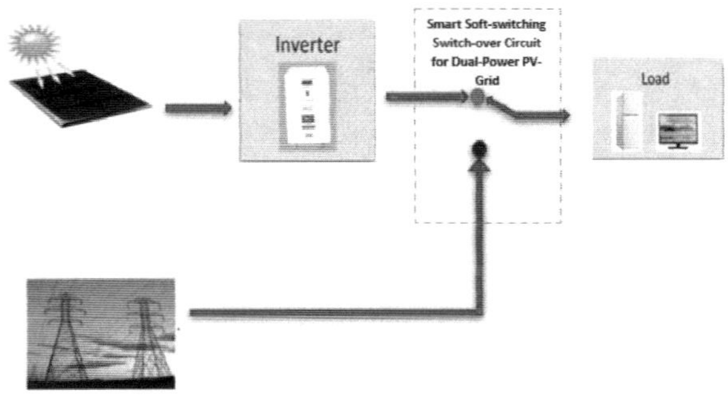

Figure 1 Dual-power PV-grid Energy System

The load is normally connected to the solar PV source. If the solar PV current is less than the present value indicates inadequate supply from the solar PV system, the load will be switched to the grid source. Due to synchronization problem of inverter output signal and grid signal, short delay was inserted between each transition of power supply as depicted by Figure 2.

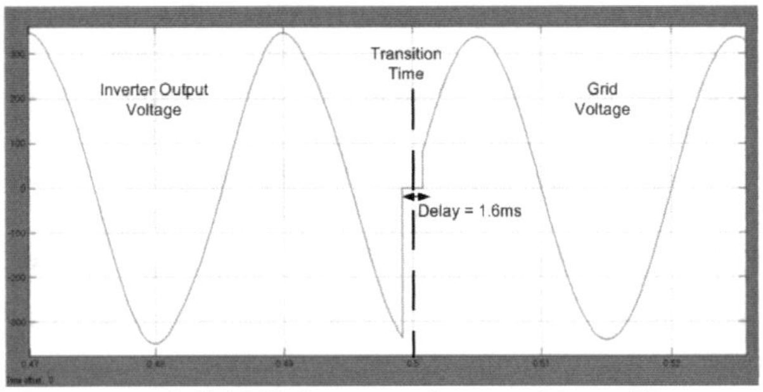

Figure 2 Inverter and Grid Voltages at the transition of Power supply

The transition delay may be very fast; however, it still could interrupt the operation of certain load such as an air condition system and computers. Switching at high voltage and current introduce sparks in the system, which consequently damage the electronic components in the system.

3. SUPERVISORY CONTROL

To overcome problems during transition of power supply, a smart supervisory control strategy is proposed as illustrated in Fig.3.

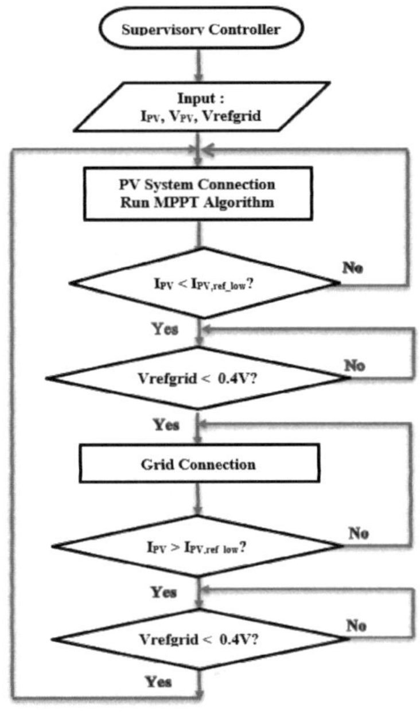

Figure 3. Supervisory Control Strategy

Under normal circumstances, the supervisory controller monitors the PV current and the load current to select which power supply source (solar PV or Grid) will be connected to the load. It will always ensure that the phase, voltage, and frequency of the power wave from an inverter are in-phase with the Grid for smooth power transition. If the current output from the PV solar system is less than the pre-set low current (meaning there is no power output from the solar PV system), it will trigger a control signal to the switchover circuit to change the power supply connection to the Grid. The zero-crossing detector circuit will wait until the Grid system voltage is less than 0.4V before connecting the supply to the Grid source.

If the current from the PV solar system returns to exceeds the pre-set level, the supervisory controller will reconnect back the load to the solar PV source. Again, a zero-cross detector circuit will be activated to ensure a reconnection is made at zero voltage point.

4.SOFT-SWITCHING SWITCH-OVER CIRCUIT

Soft-switching switchover circuit is used is to switch the source between solar PV and Grid automatically based on the detected solar PV current measured by the current sensor module.Fig.4shows the circuit of soft-switching technique transition process using the zero-crossing detector method.

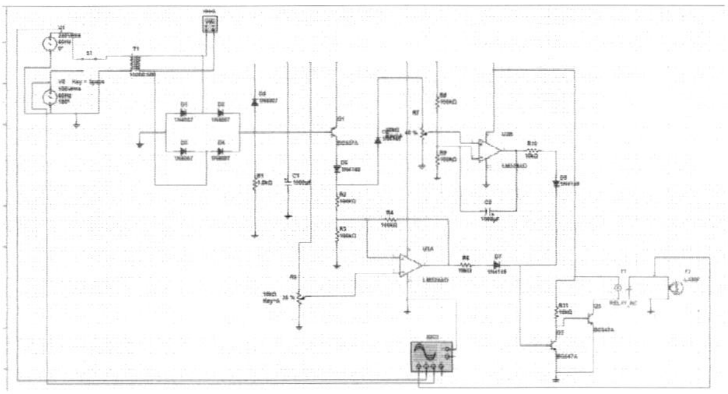

Figure 4. Soft-switching Switchover Circuit

Firstly, the 240V_{AC} voltage supply the Grid source is connected to the step-down type transformer and subsequently converted to 12V_{AC} and 12V_{DC}. 12V_{AC} is used as reference voltage ($V_{AC,ref}$) to the zero-detector circuit. Secondly, comparator (2B) is used to compare solar PV current and solar PV low voltage pre-set value. The output from the comparator will trigger the solar PV relay (normally closed). If the solar PV current is less than less the pre-set value, the solar relay will turn off and the grid relay will be triggered and supply the power to the load. This condition also simultaneously activates the zero-crossing detector circuit (ZCD) which is controlling comparator 1A. ZCD circuit only allow transition of supply when $VA_{C,ref}$ is between ±0.4V. Fig.5 illustrates 3D PCB View and hardware construction of soft-switching switchover circuit. Meanwhile, Table 1 tabulates the list of components used in the circuit.

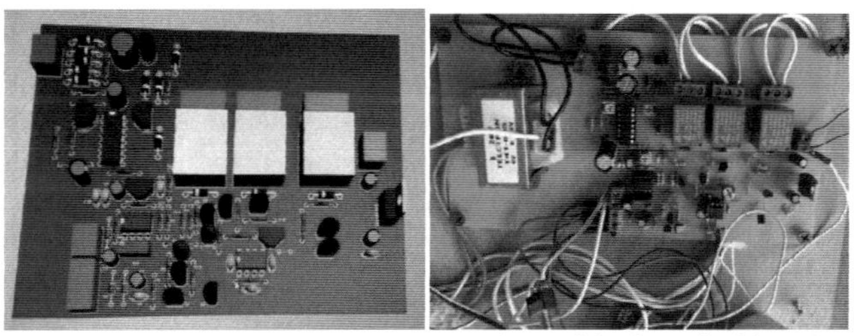

Figure 5 3D PCB View of and Hardware Construction of Soft-switching Switchover Circuit.

Table 1. List of Component of Soft-switching Switchover Circuit

Component	Value/rating
Step-down Transformer	230V to 12V
Resistor	220,330, 1k, 4.7k, 10k, 22k, 100k
Diode	1N4148, 1N4007, LEDs
Transistor	BC547, BC557, 2N3904
Comparator	LM324, LM393
Relay	12V SPSDT, Normally Close
Capacitor	10 nF, 100nF, 10μF, 100μF, 1000μF
Potentiometer (Ω)	10k, 100k
Current Sensors	ACS712ELCTR-20A
555 IC Timer	

5. RESULT AND DISCUSSION

5.1. Simulation

Figure 6 and 7 show simulation process of transition from solar PV source to Grid source and vice versa. The supervisory controller monitors the solar PV current and decides on transition of power supply based on detected solar PV current.

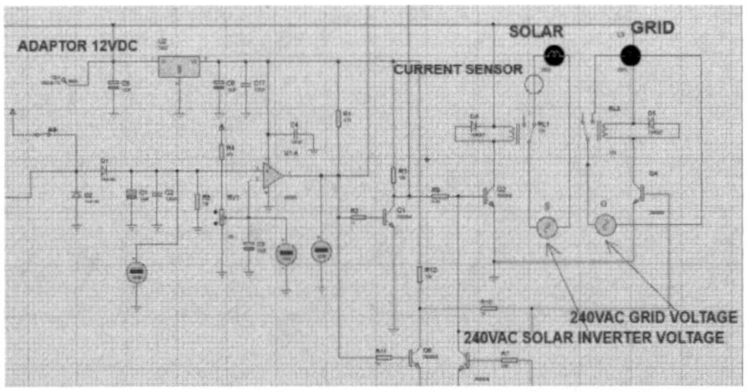

Figure 6. Solar PV Current > Pre-set Value, Solar PV Source is activated

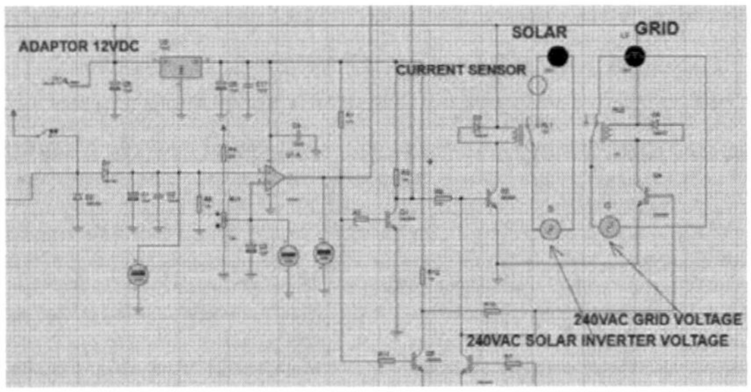

Figure 7. Solar PV Current < Pre-set Value, Grid Source is activated

During the transition, soft-switching circuit will be activated. The transition process only occurs at zero voltage as depicted in Figure 8. If both input sources are synchronized, the transition process occurred at zero voltage without any power interruption. If the input sources are unsynchronized, the switchover circuit will waits until $V_{AC,ref}$ is zero before switching the power supply. As a result, there is a delay during the transition process. Figure 9 – 11 illustrate supply power waveforms for unsynchronized power supplies.

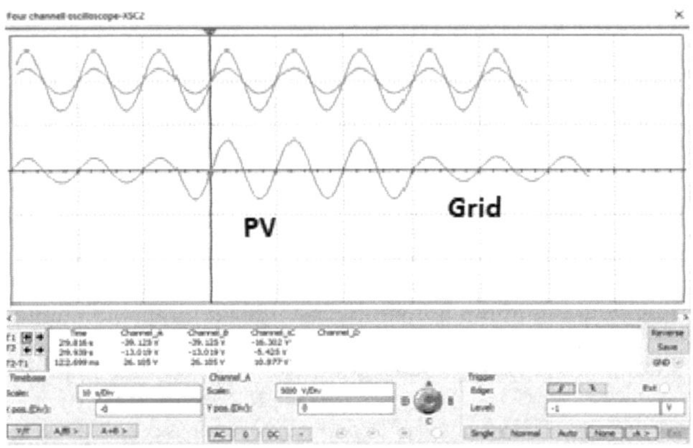

Figure 8. Synchronized Input Sources, Transition of Power Supply occurred at Zero Voltage with Negligible delay

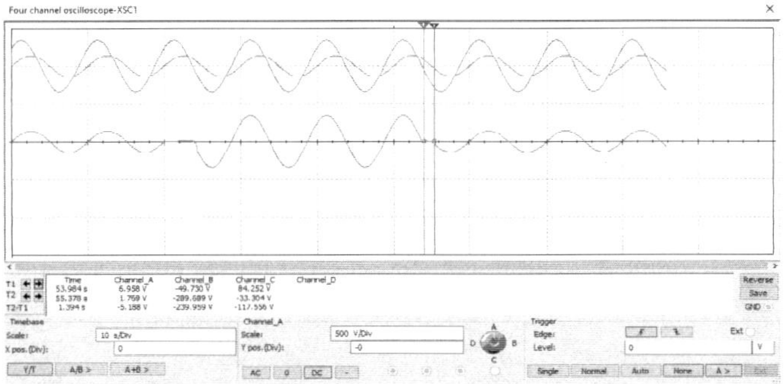

Figure 9 Input Sources out of phase by 45°,Transition of Power Supply Still occurred at Zero Voltage but with a delay

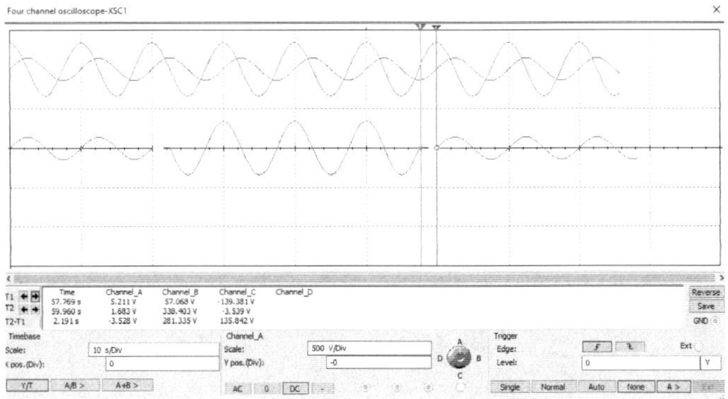

Figure 10a. Input Sources out of phase by 90°,Transition of Power Supply Still occurred at Zero Voltage but with a delay

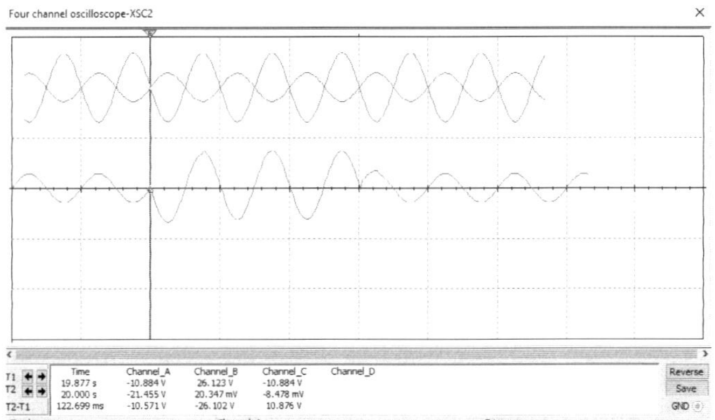

Figure 10b. Input Sources out of phase by 180°,Transition of Power Supply Still occurred at Zero Voltage but with a delay

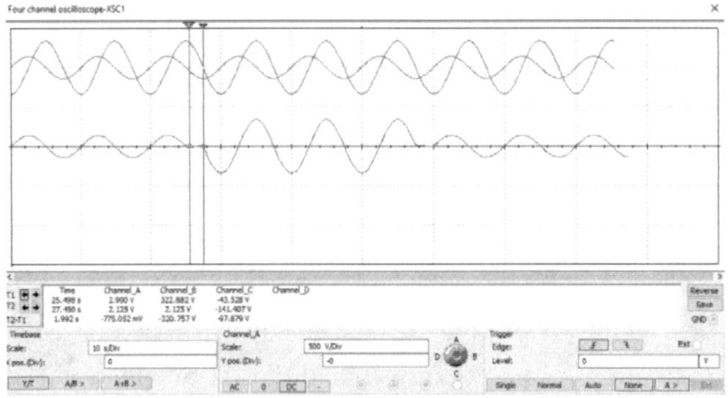

Figure 11 Input Sources out of phase by 270°,Transition of Power Supply Still occurred at Zero Voltage but with a delay

5.2 . Hardware Implementation

Hardware circuit of the soft-switching switchover circuit has been developed, tested, and analyzed to verify the simulation findings. It is observed that, without soft-switching circuit, there is a spark in the circuit during the transition of supply as shown in Figure 12. Meanwhile, Figure 13 shows smooth transition process with implementation of the soft-switching circuit.

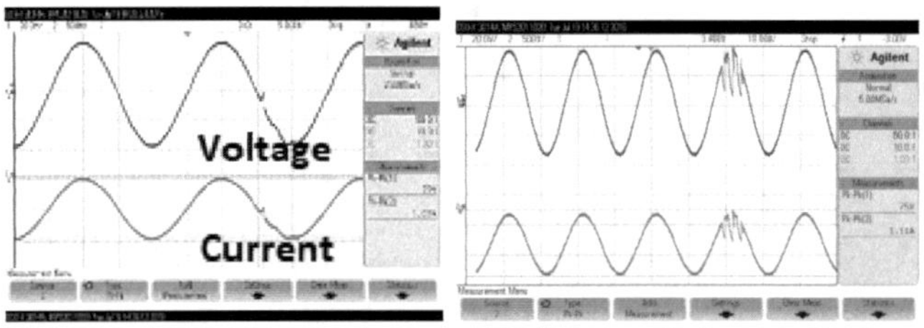

Figure 12. Without Soft-switching Circuit, there are Voltage and Current Spikes during Transition of Supply

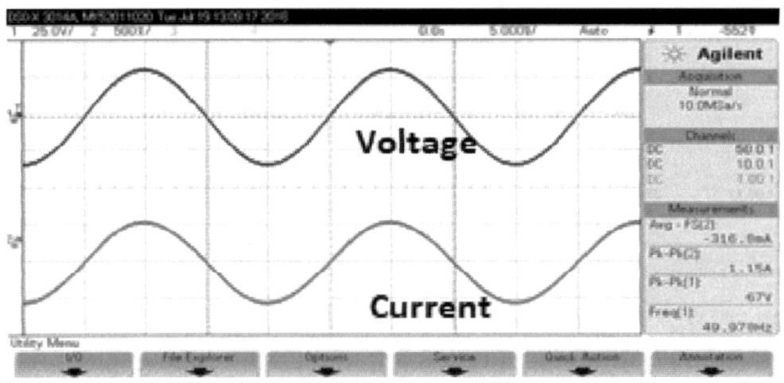

Figure. 13 Soft-switching Circuit Eliminates Voltage and Current Spikes during Transition of Supply

It is also observed that, a delay of 4ms is recorded during the transition from Grid to PV and a delay 2.2ms during transition of PV supply to Grid. This is because limitation imposed by mechanical relays.

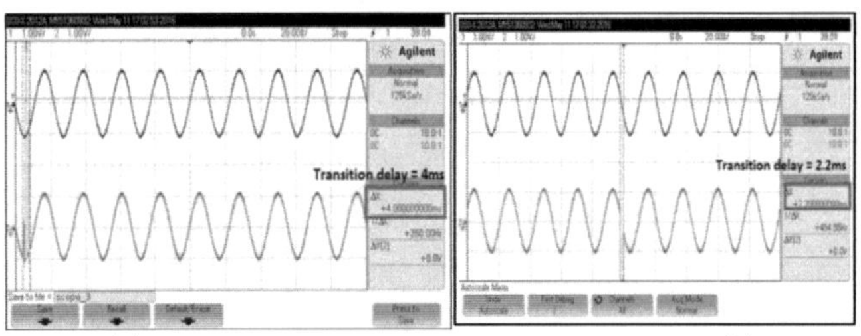

(a) (b)

Figure 14. Transition delay = 4ms for transition of Grid to PV and 2.2ms for transition of PV to Grid

Figure 15 and 16 show that whether synchronized or unsynchronized input supplies, the transition of supply process always occurs at zero voltage.

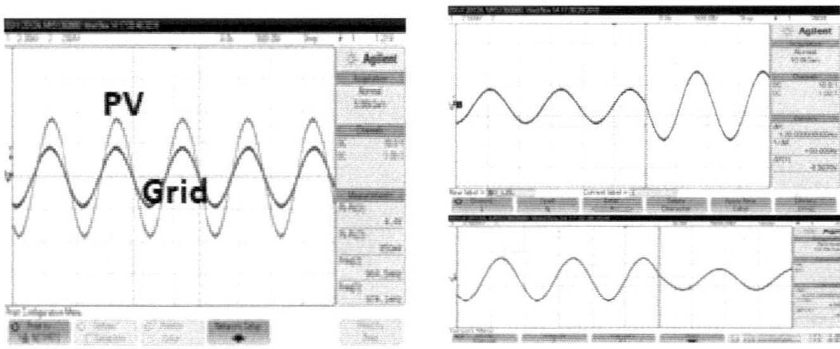

Figure 15. Synchronized Input Sources, Transition of Power Supply occurred at Zero Voltage

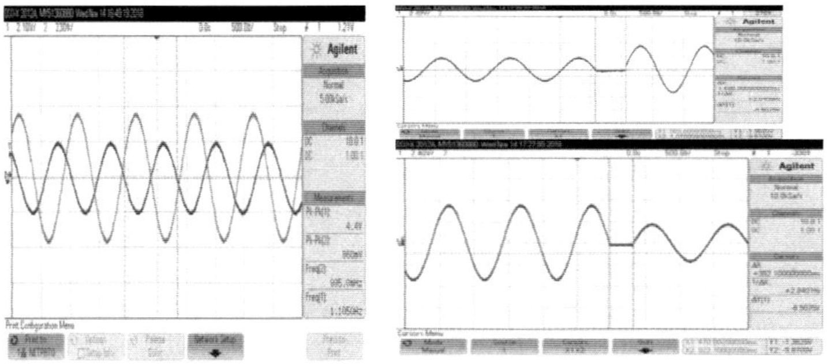

Figure 16. Input Sources out of phase by 180°,Transition of Power Supply Still occurred at Zero Voltage but with a delay

6. CONCLUSION

The dual-power PV-grid energy system is introduced to reduce the installation and maintenance cost for PV plant and to manipulate the low off-peak tariff during the night. Due synchronization issues of different power supplies, a delay is embedded during the transition of supply process. Consequently, it effects the power continuity in the demand site. The smart soft-switching switchover circuit is introduced to solve this problem. An integration of soft-switching switchover circuit and a supervisory controller is developed to monitor PV current, to decide the power supply source, and to transfer the supply power at zero voltage and current. Simulation and hardware implementation of the system proves by utilizing smart soft-switching switchover circuit, there is no spike during transition and the transition process always occurs at zero voltage thus limiting the harmful effect that might be evolved from the switching of power supply processes.

ACKNOWLEDGMENT

Technical and Financial assistance from UniversitiTeknologi MARA, Malaysia is gratefully acknowledged.

REFERENCES

[1] The Eleventh Malaysia Plan, 2016-2020, 2020 Available:https://www.talentcorp.com.my/clients/TalentCorp_2016.../RMKe-11%20Book.pdf
[2] KeTTHA. (2018). SEDA Annual Report 2017.
[3] S. Mekhilef, A. Safari, W. E. S. Mustaffa, R. Saidur, R. Omar, and M. A. A. Younis, "Solar energy in Malaysia: Current state and prospects," *Renewable and Sustainable Energy*

Reviews, vol. 16, no. 1, pp. 386-396.

[4] H. C. Ong, T. M. I. Mahlia, and H. H. Masjuki, "A review on energy scenario and sustainable energy in Malaysia," *Renewable and Sustainable Energy Reviews*, vol. 15, no. 1, pp. 639-647.

[5] P. Nema, R. Nema, and S. Rangnekar, "A current and future state of art development of hybrid energy system using wind and PV-solar: A review," *Renewable and Sustainable Energy Reviews*, vol. 13, no. 8, pp. 2096-2103, 2009.

[6] N. A. Othman, N. S. Damanhuri, I. R. Ibrahim, R. Radzali, and M. N. Mohd, "*Automated Monitoring System for Small Scale Dual-Tariff Solar PV plant in UiTM Pulau Pinang*," 2010, vol. 2.

[7] I. R. Ibrahim, Y. Mohd Ali, N. A. Othman, N. S. Damanhuri, R. Radzali, and M. N. Mohd Hussain, "*Techno-economical Study of Wind Energy Extraction in Pulau Pinang using Statistical and Optimization Analysis.*," in International Conference in Science and Technology in Education (ICSTIE 2010), Pulau Pinang, Malaysia, 2010.

[8] I. R. Ibrahim, A. M. Omar, and H. Hussain, "Power Converter for Dual-power PV-Grid Energy System Utilizing Cascaded Multilevel Inverter," *Advanced Materials Research*, vol. 925, no. 2014, pp. 505-509, 2014.

[9] I. R. Ibrahim, A. M. Omar, Z. Hussain, M. N. Hussain, Mohd, and M. Z. Noor, Siti, "Dual-Power PV-Grid Energy System: An Alternative to the Off-Grid PV Energy System," in *Advanced Materials Research*, 2014, vol. 925, pp. 661-665: Trans Tech Publ.

[10] S. Vergura, "*Scalable model of PV cell in variable environment condition based on the manufacturer datasheet for circuit simulation*," in IEEE 15th International Conference on Environment and Electrical Engineering (EEEIC), 2015, 2015, pp. 1481-1485.

[11] K. C. Wijesinghe, "*Feasibility of solar PV integration in to the grid connected telecom base stations and the ultimate challenge*," in 2017 IEEE 6th International Conference on Renewable Energy Research and Applications (ICRERA), 2017, pp. 1154-1159.

[12] I. R. Ibrahim, A. M. Omar, and Z. Hussain, "*Control strategy of power converter system in the dual-power PV-grid system energy utilizing cascaded multilevel inverter*," in 8th IEEE Conference on Industrial Electronics and Applications (ICIEA), 2013, 2013, pp. 848-853.

[13] I. R. Ibrahim, A. M. Omar, and Z. Hussain, "Power Converter for Dual-power PV-Grid Energy System Utilizing Cascaded Multilevel Inverter," i*n Advanced Materials Research*, 2014, vol. 925, pp. 505-509: Trans Tech Publ.

[14] P. M. Pandi and N. Devarajan, "*Estimation of optimized power quality in micro controller based cascaded multilevel inverter*," in Proceedings - 2010 IEEE Region 8 International Conference on Computational Technologies in Electrical and Electronics Engineering, SIBIRCON-2010, 2010, pp. 671-676.

[15] M. Sadikin, T. Senjyu, and A. Yona, "*High-frequency link DC for power quality improvement of stand-alone PV system in cascaded multilevel inverter*," in 2013 IEEE 10th International Conference on Power Electronics and Drive Systems (PEDS), 2013, pp. 597-601.

[16] S. Sirisukprasert, J. S. Lai, and T. H. Liu, "*Optimum harmonic reduction with a wide range of modulation indexes for multilevel converters*," Proceedings of Conf Rec IEEE-IAS Meeting, 2000.

[17] S. Zhang, P. Wang, M. Liu, and Z. Cheng, "*Research on the operating mode of the power supply for a telecom base station based on the peak & valley model of the power grid*," in 2017 IEEE International Telecommunications Energy Conference (INTELEC), 2017, pp. 515-520.

AUTHOR INDEX

Abrori, S A 47, 71
Agrawal, M 222
Agrawal, Vinay 183
Ahmad, F B 319
Ahmed, Mohammed Afaounodden 37
Akasapu, Akshita 129
Akmal, M H M 319
Akter, Hafsa 343
Al Mehadi, Abdullah 78
Al-Amiery, Ahmed A 65
Ali, Mohammed AH 138
Alrufaye, Faiez Musa L 21
Amri, F 71
Anand, R 191
Anshori, I 47
Aritonang, R A 47
Arora, Gurumukh Singh 119
Aruna, B 109
Awang, Mokhtar 54
Azhakath, Adresya Suresh 37
Aziz, Farah Athirah Abd 369
Baharudin, Bakhtiar Ariff 54
Bakar, Amirah Nurhafizah Abu 289
Balaji, B Aravind 312
Bhat, Pritam 145
Brahmbhatt, Bhavik 152
Chandwani, Dr. Hina 152
Chandwani, Hina 259
Chen, Jeng Yue 307
Chowdhury, Misbahul Alam 78
Chuah, L S 12
Curto, D 401
Daud, Kamarul Azhar 389
Daud, Kamarulazhar 351, 369, 379
Dejthammathorn, T 235
Deshpande, Aditya 271
Devi, R Rama 251
Fadzil, Muhammad Faris Fahmi Mohd 379
Fahmi, M Z 71
Faisal, Fahim 78, 343
Fati, Suliman Mohamed 1
Favuzza, S 401
Garg, Suyash 183
Goutham, M A 119
Hamid, Darulihsan Abdul 54
Haque, M E 28
Haron, M F Shafiee 12
Harsha, M Sri 191
Hasan, M K 28
Hasan, Nurulfadzilah 89
Hong, Lin Ann 307
Huang, Kuo Chun 307
Hung, Nguyen Thanh 171
Hussain, M N M 414
Hussain, Nur Shahira Mat 89
Huyen, Dang Thi Khanh 171
Ibrahim, I R 414
Irzaman 71
Isa, Nuratiqah Mohd 289
Islam, M Z 28
Islam, Md Minhajul 78
Ismail, Ahmad Puad 369
Ismail, Azman 54
Ismail, Nurlida 351
Joshi, Srivatsa S 271
Juang, Fuh-Shyang 307
Karim, Mohamad Shaiful Abdul 89, 289
Kasim, Nazirah Mohamat 351, 369
Khalid, Puteri Zarina Megat 54
Kumar, D Shashi 362
Kuo, Lin 307
Kusshal, M R 119
Lee, Tung Lik 54
Mahajan, Amresh 59
Mahapatra, Punyasloka 191
Manan, Muhammad Nasrullah Abdul 389
Manjula, S 251
Mhaibes, Hakeem Imad 21
Mim, Sayka Afreen 343
Misra, Pramod Kumar 162
Mohanty, Jitendra 162
Muhammad, Aisha 138
Muneer, Amgad 1
Musca, R 401
Muthalagu, Raja 37
Nagar, Ravindra 183
Nasir, M A F M 414
Navia, M Navarro 401
Neamah, Ali F 21
Nishat, Mirza Muntasir 78, 343
Nugraha, N 337
Nugraha 47
Nugroho, H S 337
Omar, A M 414
Omar, Saodah 351, 389
Omar, S 379

Osamah, Sara ... 65
Pai, Amrita B ... 119
Panatarani, C ... 337
Pandin, Veera ... 119
Patel, Hetal ... 259
Pathak, Prem ... 59
Pattananuwat, P ... 235
Poonia, Laxmi ... 215
Prabhu, Smitha ... 119
Prajwal, K T ... 145
Prasad, G ... 129
Prasertkaew, S ... 235
Prima, E C ... 337
Purushothaman, V ... 251
Qadri, Ibrahim ... 1
Rahman, Fauziah Ab ... 54
Rajak, A R Abdul ... 325
Rajan, K K ... 97, 109
Ramesh, K ... 312
Rameshkumar, K ... 297
Rao, K R ... 240
Razak, Nur Hazimah Syazana Abdul ... 89
Refantero, G ... 337
Rezki, M ... 71
Saidon, Wan Salha ... 351
Sailaja, V ... 129
Salih, Nabihah ... 54
Samat, Ahmad Asri Abd ... 379, 389
Sarif, Nurul Husna Mohd ... 289
Sarohe, Syazreena ... 351
Sasikumar, S ... 312
Septiani, N L W ... 71
Shagor, Md. Rafid Kaysar ... 343
Shah, Ahmad Syahiman Mohd ... 89, 289
Shaker, Lina M ... 65
Shanono, Ibrahim Haruna ... 138
Shant, S H ... 28
Sharma, Sanjay ... 215
Shivith, K ... 297
Siam, Muhammad Naqib Mat ... 389
Sidek, N S M ... 205
Singh, S ... 222
Smrithi, M N ... 325
Subbarayappa, Shreyanka ... 240
Suganthi, S ... 362
Suyatman ... 47, 337
Tajudin, A Idzwan ... 379
Teja, P Sai ... 191
Trinh, Le Thi Phuong ... 171
Trisno, M L A ... 47
Van Ngoc, Hoang ... 281
Van On, Vo ... 171
Verma, Rajat ... 59
Ya'acob, A ... 205
Yuliarto, B ... 47, 71, 337
Yuniasari, R ... 71
Yusof, Y ... 12
Zainol, N ... 205
Zamli, M I ... 319
Zizzo, G ... 401